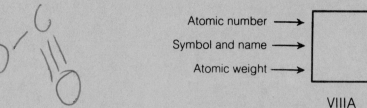

Atomic number ⟶
Symbol and name ⟶
Atomic weight ⟶

3 é gain   2 é gain   1 é gain

3 é lost

					VIIIA
					2 **He** Helium 4.003
IIIA	IVA	VA	VIA	VIIA	
5 **B** Boron 10.81	**6** **C** Carbon 12.01	**7** **N** Nitrogen 14.01	**8** **O** Oxygen 16.00	**9** **F** Fluorine 19.00	**10** **Ne** Neon 20.18
13 **Al** Aluminum 26.98	**14** **Si** Silicon 28.09	**15** **P** Phosphorus 30.97	**16** **S** Sulfur 32.06	**17** **Cl** Chlorine 35.45	**18** **Ar** Argon 39.95

IB	IIB							
28 **Ni** Nickel 58.71	**29** **Cu** Copper 63.55	**30** **Zn** Zinc 65.38	**31** **Ga** Gallium 69.72	**32** **Ge** Germanium 72.59	**33** **As** Arsenic 74.92	**34** **Se** Selenium 78.96	**35** **Br** Bromine 79.90	**36** **Kr** Krypton 83.80
46 **Pd** Palladium 106.4	**47** **Ag** Silver 107.9	**48** **Cd** Cadmium 112.4	**49** **In** Indium 114.8	**50** **Sn** Tin 118.7	**51** **Sb** Antimony 121.8	**52** **Te** Tellurium 127.6	**53** **I** Iodine 126.9	**54** **Xe** Xenon 131.3
78 **Pt** Platinum 195.1	**79** **Au** Gold 197.0	**80** **Hg** Mercury 200.6	**81** **Tl** Thallium 204.4	**82** **Pb** Lead 207.2	**83** **Bi** Bismuth 209.0	**84** **Po** Polonium (210)*	**85** **At** Astatine (210)*	**86** **Rn** Radon (222)*

63 **Eu** Europium 152.0	**64** **Gd** Gadolinium 157.2	**65** **Tb** Terbium 158.9	**66** **Dy** Dysprosium 162.5	**67** **Ho** Holmium 164.9	**68** **Er** Erbium 167.3	**69** **Tm** Thulium 168.9	**70** **Yb** Ytterbium 173.0	**71** **Lu** Lutetium 175.0
95 **Am** Americium (243)*	**96** **Cm** Curium (247)*	**97** **Bk** Berkelium (247)*	**98** **Cf** Californium (251)*	**99** **Es** Einsteinium (254)*	**100** **Fm** Fermium (257)*	**101** **Md** Mendelevium (258)*	**102** **No** Nobelium (259)*	**103** **Lr** Lawrencium (260)*

CHEMISTRY:
A FIRST COURSE

SECOND EDITION

CHEMISTRY: A FIRST COURSE

JACQUELINE I. KROSCHWITZ
Formerly of Kean College of New Jersey

MELVIN WINOKUR
Bloomfield College

McGRAW-HILL BOOK COMPANY

New York | St. Louis | San Francisco | Auckland | Bogotá | Hamburg
Johannesburg | London | Madrid | Mexico | Milan | Montreal | New Delhi
Panama | Paris | São Paulo | Singapore | Sydney | Tokyo | Toronto

CHEMISTRY: A FIRST COURSE

1234567890DOCDOC89432109876

ISBN 0-07-035539-8

This book was set in New Baskerville Book by General Graphic Services, Inc.
The editors were Karen S. Misler and Steven Tenney;
the designer was Nicholas Krenitsky;
the production supervisor was Joe Campanella.
The drawings were done by J & R Services, Inc.
R.R. Donnelley & Sons Company was printer and binder.

Library of Congress Cataloging-in-Publication Data

Kroschwitz, Jacqueline I.
  Chemistry, a first course.

  Includes index.
  1. Chemistry. I. Winokur, Melvin. II. Title.
QD31.2.K76 1987   540   86-20052
ISBN 0-07-035539-8

ABOUT THE AUTHORS

The authors bring a total of over 30 years of dedicated teaching and writing to their textbooks.

MEL WINOKUR was born and educated in New York City. After receiving his bachelor's degree from City College of New York, he went west to the California Institute of Technology where he obtained the Ph.D. in physical organic chemistry. Mel then returned to New York and taught at several branches of City University of New York, most notably at Bronx Community College. He moved to Bloomfield College and became full professor in 1984.

JACQUELINE KROSCHWITZ was born in Trenton, New Jersey, graduated from Ursinus College in Collegeville, Pennsylvania, and received the Ph.D. in physical organic chemistry from the University of Pennsylvania. Then she went west for a postdoctoral year at Caltech. Her first teaching post was at Barnard College, Columbia University. She has also taught in New York City community colleges and moved to Kean College of New Jersey in 1976.

Despite their different names, the authors are married and live in northern New Jersey with their wine cellar. Their avocation is traveling, especially among Michelin guide three-star restaurants.

CONTENTS

CHAPTER 12   335

LIQUIDS AND SOLIDS

CHAPTER 13   383

SOLUTIONS

CHAPTER 14   431

CHEMICAL EQUILIBRIUM

CHAPTER 15   455

ACIDS AND BASES

CHAPTER 16   495

OXIDATION REDUCTION

CHAPTER 17   529

NUCLEAR CHEMISTRY

CHAPTER 18   561

ORGANIC AND BIOLOGICAL MOLECULES

PREFACE

"To teach is to learn." In writing our textbooks we have attempted to apply the valuable lessons which our students have generously supplied as we have tried (with some success we think) to teach them the fundamental concepts of chemistry. Perhaps the most valuable lesson we have learned is that beginning chemistry students simply do not possess certain knowledge and reasoning skills which their science instructors generally take for granted. In this text, aimed primarily at the preparatory-level chemistry course, we take nothing for granted, neither prior scientific knowledge nor prior experience in deduction.

As we approached each topic our aim was to offer complete, logical explanations which include all necessary steps in the deductive reasoning process, rather than assuming that the student already has the ability to make deductive leaps. This approach gives students the opportunity to glimpse the complete development of scientific thought, which is usually not their natural way of thinking, and to develop their own reasoning skills as they explore increasingly complex phenomena.

We hope that students might enjoy their study of chemistry despite the fact that many are captives of a science course requirement. To this end we have used a conversational tone, a few anthropomorphic explanations, and occasionally introduced humor into some illustrations. Also, we encourage student visualization of microscopic phenomena both because such imagery is a valuable scientific skill and because it is fun.

As a convenience and learning aid, and in order to reinforce concepts and encourage necessary review, we frequently refer students to previous sections needed as a foundation for understanding the topic at hand. Cross-referencing is also used to inform students of "coming attractions" in subsequent chapters. Other convenient features and learning aids are:

Sample Exercises are very carefully worked out in stepwise detail using the **Unit Conversion Method** wherever possible and appropriate. Sample Exercises are supplied copiously.

Problems frequently immediately follow Sample Exercises within the chapter so that the student can test his or her understanding of the concept explained. At the end of each chapter there are numerous problems at varying levels so that students can reach the limits of their own capacities. In this edition problems are arranged by subsection topics for the convenience of students and instructors.

Stepwise Procedural Rules or guidelines are provided for significant manipulations; for example, see "Guidelines for writing Lewis structures" in Section 6.13.

Tables are used extensively to organize and summarize information.

Illustrations are an integral part of the explanations.

Summaries of the major points addressed conclude each chapter.

Chapter Accomplishments provide the student with the learning objectives of each chapter.

Math skills are given the status and full treatment of a chapter.

The usefulness of the periodic table is stressed and reiterated.

An entire chapter is devoted to the mole concept.

Selected answers to problems appear in Appendix 4.

Defined words are **italicized** in the index for ready location.

More applications of chemistry to students' experience or interest are offered in this edition: for example, colligative properties in Chapter 13 and blood buffers in Chapter 15.

Student reception of the first edition of this text suggests we have succeeded in writing a truly student-oriented book. Students describe the text as "self-teaching." In short, preparatory level students are able to read and understand this book and many report enjoying it.

In order to provide an early explanation for compound formation and a logical basis for remembering ionic charges, we have altered the sequence of chapters in this new edition. Electronic structure and bonding are now covered before ionic formula and nomenclature.

Instructors who wish to follow the former sequence and introduce nomenclature, formula and equation writing, and stoichiometry early in their course should have no problem covering Chapters 1, 2, 3, 4, 7, 8, 9, and 10 before returning to electronic structure and bonding in Chapters 5 and 6. An Instructor's Manual is available and provides additional comments on each chapter, suggested accompanying laboratory experiments, answers to all problems, and sample examinations for each chapter. In our Laboratory Manual each experiment is related to a specific text section.

More material is included in this book than can be covered reasonably in one semester. The core curriculum as preparation for general chemistry would be Chapters 1–11. If time permits, the instructor then has the freedom to choose additional topics based on personal preference. Chapters 12, 13, 14, 17, and 18 stand independently of one another. Chapter 15 (Acids and Bases) relies on material in Chapter 13 (Solutions) and Chapter 14 (Chemical Equilibrium). Chapter 16 presupposes coverage of Chapter 13.

We earnestly invite your comments and suggestions toward the improvement of this textbook as a learning device.

ACKNOWLEDGMENT

We would like to acknowledge and thank the many individuals who helped and encouraged us as we developed this text. First of all, there are the C.U.N.Y. community college students who were our original inspiration and the many Chemistry 110 students at Bloomfield College who have sustained the inspiration. They have used the first edition thoughtfully and offered useful suggestions for improvement.

Reviewers of our text have been very thorough and helpful, and we thank them: Ron Backus, American River Community College; Boyd Earle, University of Nevada, Las Vegas; Nancy Howard, Philadelphia College of Textile and Science; William B. Huggins, Pensacola Junior College; E. J. Kremnitz, University of Nebraska, Omaha; Steven Murov, Modesto Junior College; Raymond O'Connor, Santa Barbara City College; Donald B. Phillips, Eastern Michigan University; Gordon Parker, University of Toledo; and Ray F. Wilson, Texas Southern University.

We also thank the editors and other production staff at McGraw-Hill, who gave such careful attention to all details as the manuscript became a book.

Jacqueline I. Kroschwitz
Melvin Winokur

CHEMISTRY:
A FIRST COURSE

CLASSIFICATION OF MATTER

CHAPTER

1

H

Hydrogen

1.008

INTRODUCTION

1.1 Chemistry is the study of matter, which is everything you see around you, this book, your hand, a pencil, water, a tree, and invisible things such as air. Two characteristics define **matter:** matter occupies space and has mass.

As you begin reading this text, your view of matter resembles that of the earliest chemists who set out to study matter. That is, you probably perceive the world around you as boasting an unlimited number of different forms of matter. As it turns out, we can actually classify matter into a surprisingly small number of categories. Discovering useful classifications of matter will be our first task in this text, as it was for the earliest investigators of matter.

CLASSIFICATION OF MATTER BY PHYSICAL STATE

1.2 As citizens of the twentieth century, you have a decided advantage over the early investigators because you have absorbed some of the compiled scientific knowledge of the last 200 years in your everyday experience. For example, consider the following list of samples of matter and think about how you might group the samples into three categories based on characteristics of the samples that you have observed.

Samples of Matter (See Figure 1.1)

Steam	Iron
Gasoline	Ice
Table salt	Oxygen
Water	Carbon dioxide
Mercury	Alcohol

A useful classification system that you may have chosen involves identifying these samples as **solids, liquids,** or **gases.**

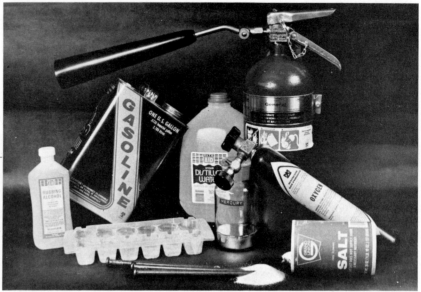

FIGURE 1.1

Matter is everything you see around you. Here are some familiar examples classified in Section 1.2. Most samples are labeled. The nails are iron, and the fire extinguisher contains carbon dioxide. Steam is not shown because of its elusiveness.
(*Photograph by Bryan Lees*)

Solids	Liquids	Gases
Table salt	Gasoline	Steam
Iron	Water	Oxygen
Ice	Mercury	Carbon dioxide
	Alcohol	

This is the classification of matter according to **physical state.** Because you have long been aware of the characteristics of the three physical states, you now recognize solids, liquids, and gases without realizing that you are examining the definiteness of the shape and volume of the sample when you make your classification. Table 1.1 summarizes the shape and volume characteristics of the three physical states.

Usually we talk about the physical state of matter at room temperature. This is so because the physical state may change as temperature changes, and the samples we consider are usually at room temperature. For example, water is a liquid at room temperature (20°C), a solid (ice) below the freezing temperature (0°C), and a gas (steam) above the boiling temperature (100°C).

CHANGES IN STATE

1.3 **Melting** is the name applied to the change between the solid and liquid physical states. Other familiar changes of state are liquids **freezing** to solids, liquids **evaporating** to gases, and gases **condensing** to liquids. Liquid-solid changes (melting or freezing) and liquid-gas changes

SHAPE AND VOLUME CHARACTERISTICS OF THE THREE PHYSICAL STATES   **TABLE 1.1**

Physical State	Shape	Volume
Solid	*Definite*—does *not* depend on the shape of the container	*Definite*—has a fixed volume (A gold bar occupies a clearly defined amount of space.)
Liquid	*Indefinite*—takes on the shape of its container	*Definite*—has a fixed volume (A glass of milk spilled on the floor spreads out, but not forever.)
Gas*	*Indefinite*—takes on the shape of its container	*Indefinite*—has no fixed volume (A small vial of gas opened in a large auditorium spreads to fill the entire room.)

* Because you are probably less familiar with this state since it is usually invisible, a whole chapter (Chapter 11) is devoted to gases. The more familiar liquids and solids are discussed in Chapter 12.
Note: Some substances, such as chocolate pudding, ice cream, etc., may not be so easily classified.

(evaporation or condensation) are the most familiar, but direct changes between the solid and gaseous states are also possible. Conversion of a solid directly into a gas is called **sublimation.** The water in wet laundry that is hung on a clothesline on a freezing winter day freezes to ice. The clothes become dry when the ice sublimes into the gaseous state and goes off into the air (see Figure 1.2).

We can summarize changes in state by the diagram below. The arrows represent the processes named, and the arrowheads indicate the direction of the change between states.

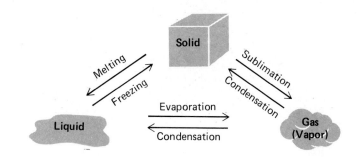

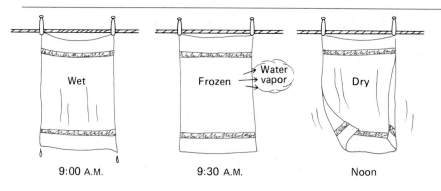

FIGURE 1.2

How your laundry dries on a cold winter day. Water in the towel freezes into ice; the ice sublimes into water vapor, and your towel is dry.

9:00 A.M.   9:30 A.M.   Noon

SAMPLE EXERCISE

1.1

If we were to heat (**a**) oxygen, (**b**) aluminum foil, (**c**) gasoline, and (**d**) ice cream until they were at the temperature of boiling water (100°C), what would their physical states be?

SOLUTION

a. Gas.
b. Aluminum remains a solid at 100°C. Remember that you can put aluminum foil in a hot oven and it will not melt. It melts at 660°C.
c. Gas. Most gasolines boil at temperatures just slightly lower than the boiling point of water.
d. Liquid.

Problem 1.1

a. At room temperature, what is the physical state of a nail, of water, of cooking oil, and of air?

b. On a very cold winter day (below freezing), what is the physical state of the substances listed in part (a)?

     The temperature at which a solid becomes a liquid is called its **melting point.** Notice from the preceding diagram that this is also the freezing point of the liquid. The temperature at which bubbles form throughout a liquid and the liquid becomes a gas is called the **boiling point.** Gas solid changes occur at the **sublimation point.**[1]

SCIENTIFIC METHOD

1.4   The classification procedure follows a line of thinking called the **scientific method.** Four basic steps are involved in the scientific method, and they can be illustrated by referring to the discovery of the classification of matter according to physical state which was just discussed.

Scientific Method	Illustration
1. Observation	The properties of the samples were observed and compared.
2. Hypothesis (suggestion)	It was suggested or hypothesized that all matter can be classified as either a solid, a liquid, or a gas.
3. Experiment or further observation	Observation of many other samples of matter were made to test the hypothesis.
4. Conclusion	Because the new observations confirm the hypothesis, we now have a means of classifying matter.

[1] Boiling points and sublimation points depend on pressure.

Accepted hypotheses are continuously tested by new experiments. Hypotheses not agreeing with experiment are modified or rejected entirely. As we proceed with the classification of matter, the scientific method is the logical framework within which we hope to fit our observations.

PURE SUBSTANCES VERSUS MIXTURES

1.5 Classification of matter into two broad categories, **pure substances** and **mixtures,** is very useful. All matter can be separated into these categories, based on the **constancy of composition** of the sample, which is reflected in the properties of the sample. **Pure substances** have a constant composition, or makeup, and this is demonstrated by the fact that they always have the same properties regardless of their origin. For example, copper metal has the same properties whether it is found in ore in the United States or in South America or is reclaimed from copper tubing in an old building. Other examples of pure substances are any pure metal, distilled water, table salt (uniodized), and refined sugar.

   Mixtures, on the other hand, display varying properties as the proportions of the components of the mixture vary (see Figure 1.3). For example, a cup of coffee is a mixture. The properties of a cup of coffee are described as strong or weak, bitter or mellow, depending on how much material has been removed from various coffee beans by water. Similarly, a piece of steak is a complex mixture. It can be juicy and tasty or dry and tasteless, depending on its particular composition. Paint is a mixture. There is no definite composition and hence no definite properties corresponding to paint. It can contain water or no water; it can contain yellow pigment or blue pigment in varying amounts. These different compositions result in different solubility and color properties.

FIGURE 1.3

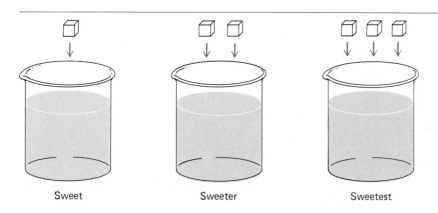

Sweet   Sweeter   Sweetest

The properties of mixtures vary with the proportions of the ingredients. Three lumps of sugar give a sweeter taste to the same volume of water than do two lumps of sugar. Also, the melting point of the solution decreases and the boiling point increases as the lumps of sugar are added.

SAMPLE EXERCISE

1.2

Are the following mixtures or pure substances?

a. Urine
b. Seawater
c. Gasoline

SOLUTION

a. Urine is a mixture of water and various bodily waste products. Its composition varies depending on what you eat and your state of health.
b. Seawater is a mixture of water, salt, seaweed, and various debris.
c. Gasoline is a mixture. The composition is variable. It can be leaded or unleaded, and different companies add different additives.

Most matter we encounter is a complex mixture.

Problem 1.2

Are the following mixtures or pure substances?

a. A page in this textbook   **c.** Pure iron metal

b. Skin   **d.** A rusty nail

A second characteristic of mixtures is that their components may be separated by physical means. For example, a mixture of sand and water can be separated by filtration. This test of the separability of components is a good experiment that distinguishes between mixtures and pure substances. In Sample Exercise 1.2 and Problem 1.2 you were asked to make the classification without doing any experimental testing, and this is possible because in each example chosen the constancy of composition is obvious through observation or through your previous everyday experiences.

Let us summarize this section by examining the scientific thinking in the classification:

Scientific Method	Illustration
1. Observation	Some samples of matter show variable properties (mixtures); others show constant properties (pure substances).
2. Hypothesis	The variation in properties is caused by variation in composition.
3. Experiment and further observation	Familiar mixtures like paint are known to show different properties as composition varies. The components of some mixtures can be readily separated and hence remixed in different proportions.
4. Conclusion	Matter can be classified into mixtures or pure substances based on constancy of composition as witnessed by constancy of properties.

PHYSICAL VERSUS CHEMICAL CHANGE

1.6 Let us explore the distinctions between changes called **physical** and those called **chemical,** since these distinctions will be necessary in understanding the subdivisions of pure substances.

Changes in state are examples of *physical changes*. A **physical change** causes no change in the basic nature or composition of pure substances in a sample of matter. Thus the freezing of water is a physical change. No change in the composition of water occurs in this process; no new kind of matter is produced. The basic nature of the matter remains the same; only the state changes; i.e., liquid water becomes solid water.

Besides changes in state, another common example of physical change is a change in size or shape of matter. For example, tearing aluminum foil into pieces represents a physical change; the chemical composition is still aluminum, but we have several smaller pieces rather than one large one.

The mixing together of two pure substances, such as iron filings and salt, *without* changing the composition of either, is a physical change. Likewise, separating the components of a mixture represents a physical change so long as the composition of the individual components remains unchanged. In Section 1.5 we mentioned separating sand and water by filtration, a physical means of separation. Notice in this case that the composition of the *mixture* is altered, but the compositions of the pure-substance components (water and sand) are *unchanged. Physical changes **never** affect the composition of pure substances.* No new pure substance is produced during a physical change.

In contrast, **chemical changes** produce a change in the basic nature or composition of pure substances in a sample of matter. Old substances are converted into new ones. Burning coal is an example of a process that produces chemical change. The solid black coal burns, giving off heat and light. Colorless carbon dioxide gas forms. This new material has a composition and a set of properties that differ from those of coal. Note that the new substance also has a physical state different from that of the old material. However, we do not call this a physical change, because the solid coal is not becoming liquid coal or gaseous coal. Rather, the different physical state arises because the new material that has been formed happens to exist in a physical state that is different from the old material, which has disappeared. That is, the new material has a different set of physical properties (Section 1.8).

Another example of a chemical change is the electrolysis of table salt (described in Table 1.2), wherein the familiar white crystals are changed into shiny sodium metal and yellowish green chlorine gas (see Figure 1.4). *Chemical changes **always** affect the composition of pure substances.* A new pure substance is always produced during a chemical change.

TABLE 1.2   EFFECTS OF EXPERIMENTS ON A SELECTION OF PURE SUBSTANCES

Pure Substances	Description	Effect of Heating (in a Vacuum)*	Effect of an Electrical Current	Conclusion and Classification
Mercuric oxide	Red powder	Turns black and a gas is given off	No further testing necessary	New materials are formed; mercuric oxide is a *compound*
Malonic acid	White solid	A gas is given off and a white solid with different properties is formed.	No further testing necessary	New materials are formed; malonic acid is a *compound*
Water	Colorless liquid	Heating produces steam; cooling restores the water	Bubbles of gas with properties other than those of steam are released	New materials are formed; water is a *compound*
Tin	Shiny solid	Can be melted and resolidified unchanged	Tin conducts electricity, but it is not changed by it	Tin remains unaltered in the experiments; tin is an *element*
Iodine	Purple solid	Sublimes and the solid reforms upon cooling	Does not conduct	No change; iodine is an *element*
Helium	Colorless gas	Remains the same	Unchanged	No change; helium is an *element*
Table salt	White solid	Melts at a very high temperature and resolidifies unchanged	Passing an electric current through melted table salt produces a shiny metal and a yellowish green gas	New materials are formed; table salt is a *compound*
Mercury	Silvery liquid	Vaporizes at high temperature, but the liquid re-forms upon cooling	Mercury conducts electricity, but it is not changed by it	No change; mercury is an *element*

* A vacuum is an environment in which essentially no other matter except the sample is present. It is necessary to exclude other matter so that it does not interfere with our observations of the sample we are studying.

SAMPLE EXERCISE

1.3

Are the following examples of physical or chemical changes?

a. A ham is sliced.
b. The sugar in grain ferments into alcohol.
c. Sugar dissolves in hot water.
d. Sugar is heated to a brownish coloration.
e. Oxygen is compressed and liquefied.

SOLUTION

a. Physical. It is still ham, but in smaller pieces.
b. Chemical. A new substance, alcohol, has been produced.
c. Physical. No new material is produced. The sugar and water have become intimately mixed but are still sugar and water.
d. Chemical. Upon heating dry sugar (sucrose), a mixture of products called caramel is formed.
e. Physical. Gaseous oxygen becomes liquid oxygen.

A mound of table salt → Electrolysis / A chemical change → A cube of shiny sodium metal + Yellow-green chlorine gas

Problem 1.3

Are the following physical or chemical changes?

a. Food is digested.

b. Cutting grass.

c. Perspiration evaporates.

d. The antiseptic and hair bleach, hydrogen peroxide, decomposes to form water and oxygen.

e. Growing of a tomato plant.

SUBDIVISIONS OF PURE SUBSTANCES

1.7 Early investigators were able to identify two categories of pure substances, **elements** and **compounds.** They made their classifications by observing that some pure substances (compounds) could be changed into distinctly different materials by certain experiments, whereas other pure substances (elements) defied attempts to alter them. Table 1.2 lists the results of two experimental tests (heating and passage of an electric current) on pure substances and shows how we use the results of these tests to classify materials as elements or compounds.

Notice that there are three possible outcomes for each experiment. One possible outcome is no change. A second possible outcome is a change in physical state. This frequently happens when the sample is heated; for example, water boils, producing steam, and tin melts. This kind of change is a physical change and results in no new substance formation. Physical changes are reversible; for example, when steam is cooled, liquid water re-forms, and when the melted tin is cooled, it resolidifies. The third possible outcome is the appearance of a new substance, as observed, for example, for mercuric oxide. This is a chemical change, and a chemical change by heating in the absence of air or other substances indicates that the pure substance tested is a compound.

Not all substances can be correctly classified by these two simple experiments of heating and passage of an electric current. However, experimentation can subdivide all pure substances into elements and compounds, and these simple tests give you an idea of how it's done.

SAMPLE EXERCISE

1.4

Classify potassium chlorate and silver as elements or compounds based on the following data:

Silver melts at 961°C and resolidifies when cooled.

Upon reheating, it again melts at 961°C.

Silver conducts electricity, but it is not changed.

When potassium chlorate is heated, a gas is released and a white solid with properties different from potassium chlorate forms.

SOLUTION

Silver is an element according to the experimental tests described because it remains unaltered, as shown by the fact that the melting point remains the same.

Potassium chlorate must be a compound because it is broken down by heat.

Summarizing this element and compound classification scheme in terms of the scientific method:

Scientific Method	Illustration
1. Observation	Some pure substances are clearly altered by experiment; others are not.
2. Hypothesis	Pure substances can be subdivided into two categories based on this difference in behavior.
3. Experiment and further observation	Tests of all known pure substances confirm these two categories.
4. Conclusion	Pure substances that can be broken down into new materials (undergo a chemical change) are classified as compounds. Pure substances that cannot be broken down are elements.

PHYSICAL AND CHEMICAL PROPERTIES

1.8 We have used the term "property" or "characteristic" in discussing classification. Properties are used to classify things into groups or categories, which is another way of saying that they serve to identify things. Pure substances can be identified on the basis of their chemical or physical properties.

Chemical properties are observed when a substance undergoes a chemical change, i.e., a change in composition. Flammability is a chemical property: some substances burn (a chemical change); others do not. Therefore, this property serves to distinguish some substances from others. Iron does not burn, but it does have the chemical property of rusting. This property uniquely identifies iron. Stainless steel, on the other hand, does not rust. Food has the chemical property of being digestible. The explosive power of TNT represents a chemical property of that substance. Whenever a chemical change is described, a chemical property of the material undergoing the change is also being described.

Substances can be identified by their chemical properties, but more often we identify substances by their physical properties. **Physical properties** can be observed and measured without *chemically* changing the composition of the substance observed. The size, color, odor, and physical state of a substance represent some of its physical properties. Whether a material conducts heat or electricity, whether it is flexible or brittle, determines other physical properties.

Problem 1.4

List some of the physical properties of a test tube or other glass apparatus commonly used in a laboratory.

Pure substances have characteristic temperatures at which they melt or boil. Table 1.3 lists the melting and boiling points of some pure substances with which you are familiar. Knowledge about the temperatures at which physical states change enables one to predict the physical state of a pure substance at any temperature.

SAMPLE EXERCISE

1.5

What is the physical state of table salt at room temperature ($\approx 20°C$), at 1000°C, and at 1500°C?

SOLUTION

Because you are familiar with table salt at room temperature, you do not need Table 1.3 to tell you that it is a solid at that temperature. If it had been unfamiliar, you would have concluded that it is solid because 20°C is well below the melting point.

On the other hand, 1000°C is above the melting point, so at that temperature salt would be in the liquid state. It would not be converted to a gas until the temperature reached 1473°C. At any temperature above 1473°C (for example, 1500°C) salt would be a gas.

Notice how the experimentally determined physical properties in Table 1.3 enable you to distinguish between look-alikes. For example, salt and sugar are both white solids at room temperature and difficult to distinguish by observation alone. However, the simple experiment

TABLE 1.3   PHYSICAL PROPERTIES OF A FEW COMMON PURE SUBSTANCES

Substance	Familiarity	Color	Odor*	Taste*	Melting Point, °C†	Boiling Point, °C†
Acetic acid	Vinegar is a 5% solution of acetic acid	Colorless	Pungent	Vinegary	16.6	118.1
Carbon tetrachloride	Dry cleaning solvent	Colorless	Like cleaning fluid	?‡	−23.0	76.5
Chlorine	Swimming pool disinfectant	Yellow-green	Irritating	?	−101.0	−34.6
Iron	Nails	Gray	Odorless	"Irony"§	1530	3000
Methane	Natural gas	Colorless	Odorless¶	?	−182.5	−161.5
Oxygen	Take a deep breath!	Colorless	Odorless	Tasteless	−218.8	−183.0
Sodium chloride	Table salt	White	Odorless	Salty	808	1473
Sucrose	Table sugar	White	Odorless	Sweet	186; decomposes	Decomposes

* Notice how odor or taste is often described in terms of our familiarity with the substance.
† The units of measurement are described in Chapter 3.
‡ Because carbon tetrachloride is a known carcinogen (cancer-producing agent), it is unlikely that the taste will be determined.
§ The characteristic taste of blood is saltiness plus the taste of iron.
¶ A substance is added to natural gas to give it a detectable odor for reasons of safety. The methane itself has no odor.

of heating the materials readily identifies them. Sugar melts easily (186°C), but salt will not melt in a match or bunsen burner flame. If you heat too vigorously, your identification will be made on the basis of a chemical property; i.e., sugar burns and salt does not. This simple example illustrates how experimentation can offer more information than observation alone.

Obviously, to distinguish between sugar and salt, we could taste them, the way we would in our kitchen. However, in the chemistry laboratory it is generally wise to refrain from tasting.

The properties of mixtures are not usually recorded because the properties vary as the proportion of ingredients in the mixture varies (see Figure 1.3).

Problem 1.5

Chloroform looks and smells like carbon tetrachloride. Its boiling point is 61°C. How might you identify two unlabeled liquids if you already know that one is chloroform and the other carbon tetrachloride?

Problem 1.6

Both methane and oxygen are colorless, odorless gases at room temperature. How might you distinguish a test tube full of methane from one containing oxygen?

ELEMENTS, COMPOUNDS, AND MIXTURES

1.9   The most useful classification of matter is into three categories: elements, compounds, or mixtures. From the foregoing discussions we can distinguish between the three categories by the following definitions:

Elements: A pure substance which *cannot* be broken down into simpler substances by ordinary chemical changes.

Compound: A pure substance which can be broken down into simpler substances (elements) by ordinary chemical changes. Compounds are composed of two or more elements combined in a fixed proportion.

Mixture: A combination of two or more pure substances in *no* fixed proportion which may be separated by a mere physical change.

Elements

Elements are the simplest substances, the building blocks from which all other matter is made up by various combinations. Whereas there are unlimited numbers of compounds and mixtures, there are only 106 known elements. Look inside the front cover of this text and you will see the names of all the elements arranged in a so-called periodic table, which we will discuss in Chapter 4. In this text we will rarely deal with more than 35 of the elements. Table 1.4 lists the names of the elements with which you need to be familiar. Notice in Table 1.4 that each element has a symbol (see Section 1.11) and that elements are classified as either **metals** or **nonmetals.**

The classification of elements as metals or nonmetals arose naturally when chemists first began studying the properties of elements. It was observed that there are two distinct sets of elements, each of which displays certain physical properties. Within one set the elements are shiny, conduct electricity, and are malleable (i.e., they bend without breaking). These substances are called **metals.** You are already familiar with metals from everyday life, and you can probably describe the properties of such metals as aluminum, copper, or gold. **Nonmetals,** on the other hand, have the opposite characteristics. They generally lack luster, do not conduct electricity, and are brittle.

As you become familiar with the elements, it is a good idea to distinguish metals from nonmetals from the very start. This classification into two types of elements which have opposite properties is nearly as important a concept in chemistry as the concept of male and female is in biology.

It is not necessary to memorize which elements are metals and which are nonmetals, because the periodic table contains this information. Look at the periodic table and note the stepped diagonal line toward the right side. This line divides metals from nonmetals. The

Element	Symbol	Element	Symbol
a. Metals		*b. Nonmetals*	
Aluminum	Al	Argon	Ar
Barium	Ba	Boron	B
Beryllium	Be	Bromine	Br
Calcium	Ca	Carbon	C
Chromium	Cr	Chlorine	Cl
Cobalt	Co	Fluorine	F
Copper	Cu	Helium	He
Gold	Au	Hydrogen	H
Iron	Fe	Iodine	I
Lead	Pb	Neon	Ne
Lithium	Li	Nitrogen	N
Magnesium	Mg	Oxygen	O
Manganese	Mn	Phosphorus	P
Mercury	Hg	Silicon	Si
Nickel	Ni	Sulfur	S
Potassium	K		
Silver	Ag		
Sodium	Na		
Tin	Sn		
Zinc	Zn		

only exception is hydrogen, which is usually treated as a nonmetal despite its being to the left of the line. Figure 1.5 shows a periodic table which emphasizes these distinctions.

Most of the human body mass is composed of the nonmetallic elements, carbon, hydrogen, oxygen, nitrogen, phosphorus, and sulfur. These elements are not found free, that is, in the elemental state, in the body; rather, they are combined in intricate ways in various compounds. Only the element oxygen is found in the elemental state in the body. Other nonmetals such as chlorine and iodine are required in the body in lesser amounts.

Many of the metallic elements are also required for good health, and their absence or insufficiency in the body leads to a deficiency symptom. Metals are always found in their combined form in the body, never free.

Compounds

Compounds are formed when two or more elements are chemically combined in a fixed ratio. *Chemical combination* means that a chemical change has occurred, that the combining elements disappear, and that

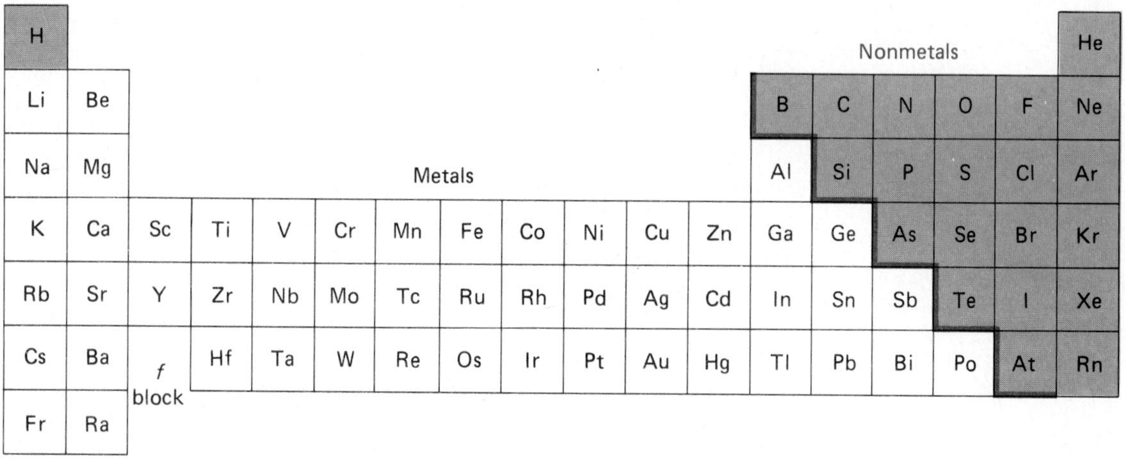

H															Nonmetals			He
Li	Be												B	C	N	O	F	Ne
Na	Mg					Metals							Al	Si	P	S	Cl	Ar
K	Ca	Sc	Ti	V	Cr	Mn	Fe	Co	Ni	Cu	Zn	Ga	Ge	As	Se	Br	Kr
Rb	Sr	Y	Zr	Nb	Mo	Tc	Ru	Rh	Pd	Ag	Cd	In	Sn	Sb	Te	I	Xe
Cs	Ba	*f* block	Hf	Ta	W	Re	Os	Ir	Pt	Au	Hg	Tl	Pb	Bi	Po	At	Rn
Fr	Ra																

FIGURE 1.5

A stepped diagonal line divides the periodic table into metals on the left and nonmetals on the right. An exception is hydrogen. Notice that there are many more metals than nonmetals.

a new substance with entirely different properties, the compound, appears. Because the elements are chemically combined to form the compound, a chemical change is necessary to separate them. Compounds are distinguished most readily from elements by experiments such as those described in Table 1.2.

The fixed-ratio notion is similar to the idea that to make a certain cake turn out the same way every time, you always follow the same recipe, with the specified amounts of each ingredient. This *constant fixed ratio* of elements in compounds gives compounds the *constancy of composition* that is a characteristic of pure substances, as described in Section 1.5. Composition is constant in elements because there is only one component; composition is constant in compounds because the ratio of component elements is fixed. This property of a compound that its component elements always appear in a fixed ratio by weight was first noticed and formalized as the **Law of Definite Composition (or Proportion)** in 1797.

Water is a common example of a compound. The elements hydrogen, a flammable gas, and oxygen, the gas needed for life, combine in the fixed proportion 1:8 by weight to give the compound water, which has the properties with which we are familiar. Only a chemical change such as that produced by an electric current can separate water into its components, as described in Table 1.2 and shown in Figure 1.6. Table salt (sodium chloride) is another compound. It is the chemical combination of the elements sodium and chlorine in the ratio 23:35.5 by weight. We looked at the very different properties of these elements and this compound in Section 1.6 and Figure 1.4. In Section 6.17 you will see how the fixed combining ratios of elements in compounds are determined.

FIGURE 1.6

(a) An electric current is being passed through water in the two glass tubes (burettes). The current breaks down the water (H_2O) into hydrogen gas and oxygen gas, bubbles of which can be observed. (b) The gases collect in the volume ratio 2 parts hydrogen (left side) to 1 part oxygen (right side) as shown in the close-up shot. This fact is reflected in the formula H_2O. (*Photographs by Bryan Lees*)

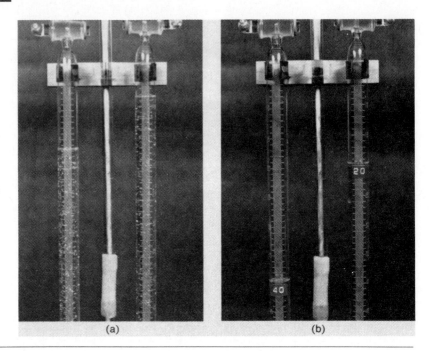

(a)   (b)

Mixtures

Mixtures are physical combinations of two or more elements, or of elements and compounds, or of two or more compounds. Because the combination is merely a physical mixing, the components of a mixture can be separated by physical means. For example, consider a mixture of iron filings (an element) and salt (a compound). We know this combination is a mixture because we can separate the components by a physical process. We can use a magnet to attract the iron away from the salt. Or we can place the mixture in water (which dissolves the salt) and filter, thereby separating the iron filings. The salt (saline) solution is also a mixture. The mixed compounds, salt and water, can be separated by the physical change of boiling the water away and thereby leaving the salt behind (see Figure 1.7).

In addition to physical separability, the other property of mixtures is the lack of a fixed proportion of components; i.e., the composition of a mixture is *not constant*. Consider the preceding mixture. We could have a 50:50 mix of iron filings and salt, or we could have more iron and less salt, or much salt and little iron. Similarly, in making a salt solution, we could dissolve just a pinch of salt in a large amount of water or we could dissolve several tablespoons. This is what is meant by no fixed proportion or nonconstant composition.

The lack of constant composition of a mixture shows up in its physical properties, as mentioned before (see Figure 1.3). As another

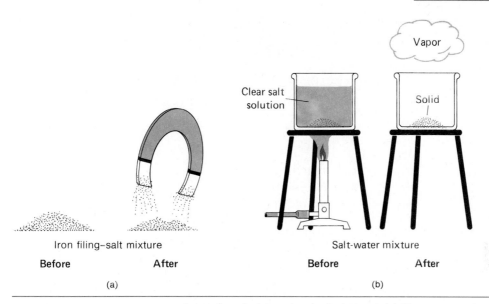

Iron filing–salt mixture

Before   After

(a)

Salt-water mixture

Before   After

(b)

FIGURE 1.7

Separation of mixtures by physical means. (a) After a magnet is applied to the iron filing–salt mixture, some of the iron has been separated from the mixture because it is attracted to the magnet while the salt is not. (b) Water can be separated from salt in a salt solution by boiling the water away into the air, thus leaving the solid salt behind in the beaker.

example, the boiling point of a salt-water solution varies with the amount of salt mixed with the water. The freezing point of the solution also varies with the composition of the salt-water mixture. In contrast, pure table salt always melts at 801°C and pure water boils at 100°C [at 1 atmosphere (atm)].

SAMPLE EXERCISE

1.6

Classify the following as elements, compounds, or mixtures:

a. Air
b. Gold
c. Iron filings and iodine crystals stirred together in a beaker
d. Iron filings and iodine chemically combined in the proportion 55.85 : 253.8 by weight
e. Toothpaste

SOLUTION

a. Mixture. The principal components are the elements oxygen and nitrogen. Air also contains trace amounts of other gases and dust and other pollutants.
b. Element. See element 79 in the periodic table.
c. Mixture. A physical combination is described.
d. Compound. A chemical combination in a fixed proportion is described.
e. Mixture. Read the ingredients on your toothpaste tube.

Problem 1.7

Classify the following as elements, compounds, or mixtures:

a. Wine
b. Chalk, a 40:12:48 (by weight) combination of calcium, carbon, and oxygen
c. Magnesium
d. Vitamin C, a 9:1:12 (by weight) combination of carbon, hydrogen, and oxygen

Figure 1.8 summarizes the overall scheme most useful for classifying matter. All matter can be classified as either mixtures or pure substances. Furthermore, pure substances can be subdivided into elements or compounds. This figure also shows us that we can classify matter as homogeneous or heterogeneous.

Problem 1.8

FIGURE 1.8

Which of the materials listed in Sample Exercise 1.6 and Problem 1.7 are pure substances?

All matter is subdivided into two large categories: pure substances and mixtures. Then there are further subdivisions as shown. Also, the properties of each subdivision are described briefly.

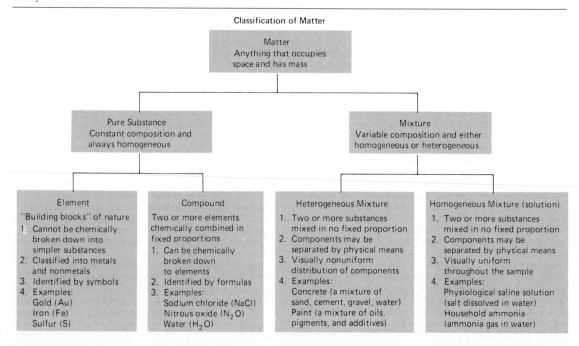

Classification of Matter

Matter
Anything that occupies space and has mass

Pure Substance
Constant composition and always homogeneous

Mixture
Variable composition and either homogeneous or heterogeneous

Element
"Building blocks" of nature
1. Cannot be chemically broken down into simpler substances
2. Classified into metals and nonmetals
3. Identified by symbols
4. Examples:
   Gold (Au)
   Iron (Fe)
   Sulfur (S)

Compound
Two or more elements chemically combined in fixed proportions
1. Can be chemically broken down to elements
2. Identified by formulas
3. Examples:
   Sodium chloride (NaCl)
   Nitrous oxide (N₂O)
   Water (H₂O)

Heterogeneous Mixture
1. Two or more substances mixed in no fixed proportion
2. Components may be separated by physical means
3. Visually nonuniform distribution of components
4. Examples:
   Concrete (a mixture of sand, cement, gravel, water)
   Paint (a mixture of oils, pigments, and additives)

Homogeneous Mixture (solution)
1. Two or more substances mixed in no fixed proportion
2. Components may be separated by physical means
3. Visually uniform throughout the sample
4. Examples:
   Physiological saline solution (salt dissolved in water)
   Household ammonia (ammonia gas in water)

Classify the following pure substances as elements or compounds:

a. Sodium bicarbonate   **c.** Chlorine
b. Carbon tetrachloride   **d.** Potassium

SOLUTION

Refer to Table 1.4 or the periodic table to identify the elements. The other pure substances must be compounds. Thus, **a** and **b** are compounds; **c** and **d** are elements.

HOMOGENEOUS VERSUS HETEROGENEOUS MATTER

1.10 **Homogeneous** matter is uniform; i.e., it has the same composition throughout. Elements are homogeneous because they consist of just one component throughout. Compounds are homogeneous because they have a uniform composition, a fixed proportion of components throughout. Thus pure substances are always chemically homogeneous.

Mixtures, on the other hand, can be either homogeneous or **heterogeneous,** i.e., *not* uniform in composition throughout. Most mixtures of solids are heterogeneous. For example, no matter how carefully we mix together iron filings and sugar, some parts of the mixture will have more iron and some parts more sugar. Mixtures of components in two different physical states are always heterogeneous. For example, a combination of iron filings and water is heterogeneous.

Solutions are homogeneous mixtures. When sugar is stirred into water and dissolves, the smallest particles intermingle and the solution is uniform throughout (having the same proportions of sugar and water). Note that the proportion of components in a solution is uniform, but not fixed or constant.

One simple way to decide whether a sample of matter is homogeneous or heterogeneous is to use your eyes to tell whether the sample is uniform throughout. If a sample is uniform in appearance throughout, it is probably homogeneous. If a sample is obviously not uniform in its composition, it is heterogeneous. Often a microscope is necessary to make this distinction.

Tell whether the materials in Sample Exercise 1.6 are homogeneous or heterogeneous.

SOLUTION

Your answers to this exercise depend to some extent on your perception of the material.

a.   The air in most metropolitan areas is definitely heterogeneous. The solid pollutants are distributed nonuniformly throughout the gases. Close inspection shows specks of solid in the gaseous air. "Clean" air (perhaps on a mountain top) may be homogeneous if it is a true solution of gases. In that case, no solid specks would be visible.

b.   Homogeneous. All elements are chemically homogeneous. Gold is a shiny solid throughout.

c.   Heterogeneous. This is a mixture of solids similar to the iron-sugar mixture described previously.

d.   Homogeneous. All compounds are chemically homogeneous. This compound is a gray-black solid throughout.

e.   Most toothpastes are heterogeneous. This is particularly obvious if you use a speckled brand. The specks do not appear uniformly. In addition, toothpaste is looser (more liquid) in spots and thicker (more solid) in other spots. Clear toothpastes may be homogeneous.

Heterogeneous matter is always a mixture, but a mixture is not always heterogeneous. Homogeneous matter may be either a mixture or a pure substance.

To test your understanding of many of the preceding ideas, think about classifying a sample of ice floating in water as homogeneous or heterogeneous. We know that water is a compound, and therefore it is chemically homogeneous (1 part H to 8 parts O by weight throughout) in both the liquid and solid states (ice). However, in a sense we have a heterogeneous mixture. The sample is not uniform throughout—one part is solid, one part is liquid. Furthermore, the solid and liquid parts can be separated by physical means.

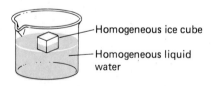
— Homogeneous ice cube

— Homogeneous liquid water

The sample as a whole is heterogeneous because we can see both a solid and a liquid part. The individual components are homogeneous. Therefore, this sample is heterogeneous because it is a mixture of two different physical states of a homogeneous compound.

Problem 1.9

Classify the materials in Problem 1.7 as homogeneous or heterogeneous.

CHEMICAL LANGUAGE

1.11 For the sake of convenience and precise communications, chemists have devised a system of abbreviations for elements and compounds with which you must become familiar. To a large extent your success in chemistry courses will depend on how well you understand the symbolic representations of substances.

Elements are represented by *symbols*. You should memorize the symbols for all the elements in Table 1.4. It would be very helpful to you to learn to think in terms of symbols. For example, when you hear the word "hydrogen," H should appear in your mind's eye. The sound of the word *iron* should conjure up Fe along with (or rather than) the letters "i r o n." Notice that the symbols always consist of either one capital letter or one capital letter followed by one small letter. These are the only correct forms. Carbon is always C, never c; calcium is always Ca, never CA, ca, or cA. Sticking to these correct forms for symbols is essential if you are to write correct representations of compounds.

Compounds are represented by *formulas*. The symbols of the combined elements are written without any space between them in a chemical formula. The formula tells not only the elements that have chemically combined to form the compound, but also the fixed proportion in which the chemical combination occurs. For example, the compound table salt has the formula NaCl, which immediately tells you sodium and chlorine have combined chemically. In Chapter 6 you will see how the formula tells you that the fixed proportion is $23:35.5$ by weight. Water is H_2O. Table sugar is $C_{12}H_{22}O_{11}$. The total meanings of these formulas will unfold in Chapters 6 and 7. For now you should realize that the formula tells you the elements that have chemically combined. Also recognize that a correctly written formula employs correct symbols, and the numbers appear as subscripts (i.e., they appear below the line on which the symbols are written).

Relating chemical language to the English language, we can see that the organization is similar. Symbols are like the letters in the alphabet. Symbols are put together to make formulas just as letters are assembled into words. Words are assembled into sentences. In Chapter 9 you will see formulas assembled into chemical equations which tell us about chemical reactions (see Figure 1.9). In your English courses you invest time and effort in order to learn forms (grammar) which increase your ability to write effectively and communicate what you really mean. In chemistry it is even more essential that you master the meaning and forms of chemical language so that there is communication between you and your instructor.

SAMPLE EXERCISE

1.9

Classify the following as elements, compounds, or mixtures:

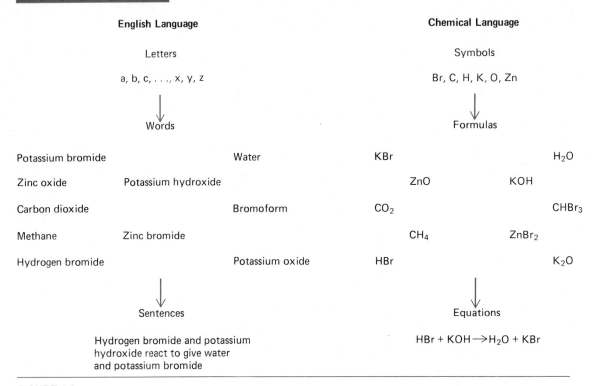

English Language

Letters

a, b, c, . . ., x, y, z

↓

Words

Potassium bromide   Water

Zinc oxide   Potassium hydroxide

Carbon dioxide   Bromoform

Methane   Zinc bromide

Hydrogen bromide   Potassium oxide

↓

Sentences

Hydrogen bromide and potassium
hydroxide react to give water
and potassium bromide

Chemical Language

Symbols

Br, C, H, K, O, Zn

↓

Formulas

KBr H_2O

ZnO   KOH

CO_2 $CHBr_3$

CH_4 $ZnBr_2$

HBr K_2O

↓

Equations

$HBr + KOH \longrightarrow H_2O + KBr$

FIGURE 1.9

Similarities in the construction
of the English language and
chemical language are
illustrated.

a. K_2S

b. K and S

c. CO_2 and H_2O

d. AgCl

e. Au

SOLUTION

a. Compound, represented by a formula
b. Mixture, the physical combination of the elements K and S
c. Mixture, the physical combination of the compounds CO_2 and H_2O
d. Compound
e. Element

ENERGY

1.12   As you learned in Section 1.1, matter is defined as anything
that occupies space and has mass. In addition to mass, all matter has
energy. **Energy** is the ability to do work, or we might say the ability
to cause some change. The change may be in such things as the
position of an object or its temperature.

There are different types or forms of energy: heat energy, electric
energy, solar energy, and atomic energy, to name a few. All forms of

energy can be classified as either kinetic or potential energy. **Kinetic energy** is the energy of matter in motion. A moving car possesses kinetic energy which lets it do the work of carrying people or pulling a trailer. Notice that this work effects a change in position.

In chemistry, we talk more about **potential energy,** which is the energy matter possesses because of its position or composition. A rock at the top of a hill has potential energy because of its position. If it begins to roll down the hill, the potential energy is converted into kinetic energy. You contain a great deal of potential energy. Move your arm. You just changed some of your potential energy into kinetic energy. Your potential energy comes from the composition of the matter in your body. All matter contains **chemical energy,** the potential energy which comes from the way the matter is put together. Gasoline contains chemical energy, which is released when it burns. Note that chemical energy and all potential energy are *stored* energy. Stored potential energy is used by changing it to a form of kinetic energy.

The energy changes that occur within an automobile cylinder offer good examples of what is meant by a change in the form of energy. The gasoline-air mixture, rich in chemical energy, is introduced into the cylinder. Electric energy produces the spark (light energy) that ignites the gasoline, converting the chemical energy of gasoline into heat energy and the mechanical energy of the moving piston. Thus the potential energy of gasoline is converted into the kinetic energy of the moving piston and moving car.

Although energy can change form, the total amount of energy in the universe is a constant. This is one way of stating the **Law of Conservation of Energy.** Another way to state this law is to say energy is neither created nor destroyed in any change in matter. Both statements tell us that we can never make new energy and we can never destroy what we have.

What is meant, then, when we hear on the news that we must save energy? The answer is that saving energy means keeping it in a useful, potential (stored) form. The message is that we must save fuels, rich sources of chemical energy. If we burn all our fuel, the universe will still contain the same amount of energy, but it will not be in a form that is useful to us.

SUMMARY

Chemistry is the study of matter and changes in matter. Matter can be classified in various useful ways (Section 1.1). For example, matter can be classified according to physical state (Section 1.2).

Classification schemes can be established by using the reasoning process known as the scientific method. This general method of science involves making observations, then offering hypotheses which are tested by experiment and further observation. Depending on the results of the experiments, conclusions are drawn in order to explain the why of the behavior of matter (Section 1.4).

An important classification of matter is based on constancy of composition. On this basis all matter can be placed into one of two categories: mixtures or pure substances (Section 1.5). Pure substances can be subdivided into classes, either elements or compounds, by experimental testing (Section 1.7).

Matter can undergo either physical or chemical changes. Changes in physical state (Section 1.3) are examples of physical changes during which no new pure substance is formed. Changes in which new pure substances are formed are called chemical changes (Section 1.6). Pure substances are characterized by noting their physical and chemical properties, which are associated with physical and chemical changes, respectively (Section 1.8).

Figure 1.8 summarizes the classification of all matter into the categories elements, compounds, or mixtures. A list of all known elements appears in the periodic table inside the front cover. Compounds are combinations of elements in fixed proportion which can be separated only by chemical change. Mixtures are combinations of substances in no fixed proportion which can be separated by physical means (Section 1.9). All pure substances are homogeneous. Mixtures may be homogeneous or heterogeneous (Section 1.10).

The chemical language represents elements by symbols, compounds by formulas, and reactions by chemical equations (Section 1.11).

All matter contains chemical energy, the potential energy which comes from the way matter is put together. Potential energy can be changed into kinetic energy, the energy of matter in motion (Section 1.12).

CHAPTER ACCOMPLISHMENTS

After completing this chapter you should be able to

Introduction
1. Define chemistry.
2. Define matter.

Classification of matter by physical state
3. State the shape and volume characteristics of the three physical states.
4. Classify common samples of matter according to physical state.

Changes in state

5. Name the processes by which matter changes physical state.

Scientific method

6. State the four basic steps of the scientific method.

Pure substances versus mixtures

7. Distinguish between the characteristics of pure substances and mixtures.

8. Classify common samples of matter as pure substances or mixtures.

Physical versus chemical change

9. Distinguish between physical and chemical changes.

Subdivisions of pure substances

10. Given experimental data, classify substances as elements or compounds.

Physical and chemical properties

11. Distinguish between physical and chemical properties.

12. Given a list of samples and their melting and boiling points, classify the substances according to their physical state at a given temperature.

13. Given a description of physical and chemical properties, distinguish between samples of matter.

Elements, compounds, and mixtures

14. Define the terms "element," "compound," and "mixture."

15. Classify common samples of matter as elements, compounds, or mixtures.

16. Given a periodic table, classify elements as metals or nonmetals.

Homogeneous versus heterogeneous matter

17. Classify common samples of matter as homogeneous or heterogeneous.

Chemical language

18. Given the name, give the symbol of each element in Table 1.4.

19. Given the symbol, give the name of each element in Table 1.4.

20. Recognize a correctly written formula.

21. Given the formula(s) of the substance(s) present, classify samples of matter as elements, compounds, or mixtures.

Energy

22. Define energy.
23. Distinguish between potential and kinetic energy.
24. Given a description of an energy transformation, identify the energy forms as kinetic or potential.
25. State the Law of Conservation of Energy.

PROBLEMS

1.1 Introduction

1.10 What does the study of chemistry involve?

1.11 What is matter?

1.2 Classification of matter by physical state

1.12 Describe experimental tests that will allow you to distinguish between the solid, liquid, and gaseous states.

1.13 What is meant by the statement "a gas has indefinite volume"?

1.14 Which of the three states of matter possess the property of fluidity?

1.15 Which of the three states of matter can be poured from one container into another?

1.16 For each of the following common substances, indicate whether it exists as a solid, liquid, or gas at room temperature:
  a. Tap water
  b. Silver
  c. Skin
  d. Mercury
  e. Carbon dioxide
  f. Molasses
  g. Chocolate pudding

1.17 A sample of matter is in the liquid state. Describe how you might convert it to a
  a. Solid
  b. Gas

1.3 Changes in state

1.18 Classify each of the materials below as a solid, liquid, or gas at room temperature and at "dry ice" temperature ($-80°C$).
  a. Ice cream
  b. Iodized table salt
  c. Cola
  d. A nickel
  e. Dry ice (frozen CO_2)
  f. Snow
  g. Methane

1.19 What name do we give to each of the following processes?
  a. Ice cubes turn to water.
  b. Steam forms water droplets on a mirror.
  c. Milk is made into ice milk.
  d. Frozen laundry "dries."
  e. Perspiration "dries."

1.20 What change in physical state occurs during the formation of the following?
  a. Rain
  b. Frost
  c. Snow
  d. Steam

1.21 A sealed glass bulb is half-filled with water, on which some ice and wood are floating. The remainder of the bulb is filled with air. How many physical states are present? Identify them.

1.4 Scientific method

1.22 Describe the steps of the scientific method as applied to a medical diagnosis by a physician.

1.5 Pure substances versus mixtures

1.23 Classify each of the materials in Problem 1.18 as a pure substance or a mixture.

1.24 Based on your everyday experience, how would you separate the following mixtures into their components?
  a. A mixture of sand and water
  b. A mixture of sugar and sand
  c. A homogeneous mixture of sugar and water

1.25 Design experiments that would prove that your classifications of cola and dry ice in Problem 1.23 were correct.

1.6 Physical versus chemical change

1.26 Classify each of the following as a physical or chemical change:

a. Water evaporates from a glass.

b. Metabolism, i.e., complex foodstuffs are broken down in the body to smaller substances, carbon dioxide and water.

c. Wood is sanded.

d. Meat is ground up into hamburger patties.

e. A firecracker explodes.

f. A leaf turns color in autumn.

g. Gasoline is burned in a car engine.

1.27 Identify the chemical and physical changes in the following sequences:

a. A lump of sugar is ground to a powder and then heated in air. It melts, then darkens, and finally bursts into flame and burns.

b. Gasoline is sprayed into the carburetor, mixed with air, converted to vapor, and burned, and the combustion products expand in the cylinder.

1.28 The electrolysis of water produces bubbles (see Figure 1.6). Boiling water produces bubbles. How does the composition of the bubbles tell us that one process is chemical and the other is physical?

1.29 Which of the following changes is physical and which chemical?

a. Melting butter

b. Iron rusting

c. Flash cube going off

d. Banging on a drum

1.7 Subdivisions of pure substances

1.30 A pure blue powder, when heated in a vacuum, releases a greenish-colored gas and leaves behind a white solid. Is the original blue powder a compound or element?

1.31 A shiny, metallic substance conducts an electric current without a change in its properties. The substance is heated until it liquefies, and then an electric current is passed through the liquid again without a change in properties. Is the substance likely to be an element or a compound?

1.32 Pure substances can be classified into how many categories? Identify them.

1.33 Classify the pure substances in Problem 1.18 as elements or compounds.

1.34 Are there more elements or compounds in the world? Explain.

1.8 Physical and chemical properties

1.35 Use the data in Table 1.3 to classify the following substances according to their physical state at the given temperature:

a. Sucrose at 78°C   **d.** Chlorine at
b. Iron at 3200°C   −121°C
c. Iron at 1700°C   **e.** Oxygen at
                                  −195°C

1.36 State three properties that may be used to identify specimens of pure substances.

1.37 What properties distinguish water from other colorless liquids such as alcohol, benzene, and acetone?

1.38 What properties distinguish white solids such as table salt and table sugar from one another?

1.39 Are the following properties of a certain metal physical or chemical?

a. It is a solid at room temperature.

b. It is easily bent into shapes.

c. It combines with water violently, giving off a gas.

d. It melts at a low temperature.

1.40 Identify the physical and chemical properties indicated in the following description. Chlorine is a greenish-yellow gas which is toxic to humans. It combines with sodium to form sodium chloride, a solid which melts at 808°C.

1.41 Compare the similarities and differences between the processes of melting sugar and dissolving sugar in water.

1.9 Elements, compounds, and mixtures

1.42 Classify the following as elements, compounds, or mixtures.

a. Sodium

b. Spaghetti with meatballs

c. Carbon particles in a hydrogen gas atmosphere

d. A cup of tea

e. Laughing gas (14 parts N to 8 parts O by weight)

f. Methane (75 percent carbon, 25 percent hydrogen by weight)

1.43 Identify the components of the following mixtures as elements or compounds:

a. Club soda (CO_2 gas in water with a pinch of salt)
b. "Clean" air
c. Aerated water

1.44 Classify the following pure substances as elements or compounds:
a. Copper
b. Iron oxide
c. Sulfur tetrafluoride
d. Fluorine

1.45 Elements *A* and *B* combine to form *AB* in a fixed proportion of 3:1 by weight. What will happen if we try to combine 4 g of *A* with 1 g of *B*?

1.46 a. Describe the basic steps of the scientific method.
   b. Apply the steps to the problem of distinguishing elements as metals and nonmetals.

1.10 Homogeneous versus heterogeneous matter

1.47 What terms do we apply to distinguish between uniform and nonuniform matter?

1.48 Classify the materials in Problems 1.16, 1.18, 1.42, and 1.43 as homogeneous or heterogeneous.

1.49 Water and white wine have in common that they are both homogeneous. How do the two materials differ?

1.11 Chemical language

1.50 Name the elements whose symbols are H, Ca, Si, C, Cl, P, and K.

1.51 Give symbols for the elements bromine, nitrogen, mercury, silver, and gold.

1.52 Classify the elements in Problems 1.50 and 1.51 as metals or nonmetals.

1.53 Tell whether the following symbols and formulas are written correctly or not. If not, tell what is wrong.
a. CL   d. f
b. C^3H^6 e. NaOH
c. Zn

1.54 Classify the following as elements, compounds, or mixtures:
a. C_2H_2 d. K_2SO_4
b. O_2 and N_2 e. H_2O and NaCl
c. Mg

1.55 Match each item on the left with as many descriptions on the right as apply.
a. C and P
b. $C_6H_{12}O_6$ (blood sugar)
c. C_2H_5OH and H_2O (alcohol in water)
d. A chocolate chip cookie
e. Oxygen dissolved in water

1. A pure compound
2. A mixture of compounds
3. An element
4. A mixture of elements
5. A homogeneous mixture
6. A heterogeneous mixture

1.12 Energy

1.56 What name do we give to the ability of matter to do work?

1.57 What are the two major classifications of energy?

1.58 For each of the following, state whether the energy described is potential or kinetic:
a. A book is poised at the edge of a table.
b. The book is falling.
c. You are walking.
d. Food.
e. A stretched rubber band.

1.59 a. Give an example of potential energy as a result of position.
   b. Give an example of potential energy as a result of composition.
   c. Can a substance have both types of potential energy, position and composition? Explain.

1.60 Give two examples of potential energy changing into kinetic energy.

1.61 State the Law of Conservation of Energy.

1.62 What is the source of energy for humans?

1.63 What happens if humans take in more energy than they use up?

MATH SKILLS

CHAPTER 2

He
Helium
4.003

INTRODUCTION

2.1 Chemists often describe laws, theories, and experimental results in quantitative (numerical) terms and therefore find it necessary to do calculations. To carry out the arithmetic calculations described in this text successfully, you will need only some very basic mathematical concepts and skills which you probably already have. We will review necessary math skills here so that you can "brush up" if you feel "rusty." We assume that you feel confident with addition, subtraction, multiplication, and division of positive decimal numbers. If not, and you need further review with any of these, please turn to Appendix 1. Also, Appendix 2 provides a brief discussion of the use of calculators for these operations. We will begin our review with addition, subtraction, multiplication, and division of signed numbers.

SIGNED NUMBERS

2.2 **A. What is a signed number?** Every number has a magnitude and a sign. The sign can be either positive or negative. If the sign is omitted, it is assumed to be positive.

Magnitude is 17.8. Magnitude is 13.2. Magnitude is 4.9.

$$+\,17.8 \qquad -\,13.2 \qquad 4.9$$

Sign is positive. Sign is negative. Sign is omitted, so we assume it is positive ($+$).

B. Multiplication of signed numbers The product of the multiplication of two factors is positive if both numbers have the same sign and negative if they have opposite signs.

In the following cases, the factors have *like signs,* so the product is *positive:*

$$(3)(6) = 18 \qquad\qquad (-3)(-6) = 18$$

Both positive Positive *Both* negative Positive

In the following cases, the factors have *opposite signs*, so the product is *negative:*

$$(-3)(6) = -18 \qquad\qquad (3)(-6) = -18$$

Negative Positive   Negative   Positive Negative   Negative

The sign of the product of more than two numbers can always be worked out two numbers at a time.

SAMPLE EXERCISE

2.1

What is the product of the following multiplication?

$$(-3)(6)(4)(5)(-2)$$

Factor   a   b   c   d   e

SOLUTION

Step 1   Work out the magnitude of the product of all the factors, which is $|720|$. The two vertical parallel lines indicate that no sign is implied.

Step 2   Work out the sign of the product, taking two signs at a time.

Taking the first two factors

$$\begin{array}{cc} a & b \\ (-)(+) & = \end{array} \quad \begin{array}{c} a \times b \\ (-) \end{array}$$

Multiplying times the third factor

$$\begin{array}{cc} a \times b & c \\ (-) & (+) \end{array} = \begin{array}{c} a \times b \times c \\ (-) \end{array}$$

Then times the fourth factor

$$\begin{array}{cc} a \times b \times c & d \\ (-) & (+) \end{array} = \begin{array}{c} a \times b \times c \times d \\ (-) \end{array}$$

Finally times the fifth factor

$$\begin{array}{cc} a \times b \times c \times d & e \\ (-) & (-) \end{array} = \begin{array}{c} a \times b \times c \times d \times e \\ (+) \end{array}$$

So the product is $+720$.

It turns out that if there are an *even number* of minus signs in a series of factors to be multiplied, the product will be positive. This is the case in Sample Exercise 2.1, in which there are two minus signs. An *odd number* of minus signs leads to a negative product.

$$(-3)(-6)(4)(-5) = -360$$

+   +

+   −

Problem 2.1

Indicate the sign of the product for the following multiplications:

a. $(3)(-6)(2)(4) =$

b. $(-4)(-5)(-3)(5) =$

C. Division of signed numbers   The sign of a quotient is determined in the same way that we determined the sign of a product. That is, the quotient of two numbers is positive if both numbers have the same sign and negative if they have opposite signs.

$$18 \div 6 = 3 \qquad -18 \div -6 = 3$$
$$-18 \div 6 = -3 \qquad 18 \div -6 = -3$$

Also as in multiplication, even numbers of minus signs lead to positive quotients and odd numbers of minus signs lead to negative quotients.

$$-18 \div 6 \div -3 = +1 \qquad -18 \div 6 \div 3 = -1$$
$$\text{Two negative signs} \qquad\qquad \text{One negative sign}$$

Where more than one division is indicated, proceed to divide from left to right two terms at a time.

D. Addition of two signed numbers

1.   In the addition of two numbers of like sign, simply add the magnitudes and attach the sign involved.

$$3 + 6 = 9$$
$$-3 + (-6) = -9$$
$$+41 + 13 = 54$$

2.   In the addition of two numbers of opposite sign, subtract the one of smaller magnitude from the one of larger magnitude and apply to the answer the sign of the number of larger magnitude.

SAMPLE EXERCISE

2.2

Add $-16 + 3$.

SOLUTION

Step 1   Recognize that the number of smaller magnitude is 3.

Step 2   Subtract 3 from 16: $16 - 3 = 13$.

Step 3   The sign of the number of larger magnitude is negative, so the answer is -13:

$$-16 + 3 = -13$$

E. Subtraction of two signed numbers The subtraction problem is converted into an addition problem by:

1. Changing the sign for the subtraction process ($-$) into the sign for the addition process ($+$)

2. Changing the sign of the subtrahend (the second number) into the opposite sign

$$14 - \qquad 6 \qquad = 14 + \qquad (-6)$$

Sign indicating Sign of Sign indicating Sign is now minus.
subtraction. subtrahend addition.
is plus.

Subtraction	Addition

$$-38 - (-14) = -38 + (+14)$$

$$14 - (-15) = 14 + (+15)$$

$$-18 - 13 = -18 + (-13)$$

This method works because subtraction is defined as addition of the negative (opposite sign) subtrahend.

Now that we have converted the problem into an addition problem, we simply follow the rules for the addition of two signed numbers.

$$14 + (-6) = 8$$

$$-38 + 14 = -24$$

$$14 + 15 = 29$$

$$-18 + (-13) = -31$$

Problem 2.2

Work out the following:

a. $(-1.4)(1.9) =$

b. $(-3.1)(-2.4) =$

c. $(6)(3)(2) =$

d. $(-6)(-3)(-2) =$

Problem 2.3

Do the following divisions:

a. $3.9 \div -1.3 =$

b. $-3.9 \div -1.3 =$

c. $-3.9 \div 1.3 =$

Problem 2.4

Calculate the following:

a. $7.4 + 3.8 =$	**d.** $7.4 - 3.8 =$
b. $7.4 + (-3.8) =$	**e.** $-7.4 - 3.8 =$
c. $-7.4 + (-3.8) =$	**f.** $-7.4 - (-3.8) =$

FRACTIONS

2.3 **A. Meaning of a fraction** A *fraction* $\frac{a}{b}$, where a is called the *numerator* and b is called the *denominator*, is an arithmetic expression meaning $a \div b$. Every fraction can be expressed as a decimal number by simply carrying out the required division.

$$\frac{3}{8} = 3 \div 8 = 0.375$$

Note that every number can be written as a fraction, with 1 as the denominator.

$$16 = \frac{16}{1} \qquad 4.2 = \frac{4.2}{1} \qquad 0.78 = \frac{0.78}{1}$$

A symbol, such as d, can also be written as $\frac{d}{1}$.

B. Equality of two fractions If two fractions $\frac{a}{b}$ and $\frac{c}{d}$ are equal, the following is true: $a \times d = c \times b$; i.e., the products of the numerator of one fraction times the denominator of the other fraction are equal. Let us look at this again.

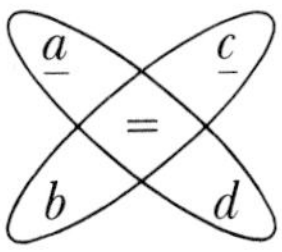

means

a	$\times$	d	$=$	c	$\times$	b
numerator (left)	$\times$	denominator (right)	$=$	numerator (right)	$\times$	denominator (left)

In Chapter 3 we will discuss a relationship $d = \frac{m}{v}$, which can be written as $\frac{d}{1} = \frac{m}{v}$ and therefore $d \times v = m \times 1$ or $m = d \times v$.

C. Multiplication of two fractions To multiply two fractions, first cancel any common factors in the numerator and the denominator, including any common units. Second, multiply the numerators together, including any uncanceled units. Then multiply the denominators, including any uncanceled units. Place the product of the numerators over the product of the denominators, as shown:

$$\left(\frac{a}{b}\right)\left(\frac{c}{d}\right) = \frac{a \times c}{b \times d}$$

Then convert fractional answers to decimal numbers by dividing the numerator of the answer by the denominator of the answer.

SAMPLE EXERCISE

2.3

Carry out the following:

$$\frac{13}{6} \times \frac{3}{4} =$$

SOLUTION

Step 1   3 is a common factor to itself $\left(\frac{3}{3} = 1\right)$ and 6 $\left(\frac{6}{3} = 2\right)$, so we may cancel out 3s.

$$\frac{13}{\cancel{6}_2} \times \frac{\cancel{3}^1}{4} = \frac{13}{2} \times \frac{1}{4}$$

There are no other common factors.

Step 2   Multiply the numerators together, then multiply the denominators.

$$\frac{13}{2} \times \frac{1}{4} = \frac{13 \times 1}{2 \times 4} = \frac{13}{8}$$

Step 3   Convert the fractional answer $\frac{13}{8}$ to a decimal number by dividing 13 by 8.

$$\frac{13}{8} = 1.625$$

SAMPLE EXERCISE

2.4

Carry out the following

$$\frac{2.54 \text{ cm}}{1.00 \text{ in}} \times \frac{12.0 \text{ in}}{1.00 \text{ ft}} =$$

SOLUTION

Step 1   The only obvious common factor is the unit inches, which we cancel:

$$\frac{2.54 \text{ cm}}{1.00 \cancel{\text{in}}} \times \frac{12.0 \cancel{\text{in}}}{1.00 \text{ ft}}$$

Step 2   Multiply together the numerators, then the denominators. Remember to put any remaining units into the numerator and denominator of the answer.

$$\frac{2.54 \text{ cm}}{1.00} \times \frac{12.0}{1.00 \text{ ft}} = \frac{2.54 \text{ cm} \times 12.0}{1.00 \times 1.00 \text{ ft}} = \frac{30.5 \text{ cm}}{1.00 \text{ ft}} = 30.5 \frac{\text{cm}}{\text{ft}}$$

D. Division of two fractions   When confronted by a division problem such as

$$\begin{array}{c} \text{Numerator} \\ \text{Denominator} \end{array} \quad \frac{a/b}{c/d} = \frac{a}{b} \div \frac{c}{d} = ?$$

$$\underset{\text{Dividend}}{} \qquad \underset{\text{Divisor}}{}$$

we solve the problem by converting the division operation into a multiplication.

To divide a fraction (a/b) by a fraction (c/d),

1.   Invert the denominator or divisor (c/d).

$$\frac{c}{d} \quad \text{inverted becomes} \quad \frac{d}{c}$$

2.   Multiply the numerator or dividend (a/b) by the inverted denominator (d/c), as in Section 2.3C.

$$\begin{array}{c} \text{Numerator} \\ \text{Denominator} \end{array} \quad \frac{a/b}{c/d} = \frac{a}{b} \div \frac{c}{d} = \frac{a}{b} \times \frac{d}{c} = \frac{a \times d}{b \times c}$$

$$\underset{\text{Dividend}}{} \qquad \underset{\text{Divisor}}{} \qquad \underset{\text{Inverted}}{}$$

SAMPLE EXERCISE

2.5

Carry out $\dfrac{3}{8} \div \dfrac{1}{4} = ?$

SOLUTION

$$\frac{3}{8} \div \frac{1}{4} = \frac{3}{8} \times \frac{4}{1} =$$

$$\frac{3}{\cancel{8}_2} \times \frac{\cancel{4}^1}{1} = \frac{3 \times 1}{2 \times 1} = \frac{3}{2} = 1.5$$

SAMPLE EXERCISE

2.6

Carry out $\dfrac{4}{9} \div 8 =$

SOLUTION

We can always write 8 as the fraction $\dfrac{8}{1}$. Then

$$\frac{4}{9} \div \frac{8}{1} = \frac{4}{9} \times \frac{1}{8} = \frac{\cancel{4}^1}{9} \times \frac{1}{\cancel{8}_2} = \frac{1 \times 1}{9 \times 2} = \frac{1}{18} = 0.055$$

SCIENTIFIC NOTATION

2.4 In scientific work it is often necessary to use extremely large and very small numbers. In this text we will work with such numbers as Avogadro's number, 602,200,000,000,000,000,000,000, and the mass of an electron, 0.00000000000000000000000000091095 g. We need a more convenient way to express such numbers. A way that eliminates the need to write all the zeros directly is desirable if for no other reason than to save ink and elbow grease. The method that is used is known as *scientific notation*. Let us review some of its basic principles.

A. Exponents and bases   You are probably already familiar with the idea of raising a number to a *power*. For example, 10^2 is read as 10 to the second power, and it is equal to 10×10, or 100. The 10 is known as the *base* and the 2 is called the *exponent*.

$$10^2 \quad \begin{matrix} \text{Exponent} \\ \text{Base} \end{matrix}$$

The exponent tells us the number of times the base is a factor in the multiplication. Review the following examples:

$$10^6 = \underbrace{10 \times 10 \times 10 \times 10 \times 10 \times 10}_{\text{Six factors of 10}} = 1{,}000{,}000$$

with Exponent and Base labels for 10^6

$$10^{10} = 10 \times 10 \times 10 \times 10 \times 10 \times 10 \times 10 \times 10 \times 10 \times 10$$

$$= 10{,}000{,}000{,}000$$

$$10^1 = 10$$

A base can also have a negative exponent, for example, 10^{-1}, 10^{-3}. We can always write a base with a negative exponent as the reciprocal, with the sign of the exponent changed to positive. The reciprocal of a number A is defined as $\dfrac{1}{A}$.

Negative exponent   Reciprocal of original

$$10^{-1} = \left(\frac{1}{10^1}\right) = 0.1$$

Same exponent with a positive sign

$$10^{-3} = \frac{1}{10^3} = \frac{1}{1000} = 0.001$$

Since $10^1 = 10$ and $10^{-1} = 0.1$, it is reasonable that $10^0 = 1$. We will prove this to you in Section 2.4D.

B. Writing numbers in scientific notation Any ordinary decimal number can be expressed as a decimal number between 1 and 10 multiplied by some power of 10. For example, the number 392,000 written in scientific notation looks like

$$\underline{3.92} \qquad \times \qquad \underline{10^5}$$

Decimal number   Power of 10
between 1 and 10

The ordinary decimal number 0.00432 written in scientific notation looks like

$$\underline{4.32} \qquad \times \qquad \underline{10^{-3}}$$

Decimal number   Power of 10
between 1 and 10

The rules for converting ordinary numbers to scientific notation are subdivided into three cases depending on the size of the number.

General rules for converting ordinary decimal numbers into scientific notation

IF THE NUMBER IS EQUAL TO OR GREATER THAN 10

1. Move the decimal point to the left, counting the number of places you must move it until you have a decimal number between 1 and 10.

2. Multiply this decimal number by 10 raised to a *positive* power equal to the number of places you moved the decimal point.

SAMPLE EXERCISE

2.7

Write 138.34 in scientific notation.

SOLUTION

We recognize that 138.34 is greater than 10. Therefore,

1. We move the decimal point to the left until we obtain a number between 1 and 10: 138.34 becomes 1.3834. This requires a movement of two places.

2. Now we write down the new number and multiply it by 10^2, because the decimal was moved two places. (The sign is positive because the number is greater than 10.)
Answer: 1.3834×10^2

IF THE NUMBER IS LESS THAN 1

1. Move the decimal point to the right, counting the number of places you must move it until you have a decimal number between 1 and 10.

2. Multiply this decimal number by 10 raised to a *negative* power equal to the number of places you moved the decimal point.

SAMPLE EXERCISE

2.8

Write 0.000108 in scientific notation.

SOLUTION

We recognize that 0.000108 is less than 1. Therefore,

1. We move the decimal point to the right until we obtain a number between 1 and 10: 0.000108 becomes 1.08. This requires a movement of four places.

2. We write down the new number and multiply it by 10^{-4}, because the decimal was moved four places. The sign is negative because the number is less than 1.

Answer: $0.000108 = 1.08 \times 10^{-4}$.

IF THE NUMBER IS EQUAL TO OR GREATER THAN 1.0 AND LESS THAN 10.0

1. Simply write down the number and multiply it by 10^0, because no (zero) movement of the decimal point is required.

SAMPLE EXERCISE

2.9

Write 3.85 in scientific notation.

SOLUTION

We recognize that the number is between 1.0 and 10.0. Therefore, we write the number down and multiply it by 10^0.

Answer: $3.85 = 3.85 \times 10^0$.

Very often 10^0 is omitted, just as the number 1 often is.

Problem 2.5

Write the following in scientific notation:

a. 173

b. 0.0029

c. 131,982

d. 0.00000401

e. 16.4

f. 8.1

C. Converting scientific notation into ordinary decimal numbers   To convert a number in scientific notation, such as 4.923×10^2, into an ordinary decimal number, we start by examining the exponent. If the exponent is positive, we move the decimal to the right a number of places equal to the value of the exponent. You may have to fill in zeros. Remember that positive exponents are associated with numbers greater than 10.

If the exponent is negative, we move the decimal to the left a number of places equal to the value of the exponent. Again, you may have to fill in zeros. Remember that negative exponents are associated with numbers that are less than 1.

SAMPLE EXERCISE

2.10

Write 4.923×10^2 as an ordinary decimal number.

SOLUTION

The exponent is positive 2, so we move the decimal point two places to the right.

$$4.923 \times 10^2 = 492.3$$

SAMPLE EXERCISE

2.11

Write 9.23×10^{-5} as an ordinary decimal number.

SOLUTION

The exponent is negative 5, so we move the decimal point five places to the left, filling in zeros.

$$00009.23 \times 10^{-5} = 0.0000923$$

Problem 2.6

Write the following as decimal numbers:

a. 7.31×10^3 **c.** 6.38×10^5

b. 1.92×10^{-4} **d.** 8.36×10^{-5}

D. Multiplying numbers in scientific notation   To multiply two numbers written in scientific notation, multiply the two decimal numbers together in the usual manner and then *add* the two exponents, being careful to treat the exponents as signed numbers. The sum is the correct power of 10 in the product.

$$(a \times 10^x)(b \times 10^y) = ab \times 10^{x+y}$$

EXAMPLES

$$(2 \times 10^4)(3 \times 10^7) = 6 \times 10^{4+7}$$
$$= 6 \times 10^{11}$$

$$(4.11 \times 10^{-1})(1.38 \times 10^3) = 5.67 \times 10^{-1+3}$$
$$= 5.67 \times 10^2$$

$$(3.39 \times 10^{-8})(2.25 \times 10^{-4}) = 7.63 \times 10^{-8+(-4)}$$
$$= 7.63 \times 10^{-12}$$

$$(9.11 \times 10^5)(2.18 \times 10^3) = 19.9 \times 10^{5+3}$$
$$= 19.9 \times 10^8$$

Notice in the last example that the decimal number of the answer turns out to be greater than 10, that is, 19.9. This answer, 19.9×10^8, is not in the correct form and must be adjusted so that the decimal number is between 1 and 10. We do this by recognizing that $19.9 = 1.99 \times 10^1$. Therefore,

$$19.9 \times 10^8 = 1.99 \times 10^{\textcircled{1}} \times 10^{\textcircled{8}} = 1.99 \times 10^9 \quad \text{Sum}$$

Summarizing the method for multiplication,

1. Multiply the decimal numbers in the usual manner.

2. Add the exponents.

3. Adjust the answer to the correct form of scientific notation.

SAMPLE EXERCISE

2.12

Carry out the following multiplication:

$$(6.42 \times 10^{-9}) \times (2.58 \times 10^2) =$$

SOLUTION

Follow the preceding summarized steps.

Step 1 $6.42 \times 2.58 = 16.6$

Step 2 $-9 + 2 = -7$

Step 3 But 16.6×10^{-7} must be adjusted because 16.6 is not between 1 and 10. Putting 16.6 in scientific notation, we get 1.66×10^1. Substituting this for 16.6, we get

$$16.6 \times 10^{-7} = 1.66 \times 10^{\textcircled{1}} \times 10^{\textcircled{-7}} = 1.66 \times 10^{-6} \quad \text{Sum}$$

Problem 2.7

Multiply the following, expressing your answer in scientific notation:

a. $(2.76 \times 10^2)(8.32 \times 10^4)$

b. $(3.37 \times 10^1)(4.89 \times 10^0)$

c. $(2.51 \times 10^{-8})(1.53 \times 10^{-6})$

d. $(8.13 \times 10^{-5})(3.89 \times 10^6)$

Now that you know how to multiply exponential numbers, you can accept the proof that $10^0 = 1$.

$$10^1 \times 10^{-1} = 10^0$$
$$10 \times 0.1 = 1$$

Therefore,

$$10^0 = 1$$

E. Division of two numbers in scientific notation   The process of division in scientific notation is similar to that of multiplication. To divide two numbers written in scientific notation, we divide one decimal number by the other in the usual manner and then we *subtract* the exponent in the denominator (divisor) from the exponent in the numerator (dividend), being careful to treat this as the subtraction of signed numbers. The difference is the correct power of 10 in the quotient.

$$\frac{a \times 10^x}{b \times 10^y} = \frac{a}{b} \times 10^{x-y}$$

EXAMPLES

$$\frac{8 \times 10^4}{4 \times 10^7} = 2 \times 10^{4-7} = 2 \times 10^{-3}$$

$$\frac{4.11 \times 10^{-1}}{1.38 \times 10^3} = 2.98 \times 10^{-1-3} = 2.98 \times 10^{-4}$$

$$\frac{3.39 \times 10^{-8}}{2.25 \times 10^{-4}} = 1.51 \times 10^{-8-(-4)} = 1.51 \times 10^{-4}$$

$$\frac{2.18 \times 10^3}{9.11 \times 10^5} = 0.239 \times 10^{3-5} = 0.239 \times 10^{-2}$$

Notice in the last example that the decimal number is not between 1 and 10, so we must rewrite this answer in the correct form. We do this by recognizing that $0.239 = 2.39 \times 10^{-1}$. Therefore,

$$0.239 \times 10^{-2} = 2.39 \times 10^{-1} \times 10^{-2} = 2.39 \times 10^{-3} \quad \text{Sum}$$

Summarizing the method for division,

1. Divide the decimal numbers in the usual manner.

2. Subtract exponents (numerator minus denominator).

3. Adjust the answer to the correct form of scientific notation.

SAMPLE EXERCISE

2.13

Carry out the division

$$\frac{1.25 \times 10^8}{6.12 \times 10^4} =$$

SOLUTION

Follow the preceding summarized steps.

Step 1 $1.25 \div 6.12 = 0.204$

Step 2 $8 - 4 = 4$

Step 3   But 0.204×10^4 must be adjusted because 0.204 is not between 1 and 10. Putting 0.204 in scientific notation, we get 2.04×10^{-1}. Substituting this for 0.204, we obtain

$$0.204 \times 10^4 = 2.04 \times 10^{-1} \times 10^{4} = 2.04 \times 10^3 \quad \text{Sum}$$

Problem 2.8

Divide the following, expressing your answer in scientific notation:

a. $\dfrac{8.0 \times 10^7}{4.0 \times 10^3} =$

c. $\dfrac{4.19 \times 10^5}{2.08 \times 10^{-3}} =$

b. $\dfrac{8.53 \times 10^{-7}}{3.12 \times 10^3} =$

d. $\dfrac{3.97 \times 10^{-8}}{4.19 \times 10^{-7}} =$

F. Addition and subtraction of numbers written in scientific notation   You will very rarely need to be able to add or subtract numbers written in scientific notation. If the need does arise, an easy way to perform the addition or subtraction is to convert to ordinary numbers and then add or subtract in the usual manner. For example, to add $(3.71 \times 10^2) + (4.550 \times 10^3)$, convert to $371 + 4550 = 4921$. The answer can then be put back into scientific notation as 4.921×10^3.

In order to add or subtract exponential numbers directly, the exponents of all the numbers must be the same. If you were asked to add $(4.5 \times 10^4) + (3.2 \times 10^4)$, simply add the decimal numbers and maintain the same base and exponent. The sum would be 7.7×10^4.

More often the numbers to be added or subtracted will have different exponents, and then the numbers must be rewritten in equivalent forms so that the exponents are the same. Let us look again at the addition $(3.71 \times 10^2) + (4.550 \times 10^3)$. The exponents are different, 2 and 3, so we cannot simply add the decimal numbers. We must either rewrite the first number so that there is a 10^3 or rewrite the second so that there is a 10^2. Let us do the latter operation. Start by "taking a 10 out of" the 10^3; that is,

$4.550 \times 10^3 = 4.550 \times 10 \times 10^2$. Now we can see that $4.550 \times 10^3 = \underbrace{4.550 \times 10}_{} \times 10^2 = 45.50 \times 10^2$. So we can do the addition:

$$\begin{array}{r} 3.71 \times 10^2 \\ 45.50 \times 10^2 \\ \hline 49.21 \times 10^2 \end{array}$$

The answer must be adjusted to correct form:

$$49.21 \times 10^2 = 4.921 \times 10^{①} \times 10^{②} = 4.921 \times 10^3 \quad \text{Sum}$$

$$(5.93 \times 10^7) - (4.0 \times 10^5) =$$

SOLUTION

The exponents are different, so before subtracting we must change one of the exponents so that it matches the other exponent. Generally it is easier to make the larger exponent smaller by "taking out tens." In this case, take out 10^2 from 5.93×10^7:

$$5.93 \times 10^7 = \underbrace{5.93 \times 10^2}_{} \times 10^5 = 593 \times 10^5$$

Now subtract:

$$\begin{array}{r} 593 \times 10^5 \\ - \quad 4 \times 10^5 \\ \hline 589 \times 10^5 \end{array}$$

Rewrite the answer in proper form:

$$589 \times 10^5 = 5.89 \times 10^2 \times 10^5 = 5.89 \times 10^7$$

To adjust negative exponents, it is easiest to "take out tenths." For example, to add $(6.94 \times 10^{-3}) + (1.920 \times 10^{-2})$, begin by adjusting the -3 exponent to -2. This is done by "taking out" 10^{-1}:

$$6.94 \times 10^{-3} = \underbrace{6.94 \times 10^{-1}} \times 10^{-2} = 0.694 \times 10^{-2}$$

Now the addition can be done:

$$
\begin{array}{r}
0.694 \times 10^{-2} \\
1.920 \times 10^{-2} \\
\hline
2.614 \times 10^{-2}
\end{array}
$$

Problem 2.9

Add or subtract and express your answer in scientific notation.

a. $(3.75 \times 10^4) + (3.15 \times 10^3) =$

b. $(2.98 \times 10^2) - (1.39 \times 10^1) =$

c. $(8.19 \times 10^{-3}) + (6.18 \times 10^{-5}) =$

d. $(5.43 \times 10^{-3}) - (3.45 \times 10^{-4}) =$

ALGEBRAIC MANIPULATIONS

2.5   Problems in this text will usually be solved by the Unit Conversion Method (Section 3.3). However, you will find that there are some problems that can be solved more easily by the use of an unknown x and the basic techniques of algebra. Only the most basic techniques of algebra will be employed.

    The key step in solving an equation for an unknown is to isolate the unknown to one side of the equation. We manipulate the equation by adding, subtracting, multiplying, and dividing, keeping in mind that *whatever is done to one side of the equation must be done to the other.*

    Let us look at some examples of simple algebraic manipulations. If we are given the equation $x + 3 = 9$ and asked to solve for x, we concentrate on getting x isolated all by itself on one side of the equation. Clearly we must remove (subtract) the 3 to get x alone. We can subtract 3 from the left side so long as we also subtract it from the right side.

$$x + \underbrace{3 - 3} = \underbrace{9 - 3}$$
$$x = 6$$

To solve the equation $6x + 5 = 29$ for x, once again we begin by subtracting from each side, that is,

$$6x + \underbrace{5 - 5} = \underbrace{29 - 5}$$
$$6x = 24$$

To complete the solution, recognize that $6x$ must be divided by 6 to isolate x, so we divide each side by 6:

$$\frac{6\!\!\!/x}{6\!\!\!/} = \frac{24}{6}$$

$$x = 4$$

Solve for t in $5t - 12 = 4 - 3t$.

SOLUTION

This problem has the unknown t on both sides of the equation. Manipulate the equation so as to isolate t to the left side.

Add $3t$ to both sides of the equation and sum together terms with the unknown t:

$$5t - 12 + 3t = 4 - 3t + 3t$$
$$8t - 12 = 4$$

Add 12 to both sides:

$$8t - 12 + 12 = 4 + 12$$
$$8t = 16$$

Divide both sides by 8:

$$\frac{8\!\!\!/t}{8\!\!\!/} = \frac{16}{8}$$

$$t = 2$$

Solve for T in $PV = nRT$.

SOLUTION

We can divide both sides of the equation by n and then divide both sides by R, *or* in one step we can divide both sides by the product nR. Let us divide both sides by nR:

$$PV = nRT$$

$$\frac{PV}{nR} = \frac{nR\!\!\!/T}{nR\!\!\!/}$$

$$\frac{PV}{nR} = T$$

It does not matter whether the unknown is isolated to the left or to the right; convenience and simplicity dictate the choice.

SAMPLE EXERCISE

2.17

Solve for V in $D = \dfrac{M}{V}$.

SOLUTION

When the unknown is in a denominator, it is usually best to begin by clearing the fraction, which is done by multiplying each side of the equation by the denominator of the fraction. In this case, multiply each side by V:

$$D = \frac{M}{V}$$

$$D \times V = \frac{M}{\cancel{V}} \times \cancel{V}$$

$$D \times V = M$$

Now divide each side by D, in order to isolate V:

$$\frac{\cancel{D} \times V}{\cancel{D}} = \frac{M}{D}$$

$$V = \frac{M}{D}$$

Refer to Section 2.3 for an alternate solution.

Problem 2.10

Solve for the unknown x in each of the following:

a. $6 - 4x = 10$

b. $6 + 16x = 33 + 7x$

c. $\dfrac{13}{x} = A$

d. $109 + 3x = 43$

PERCENTAGE CALCULATIONS

2.6 Occasionally it will be necessary to do calculations which involve percent (%). "Percent" literally means "parts per 100 parts." Therefore, 30 percent means "30 parts per 100"; 7 percent means "7 parts per 100"; etc. Because this is the definition, we can immediately write any percent as a fraction with 100 in the denominator.

$$30 \text{ percent} = \frac{30}{100}$$

$$7 \text{ percent} = \frac{7}{100}$$

$$0.45 \text{ percent} = \frac{0.45}{100}$$

From these fractions, we can easily express the percentage in decimal form by dividing through by 100.

$$30\% = \frac{30}{100} = 0.30$$

$$7\% = \frac{7}{100} = 0.07$$

$$0.45\% = \frac{0.45}{100} = 0.0045$$

Notice that in each case we get from the numerical value of percent to the decimal form by moving the (understood) decimal two places to the left.

$$.30 \text{ percent} = 0.30$$

$$.07 \text{ percent} = 0.07$$

$$.0045 \text{ percent} = 0.0045$$

Problem 2.11

Convert the following percentages to decimal form.

a. 4.5 percent

b. 33 percent

c. 0.092 percent

If you were asked to calculate the number of male students in a class of 120 students, given that the class is 30 percent male, you probably know from previous experience that to do this you "take 30 percent of 120." That is, multiply 0.30×120, which equals 36. The reasoning being followed is as follows:

$$\frac{\text{Parts}}{\text{Total parts}} = \frac{x \text{ students}}{120 \text{ students}} = \frac{30 \text{ parts}}{100 \text{ parts}}$$

$$\frac{x}{120} = \frac{30}{100}$$

Multiply each side by 120:

$$x = \frac{30}{100} \times 120$$

$$x = 0.30 \times 120 = 36$$

"Taking a percent" always involves this reasoning; you must multiply

Percent in decimal form × given total

SAMPLE EXERCISE

2.18

If water is 11 percent hydrogen by weight, calculate the number of pounds of hydrogen there are in 40 pounds (lb) of water.

SOLUTION

In decimal form 11 percent is .11, which is 0.11. The **given total** is 40 lb of water. Therefore,

Pounds of hydrogen = 0.11 × 40 = 4.4 lb

Suppose you are given data about the makeup of something that is composed of parts. For example, you are told that in a class of 75 students there are 36 female students. How do you determine the percentage of female students? The method is to divide the parts by the total parts and multiply by 100. In this example,

$$\frac{36 \text{ female students}}{75 \text{ students}} \times 100 = 48 \text{ percent}$$

The reasoning being followed is

$$\frac{\text{Parts}}{\text{Total parts}} = \frac{36 \text{ students}}{75 \text{ students}} = \frac{x \text{ parts}}{100 \text{ parts}}$$

$$\frac{36}{75} = \frac{x}{100}$$

Multiply each side by 100:

$$\frac{36}{75} \times 100 = x \qquad \text{which is the percent}$$

Determination of a percent, given numbers of parts and total parts, involves the following operation:

$$\text{Percent} = \frac{\text{parts}}{\text{total parts}} \times 100$$

SAMPLE EXERCISE

2.19

Calculate the percentage of copper in a 79.0-lb sample of copper ore in which there is 3.75 lb of copper.

SOLUTION

$$\text{Percent} = \frac{\text{parts}}{\text{total parts}} \times 100 = \frac{3.75 \text{ lb}}{79.0 \text{ lb}} \times 100 = 4.75 \text{ percent}$$

Problem 2.12

A sugar mixture is found to contain 3.50 percent water. How many pounds of water are there in 178 lb of the sugar mixture?

Problem 2.13

What is the percentage of gold in a sample in which there is 5.0 oz of gold in every 9.0 oz of sample?

GRAPHING

2.7 A graph is a visual display from which the viewer can see the variation in one quantity compared to a change in another quantity. For example, in Figure 2.1 the plotted data show the variation of inflation rate (y axis) with time (x axis) in the United States. Most common graphs and the one we will examine in this section are constructed on a grid bordered by perpendicular lines called the horizontal (x) and vertical (y) axes. Some of the guidelines below may be familiar to you, but are presented here for completeness; construction and reading of graphs are basic skills which you will find useful in many endeavors.

FIGURE 2.1

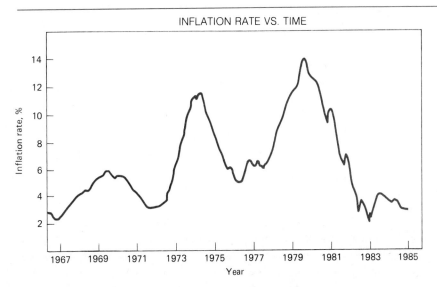

Graphs handily show the variation of one quantity with the change in another.

Guidelines for constructing a graph, given *x,y* data

1. Title the graph.

2. Draw in or darken the lines that will represent the horizontal and vertical axes and label the axes for the varying quantity, including the units of the quantity.

3. Establish appropriate and convenient scales for the axes based on the data given; that is, the data plotted on each axis must fit within the borders of the axis and the edge of the paper. The appropriate scale depends on the range of the given data (lowest to highest values) and the number of subdivisions on your graph paper. The ratio of these two quantities yields an appropriate scale value:

$$\frac{\text{Range of data}}{\text{Total number of subdivisions}} = \text{appropriate scale value}$$

For example, if the range of data in the *x* axis is 500 cm and the number of boxes from the vertical axis to approximately the edge of the paper is 50, an appropriate scale would be

$$\frac{500 \text{ cm}}{50 \text{ boxes}} = 10 \text{ cm/box}$$

In some cases, use of this formula leads to an inconvenient scale value. For example, if the range of data had been 170 cm with 50 subdivisions, the appropriate scale calculated would be 3.4 cm/box. Such a scale is less convenient than an integral value for each subdivision, and it is preferable to choose the next larger integral value. In this example, this means 4 cm/box. It is not necessary for each axis to have the same numerical scale, but once a scale for a particular axis is chosen it must be used uniformly throughout that axis.

4. Locate data points by moving along the *x* axis until the *x* value is reached and then move vertically to the *y* value. Mark each point with a dot or an X.

5. Draw the best-fitting curve through the points. Even when the graph is a straight line, the line will not generally pass through all the points. The best-fitting line will be simultaneously closest to all points; it may be that it actually goes through none of them. Although made-up data may fit a line exactly, in experimental data there is always some error in the measurement process, and thus a perfect fit is not obtained.

SAMPLE EXERCISE

2.20

Plot distance on the *y* axis against time on the *x* axis for the values in the following table:

Distance, meters (m)	Time, seconds (s)
0	0
3	1
6	2
8	3
13	4
15	5

SOLUTION

Following the guidelines above:

1. Title the graph "Distance vs. time."

2. Label the vertical axis distance (m) and the horizontal axis, time (s).

3. The range over the distance data is 15 meters (15–0), and the number of marked boxes on the vertical axis below is 20. An appropriate scale would be $\dfrac{15 \text{ meters}}{20 \text{ boxes}}$ = 0.75 meter/box. Since all the distance data are integers, however, it is more convenient to choose the higher-scale value 1.0 meter/box. The range of time data is 5 seconds, and the number of subdivisions on the horizontal axis is 20. An appropriate scale is $\dfrac{5 \text{ seconds}}{20 \text{ boxes}}$ = 0.25 second/box. Although the time data are given in integers so that a higher-scale value of 1.0 second/box could be used, this would compress the graph because the time range is small. A better display is obtained by using the 0.25 second/box scale and marking every fourth box with the appropriate integer.

4. Mark the data points in the graph. Dotted lines are drawn from the axes to the second data point to show that it is at $x = 1$ and $y = 3$.

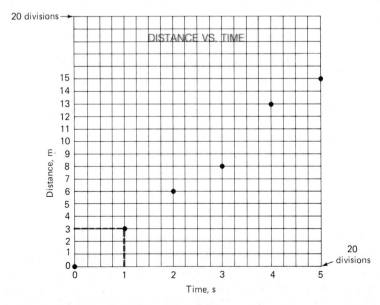

5. Draw a line so that the distance of points not touching the line on one side is balanced by the distance of points not touching on the other side.

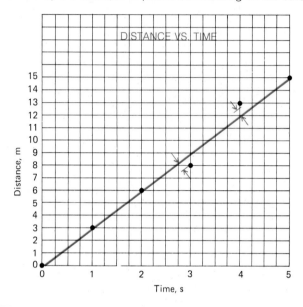

A graph can be used to determine the y (or x) value corresponding to a x (or y) value of a point not in the data table. Move along the x axis to the chosen x value and then up to the drawn curve. From this point in the curve move horizontally across to the vertical axis and obtain the y value.

SAMPLE EXERCISE

2.21

Determine the distance corresponding to a time of 2.5 seconds for the graph plotted in Sample Exercise 2.20.

SOLUTION

Move along the time axis to 2.5 seconds, then straight up to the drawn line. Moving horizontally across from the drawn line, you intersect the vertical axis at a point between 7 and 8 meters. The estimated y value is 7.6 meters. Note that the graph enabled us to determine a time-distance point that is not in the data table.

Problem 2.14

a. Make a graph from the data given in the following table.

Mass, grams (g)	Volume, milliliters (mL)
2.5	2.0
4.0	3.0
5.2	4.0
6.4	5.0
7.6	6.0
10.3	8.0

b.   Determine the mass corresponding to a volume of 6.5 mL.

SUMMARY

In order to understand many important ideas in chemistry it is necessary to have certain very basic math skills. Among these skills is the ability to deal with signed numbers, especially with respect to addition and subtraction (Section 2.2).

To use the Unit Conversion Method, an important mathematical technique which will be introduced in Chapter 3, you must understand the meaning of a fraction and be able to multiply fractions successfully (Section 2.3).

Scientific notation is a method of writing numbers as the product of a decimal number and a power of 10 (Section 2.4). It is a very convenient method for dealing with exceptionally large or small numbers. Multiplication and division of numbers written in scientific notation are frequently encountered.

Some very basic knowledge of algebra is necessary. If you understand how the unknowns are "isolated" (Section 2.5), you have enough skill to understand the algebraic manipulations encountered later in this text. In addition, percentage calculations will occasionally be necessary (Section 2.6).

Graphing and reading graphical presentations are helpful skills throughout science, and indeed, throughout many life activities (Secton 2.7).

CHAPTER ACCOMPLISHMENTS

After completing this chapter you should be able to

2.1   Introduction

2.2   Signed numbers
1.   Indicate the magnitude and sign of a signed number.
2.   Multiply a set of signed numbers.
3.   Divide one signed number by another.

4. Add any two signed numbers.
5. Subtract any two signed numbers.

2.3

6. State the meaning of a fraction in terms of the division operation.
7. Derive a multiplication expression from the equality of two fractions.
8. Multiply a set of fractions.
9. Divide one fraction by another.

2.4 Scientific notation

10. Indicate the base and exponent of an exponential number.
11. Write any decimal number in scientific notation.
12. Convert a number in scientific notation into an ordinary decimal number.
13. Multiply two numbers in scientific notation.
14. Divide two numbers in scientific notation.
15. Add two numbers in scientific notation.
16. Subtract two numbers in scientific notation.

2.5 Algebraic manipulations

17. Manipulate an algebraic equation to isolate an unknown to one side.

2.6 Percentage calculations

18. Convert a given percentage to a fraction and/or a decimal form.
19. Given the percent of one component in a given amount of the total, calculate the amount of that component.
20. Given the parts of one component in a given amount of the total, calculate the percent of that component.

2.7 Graphing

21. Construct a graph from a given data table.
22. Given a graph and a particular $x(y)$ value, determine the corresponding $y(x)$ value.

PROBLEMS

2.2 Signed numbers

2.15 Give the magnitude and sign of the following signed numbers:

a.	13.3	d.	90.5
b.	−3.13	e.	+42.3
c.	−0.0184	f.	−4

2.16 Give the product of the following multiplications:

a. $(4)(-16) =$
b. $(-1.2)(-3.9)(4.1) =$
c. $(-0.032)(-87.9)(-43.1) =$
d. $(8)(3)(-1)(-27)(4)(5) =$
e. $(-18.1)(4.5)(0.93)(-4.9)(-5.3)(8) =$

2.17 Calculate the quotient of the following divisions:

   a. $(-16) \div (4) =$
   b. $(-4) \div (-16) =$
   c. $-8.2 \div 1.9 =$
   d. $-81.9 \div -5.4 \div -3.9 =$
   e. $-81.9 \div 5.4 \div -3.9 =$

2.18 Calculate the following:

   a. $(4)(9) \div 3 =$
   b. $(-13.4)(2.1) \div 7.5 =$
   c. $(75.4 \div (4))(3) =$
   d. $(-45.7 \div (1.1))(-2.3) =$

2.19 Add the following:

   a. $-8 + (-14)$ **c.** $-49.1 + 12.5$
   b. $72 + (-11)$ **d.** $8.9 + (-14.2)$

2.20 Subtract the following:

   a. $18 - 11$ **d.** $-11 - (-18)$
   b. $11 - 18$ **e.** $-31.2 - (-49.5)$
   c. $-11 - 18$

2.3 Fractions

2.21 Show that the two fractions $\dfrac{2}{17}$ and $\dfrac{6}{51}$ are equal.

2.22 Show that the two fractions $\dfrac{3}{19}$ and $\dfrac{8}{57}$ are *not* equal.

2.23 A pizza pie is cut into eight pieces. If Jack ate three slices, what part of the pie, expressed as a decimal, did Jack eat?

2.24 Calculate the following:

   a. $\left(\dfrac{7}{2}\right)\left(\dfrac{2}{14}\right)$

   b. $\left(\dfrac{5}{8}\right)\left(\dfrac{-3}{4}\right)\left(\dfrac{2}{9}\right)$

   c. $\left(\dfrac{3 \text{ feet}}{1 \text{ meter}}\right)\left(\dfrac{1 \text{ meter}}{100 \text{ centimeters}}\right)$

   d. $\left(\dfrac{24 \text{ hours}}{1 \text{ day}}\right)\left(\dfrac{365 \text{ days}}{1 \text{ year}}\right)$

   e. $\left(\dfrac{-12}{9}\right)\left(\dfrac{-2}{3}\right)\left(\dfrac{-8}{3}\right)$

2.25 Calculate the following and express your answer in decimal form:

   a. $\dfrac{2}{3} \div 3 =$ **d.** $18 \div \dfrac{2}{3}$

   b. $\dfrac{\frac{2}{3}}{\frac{1}{3}} =$ **e.** $\dfrac{\frac{1}{4}}{\frac{3}{8}} =$

   c. $\dfrac{\frac{2}{3}}{-\frac{3}{4}} =$ **f.** $-11 \div \dfrac{9}{3} =$

2.4 Scientific Notation

2.26 Write the following numbers in scientific notation:

   a. 58.7   **d.** 3012
   b. 0.082   **e.** 73.98
   c. 631,000,000   **f.** 0.000000000718

2.27 Calculate the following and express your answer in scientific notation:

   a. $(3.0 \times 10^8)(2.0 \times 10^3)$
   b. $(8.39 \times 10^{-5})(3.21 \times 10^4)$
   c. $(2.14 \times 10^{-6})(1.39 \times 10^{-9})$
   d. $(7.15 \times 10^{-9})(3.81 \times 10^2)$
   e. $(9.6 \times 10^{16})(4.5 \times 10^{-1})$

2.28 Calculate the following and express your answer in scientific notation:

   a. $\dfrac{6.4 \times 10^6}{3.2 \times 10^2}$ **d.** $\dfrac{1.9 \times 10^4}{4.2 \times 10^5}$

   b. $\dfrac{3.2 \times 10^{-2}}{6.4 \times 10^{-6}}$ **e.** $\dfrac{2.3 \times 10^{-12}}{0.7 \times 10^{-9}}$

   c. $\dfrac{1.8 \times 10^5}{9.3 \times 10^4}$ **f.** $\dfrac{6.0 \times 10^{-23}}{3.0 \times 10^9}$

2.29 Add or subtract the following and express your answer in scientific notation:

   a. $(3.8 \times 10^4) + (1.9 \times 10^4) =$
   b. $(8.3 \times 10^4) - (2.4 \times 10^3) =$
   c. $(7.4 \times 10^4) + (8.9 \times 10^3) =$
   d. $(4.7 \times 10^7) - (5.2 \times 10^6) =$

2.5 Algebraic Manipulations

2.30 Solve for C in $A = \dfrac{B}{C}$

2.31 Solve for x in $3x + 4 = 16$

2.32 Solve for C in $A = \dfrac{B}{C} + 9$

2.33 Solve for F in $C = \dfrac{5}{9}(F - 32)$

2.34 Solve for x in $\dfrac{2x - 10}{8} = 15$

2.35 Solve for n in $PV = nRT$.

2.36 Solve for x in $0.40x + 3 = 11$

2.37 Solve for x in $(6.5)(2x) = 39$

2.38 Find the value of H, including the unit, in

$$H = (32 \text{ grams})(28°C)\frac{(1.2 \text{ calories})}{\text{grams}\cdot°C}$$

2.39 Solve for Δt, including the unit, in

$$(50 \text{ g}) (0.25 \text{ cal/g}) = \frac{1 \text{ cal}}{g\cdot°C} (5 \text{ g})(\Delta t)$$

2.6 Percentage Calculations

2.40 Calculate the percentage of nitrogen in a sample of air in which 100 ounces (oz) of air contains 76 oz of nitrogen.

2.41 A sample of ore was found to contain 1.8 percent Au. How many pounds of gold are present in 573 lb of ore?

2.42 What is the percentage of copper in 137 lb of copper ore containing 113 lb of copper?

2.43 What is the percentage of alcohol in a beer in which 12 oz of beer contains 0.39 oz of alcohol?

2.44 Calculate the percentage of sugar in a homogenous mixture containing 6.0 grams of sugar and 30.0 grams of water.

2.45 Convert 34.9 percent to a decimal number.

2.46 Which number is larger, $\dfrac{3}{8}$ or 28.9 percent?

2.7 Graphing

2.47 **a.** Make a plot of the Fahrenheit versus Celsius temperature scales from the following measured data:

Temperature, °F	Temperature, °C
−32	−35
22	−5
32	0
54	12
70	21
78	26

b. According to your graph, what is the Fahrenheit equivalent of 18°C?
c. According to your graph what is the Celsius equivalent to 68°F?

2.48 **a.** Make a plot of the following measured mass and volume data:

Mass, grams (g)	Volume, cubic centimeters (cm³)
35	12
54	18
64	21
80	27
92	31
98	33

b. What is the volume corresponding to a mass of 75 grams?
c. What is the mass corresponding to a volume of 15 cm³?
d. What is the mass corresponding to a volume of 0 cm³?

2.49 The circumference and diameter of a variety of circles were measured and found to be:

Circumference, cm	Diameter, cm
1.6	0.5
3.1	1.0
4.7	1.5
6.3	2.0
7.9	2.5
9.4	3.0

a. What is the circumference of a circle corresponding to a diameter of 1.8 cm?
b. What is the diameter of a circle which has a circumference of 8.0 cm?

2.50 Make a plot of the following measured volume and Celsius temperature data?

Volume, mL	Temperature, °C
10.5	10.0
10.7	20.0
11.1	30.0
11.5	40.0
11.9	50.0
12.2	60.0

a. What happens to the volume as the temperature increases?
b. Does the volume double when the temperature doubles?
c. What is the volume when the temperature is 25°C?

MEASUREMENT
CHAPTER

3

Li

Lithium

6.941

WHAT IS MEASUREMENT?

3.1 It is not possible to function successfully without some knowledge of weights and measures. How could you buy proper amounts of potatoes or liverwurst, decide how much fabric is needed to make a blouse, or mark out bases on a baseball diamond without understanding measurement? Because the understanding of scientific concepts is often based on measurements, it is very important that we consider exactly what the process of measurement is.

When you measure something, you are comparing it to some reference standard. For example, when you step on a bathroom scale calibrated or marked off in pounds, you are comparing your weight to the reference standard, the pound. The scale tells you how heavy you are relative to 1 lb. All systems of measurement involve this idea of comparison to a reference standard. All measurements are relative to a standard. This is an important concept which comes up repeatedly in chemistry. It is a good idea to think it through for some familiar measurements before tackling unfamiliar ones.

THE METRIC SYSTEM

3.2 The reference standards scientists use are those of the metric system, a highly logical, easy-to-use system based on multiples of 10. This system was originally developed during the French Revolution and was adopted throughout the world except for English-speaking countries. The United States is slowly converting to the metric system—new road signs in kilometers and food package weights in grams are just the beginning.

If your children grow up with the metric system, they will be able to skip this section of the text. Those of us who have been brought up with the English system must now learn the metric system. Around 1960, scientists slightly revised and updated the metric system and called it **SI** after the French name, **S**ystème **I**nternational d'Unités.

Length, the distance between two points   For **length,** the distance between two points, the reference standard is represented by a meter stick (1 meter, abbreviated m).[1] A yardstick (36 inches, abbreviated in) is somewhat shorter than a meter stick (39.37 in). Look at Figure 3.1. The man is larger than the meter stick. His measurement is 1.8 m (1.8 times a meter). The child is smaller than the meter stick, 0.9 m (0.9 times a meter).

The meter is the base unit, and all length measurements could be expressed in meters. However, it is more convenient to have some larger and smaller units (just as in the English system we have inches, feet, yards, and miles). The beauty of the metric (SI) system is that all the units are related by multiples (or divisions) of 10, so the names of the units are constructed from the base unit name preceded by a prefix which tells which multiple of 10 is involved.

$$\underbrace{\text{Prefix}} \qquad \underbrace{\text{Base unit}}$$

Tells the   Unit
multiple
of 10

For example,

$$\text{kilo} \qquad \text{meter}$$

Multiple   Base unit
of 10   of length
is 1000

---

[1] The meter is defined as the distance light will travel in $\frac{1}{299,792,458}$ of a second. This distance is unchangeable and readily measured with great accuracy by modern instruments. Meter sticks are made the length of this distance as accurately as possible.

FIGURE 3.1

The reference standard for length in the metric system is the meter. The man is 1.8 times the length of the reference standard; the boy is 0.9 times the length of the reference standard.

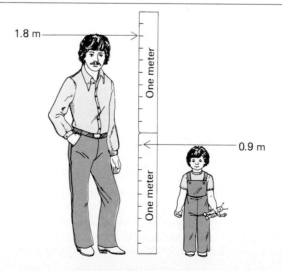

1.8 m — One meter

0.9 m

One meter

Thus *kilometer* (km) means "1000 meters."

Because the abbreviation for *kilo-* is k and for meter is m, the abbreviation for kilometer is km. Prefixes and their abbreviations appear in Table 3.1.

These prefixes and their abbreviations always have the same mathematical meaning whenever they are encountered (see Table 3.2). Now look at Figure 3.2. One thousand meter sticks laid end to end equal 1 km. Dividing the meter stick into 10 equal parts (dividing by 10 and multiplying by $\frac{1}{10}$ is the same thing) results in 10 smaller units, 10 dm. Chopping each decimeter into 10 parts results in a total of 100 centimeter divisions of the meter stick. Further dividing each centimeter into 10 parts results in 1000 mm per meter. This is all summarized in Table 3.3a.

Mass, the amount of matter in an object   The metric reference standard for mass is a metal cylinder made of a platinum-iridium alloy stored in a vault near Paris, France. The mass of this block of metal is defined as 1 kilogram (abbreviated kg). In the English system this is about 2.2 lb. The base unit of mass is the gram (g), and from the

TABLE 3.1

COMMONLY ENCOUNTERED PREFIXES OF THE METRIC SYSTEM AND THEIR MATHEMATICAL MEANINGS

		Multiply base unit by	
Prefix	Abbreviation	Common number	Exponential number
kilo-	k	1000	10^3
deci-	d	1/10, or 0.1	10^{-1}
centi-	c	1/100, or 0.01	10^{-2}
milli-	m	1/1000, or 0.001	10^{-3}
micro-	μ	1/1,000,000, or 0.000001	10^{-6}

TABLE 3.2

METRIC PREFIXES APPLIED TO THE BASE UNITS OF LENGTH AND MASS

Length	Mass
1 kilometer (km) = 1000 meters (m)	1 kilogram (kg) = 1000 grams (g)
[1 decimeter (dm) = 0.1 meter (m)]*	[1 decigram (dg) = 0.1 gram (g)]
1 centimeter (cm) = 0.01 m	[1 centigram (cg) = 0.01 g]
1 millimeter (mm) = 0.001 m	1 milligram (mg) = 0.001 g
1 micrometer (μm) = 0.000001 m	1 microgram (μg) = 0.000001 g

* Units contained in square brackets are not commonly used by scientists.

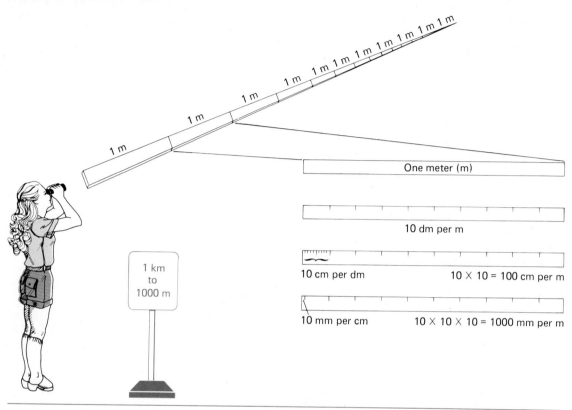

One meter (m)

10 dm per m

10 cm per dm   10 × 10 = 100 cm per m

10 mm per cm   10 × 10 × 10 = 1000 mm per m

1 km to 1000 m

FIGURE 3.2

Relationships between the common units of length in the metric system.

foregoing discussion about prefixes you can see that 1 kg must equal 1000 g. The other relationships among the units of mass are summarized in Tables 3.2 and 3.3*b*.

Scientists prefer to measure **mass,** the fixed amount of matter in an object, rather than **weight,** the amount of gravitational attraction pulling on an object, because *mass is always the same* no matter where it is measured, whereas *weight varies*. The measured value for weight depends on the distance the object being weighed is from the center

TABLE 3.3   EQUALITIES AMONG COMMON METRIC UNITS

a. Length	b. Mass	c. Volume
1 km = 1000 m	1 kg = 1000 g	
1 m = 10 dm		
1 m = 100 cm		
1 m = 1000 mm	1 g = 1000 mg	1000 cm³ = 1 liter (L)
		= 1000 milliliters (mL)
1 m = 1,000,000 μm	1 g = 1,000,000 μg	

of the earth. The weight of an object measured in Death Valley (low altitude) is slightly greater than the weight measured on top of Mount Everest (high altitude). The mass is identical in both locations.

Mass is measured by using a double-pan balance, as shown in Figure 3.3. Because gravity works on each pan equally at any altitude or on any planet, we are assured that we are actually measuring mass by comparing the unknown object with the known masses. Modern single-pan laboratory balances compensate for gravity in a less obvious way than does the double-pan balance in Figure 3.3, but the idea is the same. The crucial feature of a device for measuring mass is the knife edge on which the weighing pan is balanced.

Although mass and weight are not the same thing, it is common practice to use the words interchangeably because we use the verb "weigh" to describe the measurement of either mass or weight. Because chemists always use balances to weigh things, chemists always measure mass whether it is called mass or incorrectly called weight.

Volume, the amount of space occupied by a three-dimensional object The reference standard for **volume,** the amount of space occupied by a three-dimensional object, in the metric system is the liter, which is just slightly larger than a quart.[1] The *liter* (L) is the volume

---

[1] The metric system is the precursor of the SI system. The actual SI reference standard for volume is the cubic meter, but this unit is impractical in many settings, for example, in medical laboratories. Therefore, the older unit, liter, is maintained and discussed in this text.

FIGURE 3.3

Use of the two-pan balance.

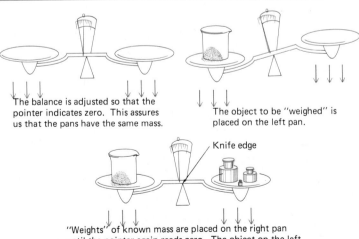

The balance is adjusted so that the pointer indicates zero. This assures us that the pans have the same mass.

The object to be "weighed" is placed on the left pan.

Knife edge

"Weights" of known mass are placed on the right pan until the pointer again reads zero. The object on the left has the same mass as the known mass on the right.

Notice that gravity (↓ ↓ ↓) is pulling on *both* pans at all times. Because it affects both pans equally we do not observe the effect of gravity and can be assured that we are truly measuring mass by comparison to some reference standard.

occupied by a perfect cube, 10 cm on each edge. Picture the cube as an empty box. We can calculate the volume of any rectangular solid in the unit cubic centimeters (cm^3) by multiplying the length × width × height. So, given that volume = l × w × h, we have in this case

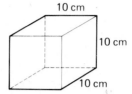

$$\text{Volume} = 10 \text{ cm} \times 10 \text{ cm} \times 10 \text{ cm} = 1000 \text{ cm}^3$$

Therefore, 1 L = 1000 cm^3. (An alternative abbreviation for cubic centimeter is cc.)

Now imagine filling up the box with water. The volume of the water is 1000 cm^3, or 1 L (see Figure 3.4). If we pour the 1 L of water into different containers, it is still 1 L, despite the fact that the shape changes. Volume is an amount of space occupied and does not depend on shape.

We are most often concerned with measuring the volumes of liquids. Chemists have invented several kinds of equipment to do this conveniently (see Figure 3.5). These devices are all calibrated in milliliters. As the prefix tells you, 1 mL is a tiny part, namely, one thousandth of a liter. There are 1000 mL in 1 L. Notice in Table 3.3, column 3 that this is the only equality listed under volume. This is because scientists do not commonly use other units. However, the wine industry does. For example, a "fifth" of wine is exactly 75 centiliters (cL).

FIGURE 3.4

One liter of water always occupies the same volume (1 L = 1000 cm^3), but the shape may change as the shape of the container changes.

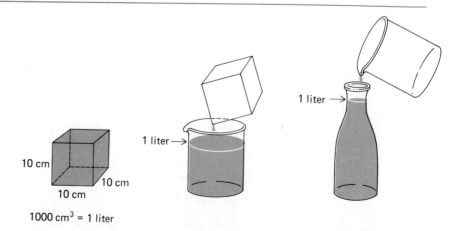

10 cm   10 cm   10 cm

1000 cm^3 = 1 liter

1 liter →   1 liter →

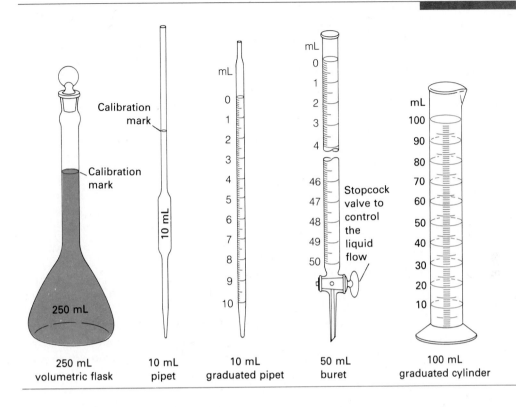

250 mL
volumetric flask

10 mL
pipet

10 mL
graduated pipet

50 mL
buret

100 mL
graduated cylinder

FIGURE 3.5

Various types of common
laboratory equipment used
to measure the volume of
liquid samples.

Because 1 L = 1000 cm³ and 1 L = 1000 mL, it must be true that

$$1000 \text{ cm}^3 = 1000 \text{ mL}$$

When each side of the equation is divided by 1000, it becomes apparent that

$$1 \text{ cm}^3 = 1 \text{ mL}$$

Thus the liquid contents of a cube 1 cm on a side will come up to the 1 mL mark of a graduated cylinder.

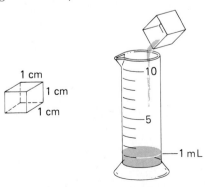

UNIT CONVERSION

3.3 If you measure the distance between the two lines marked off below, the measurement you report might be 12.7 cm, 127 mm, or 5 in, depending on the ruler used and the scale used. Notice that the number reported is meaningless without reporting the proper unit.

Measure this distance

It is frequently necessary to be able to convert the units of a measurement to different units. The method we recommend to perform these calculations is the **Unit Conversion Method.** It is well worth your time to master this method because later you can apply it to a variety of chemical problems.

Mathematical background

The Unit Conversion Method depends on two mathematical facts: (1) any equality can be used to write a fraction equal to 1; and (2) like quantities in the numerators and denominators of fractions can be "canceled out" (see Section 2.3).

A quantity divided by itself is equal to 1. For example, clearly 8 ft/8 ft = 1. The equality 1 m = 100 cm tells us that the 1 m and 100 cm represent exactly the same distance. Therefore, dividing 1 m by 100 cm is the same as dividing 1 m by itself, and therefore the fraction

$$\frac{1 \text{ m}}{100 \text{ cm}} = 1. \quad \text{Similarly,} \quad \frac{100 \text{ cm}}{1 \text{ m}} = 1.$$

Any equality can be made into a fraction equal to 1; we call that fraction a **conversion factor.** The following fraction which has a meter stick in both the numerator (100 cm) and denominator (1 m) demonstrates this visually.

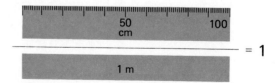

SAMPLE EXERCISE

3.1

Write conversion factors, i.e., fractions equal to 1, from the equalities

$$1 \text{ kg} = 1000 \text{ g}$$

$$1000 \text{ cm}^3 = 1000 \text{ mL}$$

SOLUTION

Both 1 kg/1000 g = 1 and 1000 g/1 kg = 1 because both 1 kg and 1000 g represent an identical mass.

Both 1000 cm³/1000 mL = 1 and 1000 mL/1000 cm³ = 1 because both 1000 cm³ and 1000 mL represent an identical volume. Notice also that 1000 can be canceled out, so that

$$\frac{\cancel{1000}\ cm^3}{\cancel{1000}\ mL} = \frac{1\ cm^3}{1\ mL}$$

Useful equalities relating the metric and English systems appear in Table 3.4.

Problem 3.1

Write conversion factors for the following equalities:

a. 1 kg = 2.2 lb

b. 1 L = 1000 mL

Multiplication by 1

As you know, multiplication by 1 does not change the quantity that is being multiplied, for example, 8 ft × 1 = 8 ft. However, units can be changed when multiplying by conversion factors. For example, if we multiply 8 kg by the factor equal to 1, that is, 1000 g/1 kg, we get 8000 g:

$$8\ \cancel{kg} \times \frac{1000\ g}{1\ \cancel{kg}} = 8000\ g$$

We can be certain that 8000 g is the same quantity as 8 kg because the multiplication is by 1, but we have done a *unit conversion*, i.e., changed the units from kilograms to grams.

For this simple example, of course, we could have used the following reasoning: if 1 kg is 1000 g, then 8 kg must be 8000 g. Fre-

a. Length	*b.* Mass	*c.* Volume
1 in = 2.54 cm	2.20 lb = 1 kg	1.06 qt = 1 L
39.4 in = 1 m	1 lb = 454 g	
1 mi = 1.61 km		

EQUALITIES BETWEEN THE METRIC AND ENGLISH SYSTEMS* **TABLE 3.4**

* Except for 1 in = 2.54 cm, these equalities are inexact and will affect the numbers of significant figures allowed in a calculated answer (see Section 3.9).

quently, however, unit conversions are more complex, and so we recommend the general method of problem solving that follows.

The Unit Conversion Method

The steps to be followed in reading a problem and setting up a calculation by the Unit Conversion Method are given here and applied in Sample Exercise 3.2:

1.   Identify the **given quantity** and **unit** and write them down.

2.   Identify the **new quantity** to be determined and write down the **new units** it is to have.

3.   Determine the conversion factor(s) that will change the given into the new quantity and unit. The factor will have given units in the denominator and new units in the numerator.

4.   Set up the calculation according to the following format:

$$\begin{array}{ccc} \textbf{Given quantity} & & \textbf{conversion} & & \textbf{new quantity} \\ \textbf{and unit} & \times & \textbf{factor(s)} & = & \textbf{and unit} \end{array}$$

$$\underline{\quad\quad} \text{given unit} \times \frac{\underline{\quad\quad} \text{ new unit}}{\underline{\quad\quad} \text{ given unit}} = \underline{\quad\quad} \text{ new unit}$$

SAMPLE EXERCISE

3.2

How many meters are there in 76 cm?

SOLUTION

Follow the steps just outlined:

Step 1   The given (old) quantity and unit is *76 cm.*

Step 2   The new quantity to be determined is numbers of meters. Thus the new unit is the *meter.*

Step 3   The equality relating meters and centimeters can be found in Table 3.3 as 1 m = 100 cm. Thus the conversion factor to be used must be either 1 m/100 cm or 100 cm/1 m. Choose the one that has the given units in the denominator, $\dfrac{\textbf{new} \text{ units}}{\textbf{given} \text{ units}}$, which in this case is 1 m/100 cm. This conversion factor is chosen so that the given units will cancel out.

Step 4   Set up in the proper format:

Given quantity and unit × conversion factor = new quantity and unit

$$76 \text{ cm} \times \frac{1 \text{ m}}{100 \text{ cm}} = \underline{\quad\quad} \text{ m}$$

Notice that the given units cancel to give the desired new unit, meters. The setup of the fractions tells you to divide 76 by 100, that is,

$$76 \text{ cm} \times \frac{1 \text{ m}}{100 \text{ cm}} = \frac{76 \text{ m}}{100} = 0.76 \text{ m}$$

Multistep conversions

This method can be employed for more complex conversions in which equalities may not be given directly in the tables. The idea in such a case is to examine the table and assemble all needed equalities in order to establish the one you want.

SAMPLE EXERCISE

3.3

How many milligrams are there in 0.53 kg?

Solution

Follow the preceding steps.

Step 1 The given quantity and unit is *0.53 kg.*

Step 2 The new unit to be determined is *milligrams.*

Step 3 Table 3.3 does not offer a direct equality between kilograms and milligrams. However, it does relate kilograms to grams and grams to milligrams. Therefore, if we use the conversion factors corresponding to these equalities, we should be able to convert kilograms to grams to milligrams. The factors are always in the form new unit/given (old) unit, so in this case we have 1000 g/1 kg and 1000 mg/1 g.

Step 4 The setup is then

Given quantity and unit × conversion factors = new quantity and unit

$$0.53 \text{ kg} \times \frac{1000 \text{ g}}{1 \text{ kg}} \times \frac{1000 \text{ mg}}{1 \text{ g}} = 530,000 \text{ mg}$$
$$(5.3 \times 10^5 \text{ mg})$$

The units cancel to give milligrams, and the setup tells you to multiply 0.53 × 1000 × 1000.

Problem 3.2

How many milliliters are in 0.439 L?

With the help of Table 3.4 one can also do conversions between the metric and English systems. For example, you can now prove that the three measurements reported at the very beginning of Section 3.3 are all correct. The three values given for the distance between the two lines were 12.7 cm, 127 mm, and 5 in.

$$12.7 \ \cancel{cm} \times \frac{1 \ \cancel{m}}{100 \ \cancel{cm}} \times \frac{1000 \ mm}{1 \ \cancel{m}} = \frac{12.7 \times 1000 \ mm}{100} = 127 \ mm$$

$$12.7 \ \cancel{cm} \times \frac{1 \ in}{2.54 \ \cancel{cm}} = \frac{12.7 \times 1 \ in}{2.54} = 5.00 \ in$$

SAMPLE EXERCISE

3.4

The distance from Reims, France to Strasbourg, France is 346 km. What is this distance in miles?

SOLUTION

Following the above guidelines:

Step 1 The given quantity and unit is 346 km.

Step 2 The new unit to be determined is miles.

Step 3 The equality relating kilometers and miles can be found in Table 3.4 as 1 mi = 1.61 km. In this case, selection of the conversion factor with new units/given units yields 1 mi/1.61 km.

Step 4 The setup is then

Given quantity and unit × conversion factors = new quantity and unit

$$346 \ \cancel{km} \times \frac{1 \ mi}{1.61 \ \cancel{km}} = 215 \ mi$$

Problem 3.3

How many grams are in 1.21 lb?

The bottom line

The Unit Conversion Method is a powerful tool. Remember,

1. Whenever you have an equality, you can construct a fraction equal to 1.

2. Fractions equal to 1 can be used as conversion factors to convert units.

We will continue to use the Unit Conversion Method throughout this book.

DENSITY

3.4 Neither the mass nor the volume of a substance is a characteristic property because both mass and volume vary with the size of the

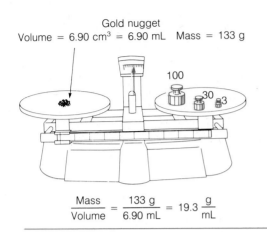

Gold nugget
Volume = 6.90 cm³ = 6.90 mL   Mass = 133 g

100
30
3

$$\frac{Mass}{Volume} = \frac{133 \text{ g}}{6.90 \text{ mL}} = 19.3 \frac{\text{g}}{\text{mL}}$$

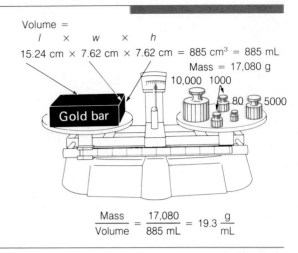

Volume =
$l \times w \times h$
15.24 cm × 7.62 cm × 7.62 cm = 885 cm³ = 885 mL

Mass = 17,080 g
10,000 1000

Gold bar

80   5000

$$\frac{Mass}{Volume} = \frac{17,080}{885 \text{ mL}} = 19.3 \frac{\text{g}}{\text{mL}}$$

FIGURE 3.6

The masses and volumes of the two gold samples are very different, but the ratio $\frac{mass}{volume}$ is a constant characteristic property. This property is called **density**.

sample. For example, consider two samples of gold, one a small gold nugget and the other a gold bar from Fort Knox (see Figure 3.6). Clearly the gold bar has a larger volume (occupies more space) and weighs more (has a larger mass), but it turns out that the ratio of the mass to the volume, i.e., mass/volume, is a constant, characteristic property. This ratio is called the **density.**

Density is determined in the laboratory by measuring both the mass and the volume of a sample and dividing the mass by the volume.

$$\text{Density} = \frac{\text{mass}}{\text{volume}}$$

Because the mass is usually measured in grams and the volume in milliliters, density has the units grams per milliliter, g/mL or $\frac{\text{g}}{\text{mL}}$.

If you were to measure the masses and volumes of the two gold samples shown in Figure 3.6, you would find very different values, as you would expect.

	Gold Nugget	Gold Bar
Mass	133 g	17,080 g
Volume	6.90 cm³ = 6.90 mL	885 cm³ = 885 mL

But consider the mass/volume ratios for the two samples:

$$\text{Density} = \frac{\text{mass}}{\text{volume}} = \frac{133 \text{ g}}{6.90 \text{ mL}} = 19.3 \frac{\text{g}}{\text{mL}} \qquad \frac{17,080 \text{ g}}{885 \text{ mL}} = 19.3 \frac{\text{g}}{\text{mL}}$$

We see that the density is a constant. A density of 19.3 g/mL is a characteristic property of any sample of gold of any size.

SAMPLE EXERCISE

3.5

A student wants to know the density of ethyl alcohol. She carefully measures out 100. mL of the liquid in a graduated cylinder. Then she carefully weighs the 100. mL and finds the mass to be 79 g (see Figure 3.7). What is the density?

SOLUTION

To find density, divide the mass of a sample by its volume. In this case,

$$\text{Density} = \frac{79\ g}{100.\ mL} = \frac{0.79\ g}{mL}$$

FIGURE 3.7

All matter has both mass and volume. Mass is measured by the double-pan balance. For liquids, one way to measure volume is to use a graduated cylinder calibrated in milliliters.

Problem 3.4

What is the density of olive oil if 21 mL weighs 19 g?

Density varies slightly with temperature. Therefore, when densities are reported, the temperature at which the measurement was

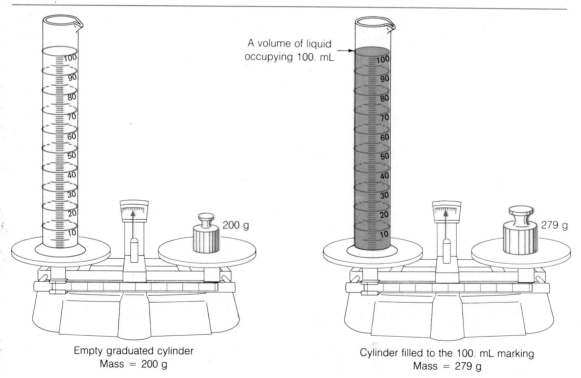

Empty graduated cylinder
Mass = 200 g

A volume of liquid occupying 100. mL

Cylinder filled to the 100. mL marking
Mass = 279 g

200 g

279 g

Mass of liquid = 279 g − 200 g = 79 g
Volume of liquid = 100. mL

$$\text{Density} = \frac{\text{Mass}}{\text{Volume}} = \frac{79\ g}{100.\ mL} = 0.79\ \frac{g}{mL}$$

DENSITIES OF COMMON MATERIALS   **TABLE 3.5**

Solids	Density, g/mL, 20°C	Liquids	Density, g/mL, 20°C	Gases	Density, g/L, 0°C*
Gold	19.3	Water	1.00	Air	1.29
Lead	11.3	Gasoline	0.67	Oxygen	1.43
Copper	8.92	Milk	1.03	Hydrogen	0.090
Iron	7.86	Seawater	1.03	Helium	0.178
Aluminum	2.70	Blood	1.06	Carbon dioxide	1.96
Salt	2.16	Mercury	13.6		
Paper	0.70	Olive oil	0.92		
Balsawood	0.20	Alcohol	0.79		
Redwood	0.44	Vinegar	1.01		
Rubber	1.1	Ether	0.70		
Ice	0.92	Carbon tetrachloride	1.59		

* Notice that for gases the units are grams per liter and the temperature is 0°C. This is because gases are much less dense than liquids or solids and change more dramatically with temperature. Density increases as temperature decreases.

done is usually noted. Table 3.5 lists the densities of some common materials.

When working in the laboratory, you will find that it is particularly useful to know the densities of liquids. When you wish to know the mass of a liquid, it is generally easier to measure the volume and use the known density to calculate the mass. Density relates mass and volume and allows the calculation of one from the other either by algebraic manipulation of the equation density = mass/volume or by regarding density as a conversion factor relating mass and volume units.

Density is a conversion factor for converting units of mass into volume, or vice versa. Notice in Sample Exercise 3.5 that 100 mL of ethyl alcohol and 79 g of ethyl alcohol represent the identical sample; therefore, the fraction 79 g/100 mL, or the reduced form 0.79 g/mL, is a fraction equivalent to 1 and can be used as a conversion factor. This is true of all densities.

SAMPLE EXERCISE

3.6

What is the mass of 3.0 mL of ether?

SOLUTION

Approach this by the Unit Conversion Method.

Step 1   The given quantity is 3.0 mL.

Step 2   The new quantity to be determined is the mass, i.e., the number of grams of sample.

Step 3   Density relates grams and milliliters. The density of ether from Table 3.5 is equal to 0.70 g/mL.

Step 4   Set up in the proper format.

Given quantity and unit × conversion factor = quantity and unit

$$3.0 \; \text{mL} \times \quad \frac{0.70 \text{ g}}{1.0 \text{ mL}} \quad = \frac{3.0 \times 0.70 \text{ g}}{1.0} = 2.1 \text{ g}$$

SAMPLE EXERCISE

3.7

What is the volume of 27.2 g of mercury?

SOLUTION

Step 1   The given is 27.2 g.

Step 2   The new quantity is the volume, i.e., the number of milliliters.

Step 3   Density relates grams and milliliters. The density of mercury is 13.6 g/mL, which can also be written as 1.00 mL/13.6 g.

Step 4.   The setup is

Given quantity and unit × conversion factor = new quantity and unit

$$27.2 \; \text{g} \times \quad \frac{1.00 \text{ mL}}{13.6 \text{ g}} \quad = 2.00 \text{ mL}$$

Problem 3.5

What volume does 84 g of carbon tetrachloride occupy?

Often densities of liquids and solids are compared with that of water (1.0 g/mL). Anything less dense than water will float, anything more dense than water will sink. Oil and gasoline both float on water, as you have probably observed at some point.

Densities of gases are compared with air. A gas less dense than air will rise in air; one more dense will sink. Balloons filled with helium will rise in air because helium has a low density. A balloon filled with pure oxygen will sink.

TEMPERATURE

3.5   **Temperature** is a measure of the hotness or coldness of an object. There is a difference between temperature and heat which will unfold in this section and the next. Weather reports now commonly give two values for the daily temperature. One is the familiar Fahrenheit temperature (°F); the other is the metric system equivalent, i.e., the Celsius temperature (°C). By choosing two fixed points, namely, the normal boiling point (bp) and freezing point (fp) of water, you can determine a mathematical relationship between degrees Fahrenheit and degrees Celsius (see Figure 3.8). Table 3.6 summarizes the data gained from the Fahrenheit and Celsius thermometers shown in Figure 3.8.

From the data in Table 3.6 you can construct an equation relating degrees Fahrenheit and Celsius. One form of the equation is

$$°F = 1.8°C + 32 \qquad (3.1)$$

This form is convenient to use when you are given a Celsius temperature and asked to calculate a Fahrenheit temperature because the unknown (°F) is isolated.

COMPARISON OF FAHRENHEIT AND CELSIUS TEMPERATURE SCALES **TABLE 3.6**

	Normal Boiling Point (BP)	Normal Freezing Point (FP)	BP − FP	Degree Ratio
Fahrenheit	212°F	32°F	212° − 32° = 180°	$\frac{180}{100} = 1.8$
Celsius	100°C	0°C	100° − 0° = 100°	

FIGURE 3.8

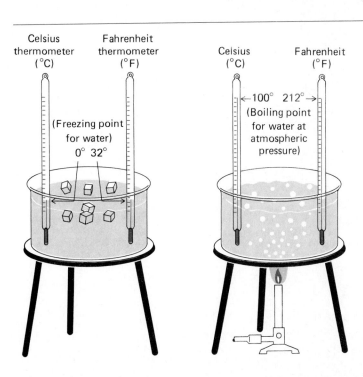

Celsius thermometer (°C)   Fahrenheit thermometer (°F)   Celsius (°C)   Fahrenheit (°F)

(Freezing point for water)
0° 32°

←100° 212°→
(Boiling point for water at atmospheric pressure)

The relationship between the Celsius and Fahrenheit temperature scales can be determined by measuring the freezing point and boiling point of water with both kinds of thermometers.

SAMPLE EXERCISE

3.8

A scientist measures the temperature of her laboratory as 25°C. What is the temperature in degrees Fahrenheit?

SOLUTION

We are given the number of degrees Celsius; therefore, we substitute this number in Equation (3.1):

$$°F = 1.8°C + 32 \qquad (3.1)$$

Substituting, we have

$$°F = 1.8(25) + 32$$

To complete this calculation, we must multiply before adding because the plus sign is outside the parentheses. Thus

$$°F = 45 + 32 = 77°$$

Equation (3.1) can be rearranged to a form that is more convenient for calculating degrees Celsius, given degrees Fahrenheit. Let us do the rearrangements step by step.
Subtract 32 from each side:

$$
\begin{aligned}
°F &= 1.8°C + 32 \\
-32 & \qquad\quad -32 \\
\hline
°F - 32 &= 1.8°C
\end{aligned}
\qquad (3.1)
$$

Divide each side by 1.8:

$$\frac{(°F - 32)}{1.8} = \frac{\cancel{1.8}°C}{\cancel{1.8}} = °C$$

Rewrite the equation putting degrees Celsius on the left:

$$°C = \frac{(°F - 32)}{1.8} \qquad (3.2)$$

SAMPLE EXERCISE

3.9

Body temperature is 98.6°F. What is body temperature in degrees Celsius?

SOLUTION

You are given the number of degrees Fahrenheit; therefore, substitute the number in Equation (3.2):

$$°C = \frac{(°F - 32)}{1.8} \qquad (3.2)$$

Substitution yields

$$°C = \frac{(98.6 - 32)}{1.8}$$

To complete this calculation, you must subtract before dividing because the minus sign is inside the parentheses.

$$°C = \frac{66.6}{1.8} = 37°$$

ALTERNATE SOLUTION

We really need to know only one equation to solve for temperature. We can use Equation (3.1) to solve this problem.

$$°F = 1.8°C + 32 \qquad (3.1)$$

Substitute 98.6 for degrees Fahrenheit because that is the given:

$$98.6 = 1.8°C + 32$$

Subtract 32 from each side:

$$66.6 = 1.8°C$$

Divide each side by 1.8:

$$37 = °C$$

Your choice of solution will depend on how comfortable you feel with algebraic manipulations.

The general method of problem-solving for temperature conversions is:

1. Identify the kind of degrees given.

2. Select a correct equation relating degrees Celsius and degrees Fahrenheit, and substitute for the degrees given.

3. Complete the calculation, being careful to multiply or add, subtract or divide in the proper order.

Notice that temperature conversion cannot be done by the Unit Conversion Method because the relationship between degrees Celsius and Fahrenheit is not strictly multiplicative. One must always add or subtract 32 at some point in the calculation.

Problem 3.6

A house thermostat reads 72°F. What is this temperature in degrees Celsius?

SAMPLE EXERCISE

3.10

One day last winter the temperature was reported as −10°C. What was the Fahrenheit reading?

SOLUTION

Use the preceding method, being careful to treat the minus sign correctly.

Step 1   The given is degrees Celsius.

Step 2   Therefore, we select Equation (3.1) and substitute for degrees Celsius.

$$°F = 1.8°C + 32 \tag{3.1}$$

Substituting, we have

$$°F = 1.8(-10) + 32$$

Multiply first:

$$°F = -18 + 32$$

Then add:

$$°F = +14$$

UNITS OF ENERGY

3.6   Heat is a form of energy. Temperature is an indicator of the tendency of heat energy to be transferred. Heat energy flows from objects of higher temperature to objects of lower temperature.

We saw in Section 1.12 that there are different forms of energy. Historically for chemists, the energy changes that were most useful and most easily measured experimentally were those involving heat energy. For this reason chemists traditionally use calories (abbreviated cal), units of heat energy, for all energy forms.[1]

One calorie is the amount of heat necessary to raise the temperature of one gram of water by one degree Celsius.[2]

Notice that temperature does *not* measure heat energy. The amount of heat energy contained in a body depends on:

1. Nature of the substance

2. Mass of the substance

3. Temperature

---

[1] This practice is slowly being abandoned. The SI unit of energy is the joule.
[2] To be absolutely accurate, the calorie is defined as the heat required to raise one gram of water from 14.5 to 15.5°C.

These three items are all reflected in the definition of a calorie:

1 cal = amount of heat to raise 1 g of water by 1°C

           ↑   ↑   ↑

         Mass   Nature Temperature

For every pure substance we can measure a physical property called the **specific heat** of that substance. The specific heat tells us the amount of heat necessary to raise the temperature of one gram of material by one degree Celsius. Table 3.7 lists specific heats for some common materials. Notice that the units of specific heat are cal/(g·°C). The specific heat of water is 1.0 cal/(g·°C); that is, it takes 1 cal to raise 1 g of water by 1°C. For iron the specific heat is 0.11 cal/(g·°C); that is, it takes only 0.11 cal to raise 1 g of iron by 1°C.

SAMPLE EXERCISE

3.11

How much heat energy must be applied to raise the temperataure of 10 g of water by 3°C?

SOLUTION

It requires 1 cal to raise the temperature of 1 g of water by 1°C. Therefore, 10 cal must be applied to raise the temperature of 10 g of water by 1°C, but we want the temperature to go up 3°C; therefore, we need 3 times the 10 cal, or 30 cal.

We can also do this problem by the Unit Conversion Method. Follow the usual steps:

Step 1 The given quantities and units are 10 g of water and 3°C change in temperature.

SPECIFIC HEATS OF SOME COMMON SUBSTANCES   **TABLE 3.7**

Substance	Specific Heat, cal/(g × °C)
Aluminum	0.22
Copper	0.093
Ethyl alcohol	0.51
Iron	0.11
Lead	0.031
Olive oil	0.50
Silver	0.056
Table salt	0.21
Water	1.0

Step 2   The new unit to be determined is calories, the unit of heat energy.

Step 3   The conversion factor relating calories, grams, and degrees Celsius is the specific heat of the given substance, in this case water.

Step 4   The setup is therefore

Given quantities and unit × conversion factor = new quantity and unit

$$10\,\cancel{g}\ \text{water} \times 3\cancel{^\circ C} \times \frac{1\ \text{cal}}{\cancel{g} \times \cancel{^\circ C}} = 30\ \text{cal}$$

In general the heat energy change for a given mass of substance m undergoing a temperature change t_1 to t_2 is given by the equation

$$m \times (t_2 - t_1) \times \text{specific heat} = \text{heat energy change}$$

$$\cancel{\text{grams}} \times \cancel{^\circ C} \times \frac{\text{calories}}{\cancel{g} \times \cancel{^\circ C}} = \text{calories}$$

Problem 3.7

How much heat energy must be applied to raise the temperature of 36 g of copper 18°C?

To determine the specific heat of a substance experimentally, you measure the temperature change that occurs when a known amount of heat is applied to a known mass of material. The experimental data is then used in the equation

$$\text{Specific heat} = \frac{\text{heat energy in calories}}{\text{mass in g} \times \text{temperature change in °C}}$$

SAMPLE EXERCISE

3.12

When 30.0 g of gold absorbs 4.78 cal, the temperature of the gold changes from 21.3 to 26.4°C. What is the specific heat of gold?

SOLUTION

Use the preceding equation for specific heat.

$$\text{Specific heat} = \frac{4.78\ \text{cal}}{30.0\ \text{g} \times 5.1°C}$$

(Use the change in temperature 26.4 − 21.3)

Doing the indicated multiplication and division, we get

$$\text{Specific heat of gold} = 0.031\ \frac{\text{cal}}{\text{g} \times °C}$$

The prefix "kilo-" is used with units of energy in the usual manner, that is, 1 kilocalorie = 1000 cal. When nutritionists speak of Calories (with a capital C, abbreviated Cal) as applied to foods, they mean kilocalories (abbreviated kcal). Because energy can change form, we can use calories and kilocalories as the units for any form of energy, even though they have been defined only in terms of heat energy.

The energy content (caloric content) in food is measured by totally burning the food and measuring the heat given off by the burning food. This amount of heat energy is exactly equivalent to the amount of energy the food can potentially provide to the body for moving muscles, conducting electrical impulses along nerves, maintaining body temperature, etc. Table 3.8 shows the caloric content of some common foods.

CALORIC VALUE OF SOME COMMON FOODS   **TABLE 3.8**

Food	Typical Serving Size	Caloric Content of Serving
Apples (raw)	1 apple (~175 g)	105
Beer	12 oz	150
Bread (white, enriched)	1 slice	68
Cheese (cheddar)	1 oz	133
Chocolate	1 oz	156
Eggs	1 egg	80
Hamburger	$\frac{1}{4}$ lb	409
Milk	1 glass	160
Peanuts	1 oz	159
Sugar	2 tsp	25

WHAT ARE SIGNIFICANT FIGURES?

3.7 Whenever you make a measurement, there is always some uncertainty associated with the measurement. The uncertainty comes about because of the nature of the object measured and the limitations of the measuring device. For example, consider the attempts at measuring the length of a common ball-point pen shown in Figure 3.9. There is an uncertainty because the shape of the pen makes it difficult to line up the ends for measurement accurately. This is an unavoidable uncertainty, but we can see that the measuring device also limits our knowledge. In Figure 3.9a, the use of a meter stick calibrated only in decimeters allows us to say with certainty that the pen measures between 0.1 and 0.2 m. We can estimate that it is about 0.14 m.

In Figure 3.9b, using a meter stick calibrated in centimeters allows us to say with certainty that the pen measures between 14 cm

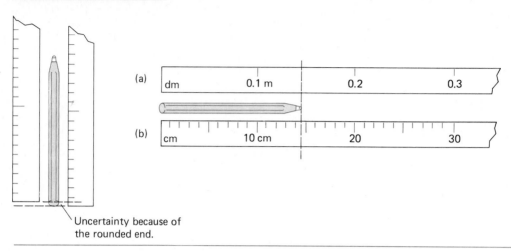

Uncertainty because of
the rounded end.

FIGURE 3.9

Uncertainty in measurement.
The nature of the measured
object introduces uncertainty.
The measuring device has
limitations. (a) With the meter
ruler, one can say with
certainty that the pen
measures between 0.1 and
0.2 m. We can estimate that
it is 0.14 m. (b) With the
centimeter ruler, one can say
with certainty that the pen
measures between 14 and
15 cm (between 0.14 and
0.15 m). We can estimate
that it is 14.5 cm or 0.145 m.

(0.14 m) and 15 cm (0.15 m). We can estimate that it is about
14.5 cm (0.145 m).

In the first measurement (Figure 3.9*a*), we say that the measurement is good to two significant figures (2 sig figs), one we know for certain and the one we guessed.

0.14 m

Certain **Good guess**

2 sig figs

In the second measurement, using a more sensitive measuring device, the measurement is good to 3 sig figs:

0.145 m

Certain **Good guess**

3 sig figs

*For any measurement, the number of **significant figures** (we will use sig figs for short) that can be reported is the number of figures that can be read accurately from the measuring device plus one more figure that must be estimated.*

SAMPLE EXERCISE

3.13

How many sig figs can be reported in each of the following measurements?

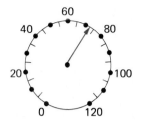

a. Bathroom scale calibrated in pounds

b. Speedometer calibrated in kilometers per hour

c. Graduated cylinder calibrated in milliliters

SOLUTION

a. We can tell with accuracy that the weight is between 136 and 137 lb. Then we can estimate that the pointer is midway between 136 and 137 and report 4 sig figs:

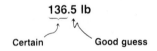

136.5 lb

Certain   Good guess

b. The speed is definitely between 70 and 75 km/hour (h). Because there are no calibration lines between 70 and 75, we must estimate the nearest kilometer. Our guess is 72 km/h. Therefore, we are allowed 2 sig figs: the 7 is certain, the 2 is a good guess.

c. The volume is between 6 and 7 mL. We estimate that it is 6.8 mL. Two sig figs are allowed.

When we actually do or see the measurement, as in Sample Exercise 3.13, it is easy to determine how many sig figs are allowed and are represented by the reported number. More often you are given a measured number and asked to tell how many sig figs it contains. To do this requires a set of rules.

Rules for counting sig figs in a given number

1. All *nonzero* figures are significant.

2. All *zeros between nonzero* figures are significant.

3. When a decimal point is shown, *zeros to the right of nonzero* figures are significant. (When a decimal point is not shown, the number is ambiguous and the number of sig figs cannot be determined.)

4. *Zeroes to the left* of the first nonzero figure are *not* significant.

An example demonstrating all the rules is

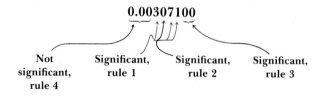

Total of 6 sig figs.

SAMPLE EXERCISE

3.14

For each of the following measurements, tell how many sig figs are represented.

a. 4.065 m

b. 0.32 g

c. 57.98 mL

d. 20.00 s

e. 0.00040 km

f. 604.0820 kg

SOLUTION

a. 4 sig figs, rules 1 and 2
b. 2 sig figs, rules 1 and 4
c. 4 sig figs, rule 1
d. 4 sig figs, rules 1 and 3
e. 2 sig figs, rules 1, 3, and 4

$$0.00040$$

Not significant   Significant

f. 7 sig figs, rules 1, 2, and 3

Problem 3.8

Tell the number of sig figs in each of the following:

a. 21.94 mL

b. 3.90 g

c. 0.08040 cm

d. 0.9000 g

SAMPLE EXERCISE

3.15

How many sig figs are represented by each of the following measurements?

a. 3.24×10^{-8} cm   **c.** 7.0300×10^2 mL

b. 5.0×10^3 g   **d.** 4.705×10^{-4} km

SOLUTION

To determine the number of sig figs in a number written in scientific notation, apply the rules to the number preceding the times sign. You will find that if the number is in the correct notation form, then all digits are significant.

a.   3   **c.**   5

b.   2   **d.**   4

ROUNDING OFF

3.8   In doing calculations with sig figs you will frequently need to round off answers to smaller numbers of digits. Use the following rules:

Rules for rounding off

1.   If the first figure to be dropped is less than 5, the preceding digit is not changed. Therefore, 1.234 is rounded to 1.23 (3 sig figs) or 1.2 (2 sig figs) or 1. (1 sig fig).

2.   If the first figure to be dropped is 5 or more, the preceding digit is increased by 1. Therefore, 1.756 is rounded to 1.76 (3 sig figs) or 1.8 (2 sig figs) or 2. (1 sig fig).

SIG FIGS IN CALCULATIONS

3.9   Multiplication and division   If you multiply the measured numbers 2.3×5.69, the calculator will read out 13.087. But this is not a proper answer to the multiplication because in multiplying or dividing measurements, the answer should contain no more *sig figs* than the measurement with the fewest number of *sig figs*. Therefore, in multiplying 2.3 (2 sig figs) $\times$ 5.69 (3 sig figs), we are allowed only 2 sig figs in the answer. So we must round off to 2 sig figs:

$$2.3 \times 5.69 = 13.087 = 13$$

SAMPLE EXERCISE

3.16

The mass and volume of a sample of blood are measured and found to be 12.193 g and 11.5 mL, respectively. Determine the density and express the answer in the proper number of sig figs.

SOLUTION

$$\text{Density} = \frac{\text{mass}}{\text{volume}}$$

Using a calculator,

$$\text{Density} = \frac{12.193 \text{ g}}{11.5 \text{ mL}} = 1.0602609 \text{ g/mL}$$

But because we know the volume to only 3 sig figs, we are allowed only 3 sig figs in the final answer. Therefore, we round off and report the density as 1.06 g/mL.

Even when there is a series of multiplications and divisions, the measurement with the smallest number of sig figs governs the number of sig figs in the answer. Consider doing the following operations with a calculator.

$$\frac{0.04 \times 2.546 \times (3.12 \times 10^6)}{7.0012} = 4.5384 \times 10^4$$

Calculator answer

Because one of the multipliers (0.04) has only 1 sig fig, the correct answer to report must have only 1 sig fig. In this case, the answer would be 5×10^4.

Problem 3.9

Assuming that the following numbers are measurements, perform the calculation and report the answer to the correct number of sig figs:

$$\frac{3.8 \times 48.20}{7.81} =$$

When rounding off larger numbers to a small number of sig figs, it is best to express the answer in scientific notation. For example, 1903 rounded to 2 sig figs is 1.9×10^2. If you were to express the answer as 1900, it would be ambiguous. The number 1900 may indicate 2, 3, or 4 sig figs; 1900. indicates 4 sig figs. As you can see, scientific notation offers the best expression.

The idea of sig figs applies only to *measured* numbers. Some numbers are **defined** or **exact** numbers and when used in a calculation do not affect the numbers of sig figs allowed in the answer. For example, 1 dozen is defined as 12 eggs. To determine the number of eggs in 11 dozen, we multiply 11 dozen × 12 eggs/1 dozen = 132 eggs. We do not round off here because we are dealing with exact numbers, not measured numbers.

All the equalities given in Tables 3.2 and 3.3 are exact or defined.

For example, 1 m is defined as exactly 100 cm, just as 1 dozen is defined as 12 eggs, so the use of conversion factors involving meters and centimeters will not affect the number of sig figs allowed. The equalities in Table 3.4 are inexact because they are measured and must be considered in determining sig figs allowed. They are expressed in such a way that the 1 is exact, but the other number is good only to the number of sig figs shown. For example, for 1 lb = 454 g, the 1 is exact or defined and will not affect a calculation, 454 is measured and will limit a calculated answer to 3 sig figs.

To summarize, only **measured** *numbers* and **inexact** *equalities* limit the number of sig figs in a calculated answer.

SAMPLE EXERCISE

3.17

A student's height is measured as 69.5 in and the student wishes to convert this measurement to kilometers. Perform this conversion and report the answer to the proper number of sig figs.

SOLUTION

Following the usual Unit Conversion Method, we will get the following setup:

Given quantity and unit × conversion factors = new quantity and unit

$$69.5 \text{ in} \times \frac{1 \text{ m}}{39.4 \text{ in}} \times \frac{1 \text{ km}}{1000 \text{ m}} = 0.00176396 \text{ km}$$

Calculator answer

To determine the numbers of sig figs allowed, we must identify measured, exact, and inexact numbers. From the preceding discussion realize that

$$69.5 \text{ in} \times \frac{1 \text{ m}}{39.4 \text{ in}} \times \frac{1 \text{ km}}{1000 \text{ m}}$$

Exact

Measured   Inexact

Therefore, the number of sig figs allowed will be 3, the minimum number of sig figs in the measured and inexact numbers. The answer is 0.00176 km.

In practice, tables of equalities are constructed so that they do not limit the number of sig figs allowed any more than the measured number does. This was the case in Sample Exercise 3.17. The measured value was good to 3 sig figs and so was the conversion factor.

Addition and subtraction Addition and subtraction of measured numbers follows a different rule. In adding and subtracting measurements, the final answer must not contain any more digits to the right of the decimal point than does the measured number with the *least number of digits to the right of the decimal point.*

For example, in adding 937.3 g + 15.224 g + 71.04 g, the number 937.3 governs the sig figs in the answer—there may be only *one digit* to the right of the decimal point.

$$
\begin{array}{r}
937.3 \text{ g} \\
15.224 \text{ g} \\
\underline{71.04 \text{ g}} \\
1023.564 \text{ g} = 1023.6 \text{ g}
\end{array}
$$

Because one mass is known only to the nearest tenth, the answer must be rounded off to tenths.

SAMPLE EXERCISE

3.18

Report the result of the following subtraction to the correct number of sig figs:

$$2.572 \text{ m} - 0.41 \text{ m} = ?$$

SOLUTION

$$
\begin{array}{l}
2.572 \text{ m} \\
\underline{-0.41 \text{ m}} \leftarrow \text{This is the limiting number because there are fewer digits} \\
2.162 \text{ m} \quad \text{to the right of the decimal point.}
\end{array}
$$

The answer must be rounded off to 2.16 m.

Problem 3.10

Add up the following weights and report the sum to the proper number of sig figs:

$$2.2332 \text{ g} + 1004.12 \text{ g} + 26.557 \text{ g} = ?$$

Comment on the use of sig figs in this book

The solutions to all the sample exercises in this book from Chapter 3 on and the solutions to the problems given in Appendix 4 are reported to the correct number of sig figs. You should always round off all your final answers to the proper number of sig figs.

SUMMARY

Measurement is comparison to some reference standard (Section 3.1). Scientists use the metric system, which is based on multiples and divisions by 10. The base units of the metric system are meters for length, grams for mass, and liters for volume (Section 3.2). Chemists measure energy in calorie units which are related to the nutritionist's Calorie (Section 3.6). The Celsius scale (°C) is used for temperature measurement in the metric system. The Celsius scale is related to the Fahrenheit scale (°F) by a formula derived from the boiling and freezing points of water (Section 3.5).

The Unit Conversion Method is a powerful tool for doing many types of calculations. In this chapter it is used to change the units of metric measurements (Section 3.3). Other applications will be discussed in later chapters.

Density is the ratio of the mass of a substance to its volume and is a constant characteristic property of pure substances (Section 3.4). Another property of a pure substance is its specific heat (Section 3.6).

Measurements are reported to the correct number of significant figures in order to indicate the uncertainty in the measurement (Section 3.7). In the multiplication and division of measurements, the number of sig figs allowed in the answer depends on the measurement with the fewest number of sig figs. In addition and subtraction, the number of sig figs in the answer depends on the measurement with the least number of digits to the right of the decimal point (Section 3.9). Working with sig figs frequently requires "rounding off" (Section 3.8).

CHAPTER ACCOMPLISHMENTS

After completing this chapter you should be able to

3.1 What is measurement?
1. State the role of a reference standard in measurement.

3.2 The metric system
2. State the metric base units used for length, mass, and volume measurements.
3. State the meanings of the common metric prefixes (Table 3.1).
4. State the equalities between common metric units (Table 3.3).
5. Distinguish between mass and weight.
6. State the advantage of measuring mass rather than weight.

3.3 Unit conversion
7. Construct conversion factors from equalities.

8. Beginning with a given quantity and unit, use conversion factors to calculate some new quantity and unit.

3.4 Density

9. Define density.
10. State the usual units in which density is expressed.
11. Given the mass and volume of a sample, calculate the density of the sample.
12. Recognize density as a conversion factor.
13. Given the density, convert the volume of a sample to the mass of the sample.
14. Given the density, convert the mass of a sample to the volume of the sample.

3.5 Temperature

15. Given a temperature in degrees Fahrenheit or degrees Celsius, convert from the given scale to the other scale.

3.6 Units of energy

16. State the unit of energy most commonly used by chemists.
17. Define calorie.
18. Define specific heat.
19. Given a table of specific heats, calculate the amount of heat energy necessary to raise the temperature of a given mass a given number of degrees.
20. Given appropriate experimental data, calculate the specific heat of a substance.

3.7 What are significant figures?

21. Given a measuring device, state the number of sig figs that can be reported in a measurement.
22. Given a measured number, state the number of sig figs it has.

3.8 Rounding off

23. Round off numbers to some indicated number of sig figs.

3.9 Sig figs in calculations

24. Express the result of the multiplication or division of measured numbers to the correct number of sig figs.
25. Distinguish between measured numbers and defined or exact numbers.
26. Express the result of the addition or subtraction of measured numbers to the correct number of sig figs.

PROBLEMS

3.1   What is measurement?

3.11   Describe in your own words what is meant by the measurement process.

3.2   The metric system

3.12   Give five common prefixes of the metric system and their mathematical meanings.

3.13   State the equalities between the following metric units:
  a.   Grams and kilograms
  b.   Liters and milliliters
  c.   Meters and centimeters
  d.   Micrometers and meters
  e.   Milliliters and cubic centimeters

3.14   Give the names of the metric system units having the following abbreviations:
  a.   mm   d.   mg
  b.   mL   e.   km
  c.   g

3.15   Arrange each of the following units in an increasing sequence from smallest to largest:
  a.   Meter, centimeter, kilometer
  b.   Milligram, microgram, gram
  c.   Liter, milliliter, cubic centimeter

3.16   Express the following metric prefixes as a power of 10:
  a.   Centi   c.   Milli
  b.   Kilo

3.3   Unit conversion

3.17   Using the unit conversion method, carry out the following one-step conversions within the metric system:
  a.   18.9 m to kilometers
  b.   37.5 cm to meters
  c.   0.145 L to milliliters
  d.   452 mL to liters
  e.   645 cm^3 to milliliters
  f.   0.089 kg to grams
  g.   1.9 L to milliliters
  h.   38.4 g to milligrams

3.18   Using Table 3.4, carry out these one-step conversions between the English and metric systems:
  a.   26.0 mi to kilometers
  b.   3.6 kg to pounds
  c.   5.00 in to centimeters
  d.   4.00 L to quarts
  e.   1.92 lb to kilograms
  f.   125 g to pounds
  g.   1.00 qt to liters

3.19   Carry out the following two-step conversions within the metric system:
  a.   0.053 km to centimeters
  b.   38.5 cm to kilometers
  c.   3190. mg to kilograms
  d.   82 mm to centimeters

3.20   Using Table 3.4, carry out the following two-step conversions between the English and metric systems:
  a.   3.00 ft to meters
  b.   1.42 qt to milliliters
  c.   78 mm to inches

3.21   Carry out the following multistep conversions:
  a.   0.52 mi to centimeters
  b.   400. mL to gallons
  c.   1.50 oz to grams (16 oz = 1 lb)

3.22   Do the following conversions:
  a. $\frac{1}{4}$ lb of salami to kg salami
  b. 9.3×10^7 mi to kilometers
  c.   4.50 yd to millimeters
  d.   42.0 in to centimeters
  e.   6.80 μg to kilograms
  f.   750. mL to quarts

3.23   A run of 100. m corresponds to how many yards?

3.24   Which is more gold, 1.9 oz or 39 g?

3.25   A highway sign tells you that you are 341 km away from home. How many miles away are you?

3.26   How long is a standard football field (100. yd) in meters?

3.27 Men's shirt sizes are based on the circumference of men's necks measured in inches. Size 16 means the neckband measures 16 in. What is the metric size (in centimeters) corresponding to size 16?

3.28 Trouser sizes are based on waist and inseam (leg length) measurements in inches. What is the metric size (in centimeters) corresponding to 38 waist, 32 length?

3.29 The distance from home plate to the pitcher's mound on a standard baseball field is 60 ft 6 in. How far is it in meters?

3.30 A basketball must weigh at least 20. oz. What is the minimum weight in grams (16 oz = 1 lb)?

3.31 A swimmer swims 50. lengths of a pool each day. The pool is 30. m long. What distance does she swim each day (**a**) in meters, (**b**) in miles?

3.32 In France gasoline is sold by the liter. If you wanted 12 gallons (gal) of gasoline, how many liters should you ask for (1 gal = 4 qt)?

3.33 If gasoline costs $1.14 per gallon, what is the price in cents per liter?

3.34 A speeder is found to be moving at 82 mi/h. What is his speed in kilometers per hour?

3.35 If your car gets 26 mi/gal, how many kilometers per liter does it get?

3.36 On a long trip you travel 565 mi in 14 h and use 19 gal of gasoline.
    a What is your average speed in miles per hour?
    b. What is your gas mileage in miles per gallon?

3.37 On another trip you travel 308 km in 3.5 h and use 32 L of gasoline.
    a. What is your average speed in kilometers per hour?
    b. What is your gas mileage in kilometers per liter?

3.38 Reconsider the trip described in Problem 3.36.
    a. What is the average speed in kilometers per hour?
    b. What is the gas mileage in kilometers per liter?

3.39 Reconsider the trip described in Problem 3.37.
    a. What is the average speed in miles per hour?
    b. What is the gas mileage in miles per gallon?

3.40 A person's body measurements are weight = 152 lb, height = 5 ft 8 in, chest = 39 in, and waist = 34 in. Convert the weight measurement to kilograms and the length measurements to centimeters.

3.41 The moon is 238,860. mi from the earth.
    a. How far away is this in meters? (Use scientific notation.)
    b. Given that a radio signal travels 3.00×10^8 m/s, how many seconds does it take for the radio signal to travel from the moon to the earth?

3.42 A fish tank is 30. in long, 20. in deep, and 12 in high.
    a. How many liters of water are required to fill it?
    b. How many gallons are needed?

3.4 Density

3.43 Calculate the density of a liquid, 31.90 mL of which weighs 22.38 g.

3.44 A piece of metal weighs 59.24 g and occupies a volume of 6.64 mL.
    a. What is the metal's density?
    b. Consult Table 3.5 and identify the metal.

3.45 Use Table 3.5 to calculate the mass in grams of each of the following samples:
    a. 35.0 mL of gasoline
    b. 989.0 mL of vinegar
    c. 18.0 mL of olive oil

3.46 Use Table 3.5 to calculate the volume of each of the following samples:
    a. 65 g of alcohol
    b. 98.2 g of gold
    c. 454 g of mercury

3.47 How much does 1 pt of milk weigh in pounds (see Table 3.5; 1 qt = 2 pt)?

3.48 Which liquid sample will occupy the greater volume, 100. g of water or 100. g of carbon tetrachloride?

3.49 Consult Table 3.5 and calculate the mass in pounds of 1.0 qt of
  a. blood
  b. gasoline

3.50 You are given three liquid samples, *A*, *B*, and *C*, and told that one is water, one is alcohol, and one is ether. Each sample is 10.00 mL. The masses of the samples are *A*, 10 g; *B*, 7 g; and *C*, 8 g. Identify *A*, *B*, and *C*.

3.51 A cube of lead, measures 2.50 cm on each edge and has a mass of 176 g. Calculate the density of lead.

3.52 A flask has a mass of 54.12 g when empty and 180.27 g when completely filled with water.
  a. Using the density of water at 20°C (1.00 g/mL), calculate the volume of the flask.
  b. When the same flask is filled with an unknown liquid, the total mass is 142.48 g. What is the density of the unknown liquid?
  c. Identify the unknown liquid by consulting Table 3.5.

3.53 What is the mass of air in a room measuring 10. m × 5.0 m × 3.0 m?

3.54 A shiny yellow stone of volume 7.0 mL has a mass of 56.0 g. Could this stone be gold? Explain.

3.5 Temperature

3.55 Do the following temperature conversions:
  a. 75°F to degrees Celsius
  b. −5.0°F to degrees Celsius
  c. 18°C to degrees Fahrenheit
  d. −80.°C to degrees Fahrenheit
  e. 15.45°F to degrees Celsius

3.56 Which is the higher temperature, −25°C or −15°F?

3.57 A typical antifreeze mixture raises the boiling point of the water in your radiator to 240.°F. What is this in degrees Celsius?

3.58 The highest recorded weather temperature on the earth is 136.4°F. What is this in Celsius?

3.59 Which is warmer, 265°C or 425°F?

3.6 Units of energy

3.60 What does the amount of heat energy within a sample depend on?

3.61 How many calories are there in 43.5 kcal?

3.62 What is 1032 cal expressed in kilocalories?

3.63 How much heat must be applied to raise the temperature of 84 g of olive oil 75°C (see Table 3.7)?

3.64 How much heat must be removed to cool 76 g of iron from 95 to 16°C (see Table 3.7)?

3.65 1.50 kcal is applied to a sample of water and the temperature rises from 22.3 to 28.9°C. What is the mass of the water sample?

3.66 When a piece of graphite weighing 3.00 g absorbs 15.0 cal, the temperature changes from 25.0 to 54.0°C. What is the specific heat of graphite?

3.67 When a 75.4-g sample of Ni loses 341 cal, its temperature falls from 88.5 to 45.8°C. What is the specific heat of Ni?

3.68 Consult Table 3.7 and decide which sample below is the hottest (highest temperature). Both materials were originally at room temperature.
  a. 10 g of water absorbs 75 cal
  b. 10 g of iron absorbs 75 cal

3.69 Consult Table 3.7 and decide which sample is the coldest. Both materials were originally at room temperature.
  a. 2 g of olive oil loses 0.02 kcal
  b. 5 g of aluminum loses 0.03 kcal

3.70 A cup of yogurt contains 150 Calories.
  a. How many calories is this?
  b. If this energy were used to heat 100. lb of water originally at 98.6°F, what would be the final temperature of the water?

3.7 What are significant figures?

3.71 How many sig figs are in each of the following measured numbers?
  a. 1.09   **f.** 0.000001
  b. 0.1   **g.** 11×10^7
  c. 0.3040   **h.** 0.0230
  d. 9.2×10^{-3} **i.** 6.023×10^{23}
  e. 901.0   **j.** 100.

3.8 Rounding off

3.72 Round off each of the following to 4 sig figs:
  a. 235.674 g
  b. 10.528 mL
  c. 4.0534×10^{-4} liters
  d. 7.2457 g/mL
  e. 0.000328730 kg

3.9 Sig figs in calculation

3.73 Do the following multiplications or divisions and express your answer to the proper number of sig figs:
  a. 100.1×0.094
  b. 3×2.95
  c. $\dfrac{0.04}{18.1}$
  d. $\dfrac{2.00 \times 10^1}{3.9 \times 10^3}$
  e. 0.00903×21.18

3.74 Do the following additions or subtractions and express your answer to the proper number of sig figs:
  a. $12.0140 + 11.91$
  b. $84.938 + 7.452 + 0.9$
  c. $28.75 - 1.4$
  d. $0.9523 - 6.540$

3.75 Do the following calculations and report the answers to the proper number of sig figs. Convert 3182 g to pounds.
  a. Given: 1 lb = 453.6 g
  b. Given: 1 lb = 454 g

3.76 Calculate the density of ethanol to the proper number of sig figs, given the data that for a certain sample, mass = 15.79 g and volume = 20.00 mL.

3.77 What is the total volume accumulated in a beaker by successive additions of liquid in the amounts 50.40 mL, 1.5 mL, 8.9 mL and 0.123 L? Report the answer to the proper number of sig figs.

3.78 A student is given a liquid sample to identify. He performs the following measurements to do the identification:
  a. Weighs a flask and stopper (mass = 74.931 g)
  b. Adds exactly 10.00 mL of the unknown liquid to the flask and stoppers it again
  c. Weighs the flask and stopper and liquid (mass = 82.069 g)

Calculate the density to the correct number of sig figs. What is the liquid?

ELEMENTS AND THEIR INVISIBLE STRUCTURES

CHAPTER

4

Be

Beryllium

9.012

4.1 Thus far the discussion of matter has been in *macroscopic* terms. That is, we have explained the classification of matter by examining large (*macro-* means "large") "chunks" of matter which we can see and hold. For a true and clearer understanding of matter, it must be explored in *microscopic* (*micro-* means "small") terms. We should really say *sub*microscopic terms, because the smallest particles of matter cannot be seen even by the most sensitive electron microscopes. As soon as you learn the basic makeup of **atoms,** the tiniest particles of matter, you will be able to understand such topics of everyday interest as radioactivity, nuclear medicine, and nuclear power plants.

ATOMIC THEORY

4.2 We assume that you have already heard that all matter is made up of tiny particles called atoms because of the publicity Atomic Theory receives through such things as the atomic bomb and atomic energy. Philosophically there are only two ways matter could be constructed: (1) it might be *continuous,* i.e., divisible into ever smaller parts so that no smallest particle exists; or (2) it might be *discrete,* i.e., made up of characteristic small particles which if further divided no longer have the properties of the matter which we are examining. Experiment indicates that matter is discrete, i.e., made up of atoms.

The concept of the atom is an old one. The word was coined by the Greek Democritus, who first suggested the atomic idea in 400 B.C. Scientists and philosophers generally did not believe in atoms until the early nineteenth century. At that time an English school teacher, John Dalton, stated the Atomic Theory of Matter in a clear manner which explained many known scientific laws and experiments and also accounted for the classification of matter into elements, compounds, and mixtures. Some of the key statements of Dalton's theory appear below:

1. All matter is made up of atoms which are indestructible by ordinary means.

2. All atoms of a given element are identical in chemical properties.

3. Atoms of different elements have different chemical properties.

4. Atoms of different elements can chemically combine in simple, whole-number ratios to form compounds.

Now we can define elements, compounds, and mixtures in terms of their atomic makeup.

Element: Matter which is composed of one kind of atom.

Compound: Matter which is composed of different kinds of atoms chemically combined in simple, whole-number ratios.

Mixture: Matter which is a physical combination of the particles of elements or compounds.

Compare these definitions with those given in Section 1.9.

SAMPLE EXERCISE

4.1

How does this new atomic definition of element explain the former definition given in Section 1.9, i.e., that an element is "a pure substance which cannot be broken down into simpler substances by ordinary chemical changes?"

SOLUTION

Elements are pure substances, which means that they have a constant composition and unique set of properties because they are collections of just one kind of atom. Elements cannot be broken down because atoms are essentially indestructible.

Problem 4.1

Explain how the new atomic definition of compound relates to the former definition given in Section 1.9.

In this chapter we will discuss various elements and look more closely at how atoms are constructed.

PICTURING ATOMS AND ELEMENTS

4.3 Dalton pictured atoms as tiny spherical particles (see Figure 4.1). We will find that this is not a completely accurate picture, but it pro-

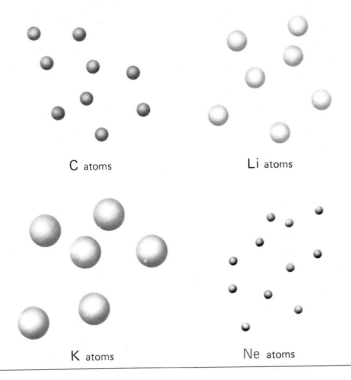

C atoms   Li atoms

K atoms   Ne atoms

FIGURE 4.1

Elements are made up of one
kind of atom. The atoms of
different elements have
different properties. Note the
different sizes and illustrative
colors of the examples.

vides a useful mental image for understanding the composition of
matter.

It is difficult to picture how very, very small an atom is and thus
to realize the extremely large number of atoms which are present in
a sample of an element. For example, the largest atoms are those of
cesium, but one cesium atom has a diameter of only 0.0000000524
cm. This means that between adjacent millimeter-marking lines on a
meter stick we could line up 1,910,000 Cs atoms. A 1-cm cube of
cesium metal (smaller than a sugar cube) would contain approximately
6,950,000,000,000,000,000,000 Cs atoms. Now you can see one reason
why we need to be able to express numbers in scientific notation.
Discussing sizes and numbers of atoms involves very small and very
large numbers (Figure 4.2).

Problem 4.2

Refer to the preceding paragraph. Express the diameter of a cesium atom,
the number of cesium atoms along a 1-mm line, and the number in a 1-cm
cube in scientific notation.

Of course, because atoms are so very small in size, the mass of
one atom is very small. One Cs atom weighs only 2.21×10^{-22} g. If
you weigh 130 lb, then 2.67×10^{26} Cs atoms correspond to your
mass.

Diameter = 0.0000000524 cm

In this tiny distance (1 mm) 1,910,000 Cs atoms can line up.

mm 1   2   3   4   5

Metric ruler

(a) This drawing has been scaled up almost *40 million* times the actual size of the Cs atom.

(b) To line up 1,910,000 spheres with diameter 2 cm [approximately the size shown in (a)] would require 38 km.

1 cm

—6,950,000,000,000,000,000,000 Cs atoms will fit within this cube.

FIGURE 4.2

Atomic size: Cs atoms are the largest of the atoms, but they are still very, very tiny.

(c) To fit this number of spheres of the size shown in (a) would require a cube 380 km on its edge. The volume of the cube (5.5×10^7 km^3) then would be a volume corresponding to the water content of 15 Mediterranean Seas!

INSIDE THE ATOM

4.4   Atoms are not really hard little balls as they are shown in Figure 4.1. Actually, atoms are mostly empty space because of the arrangement of the **subatomic** particles of which they are made. That is, by the end of the nineteenth century it had become clear that atoms were not the smallest particles. Rather, atoms can be subdivided into smaller subatomic particles. By today, atomic physicists have identified hundreds of indescribably small and often short-lived particles. Luckily, to understand chemistry, one need be aware of only the three major types of subatomic particles: protons, electrons, and neutrons. Atoms of different elements differ in the *numbers* of these particles contained. Figure 4.3 shows the arrangement of these particles and explains the statement that atoms are mostly empty space.

Characteristics of subatomic particles   The important characteristics of protons, electrons, and neutrons are their relative masses and electrical charges. These are summarized in Table 4.1.
       Protons and neutrons have approximately the same mass. Electrons are much smaller and lighter; the electron mass is only roughly $\frac{1}{1840}$ the mass of a proton. To express these mass relationships conveniently we assign the proton a mass of 1, then the neutron mass must be 1, and the electron mass must be $\frac{1}{1840}$. Because the electron mass is so much less than the proton or neutron, we can ignore it. Electrons contribute practically no mass to an atom. Therefore, we

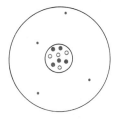

(a) A cross section of an atom shows the protons (○) and neutrons (●) clustered in the nucleus and the electrons outside the nucleus and fairly far apart.

(b) A better representation of the atom is a fuzzy cloud for the electrons to show their constant motion and to better show that atoms are three-dimensional. Cutting the cloud in half reveals the nucleus in the center.

FIGURE 4.3

The arrangement of subatomic particles in the atom.

CHARACTERISTICS OF SUBATOMIC PARTICLES   **TABLE 4.1**

Particle	Symbol	Relative Charge	Relative Mass
Proton	p^+	$+1$	1
Electron	e^-	-1	0 (1/1840)
Neutron	n^0	0	1

assign the electron a relative mass of zero. In Section 4.9 we will assign units to these relative masses.

You probably are familiar with the positive ($+$) and negative ($-$) character of electricity. Look at any battery; there is a positive pole and a negative pole. If you study physics, you will find that electricity is a flow of electrons and that electrons have a negative ($-$) charge. Atoms are electrically neutral (no net charge). This is so because for every electron with a -1 charge, there is a proton which has a $+1$ charge. Just as adding signed numbers of equal magnitude and opposite signs results in zero, equal numbers of positive protons and negative electrons result in zero charge (electric neutrality) (see Figure 4.4). Neutrons are so named because they are electrically neutral (zero charge).

Arrangement of subatomic particles   The arrangement of protons, electrons, and neutrons is shown in Figure 4.3. Notice that the mass of an atom is concentrated in the center because the protons and neutrons are clustered there. The center of the atom occupied by protons and neutrons is called the **nucleus.** The nucleus is described as being dense or compact because most of the mass of the atom is contained in a small volume. In addition, the nucleus bears a positive charge because of the positive protons.

Sprinkled around the dense, positive nucleus are the "weightless,"

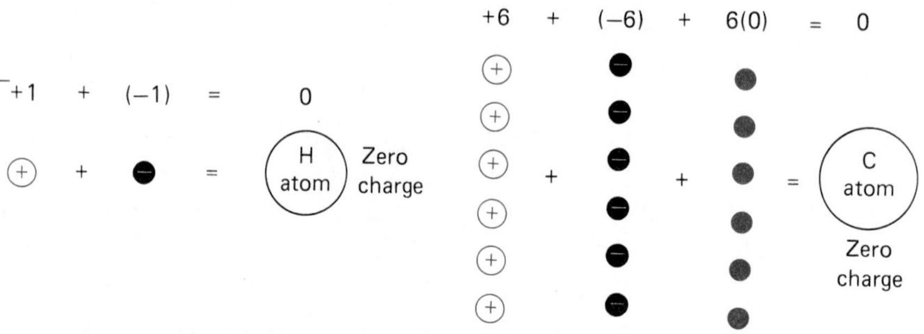

FIGURE 4.4

Proton ($+$) and electron ($-$) charges are added in the same way that signed numbers are added. Neutrons ($\bullet$) have zero charge. In the neutral atom the sum of the charges is zero.

$$+1 \quad + \quad (-1) \quad = \quad 0$$

$$+6 \quad + \quad (-6) \quad + \quad 6(0) \quad = \quad 0$$

flighty negative electrons. In Chapter 5 you will learn more details about the arrangement of electrons. For now, it is sufficient to know the following:

1. Electrons are outside the nucleus.

2. Electrons are constantly moving.

3. Because of the spaces between electrons and between the electrons and the nucleus, atoms are mostly empty space.

4. Because electrons bear negative charges and are constantly moving, they appear to be a negatively charged cloud with fuzzy borders.

Look out your window at the clouds. Pick out a nice round or egg-shaped cloud. Imagine that the cloud is negatively charged and that buried deep within the cloud is a tiny positively charged pebble. You have just developed a mental picture of an atom.

ATOMIC NUMBER

4.5 The number of protons in an atom is called the **atomic number.** Atoms and elements are identified by their atomic number. Look at the periodic table inside the front cover. The numbers above the symbols are the atomic numbers, that is, the number of protons in atoms of that element. Any atom with six protons must be carbon. An atom with just one proton must be hydrogen. For an element to be gold, there must be 79 protons in the nucleus of each atom.

Problem 4.3

a. What are the atomic numbers of Mg, S, and Ag?

b. How many protons are there in Mg, S, and Ag?

Problem 4.4

Identify the elements for which the following information is given:

a. Atomic number, 17

b. Number of protons, 20

Atoms are electrically neutral because for every proton $(+1)$ there is an electron (-1). Therefore, the atomic number also tells you the number of electrons in a neutral atom. Carbon atoms with atomic number 6 have six protons and six electrons.

SAMPLE EXERCISE

4.2

A neutral atom is known to have 14 electrons.

a. How many protons does it have?
b. What is its atomic number?
c. What is the name of the element with this atomic number?

SOLUTION

a. Fourteen protons because the number of protons equals the number of electrons.
b. Fourteen because atomic number equals the number of protons.
c. Silicon; only the element silicon is made up of atoms with atomic number 14.

Problem 4.5

How many electrons are there in neutral atoms of Al, K, and Cu?

Problem 4.6

a. What are the atomic numbers of elements with 3, 18, and 53 electrons in their neutral atoms?

b. What are the names of these atoms?

ISOTOPES

4.6 Whereas the number of protons (atomic number) identifies atoms of a particular element, the atoms of the same element may contain a different number of neutrons. For example, *all* carbon atoms have six protons and most carbon atoms have six neutrons. But about 1 percent of carbon atoms contain seven neutrons, and there are some carbon atoms with eight neutrons. The six protons in the atom make that atom the element carbon; the number of neutrons may vary. Atoms with the same number of protons but different numbers of neutrons are called **isotopes.**

There are three isotopes of hydrogen (Figure 4.5). All isotopes of hydrogen have one proton in the nucleus. Most isotopes of hydro-

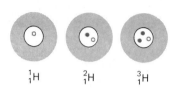

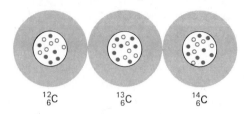

$^{1}_{1}H$ $^{2}_{1}H$ $^{3}_{1}H$

$^{12}_{6}C$ $^{13}_{6}C$ $^{14}_{6}C$

(a) Three isotopes of hydrogen with 0, 1, and 2 neutrons (•) respectively. Note in each case the atom has 1 proton (○); this makes it hydrogen.

(b) Three isotopes of carbon with 6, 7, and 8 neutrons (•) respectively. Note in each case the atom has 6 protons (○); this makes it carbon.

FIGURE 4.5

Isotopes of an element differ in the numbers of neutrons they possess. This also means that the mass number will vary.

gen have no neutrons, some have one neutron, and a few have two neutrons.

"Isotope" is really just another word for "atom." The two words can often be used interchangeably. Notice: "All *atoms* of nitrogen contain seven protons. All *isotopes* of nitrogen contain seven protons." Keep this idea in mind as further discussions of isotopes arise.

Problem 4.7

How many protons are in the nucleus of all isotopes of magnesium?

MASS NUMBER

4.7 The sum of the number of protons and the number of neutrons in an atom is called the **mass number.** Remember that the mass of the atom is almost totally accounted for by the protons and neutrons because the mass of an electron is effectively zero.

If you know the atomic number and the mass number for an isotope, you can figure out the number of protons, electrons, and neutrons in that atom. For a neutral atom,

Atomic number = number of protons = number of electrons

Mass number = number of protons + number of neutrons

or we can substitute **atomic number** for number of protons. Then

Mass number = atomic number + number of neutrons

Now subtract **atomic number** from each side of the equation, and you get

Mass number − atomic number = number of neutrons

SAMPLE EXERCISE

4.3

How many protons, electrons, and neutrons are there in isotopes of sodium which have atomic number 11 and mass number 23?

SOLUTION

Atomic number = number of protons = number of electrons

11 = number of protons = number of electrons

Mass number − **atomic number** = number of neutrons

23 − 11 = number of neutrons

12 = number of neutrons

Problem 4.8

How many protons, neutrons, and electrons are there in neutral atoms which have:

a.   Atomic number 8, mass number 17

b.   Atomic number 15, mass number 31

c.   Atomic number 27, mass number 60

Atomic number and mass number data are conveniently given in shorthand symbols. For example, the symbol that represents neutral atoms of carbon with six protons and six neutrons is $^{12}_{6}C$. The atomic number is written as a *sub*script to the left of the symbol for carbon. The mass number is written as a *super*script to the left of the symbol. This method can be represented in general by the following illustration, in which Sy stands for any chemical symbol:

$$^{\text{mass number}}_{\text{atomic number}}Sy$$

The symbol $^{12}_{6}C$ can also be read "carbon-12." "Carbon-13" is $^{13}_{6}C$. Remember, all atoms of carbon have the same atomic number. But different isotopes of carbon will have different mass numbers because the number of neutrons varies. We can also define *isotopes* as atoms with the same atomic number but different mass numbers.

Often a particular isotope of an element is indicated only by its name or symbol and mass number, e.g., carbon-12, ^{12}C, or uranium-238, ^{238}U. It is not necessary to include the atomic number since the name or symbol is sufficient to identify the element with a unique atomic number.

SAMPLE EXERCISE

4.4

Determine the number of protons (p^+), electrons (e^-), and neutrons (n^0) in the neutral atoms indicated. Also, identify the atoms.

a. $^{65}_{30}X$ d. $^{28}_{14}X$

b. $^{27}_{13}X$ e. $^{64}_{29}X$

c. $^{28}_{13}X$

SOLUTION

$^{\text{mass number}}_{\text{atomic number}} Sy$	Atomic number $= p^+ = e^-$	Mass number $-$ atomic number $= n^0$
a. $^{65}_{30}X$	$30 = p^+ = e^-$	$65 - 30 = n^0$ $35 = n^0$
b. $^{27}_{13}X$	$13 = p^+ = e^-$	$27 - 13 = n^0$ $14 = n^0$
c. $^{28}_{13}X$	$13 = p^+ = e^-$	$28 - 13 = n^0$ $15 = n^0$
d. $^{28}_{14}X$	$14 = p^+ = e^-$	$28 - 14 = n^0$ $14 = n^0$
e. $^{64}_{29}X$	$29 = p^+ = e^-$	$64 - 29 = n^0$ $35 = n^0$

The atomic number identifies the atoms. Consult the periodic table to see that the atoms are Zn, Al, Al, Si, and Cu in the order they are given.

SAMPLE EXERCISE

4.5

Refer to Sample Exercise 4.4. Which of the atoms are isotopes of the same element?

SOLUTION

$^{27}_{13}X$ and $^{28}_{13}X$ are isotopes of Al. They are atoms with the same number of protons and different numbers of neutrons. They are atoms with the same atomic number and different mass numbers.

The basic arrangement of the subatomic particles within the atom can be indicated by the use of diagrams which show the *numbers of protons and neutrons in a small circle* surrounded by a larger circle on which the number of electrons is indicated. For example, for $^{65}_{30}$Zn the diagram would be

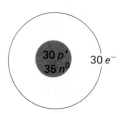

The small central circle emphasizes the idea that the protons and neutrons are clustered in the nucleus, the center of the atom. The electrons are located outside of the nucleus, and the atom as a whole is largely empty space.

Show a pictorial diagram of the two isotopes of aluminum ^{27}Al and ^{28}Al.

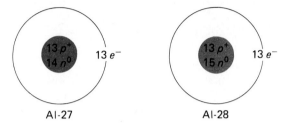

Al-27   Al-28

SOLUTION

Note that the only difference between these two isotopes is the number of neutrons in the nucleus.

Problem 4.9

Consider the neutral atoms $^{32}_{16}Y$, $^{32}_{15}Y$, $^{127}_{53}Y$, $^{31}_{15}Y$, and $^{130}_{53}Y$.

a. Indicate the numbers of protons, neutrons, and electrons in each.

b. Identify each.

c. Show a pictorial diagram of each.

d. Pick out any sets of isotopes.

CHARGED ATOMS

4.8 Eventually you will see that atoms often gain or lose electrons. When this happens, charged atoms called **ions** result because the number of protons and the number of electrons are no longer equal. It is important to realize that *atoms* and *ions* of an element are very different in physical, chemical, and biological properties.

For example, chlorine atoms (atomic number 17) often gain one electron and become so-called chloride ions. Pictorially:

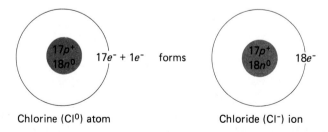

Chlorine (Cl^0) atom   Chloride (Cl^-) ion

Whereas chlorine atoms are electrically neutral because the $+17$ charge from 17 protons is just canceled by the -17 charge from electrons, chloride ions have a net charge of -1. This is because of the one extra electron in the chloride ion $(-18 + 17 = -1)$. *The properties of chlorine atoms and chloride ions are very different.*

Ions with a **negative** charge are called **anions.** Anions are formed when a neutral atom *gains* electrons. Remember, gaining electrons is gaining negative charge.

The charge on an ion is written as a superscript to the right of the symbol, with the magnitude first and the sign following, for example, Cl^{1-}. When the magnitude is 1, the 1 can be omitted, Cl^{-}.

Sodium atoms (atomic number 11) tend to lose one electron and become sodium ions. Sodium ions have a $+1$ charge because 11 protons $(+11)$ and 10 electrons (-10) result in an overall $+1$. Ions with a **positive** charge are called **cations.** Cations are formed when a neutral atom *loses* electrons. Remember, losing electrons is losing negative charge; therefore the ion is positive because the number of protons is greater than the number of electrons. Pictorially, sodium atoms become cations in the following way:

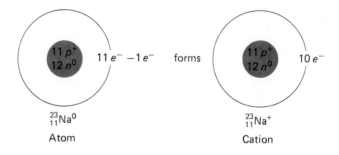

| $^{23}_{11}Na^0$ | | $^{23}_{11}Na^+$ |
| Atom | | Cation |

SAMPLE EXERCISE

4.7

Calculate the charges on the following ions, and indicate whether the ions are cations or anions:

Ion	Protons	Electrons
X	35	36
Y	3	2
Z	20	18

SOLUTION

To find the charge, take the algebraic sum of the number of protons assigned a plus sign and the number of electrons assigned a minus sign.

X: Charge $= +35 + (-36) = -1$ anion (negative charge)

Y: Charge $= +3 + (-2) = +1$ cation (positive charge)

Z: Charge $= +20 + (-18) = +2$ cation (positive charge)

Problem 4.10

Calculate the charges on ions with the following numbers of protons and electrons, and indicate whether the ions are cations or anions.

Ion	Protons	Electrons
Q	21	18
R	12	10
T	9	10

Remember the symbolism we have seen so far:

$$\text{mass number} \atop \text{atomic number} Sy \, ^{\text{charge}}$$

Atomic number = number of protons
Mass number = number of protons + number of neutrons
Charge = algebraic sum of the protons, assigned a plus sign and the electrons, assigned a negative sign

For neutral atoms, in which the number of p^+ = the number of e^- so that the charge = 0, the 0 is frequently omitted.

SAMPLE EXERCISE

4.8

Calculate the number of protons, neutrons, and electrons in the following ions: $^{24}_{12}Mg^{2+}$ and $^{14}_{7}N^{3-}$.

SOLUTION

Refer to the immediately preceding section, which reviews this symbolism.

	$^{24}_{12}Mg^{2+}$	$^{14}_{7}N^{3-}$
Number of p^+ = atomic number	= 12	7
Number of n^0 = mass number − atomic number	= 12	7

Saying that the charge is the algebraic sum of the number of protons, assigned a + (because they are positively charged), and the number of electrons, assigned a − (because they are negatively charged), can be most conveniently expressed as

Charge = number of protons − number of electrons

Subtracting the number of protons from each side:

Charge − number of p^+ = − number of electrons

Therefore, for Mg^{2+}

$$+2 - 12 = -10$$
10 electrons in Mg^{2+}

and for N^{3-}

$$-3 - 7 = -10$$
10 electrons in N^{3-}

RELATIVE ATOMIC MASS

4.9   A single atom is much too tiny to be weighed. Fortunately, scientists realized a long time ago that it is not necessary to weigh individual atoms. What we need to know are the masses of atoms in one element *relative* to atoms in other elements. Knowing which atoms are heavier or lighter than others and by what factor enables us to set up all needed weight relationships for elements and compounds.

In Section 3.1 we discussed the idea that all measurements are made relative to some standard. (Refer also to Section 3.2, which discussed the various standards in the metric system.) The standard chosen for the atomic-weight scale is the most abundant isotope of carbon, $^{12}_{6}C$ (carbon-12). In 1961 carbon-12 was assigned a mass of 12 **a**tomic **m**ass **u**nits (amu). Then it was experimentally determined how much heavier or lighter atoms of other elements were. From these experiments, the atomic-weight scale was established.

For example, it was found that helium atoms are $\frac{1}{3}$ as heavy as carbon-12 atoms and titanium atoms are 4 times as heavy. This means that the mass of a helium atom is

$$\tfrac{1}{3} \times 12 \text{ amu} = 4 \text{ amu}$$

and the mass of a titanium atom is

$$4 \times 12 \text{ amu} = 48 \text{ amu}$$

These multiplying factors can be found for all atoms, and thus relative atomic masses can be assigned to all atoms.

SAMPLE EXERCISE

4.9

Atoms of elements *X* and *Y* are, respectively, 1/2 and 6.25 times as heavy as carbon-12 atoms. What are the atomic masses of these atoms?

SOLUTION

Multiply the relative weight factor times 12 amu, the mass of $^{12}_{6}C$. For *X* atoms, $\frac{1}{2} \times 12$ amu = 6 amu. For *Y* atoms, 6.25×12 amu = 75 amu.

Problem 4.11

Atoms of element *Q* are $1\frac{1}{3}$ times as heavy as ^{12}C atoms. What is the atomic mass of these atoms?

On the carbon-12 scale, the protons and neutrons weigh approximately 1 amu. Because this is true, the mass number of an atom is approximately equal to its atomic mass in atomic mass units. The approximation is usually good to at least 3 significant figures.

SAMPLE EXERCISE

4.10

What are the approximate atomic masses of the isotopes ^1_1H, $^{19}_9F$, and $^{11}_5B$?

SOLUTION

$$\text{Mass number} = \text{protons} + \text{neutrons}$$

Because each proton and neutron weighs approximately 1 amu, the mass number also gives the approximate mass of the atom in atomic mass units. Therefore,

^1_1H	1 amu
$^{19}_9F$	19 amu
$^{11}_5B$	11 amu

AVERAGE ATOMIC WEIGHT

4.10 Look at the square devoted to the element carbon in the periodic table. Above the symbol, the number 6 is the atomic number. Below the symbol is the **average atomic weight,** 12.01. The average atomic weight reflects the fact that elements are made up of atoms with different masses, i.e., isotopes. In the case of carbon, most carbon atoms are carbon-12, mass = 12 amu. But about 1 percent of carbon atoms are carbon-13, mass = 13 amu, and there are a very small number of carbon-14 atoms, mass = 14 amu. The weighted average turns out to be 12.01 amu. The weighted average takes into account the amount of each isotope as well as the weight of each isotope.

Notice the other atomic weights (the numbers below the symbols) in the periodic table. They are not whole numbers because they are weighted averages of the masses of all isotopes of that element.

How do we determine an average atomic weight? To do this we need to know (1) which isotopes exist for a given element and (2) the percentage abundance of each isotope.

For example, the element magnesium is made up of three isotopes: 78.7 percent ^{24}Mg, 10.1 percent ^{25}Mg, and 11.2 percent ^{26}Mg. The weight contributed by each isotope is its fractional abundance [i.e., the percentage in fractional or decimal form (see Section 2.6)] multiplied by the isotopic mass. Therefore, the weight contributions for magnesium are:

	Fractional abundance	$\times$	isotopic mass	$=$	weight contribution
^{24}Mg contribution:	78.7/100	$\times$	24.0	$=$	18.9
^{25}Mg contribution:	10.1/100	$\times$	25.0	$=$	2.53
^{26}Mg contribution:	11.2/100	$\times$	26.0	$=$	2.91
					24.34

The average atomic weight is the sum of the contributions. In this example, the average atomic weight of Mg is 24.3 amu. (Note that 3 sig figs are all that we are allowed in the final sum.)

SAMPLE EXERCISE

4.11

Calculate the average atomic weight of potassium, given that potassium consists of 93.1 percent ^{39}K and 6.9 percent ^{40}K isotopes.

SOLUTION

Fractional abundance is the percentage divided by 100. Therefore, the fractional abundances are 0.931 for ^{39}K and 0.069 for ^{40}K.

^{39}K contribution:	$0.931 \times 39.0 =$	36.3
^{40}K contribution:	$0.069 \times 40.0 =$	2.8
Average atomic weight	$=$	39.1

Problem 4.12

Calculate the average atomic weight of lithium. The element lithium is 7.42 percent ^{6}Li and 92.58 percent ^{7}Li.

PERIODIC TABLE

4.11   The periodic table contains a wealth of information which will unfold chapter by chapter. In Chapter 1 you saw how to identify metals and nonmetals (Section 1.9).

The elements in the periodic table are arranged in order of increasing atomic number, the number above the symbol. The average atomic weight appears below the symbol. Arranging the elements in this order produces the surprising and useful result that elements appearing in the same vertical column have similar physical and chemical properties. Vertical columns are called **groups,** or **families.** The horizontal rows are termed **periods.**

Notice that the groups are designated by Roman numerals and a capital letter A or B. We will consider mostly A Group elements, which are more regular in their behavior. There are special names for some of the groups. The names of the A groups appear in the table in the left margin. Scientists have proposed alternative designations and numbering of groups in the periodic table. The desig-

Group	Name
IA	Alkali metals
IIA	Alkaline earth metals
IIIA	Boron family
IVA	Carbon family
VA	Nitrogen family
VIA	Oxygen family (chalcogens)
VIIA	Halogens
VIIIA	Inert gases

nations in the periodic table in this book are most convenient for beginning students.

Periods are numbered to the left of each row. The first period contains only hydrogen and helium. The second period begins with lithium in Group IA and ends with neon in Group VIIIA. The third period sequence begins again in IA and terminates with argon in VIIIA.

Because the members of a group show similar properties, learning about one member of a group means that you have learned something about the other members. For example, when you learn how sodium reacts, you will then know that lithium and potassium will react similarly. This is how the periodic table can help you keep track of many elements more efficiently.

PERIODIC GROUPS

4.12   Let us look at some of the characteristics of the A Groups.

Alkali metals

With the exception of hydrogen, all the elements in Group IA are soft metals which react violently with water, producing hydrogen gas and another compound, a metal hydroxide. Because of this vigorous reaction, the alkali metals are stored under oil or kerosene to protect them from moist air. Because of their high reactivity, the alkali metals are never found free in nature, but always as an ion combined in a compound. Sodium and potassium are the most abundant alkali metals and the most important in the human body, where they occur as ions (Na^+ and K^+) which play an important role in nerve transmission and maintenance of proper fluid balance between the cells and tissues of the body.

Hydrogen, although in Group IA, is not a metal, but it does react and form compounds of a composition similar to that of the alkali metals. Hydrogen, a gas at room temperature, is the most abundant element in the universe. The energy radiated from the sun is due to a nuclear reaction fusing together hydrogen nuclei to form helium nuclei.

Alkaline earth metals

The alkaline earth metals are all less chemically reactive than the alkali metals. They do react with oxygen in the air, but the compound formed (an oxide) as a coating protects the metal from further attack. Thus elemental magnesium can be used in a mixture with other metals as a low-density, structural material. Calcium is the most abundant metal in the body, occurring in the ionic form Ca^{2+}, principally in bones and teeth. Barium may be familiar to you because of its presence in the "barium cocktail," a milky suspension which patients drink to make the gastrointestinal tract opaque to x-rays.

Boron family

At the top of this periodic group is the nonmetal boron; proceeding down the group, the other four members, aluminum, gallium, indium, and thallium, are metals. This reflects a general feature of all groups, namely, that metallic character increases as we proceed down a group.

Carbon family

Carbon, a nonmetal, is found in the elemental form in graphite, the "lead" of pencils, and diamond. The six million or so compounds of carbon make up the field of organic chemistry which includes all the compounds found in living systems. The metalloid character of silicon confers special electrical, semiconducting properties, making it a basic component of transistors and integrated circuits. Again, moving down the group, metallic character tends to increase, with lead displaying all the common properties of a metal.

Nitrogen family

Nitrogen, a diatomic gas (N_2) (see Sections 4.13 and 6.11) at room temperature, is a nonmetal, which makes up about 80 percent of air. Nitrogen and phosphorus are key components of many physiological compounds. Bismuth has metallic character.

Oxygen family

Oxygen is the most abundant element on earth. It combines with most metals to form compounds called **oxides.** Elemental oxygen is a diatomic gas which is inhaled in that form by the human body and eventually combines with hydrogen ions to form water and supply the organism with energy. Sulfur, a nonmetallic solid, has many industrial applications, including vulcanization, the addition of sulfur to rubber to make rubber less brittle.

Halogens

All the halogens are diatomic. Fluorine and chlorine are gases, bromine is a liquid, and iodine is a solid. At this end of the periodic table all the elements within the group are nonmetals. Chlorine is present as an anion (Cl^-) with the sodium cation (Na^+) in the familiar substance, table salt or sodium chloride (NaCl). Iodine sublimes to yield a characteristic purplish vapor; a homogeneous mixture of iodine in alcohol has been used as a disinfectant.

Inert gases

All the elements of this group are gases, and all display a lack of chemical reactivity.

ELEMENTAL MAKEUP

4.13   The inert-gas elements (Group VIIIA) are made up of uncombined atoms, as we have shown in the simplified drawings in Figure 4.1. But very few other elements are actually made up of separate individual atoms.

Most metals are better represented as positive nuclei in a sea of electrons (Figure 4.6). The element carbon comes in two common forms, graphite and diamond, both of which involve carbon atoms attached to one another in different arrangements (Figure 4.7).

Many common, frequently encountered elements exist as **diatomic molecules.** A molecule is several atoms hooked together. The prefix **di-** always means "two." Therefore, a diatomic molecule is two atoms hooked together. Figure 4.8 gives an accurate picture of the element hydrogen, which exists as diatomic molecules with the formula H_2. The formula stands for two hydrogen atoms hooked together. The other elements which exist as diatomic molecules are nitrogen, oxygen, and the halogens, Group VIIA.

SAMPLE EXERCISE

4.12

Write proper formulas for the elements chlorine, bromine, and nitrogen.

SOLUTION

Because these elements exist as diatomic molecules, we show this with the formulas Cl_2, Br_2, and N_2.

FIGURE 4.6

Electron sea model of metals. (*From William L. Masterson and Emil J. Slowinski, Chemical Principles, 1977, W. B. Saunders Co.*)

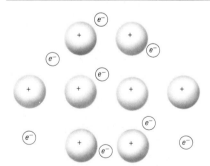

(a) For the metal sodium, one electron per atom is "loose" in the "electron sea." This leaves a net +1 charge on the sodium ion.

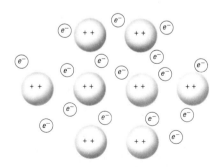

(b) For the metal magnesium, two electrons per atom are "loose" in the "electron sea." This leaves a net charge of +2 on the magnesium ion.

Graphite layer

Diamond crystal

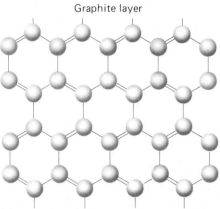

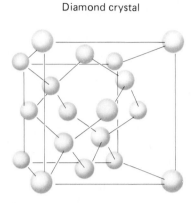

(a) In graphite, C atoms form overlapping layers of a chicken-wire-fence pattern.

(b) In diamond, C atoms are arranged in a three-dimensional array.

FIGURE 4.7

Arrangements of C atoms in different forms of the element carbon. Above: atomic models. Below: actual samples. (*From William L. Masterson and Emil J. Slowinski, Chemical Principles, 1977, W. B. Saunders Co.*) (*Photos [trans. nos. 3067, K13784] courtesy Department Library Services, American Museum of Natural History. Lower left photo by O. Bauer.*)

FIGURE 4.8

The element hydrogen exists as a diatomic molecule; that is, two H atoms are stuck together. Similar pictures could be shown for O_2, N_2, and the halogens.

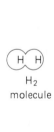

H_2
molecule

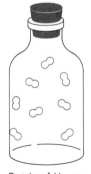

Bottle of H_2 gas.

Problem 4.13

What is meant by the formulas F_2 and O_2?

It is possible to treat most elements as if they existed as individual atoms, as shown in Figure 4.1. But very often we must remember the small list of elements that exist as diatomic molecules, H_2, N_2, O_2, F_2, Cl_2, Br_2, I_2, and picture them as in Figure 4.9. You will find out why these atoms "double up" in Section 6.11.

SUMMARY

All matter is made up of atoms. Different elements are made up of different atoms. Compounds are made up of different atoms combined in various ways (Section 4.2). Atoms are extremely tiny (Section 4.3) and are actually mostly empty space because of the arrangement of the subatomic particles, the protons, electrons, and neutrons (Section 4.4).

The atomic number tells you the number of protons in an atom. In a neutral atom the number of protons equals the number of electrons (Section 4.5). If the number of protons and electrons in an atom are not equal, the atom has a charge and is called an ion (Section 4.8). The mass number tells you the sum of the numbers of protons and neutrons in an atom (Section 4.7). Isotopes have identical atomic numbers but different mass numbers (Section 4.6).

Carbon-12 is the standard for the atomic-weight scale and is assigned a mass of 12 atomic mass units. Other atoms are assigned masses relative to carbon-12 (Section 4.9). The average atomic weight of an element is the weighted average of the masses of the isotopes of which the element is composed (Section 4.10).

The periodic table contains a great deal of information, including atomic numbers and atomic weights. Within the vertical columns, called groups, are elements with similar properties (Sections 4.11 and 4.12).

Elements rarely consist of individual atoms. For example, it is most important to bear in mind that hydrogen, oxygen, nitrogen, fluorine, chlorine, bromine, and iodine exist as diatomic molecules (Section 4.13).

CHAPTER ACCOMPLISHMENTS

After completing this chapter you should be able to

4.1 Introduction

4.2 Atomic theory

1. Define elements, compounds, and mixtures in terms of their atomic makeup.

4.3   Picturing atoms and elements

2.   Form a mental picture of atoms and elements.

4.4   Inside the atom

3.   State the names and symbols of the three subatomic particles.
4.   State the relative charges and masses of the subatomic particles.
5.   Form a mental picture of atoms, showing the arrangement of the subatomic particles.

4.5   Atomic number

6.   Define atomic number.
7.   Given the atomic number, determine the number of protons and electrons in a neutral atom.
8.   Given the number of protons or electrons in a neutral atom, write the atomic number.

4.6   Isotopes

9.   Define isotope.
10.   Given the number of protons and neutrons in two atoms, indicate whether the atoms are isotopes.

4.7   Mass number

11.   Define mass number.
12.   Given the atomic and mass numbers, determine the number of subatomic particles in an atom.
13.   Given the number of protons and neutrons in an atom and a periodic table, write a correct symbol showing the atomic and mass numbers.
14.   Given the atomic and mass numbers of a set of atoms, indicate which atoms are isotopes.

4.8   Charged atoms

15.   Define ion.
16.   Define anion and cation.
17.   Indicate how anions are formed.
18.   Indicate how cations are formed.
19.   Given the number of protons and electrons in an ion, calculate the charge on the ion.
20.   Given the symbol for an ion, including the mass number, charge, and atomic number, calculate the number of protons, neutrons, and electrons in the ion.
21.   Given the number of protons, neutrons, and electrons in an ion, write a correct symbol for the ion, including the mass number, atomic number, and charge.

4.9 Relative atomic mass

22. Given the relative weight factor between an atom X and a ^{12}C atom, calculate the atomic mass of atom X.

4.10 Average atomic weight

23. Calculate the average atomic weight of an element given the percentage abundances for the isotopes of that element.

4.11 Periodic table

24. Use the periodic table to determine atomic numbers and atomic weights.
25. Explain the significance of elements appearing in the same group in the periodic table.
26. Given a periodic table, indicate the name of the family for an A Group element.

4.12 Periodic groups

27. Describe the trend of nonmetallic or metallic character of elements within a group.

4.13 Elemental makeup

28. List the elements that exist as diatomic molecules.

PROBLEMS

4.2 Atomic theory

4.14 In what way do atoms of different elements differ?

4.15 In what way are atoms of different elements similar?

4.16 How do elements and compounds differ in terms of Atomic Theory?

4.3 Picturing atoms and elements

4.17 Draw a picture of an atom the way you visualize it.

4.4 Inside the atom

4.18 Name the three subatomic particles.

4.19 Which subatomic particle has zero mass?

4.20 Which subatomic particle has zero charge?

4.21 Compare the relative masses of the three subatomic particles.

4.22 Compare the relative charges of the three subatomic particles.

4.23 Which subatomic particles account for the charge on the nucleus?

4.24 How can an atom which contains charged subatomic particles be neutral?

4.25 Describe the location of each of the three subatomic particles.

4.26 What is between the nucleus and the electrons of an atom?

4.5 Atomic number

4.27 What is the difference between the atomic number and the number of electrons in an atom?

4.28 An atom has 12 protons and 12 neutrons.
  a. What is its atomic number?
  b. What is the number of electrons?
  c. What is the name of this atom?

4.29 What is the name of the element with the following atomic number?
  a. 7   **c.** 18
  b. 11   **d.** 2

4.30 What is the name of the element with the following number of electrons in their neutral atoms?
  a. 14   **c.** 20
  b. 13   **d.** 35

4.6 Isotopes

4.31 Two isotopes of the same element are alike in what way? In what way do they differ?

4.32 Give a possible set of protons, electrons, and neutrons for an atom that is an isotope of:
  a. A carbon atom with 6 neutrons
  b. An atom with 11 protons, 11 electrons and 12 neutrons
  c. A neutral atom with 19 electrons and 20 neutrons

4.33 What is the restriction on the number of protons for two isotopes of the same element?

4.7 Mass number

4.34 Complete the following table for the neutral atoms indicated:

Element Name	Atomic Number	Mass Number	Number of Protons	Number of Neutrons	Number of Electrons
	35	80			
	4			5	
		40	51		
		59			28

4.35 Tell the numbers of protons, neutrons, and electrons in neutral atoms which have

	Mass number	Number of Neutrons
a.	51	28
b.	28	14
c.	19	10

4.36 With the help of the periodic table, tell the number of protons, neutrons, and electrons in the following isotopes:
  a. Carbon-13   **c.** Bromine-79
  b. Cobalt-59   **d.** Strontium-87

4.37 Complete the following table:

Symbol	Protons	Neutrons	Electrons
$^{9}_{4}Be$			
$^{127}_{53}I$			
$^{31}_{15}P$			
$^{40}_{18}Ar$			

4.38 Write symbols for atoms with:

	Number of Protons	Number of Neutrons
a.	11	12
b.	2	2
c.	17	18

4.39 Which of the following atoms are isotopes of each other?
  a. $^{28}_{14}X$ **d.** $^{15}_{7}X$
  b. $^{14}_{7}X$ **e.** $^{31}_{15}X$
  c. $^{45}_{21}X$ **f.** $^{30}_{14}X$

4.40 Consult the periodic table and identify each of the elements represented in Problem 4.39.

4.41 a. An atom contains 19 protons, 20 neutrons, and 19 electrons. Write the symbol for this atom.

b. Write the symbol for a possible isotope of this atom.

4.42 An isotope of neon has 11 neutrons in its nucleus. Write the symbol of the isotope.

4 43 The mass number of an isotope of nitrogen is 15. Write the symbol for the isotope.

4.44 Write the symbol for an isotope of $^{16}_{8}O$ that contains:

   a. One more neutron
   b. Two more neutrons

4.45 Is there any difference between the representations $^{13}_{6}C$ and carbon-12? Between $^{13}_{6}C$ and carbon-13? Explain.

	Protons	Neutrons	Electrons
a.	15	16	18
b.	19	20	18
c.	1	0	0
d.	9	10	10
e.	37	48	36

4.50 Explain how cations are formed from atoms.

4.51 Explain how anions are formed from atoms.

(Problems continue on p. 118.)

4.3 Charged atoms

4.46 Calculate the charges on the following ions, and indicate whether the ion is a cation or anion.

	Ion	Protons	Electrons	Charge	Cation or Anion
a.	F ion	9	10		
b.	Al ion	13	10		
c.	P ion	15	18		
d.	Ag ion	47	46		

4.47 Calculate the numbers of protons, neutrons, and electrons in the following ions:

	Ion	Protons	Neutrons	Electrons
a.	$^{88}_{38}Sr^{2+}$			
b.	$^{23}_{11}Na^{+}$			
c.	$^{14}_{7}N^{3-}$			
d.	$^{127}_{53}I^{-}$			

4.48 Consult the periodic table, and name the ions that have

   a. 35 protons and 36 electrons
   b. 30 protons and 28 electrons

4.49 Write a symbol, including mass number, atomic number, and charge, for ions containing the following numbers of protons, neutrons, and electrons.

4.52 Complete the following table:

Isotope	Symbol	Atomic Number	Mass Number	Protons	Neutrons	Electrons
Lithium-7						3
	$^{59}_{27}Co$					
		26			30	
			40	18		
	$^{16}_{8}O^{2-}$					
	$^{138}_{56}Ba^{2+}$					

4.53 Give the symbol and charge of an ion having 10 electrons and:
a. 9 protons   **c.** 11 protons   **e.** 7 protons
b. 12 protons   **d.** 8 protons

4.54 What is the difference between the representations Mg and Mg^{2+}?

4.9   Relative atomic mass

4.55 A certain isotope of Co is 5 times as heavy as ^{12}C. What is the atomic mass of this isotope?

4.56 What are the approximate atomic masses of the atoms symbolized in Problem 4.39?

4.57 What is the weight ratio of a magnesium-24 atom to a carbon-12 atom?

4.58 If a carbon atom had a mass of 6 instead of 12 amu, what would be the mass of a helium atom? Titanium?

4.10   Average atomic weight

4.59 The element sulfur is made up of 95.0 percent sulfur-32, 0.76 percent sulfur-33, and 4.22 percent sulfur-34. What are the fractional abundances of the three isotopes?

4.60 Refer to Problem 4.59 and calculate the average atomic weight of sulfur.

4.61 Neon is 90.9 percent ^{20}Ne, 0.257 percent ^{21}Ne, and 8.82 percent ^{22}Ne. Calculate the atomic weight of neon to 3 sig figs.

4.62 The element iron is made up of the isotopes ^{54}Fe (5.82 percent), ^{56}Fe (91.7 percent), ^{57}Fe (2.19 percent), and ^{58}Fe (0.330 percent). Calculate the average atomic weight of iron.

4.11 and 4.12   Periodic table and groups

4.63 Consult the periodic table and arrange the following elements in groups: oxygen, sodium, sulfur, phosphorus, potassium, arsenic, lithium.

4.64 Give the number and name of the group containing iodine.

4.65 Give the symbol of the element that is found in the following location in the periodic table:
a. Period 2, Group IIA
b. Period 5, Group VIA
c. Period 2, Group VIIA
d. Period 4, Group VIIIA

4.66 Give the symbol of an element that would have properties similar to
   a. Al   **b.** N   **c.** Si   **d.** Ar

4.67 Which of the following elements would you predict to have similar properties?
a. Atomic number 13   **c.** Atomic number 11
b. Atomic number 37   **d.** Atomic number 55

4.68 Elements in Groups IA, IIA, and IIIA form compounds with oxygen that have the following general formulas:

IA	IIA	IIIA
M_2O	MO	M_2O_3

a. Write a formula for the compound of calcium and oxygen.
b. Write a formula for a compound of aluminum and oxygen.

4.69 Which element in Group IIIA do you predict to be most nonmetallic? Which is the most metallic?

ELECTRONIC STRUCTURE OF THE ATOM

CHAPTER 5

B

Boron

10.81

INTRODUCTION

5.1 As you learned in Chapter 4, an atom has a central nucleus with surrounding electrons. The chemical reactions that we encounter in daily life involve electrons and electrons only; the nucleus is unaltered. In Chapter 17 nuclear reactions will be considered, but until then the chemistry we discuss is a consequence of electronic changes.

In order to understand why compounds form and chemical reactions occur, you must know more of the details of how the electrons are arranged about the nucleus. This arrangement of electrons is called **electron structure.** As soon as you know about electron structure, you will be able to understand how elements combine to form compounds. Compound formation will be discussed in Chapter 6, and the discussions rely heavily on principles in this chapter. An understanding of the trends within the periodic table will also unfold in this chapter.

ENERGY REVISITED

5.2 In Section 1.12 **energy** was defined as the ability to do work, and we discussed the idea that energy can be witnessed only when work is done and some *energy change* occurs. It was also pointed out that all matter has energy stored within it. The amount of stored energy depends on how the sample of matter is constructed. One important aspect of the construction is the arrangement of the electrons in atoms.

Different amounts of energy are associated with different electron arrangements. As electron arrangements change, we can observe changes in the energy content of samples of matter. It is impossible to consider electron arrangements without considering the energies of such arrangements.

It would probably be a good idea to reread Section 1.12 before continuing to read this chapter.

CONCEPT OF MINIMUM ENERGY

5.3 Scientists talk about **energy states.** It is hoped that you approach your studies in an *excited* energy state, because this is a high-energy condition. Matter in a high-energy state has a great deal of energy. If a sample of matter loses energy, then it will be in a lower-energy state. So, modification of the term *state* is used to describe how much energy something has.

It is a principle of nature that all things try to reach a minimum-(lowest-) energy state. For example, lay your pencil at the edge of your desk. You know that there is a natural tendency for that pencil to roll off and fall to the floor. In terms of energy states, the pencil on the desk is in a high-potential-energy state. When the pencil hits the floor, it has reached a minimum-energy state for the room and in so doing has lost the higher potential energy it possessed when it hovered at the edge of the desk. It still has some smaller amount of potential energy—for example, if you were to cut a hole in the floor, the pencil would fall through. In order for the pencil to return to the higher-energy state that it occupied while it lingered at the edge of the desk, energy must be put in. If you were to scoop the pencil up, you would be transferring the energy from yourself to the pencil. When the pencil is back on the desk, it has the same tendency to fall again to the lower state.

This natural tendency toward minimum energy is true for the chemical energy of matter also. Matter is constructed so that it is in a low-energy state. Although it can be excited into a higher-energy state temporarily through the addition of outside energy, it exhibits a natural tendency to return to the low-energy state (called the **ground state**). Because matter will stay in the ground state if it is left alone, its condition in this state is labeled a **stable** condition. Something that is stable or has **stability** is in a low-potential-energy state. Higher-energy states are less stable. This idea that **high energy corresponds to low stability and low energy corresponds to high stability** is very important but can be confusing. You will find it well worth the effort

FIGURE 5.1

The pencil has a natural tendency to assume the lower potential energy, more stable state.

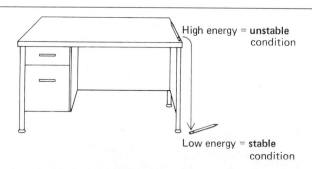

High energy = **unstable** condition

Low energy = **stable** condition

to clarify these terms for yourself now to save confusion later. Think about the pencil at the edge of the desk. The pencil is **unstable** (exhibits low stability) because it has high potential energy. It will fall to the ground state, or lowest-energy state, and become more stable (see Figure 5.1).

SAMPLE EXERCISE

5.1

Consider a woman who is standing on a diving board.

a. Does she possess more potential energy before or after diving?
b. Is she in a more stable condition when standing on the diving board or after hitting the water?

SOLUTION

a. Her potential energy is higher when she is poised on the board. When she dives, some potential energy is converted into kinetic energy, so that she has less potential energy when she hits the water.
b. While she is in the water, her condition is more stable since it is a lower-energy state.

SAMPLE EXERCISE

5.2

When gasoline burns, the products of the combustion are carbon dioxide (CO_2) and water (H_2O), which are more stable than gasoline. Describe the relative energy contents of gasoline, CO_2, and H_2O.

SOLUTION

We are told that gasoline is *less stable* than CO_2 and H_2O; therefore, it must possess *more energy* than CO_2 and H_2O.

Problem 5.1

Food contains more energy than the products of metabolism into which it is converted in the body. Is food more or less stable than its metabolic products?

MINIMUM ENERGY IN THE ATOM

5.4 In terms of energy, electrons in atoms obey the same rules as all other matter in the universe; they prefer to be arranged in a condition (state) of minimum (lowest) energy and hence maximum (highest) stability.

Electrostatics

Let us divert our attention from atoms for a moment to discuss some aspects of electrostatics that will be useful in discovering the minimum-energy electron structure (arrangement) of the atom. **Electrostatics** deals with interactions between charged particles. In terms of "male-female electricity" the concept of "opposites attract and likes repel" is well established, although there are exceptions. In electrostatics, *without exception, objects with opposite electrical charge (one + and one −) attract each other; objects with identical charge (both + or both −) repel each other.* The attraction or repulsion increases as the objects come closer together.

In terms of energy states, oppositely charged bodies are in a low-energy stable state when they remain close to each other; work must be performed and energy thereby increased in order for them to move farther apart. The low-energy stable condition for identically charged bodies occurs when they are far apart. Work must be performed and energy increased in order to push them closer together.

SAMPLE EXERCISE

5.3

Two positively (+) charged tennis balls are held at opposite ends of a room 3 m long, 3 m wide, and 3 m high (volume = 3 m × 3 m × 3 m = 27 m³). Is their energy state in this room higher or lower than when they were in a cubic container that measured 10 cm along an edge?

SOLUTION

The energy state in the room is lower, because the two like-charged objects are farther apart there than when they were in the container with the smaller volume.

SAMPLE EXERCISE

5.4

If the tennis balls in Sample Exercise 5.3 were oppositely charged (one + and one −), then would their energy be lower in the room or in the container?

SOLUTION

The energy state in the container would be lower because the positively charged ball and the negatively charged ball can get closer together.

The low-energy arrangement of electrons in an atom must be based on the ideas that:

1. The close approach of a negative electron to the positive nucleus is a low-energy condition.

2. The far separation of two negative electrons is a low-energy condition.

To obtain the minimum energy condition for the atom as a whole, these two conditions must be met simultaneously. That is, electrons prefer to be close to the nucleus, but they must avoid other electrons. These simple ideas lay the entire foundation for the following discussion of the electronic arrangements in atoms.

Problem 5.2

In which case would the overall energy of the system be lower?

Case A: Two electrons in a box of volume $1 \times 10^{-12} \, \mu m^3$

Case B: Two electrons in a box of volume $1 \times 10^{-13} \, \mu m^3$

Hydrogen, the Smallest Atom

Consider the H atom with its one (+) proton in the nucleus and one (−) electron outside the nucleus. From the foregoing discussions you might immediately conclude that the (−) electron must "fall into" the nucleus, i.e., draw as close as possible to the (+) proton. But we must accept a puzzling but firmly established fact: although the electron may get very close to the nucleus, it never "falls in." This contradiction of classical electrostatics really had scientists baffled early in this century. The person who offered the first solution to the problem was Niels Bohr, the Danish physicist who eventually won the Nobel Prize for this and other brilliant work.

Bohr said that electrons could only have certain definite energies and therefore could only occupy certain definite orbits around the nucleus, just as planets orbit around the sun. The Bohr atom is shown in Figure 5.2a. The lowest energy state for the hydrogen atom would occur when the electron was in orbit number 1, the one closest to the nucleus. If the electron were in orbits farther from the nucleus, the atom would exhibit higher energy and be in an excited state. The electron can never be in the regions between the orbits.

Bohr's model of the atom is not entirely correct. The Bohr model did not agree with experimental observations for atoms containing more than one electron and hence required further modification. However, this idea that electrons possess only certain definite energies

FIGURE 5.2

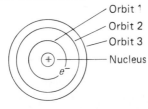

This is a Bohr H atom!

(a)

This is a real H atom!

(b)

(a) The Bohr atom restricts electrons to definite orbits. The figure shows hydrogen's one electron in the lowest energy orbit closest to the nucleus. (b) In the real atom the electron moves about the nucleus within a sphere which has the appearance of a negatively charged cloud.

associated with certain regions in space is correct and is very important. This is the idea of **quantization.** Electron energies in atoms are quantized; that is, there are definite allowed energy states that electrons may occupy. Electrons never have intermediate energies. Bohr's introduction of quantization eventually led to the development of **quantum mechanics,** which fully describes electronic energies and arrangements mathematically. It turns out from this mathematical description that it is not possible to predict the precise positions of electrons, but rather one can only predict the likelihood of electrons being within some volume of space.

Whereas the mathematical description is the only precise one, most people find it easier and more useful to develop a mental picture of the atom. Happily, it is possible to do so by sticking to the ideas already discussed. To recap, (1) the close approach of a negative electron to the positive nucleus is a low-energy condition, and (2) the far separation of two negative electrons is a low-energy condition. The only exception to these two principles that we need to remember is that electrons do not enter the nucleus.[1] If we picture electrons as continuously moving rapidly, then a snapshot would capture them as a fuzzy cloud. The cloud is negatively charged because electrons bear negative charges. Figure 5.2*b* shows the picture of a hydrogen atom implied by these ideas. The one negative electron has a definite lowest-energy state which confines it to movement within a region close to the positive nucleus.

The fuzzy cloud, or more scientifically, the volume, in which the electron is likely to be found is called an **orbital.** An orbital in general is defined as a volume in space in which an electron or a pair of electrons is found. As you continue reading you will find that no more than two electrons can occupy an orbital. There is a specific energy associated with every orbital.

Helium Atoms

Let us go on to the helium atom, with its two-proton ($+2$) nucleus and its two electrons ($-$). There are attractive forces between the nucleus and each of the electrons, and the attractive forces will attempt to keep each electron as close to the nucleus as possible. However, there is an additional force in this system of charges: the repulsive force operating between the two like-charged ($-$) electrons. To answer the question of how the electrons are arranged in the helium atom, one must ask: What is the arrangement of minimum energy of this

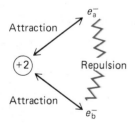

Attraction

e_a^-

($+2$)   Repulsion

Attraction

e_b^-

---

[1] You may be wondering how so many positive particles (the protons) can cluster together in the nucleus. This question is one of the major puzzles that physicists are trying to unravel. Fortunately for us, we need to know only that protons do cluster, and we can regard the nucleus as one clump of positive charge.

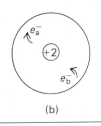

The 1s orbital for He: (a) spherical volume occupied by the two electrons; (b) cross section of the orbital.

(a)   (b)

FIGURE 5.3

three-component system which includes the two negative electrons and the positive nucleus?

It turns out that the volume of space within which the electron of the hydrogen atom is found is large enough to accommodate another electron. Placing the two electrons in a spherical volume, or orbital, surrounding the nucleus satisfies the two principles which lead to a minimum-energy condition, namely, proximity to the nucleus and the ability to avoid one another (see Figure 5.3). This sphere within which the two electrons of lowest energy in any atom reside is called the **1s orbital.**

Li and Be

What is the minimum-energy picture for an atom containing three electrons? If all three electrons are put in the 1s spherical volume (orbital), calculations show that the repulsion between electrons in that small volume is greater than the attractive energy from the pull of the nucleus on the electrons. Minimum energy can be obtained only when the *third* electron is located in a spherical volume around the nucleus that is larger than the volume of the 1s orbital. This new spherical volume is called the 2s orbital. You can envision the relationship between the 1s and 2s orbitals as a softball (1s) within a basketball (2s). The 2s orbital, like the 1s, can contain a maximum of two electrons (see Figure 5.4).

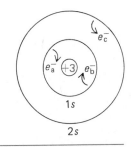

FIGURE 5.4

Cross section of the Li atom. The third electron occupies the 2s orbital to minimize electron repulsion. The arrows denote the property of spin, which is discussed in Section 5.4.

SAMPLE EXERCISE

5.5

How are the four electrons of Be arranged?

SOLUTION

Two electrons are in the 1s and two are in the 2s. This maintains the electrons close to the nucleus, but does not exceed the limit of two electrons per orbital.

Orbitals Have Various Shapes

What is the minimum-energy picture for an atom containing five electrons? The fifth electron cannot be in the 2s orbital because of repulsive forces with the two electrons already in that orbital. How-

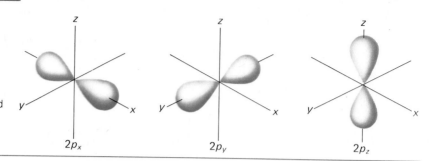

The three 2p orbitals directed along the three mutually perpendicular Cartesian axes.

$2p_x$ $2p_y$ $2p_z$

FIGURE 5.5

FIGURE 5.6

(a) The volume corresponding to the first energy level is large enough to accommodate only one orbital, the 1s. (b) The volume corresponding to the second energy level contains the 2s orbital and the three perpendicular 2p orbitals. Notice the 1s "buried" inside the 2s.

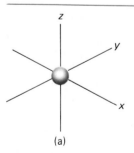

(a)

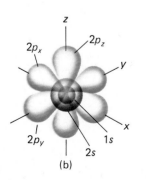

(b)

ever, to minimize energy, as always it is desirable to keep electrons close to the nucleus rather than move them significantly farther away than the 2s orbital. Orbitals of a different shape accomplish the juggling act of keeping the fifth electron close to the nucleus, but at the same time away from other electrons. These new orbitals are called 2p and are shaped like a long balloon pinched in the middle or an iceskater's figure eights (see Figure 5.5). We can think of electrons in these 2p orbitals as being only slightly farther from the nucleus, on the average, than they would be if they were in the 2s. However, repulsion between the 2s electrons and them is minimized by the fact that they are located in different three-dimensional volumes.

The three dimensions of space are represented mathematically by three mutually perpendicular axes, the x, y, and z (Figure 5.5). There are three 2p orbitals: $2p_x$, $2p_y$, and $2p_z$, where the subscript tells us along which axis in space the orbital is oriented. Each directed 2p orbital can hold a maximum of two electrons, just as the 1s and 2s orbitals each hold a maximum of two electrons.

Orbitals and Energy Levels

Perhaps you have begun to see the idea of energy levels surrounding the nucleus. By far the lowest-energy level, because it allows electrons the closest proximity to the nucleus, is the first energy level. The volume corresponding to the first energy level is only large enough to accommodate one orbital, the 1s. Because this is so, only two electrons in any atom can have this lowest-energy state.

Next in energy is a larger level, in which electrons can travel in a spherical volume, the 2s orbital, or in figure eights, the 2p orbitals. Because this second energy level contains two types of orbitals of slightly different energies, it is said to consist of sublevels, with the lower-energy 2s sublevel consisting of one 2s orbital, and the slightly higher-energy 2p sublevel consisting of the three perpendicular 2p orbitals (see Figure 5.6). You will see shortly that a third level contains three sublevels, the 3s, 3p, and 3d. Figure 5.7 summarizes the maximum numbers of electrons that can occupy the main energy levels.

With the information given so far about $1s$, $2s$, and $2p$ orbitals, it is possible to predict the proper location for the 10 electrons of lowest energy in any atom. First, two electrons occupy the $1s$ sublevel, or orbital; then electrons three and four fit into the $2s$ sublevel or orbital; then electrons 5 to 10 occupy the $2p$ sublevel. The $2p$ sublevel can hold six electrons, with two in each of the three $2p$ orbitals. Try to visualize the arrangement of the 10 electrons in a neon atom by looking at Figure 5.6b. Two electrons are in the $1s$ orbital, two in the $2s$, and two each in the $2p_x$, $2p_y$, and $2p_z$ orbitals.

The Third Energy Level

Electrons 11 and 12 in any atom are found to occupy a $3s$ *orbital*, another spherical orbital similar in shape to the $1s$ and $2s$, but farther away from the nucleus. Electrons 13 to 18 are placed in a $3p$ sublevel (consisting of $3p_x$, $3p_y$, and $3p_z$ orbitals). The $3p$ orbitals have the same shape as the $2p$, but on the average they are farther away from the nucleus. Within the third energy level there is another set of orbitals

FIGURE 5.7

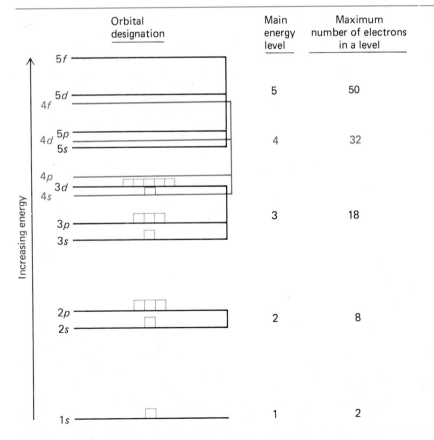

	Orbital designation	Main energy level	Maximum number of electrons in a level
5f		5	50
5d 4f			
4d 5p 5s		4	32
4p 3d 4s		3	18
3p			
3s			
2p		2	8
2s			
1s		1	2

Increasing energy

Electrons are in orbitals (▢) within s, p, d, or f sublevels within numbered main levels. The energy of the main levels begins to overlap at level 4. The spacings between the levels vary for different elements.

with unique shapes, called 3d orbitals. However, after electron 18 completes the filling of the 3p orbitals, electrons 19 and 20 go into the 4s orbital. The 4s orbital penetrates closer to the nucleus than the 3d, making the overall energy of an electron in a 4s orbital lower than one in a 3d orbital. Consequently, the total energy of an atomic system is found to be at a minimum when electrons 19 and 20 are contained in a 4s orbital. Another way of saying this is that the third and fourth energy levels overlap, as shown in Figure 5.7. It is this overlap that allows two electrons to enter the fourth energy level (4s sublevel) before the third energy level is filled.

Electrons 21 to 30 occupy the five 3d orbitals. Once again we find that a maximum of two electrons can fit into each 3d orbital. As you can see in Figure 5.8, the shapes of these orbitals are quite complex. These shapes need not concern you. The important idea is that there are five d orbitals in the d sublevel and that each orbital can hold a maximum of two electrons. There are also f and g sublevels.

SAMPLE EXERCISE

5.6

For each of the following atoms, tell how the electrons are arranged.

a. Oxygen (8 electrons)
b. Aluminum (13 electrons)
c. Calcium (20 electrons)

FIGURE 5.8

The shapes of the five 3d orbitals.

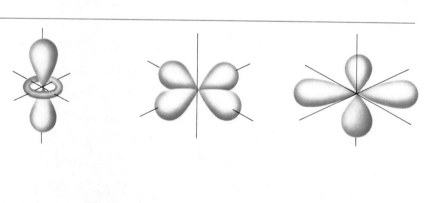

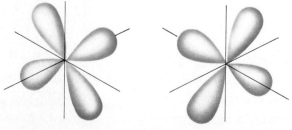

SOLUTION

a. The 1s and 2s sublevels can each hold two electrons for a total of four. That leaves four more electrons, which can occupy the 2p sublevel since the 2p can hold up to a total of six electrons.

b. Two electrons in the 1s, two in the 2s, six in the 2p, and two in the 3s account for 12 electrons (2 + 2 + 6 + 2 = 12). The thirteenth electron must go into the 3p.

c. Two in the 1s, two in the 2s, six in the 2p, two in the 3s, and six in the 3p account for 18 electrons (2 + 2 + 6 + 2 + 6 = 18). The nineteenth and twentieth electrons must go into the 4s.

Problem 5.3

For each of the following atoms, tell how the electrons are arranged.

a. Sulfur (16 electrons)

b. Phosphorus (15 electrons)

c. Magnesium (12 electrons)

Spin

Scientists have found that electrons give off very small magnetic fields and act almost as if they were tiny bar magnets. This property leads us to believe that electrons are spinning, either clockwise or counterclockwise, because it has been known for a long time that any charged particle which spins gives off a magnetic field. A spin in one direction, say, clockwise, would give off a magnetic field like the positive pole of a magnet, whereas a spin in the other direction would give off the opposite field, the negative pole. The ideas of electrostatics apply here as well as to the preceding material. Opposite (+ and −) poles attract; like poles repel. In energy terms, two magnets have lower energy when opposite poles (+ and −) are placed near each other than they do when similar poles (+ and + or − and −) are placed together. *When two electrons are placed in one orbital, energy will be lower when the two electrons have opposite spin rather than when their spin is the same.* Return to Figures 5.3 and 5.4 and notice that each electron has been assigned a spin which is indicated by a curved arrow denoting either clockwise or counterclockwise motion. This requirement of opposite spins to assure low energy also accounts for the existence of a maximum of two electrons per orbital. If more than two electrons are placed in an orbital, then two electrons would necessarily have the same spin.

Summary of Electron Arrangements

Let us try to summarize briefly the picture of electrons in atoms that you should have developed by now.

For (−) electrons, low energy is linked with being close to

a ($+$) nucleus. However, at the same time, electrons must avoid one another. Thus, they are located within different three-dimensional patterns (orbitals) as close to the nucleus as their need to avoid other electrons allows. Each orbital can contain two electrons with opposite spins.

ELECTRON CONFIGURATION NOTATION

5.5 Chemists employ a system of shorthand notation to show the **electron configuration** of an atom, i.e., how the electrons within a given atom are arranged in orbitals within energy sublevels. For example, the fact that the one electron of hydrogen occupies the lowest-energy $1s$ orbital is abbreviated $1s^1$. The superscript 1 refers to the one electron in the $1s$ orbital. Helium has two electrons in the $1s$, and the superscript 2 is used:

Energy-level designation $1s^2$ Number of electrons in the orbital

Orbital type (shape)

Now consider fluorine, which has nine electrons to be arranged. The lowest-energy arrangement calls for two electrons in the $1s$ orbital, $1s^2$, and two electrons in the $2s$ orbital, $2s^2$, and the last five electrons distributed among the $2p_x$, $2p_y$, and $2p_z$ orbitals (abbreviated $2p^5$). The full fluorine configuration is abbreviated $1s^2 2s^2 2p^5$.

To write an electron configuration for an atom, begin by determining the atom's atomic number. Because in a neutral atom the number of protons always equals the number of electrons, the atomic number tells you the number of electrons in a neutral atom as well as the number of protons. Place the electrons in orbitals of increasing energy (Figure 5.7) until the total number of electrons is accounted for. In this notation system, p designates the set of three p orbitals (p_x, p_y, p_z), or more rigorously, the sublevel of energy which distinguishes s orbitals from p orbitals. Consequently, the superscript of p may be as large as 6 if all three p orbitals are occupied. Similarly, the superscript for d may be as high as 10 because d designates the set of five d orbitals. Table 5.1 shows the lowest-energy arrangement of electrons for the first 18 elements.

SAMPLE EXERCISE

5.7

Write out the electron configuration of sodium.

SOLUTION

The atomic number of sodium is 11, which indicates that there are 11 electrons in the sodium atom. Place these electrons in orbitals in order of increasing

ELECTRON CONFIGURATION OF FIRST
18 ELEMENTS IN THE PERIODIC TABLE

TABLE 5.1

Element	Atomic Number	Notation
Hydrogen	1	$1s^1$
Helium	2	$1s^2$
Lithium	3	$1s^2 2s^1$
Beryllium	4	$1s^2 2s^2$
Boron	5	$1s^2 2s^2 2p^1$
Carbon	6	$1s^2 2s^2 2p^2$
Nitrogen	7	$1s^2 2s^2 2p^3$
Oxygen	8	$1s^2 2s^2 2p^4$
Fluorine	9	$1s^2 2s^2 2p^5$
Neon	10	$1s^2 2s^2 2p^6$
Sodium	11	$1s^2 2s^2 2p^6 3s^1$
Magnesium	12	$1s^2 2s^2 2p^6 3s^2$
Aluminum	13	$1s^2 2s^2 2p^6 3s^2 3p^1$
Silicon	14	$1s^2 2s^2 2p^6 3s^2 3p^2$
Phosphorus	15	$1s^2 2s^2 2p^6 3s^2 3p^3$
Sulfur	16	$1s^2 2s^2 2p^6 3s^2 3p^4$
Chlorine	17	$1s^2 2s^2 2p^6 3s^2 3p^5$
Argon	18	$1s^2 2s^2 2p^6 3s^2 3p^6$

energy, in accord with the maximum number that can fit into each sublevel: two electrons in s; six electrons in p; and 10 electrons in d when necessary.

$$\text{Na } 1s^2 2s^2 2p^6 3s^1$$

The total of the superscripts should equal the number of electrons in the atom, in this case 11.

Problem 5.4

Write out the electron configuration for silicon (without looking at Table 5.1).

Remembering the orbitals or sublevels in order of increasing energy is important; fortunately, there is a device (shown in Figure 5.9) for easily determining the energy order.

Aufbau Principle

The orderly placement of electrons, first in lower-energy orbitals and then in higher-energy orbitals, is known as the **Aufbau** principle. *Aufbau* means "buildup" in German. Another way of showing the

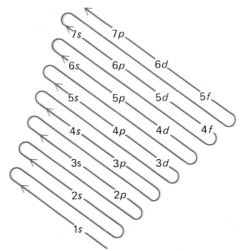

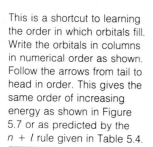

This is a shortcut to learning the order in which orbitals fill. Write the orbitals in columns in numerical order as shown. Follow the arrows from tail to head in order. This gives the same order of increasing energy as shown in Figure 5.7 or as predicted by the $n + l$ rule given in Table 5.4.

FIGURE 5.9

electron configuration is by using boxes for orbitals and arrows for electrons. Consider Na, atomic number 11, as a first example.

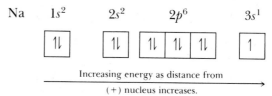

The arrows pointing in opposite directions show that the two electrons in an orbital have opposite spins. The separation of boxes indicates differences in energy. The drawing of the three $2p$ boxes together indicates that the three orbitals all have the same energy, i.e., the $2p$ sublevel contains three orbitals all of the same energy. Consider Cl, atomic number 17, as another example.

Cl $1s^2$ $2s^2$ $2p^6$ $3s^2$ $3p^5$

N, atomic number 7, presents us with another consideration:

N $1s^2$ $2s^2$ $2p^3$

Notice that the electrons fill the $2p$ orbitals singly, with the same spin, before any pairing takes place. This is completely consistent with our

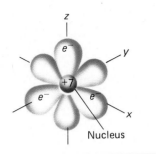

N $1s^2$ $2s^2$ $2p^3$

The three p-electrons are all the same distance from the nucleus and thus feel the same attractive force but are far apart from one another.

FIGURE 5.10

idea that low energy is associated with maximum separation of electrons, but at the minimum distance from the nucleus. This lowest-energy arrangement for the three electrons in the p orbitals of nitrogen is shown in Figure 5.10.

This principle that electrons enter orbitals of equal energy singly before they become paired is called **Hund's rule.**

At this point you should be able to make a diagram, complete with boxes and arrows, for the electronic configuration of any A Group element. Remember:

1. The atomic number tells you the total number of electrons in the atom.

2. Lower-energy sublevels are completely filled with electrons before electrons enter higher-energy sublevels.

3. Use Hund's rule when placing electrons in p orbitals and d orbitals.

SAMPLE EXERCISE

5.8

How many single (unpaired) electrons are there in oxygen?

SOLUTION

The distribution of electrons in oxygen is described in Sample Exercise 5.6a and shown in Table 5.1 to be $1s^2 2s^2 2p^4$. Showing the boxes-and-arrows diagram and following Hund's rule, we see

1s   2s   2p

Thus we find *two* single (unpaired) electrons in the $2p$ sublevel.

Problem 5.5

How many single (unpaired) electrons are there in phosphorus?

QUANTUM NUMBERS

5.6 Everyone is familiar with being identified by a set of numbers. For example, no doubt you have a Social Security Number. Electrons are identified by a set of four **quantum numbers** which come from the quantum mechanics we mentioned in Section 5.4 as the basic theory of electron arrangements. Every electron in an atom has a set of four quantum numbers, each of which reveals specific information about a different aspect of the electron. Most importantly, quantum numbers offer a method of predicting the energy rankings of sublevels without memorization (Figure 5.7) or ordering diagrams (Figure 5.9).

Quantum numbers are a sort of numerical code that describes the orbital that the electron occupies and the spin of the electron. Familiarity with the code enables chemists to "translate" back and forth from quantum numbers into specific information about electron arrangements and from data about electron arrangements into specific sets of quantum numbers. Table 5.2 summarizes the representations, meanings, and typical values of the quantum numbers.

The Pauli exclusion principle tells us that *no two* electrons in any given atom can have the same four quantum numbers, or in different words, that every electron in an atom must differ by at least one of its four quantum numbers from every other electron in that atom. This enables an electron to be uniquely identified by its set of quantum numbers, just as every Social Security Number is unique.

Principal Quantum Number *n*

We have been discussing electrons in orbitals in terms of their distance from the nucleus and their relative energy. In general, the farther an orbital is from the nucleus, the higher is its energy, because of the smaller attraction between the positive nucleus and the negative electron. You have also seen the notion that orbitals are grouped within

TABLE 5.2   ALLOWED VALUES OF QUANTUM NUMBERS

Name and Symbol	Meaning	Possible Value
Principal quantum number *n*	Distance from nucleus	$n = 1,2,3,4,5$, etc. [only positive integers (whole numbers)]
Orbital quantum number *l*	Type of volume in which electron is moving	$l = 0$ to $n - 1$
Orientation quantum number *m*	Orientation of volume	$m = -l$ to $+l$ (all integers and zero in between $-l$ and $+l$)
Spin quantum number *S*	Direction of spin	$S = +\frac{1}{2}$ or $-\frac{1}{2}$

levels around the nucleus. There is the innermost $1s$ level, which contains one orbital ($1s$); the next level, which contains the $2s$ and $2p$ orbitals; then a still larger level, which contains the $3s$, $3p$, and $3d$ orbitals. In each case the numerical coefficients preceding the orbital or sublevel symbol designate relative distances from the nucleus and relative energies. That is, a smaller number indicates a smaller distance from the nucleus and in general a lower energy. This number which indicates distance from the nucleus and the approximate relative energy of the orbital or sublevel is the principal quantum number, which is given the symbol n.

The lowest possible value of n is 1; from there the values increase in *integral* (whole-number) quantities. As n increases, the orbital is found farther from the nucleus. *Thus the first quantum number assigned to an electron indicates its relative distance from the nucleus.* The first quantum number indicates which main level the electron occupies.

Orbital Quantum Number *l*

For a given distance or within a specified main level, an electron can reside in variously shaped volumes around the nucleus. The types of three-dimensional shapes have been described with such symbols as s, p, and d. There is also a fourth type of shape, f, which will not be discussed further in this text. *The orbital quantum number (l) tells the shape of the volume within which the electron is located.* The numerical value is directly related to the s, p, d, and f designations as shown in Table 5.3.

Indicates the main level, $n = 3$ →→ **3d** ←← Indicates the orbital or sublevel type — For d orbitals, $l = 2$

We have already observed that as n is increased, say, from $n = 1$ to $n = 2$, the number of ways in which the electrons can exist around the nucleus without interfering with one another increases. Recall that in the first level there was only one type of volume shape, s, but in the second level there are two different shapes, s and p. In the third level, there are three shapes, s, p, and d. What we are saying, then, is that the possible volume shapes and thus the values of l depend on the values of n. The general scientific rule is that l can possess integral values from 0 to $n - 1$.

For example, if $n = 2$, l can equal 0 or 1 ($n - 1 = 1$ if $n = 2$). This is just the numerical code way of telling us that in the second level we can have ($l = 0$) an s sublevel (s orbital) and ($l = 1$) a p sublevel (p orbitals). No other values of l are allowed and thus only the s and p sublevels are in the $n = 2$ level. For a given value of n, an electron on the average will be slightly closer to the nucleus in an s sublevel

TABLE 5.3
SUBLEVEL TYPE AND VALUE OF *l*

$l = 0$	s sublevel
$l = 1$	p sublevel
$l = 2$	d sublevel
$l = 3$	f sublevel

Notice that the value of l tells the sublevel type and the orbital type.

than in a *p* sublevel, which in turn would be slightly closer than a *d* sublevel, which would be slightly closer than an *f* sublevel.

SAMPLE EXERCISE

5.9

The quantum numbers of an electron are $n = 3$ and $l = 1$. What sublevel does it occupy?

SOLUTION

$3p$, because n gives the main level and $l = 1$ corresponds to a p sublevel.

SAMPLE EXERCISE

5.10

What are the values of n and l for an electron in the $4s$ sublevel? -

SOLUTION

$n = 4$, the main level, and $l = 0$, which corresponds to an s sublevel.

Problem 5.6

An electron's quantum numbers are $n = 4$ and $l = 1$. What sublevel does it occupy?

Orientation Quantum Number *m*

Keeping in mind that the space in which an electron is located is three-dimensional, we see that there can be more than one orientation for some orbitals or volume shapes. An *s* orbital is spherical, and there is only one possible orientation of a sphere in space; that is, no matter how you turn a ball it always looks the same. However, an electron in a "figure eight" volume shape, i.e., a *p* orbital, can take three possible orientations, along the *x* axis, *y* axis, or *z* axis (see Figure 5.5). All three *p* orbitals (p_x, p_y, and p_z) are the same distance from the nucleus and have the same relationship to other electrons in the atom. Hence these orbitals have the same energy. When two or more orbitals possess the same energy, they are said to be **degenerate.** *The orientation quantum number m is introduced to account for the orientation factor.* It turns out that the rule for allowable values for *m* is quite simple: $m = -l$ to $+l$ (including 0). For a *p* orbital for which $l = 1$, $m = -1$, 0, or $+1$, which correspond to the three *p* orbitals, p_x, p_y, and p_z. For a *d* orbital, $l = 2$ and there are five possible *m* values, $m = -2, -1, 0, +1$, or $+2$, which in turn means that there are five degenerate *d* orbitals.

We can now describe the relative distance of an electron (*n*), the shape of the volume within which it resides (*l*), and the orientation of that volume in space (*m*).

Spin Quantum Number S

We must also assign a number to describe the electron's spin. A fourth quantum number, the spin quantum number S, is used to account for spin. S can have the values $S = +\frac{1}{2}$ or $-\frac{1}{2}$, corresponding to either a clockwise (CW) or counterclockwise (CCW) spin. (There are no other possible directions.)

SAMPLE EXERCISE

5.11

Assign quantum numbers to the two electrons in a 2s orbital.

SOLUTION

	n	l	m	S
e_1^-	2	0	0	$+\frac{1}{2}$
e_2^-	2	0	0	$-\frac{1}{2}$

Main level is 2; $l = 0$ codes for an s orbital; m can only be 0 because l is 0; and one electron spins CW and one CCW.

SAMPLE EXERCISE

5.12

There are six electrons in the 2p sublevel. Below are the quantum numbers for four of those electrons. Write the two missing sets.

	n	l	m	S
e_1^-	2	1	-1	$+\frac{1}{2}$
e_2^-	2	1	0	$-\frac{1}{2}$
e_3^-	2	1	$+1$	$+\frac{1}{2}$
e_4^-	2	1	$+1$	$-\frac{1}{2}$

SOLUTION

	n	l	m	S
e_5^-	2	1	-1	$-\frac{1}{2}$
e_6^-	2	1	0	$+\frac{1}{2}$

All 2p electrons have $n = 2$, $l = 1$; m can be -1, 0, or $+1$, while $S = \pm\frac{1}{2}$.

Problem 5.7

Assign quantum numbers to all the electrons in magnesium.

Quantum numbers not only uniquely identify electrons, they also provide other useful information for organizing our knowledge of electron configuration. For example, consider n, the principal quantum number:

n indicates main level or distance.

n^2 indicates number of orbitals at that level.

If $n = 1$, $n^2 = 1$ and there is one orbital, the $1s$.

If $n = 2$, $n^2 = 4$ and there are four orbitals, the $2s$ and three $2p$ orbitals

If $n = 3$, $n^2 = 9$ and there are nine orbitals, the $3s$, three $3p$, and five $3d$.

$2n^2$ indicates maximum number of electrons that may occupy the main level.

Because each orbital can contain a maximum of two electrons and the number of orbitals is n^2, the maximum number of electrons in a main level is $2n^2$.

$n + l$ Rule

The primary rule in the *Aufbau* process is to fill sublevels of lower energy before those of higher energy. We have stated that the order of increasing energy is $1s$, $2s$, $2p$, $3s$, $3p$, $4s$, and $3d$. It is not necessary to memorize this order. Sublevels can be ranked in order of increasing energy by realizing that lower energies are associated with smaller values for the sum of the n and l quantum numbers. Thus an electron in the $1s$ sublevel has $n = 1$, $l = 0$, and $n + l = 1$. This sum is less than for an electron in the $2s$ sublevel, for which $n = 2$, $l = 0$, $n + l = 2$. Therefore, we know that the $1s$ sublevel ($n + l = 1$) is of lower energy than the $2s$ ($n + l = 2$).

For electrons in the $2p$ sublevel, $n = 2$, $l = 1$, and $n + l = 3$, and thus these electrons are of higher energy than $2s$ electrons. Electrons in the $3s$ also have $n + l = 3$ ($n = 3$, $l = 0$). The $2p$ sublevel is lower in energy than the $3s$ because of the rule that if electrons in two sublevels have the same value of $n + l$, then the one of lower n is lower in energy. Thus $2p$ is lower in energy than $3s$. Table 5.4 lists sublevel energies in order and gives the values of $n + l$. Verify for yourself that the ranking is correct by referring to Figure 5.7 or 5.9. In comparing the energy of electrons in two sublevels, remember that:

1. Lower $n + l$ means lower energy.

2. When the sums of $n + l$ are identical, lower n means lower energy.

TABLE 5.4 ORDERING SUBLEVEL ENERGIES

	n	l	$n + l$
$7p$	7	1	8
$6d$	6	2	8
$5f$	5	3	8
$7s$	7	0	7
$6p$	6	1	7
$5d$	5	2	7
$4f$	4	3	7
$6s$	6	0	6
$5p$	5	1	6
$4d$	4	2	6
$5s$	5	0	5
$4p$	4	1	5
$3d$	3	2	5
$4s$	4	0	4
$3p$	3	1	4
$3s$	3	0	3
$2p$	2	1	3
$2s$	2	0	2
$1s$	1	0	1

Increasing energy ↑

SAMPLE EXERCISE 5.13

How do we know that the 5*s* sublevel is lower in energy than the 4*d* sublevel?

SOLUTION

Examine the *n* and *l* quantum numbers for each sublevel.

	n	l	$n + l$
5*s*	5	0	5
4*d*	4	2	6

Because $n + l$ is less for the 5*s*, the 5*s* is of lower energy.

SAMPLE EXERCISE 5.14

How do we know that the 5*d* sublevel is lower in energy than the 6*p* sublevel?

SOLUTION

Examine the *n* and *l* quantum numbers for each sublevel:

	n	l	$n + l$
5*d*	5	2	7
6*p*	6	1	7

Because the sums of $n + l$ are identical in this case, we decide on the lower energy sublevel by considering values of *n*. Therefore, 5*d* is of lower energy.

Problem 5.8

a. How do we know that the 6*s* sublevel is lower in energy than the 4*f* sublevel?

b. How do we know that the 4*f* sublevel is lower in energy than the 5*d* sublevel?

The $n + l$ rule or the scheme in Figure 5.9 can be used to order sublevels by increasing energy, thus allowing you to write the electron configuration for any representative (A Group) element.

SAMPLE EXERCISE 5.15

Write out the electron configuration for rubidium (Rb).

SOLUTION

The atomic number of rubidium is 37, which indicates that there are 37 electrons in the rubidium atom. Place electrons into the sublevels in order of increasing energy until 37 electrons have been accounted for.

$$\text{Rb} \qquad 1s^2 2s^2 2p^6 3s^2 3p^6 4s^2 3d^{10} 4p^6 5s^1$$

Problem 5.9

Write out the electron configuration for cesium.

EVIDENCE AND USES OF THE ELECTRON ENERGY LEVELS

5.7   You have now seen that the arrangement of electrons in an atom is based on the buildup of electrons in discrete energy levels, with low-energy levels close to the nucleus and higher-energy levels as we move farther from the nucleus. In the undisturbed ground state, the electrons occupy the lowest-energy levels consistent with the ideas presented in Section 5.4.

However, energy can be put into an atom. For example, if we heat atoms in a flame, we are adding heat energy. As a result of this added energy, one or more electrons may be raised to a higher-energy level and the atom is then said to be in an excited state. In time the electrons will fall back to the lower levels of the ground state when the excess energy is given off. Figure 5.11 summarizes the excitement of a He atom by electrical energy and its relaxation when light energy is emitted.

Light is a form of radiant energy or radiation. You know it is energy because it is capable of doing work or effecting a change. For example, light provides you with a suntan or sunburn, or it can develop a photographic film. Visible light from the sun or a light bulb is usually white light. A prism can divide white light into a rainbow. Scientists call the array of colors in the rainbow a *spectrum*. (Raindrops in the atmosphere sometimes act as prisms in the sky and produce a rainbow.) What the prism and rainbow (spectrum) tell us is that white light is a mixture of light rays of different colors. Each color of visible light has associated with it a distinct energy, the shades of violet corresponding to the higher energies and red to the lower energies. The actual definition of a **spectrum** is that it is an array of energies.

Excited atoms do not emit white light, i.e., the total spectrum of energies; rather, they emit a spectrum of energies unique to that atom. This is so because the energy levels of each element (sodium, potassium, lead, etc.) are uniquely spaced and electrons can occupy only these states; therefore, a particular excited element will emit its own unique spectrum of energies as electrons fall back to the ground state. This, in turn, means that there will be a unique spectrum of discrete colors in the visible light that is emitted. This emitted radiation can be precisely analyzed with an instrument known as a **spectroscope.** The presence or absence of a particular element can be confirmed by

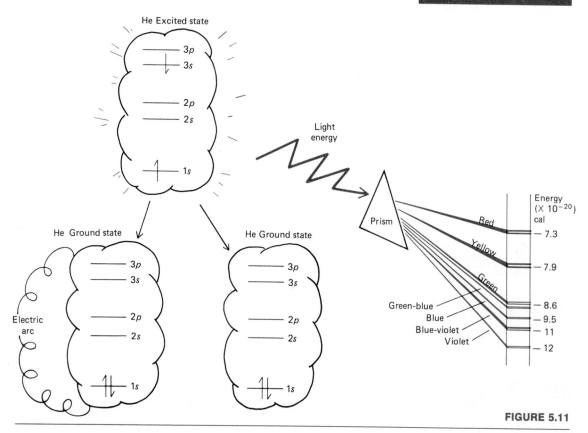

FIGURE 5.11

Electrical energy promotes electrons to higher-energy levels. In this representative case we see an electron promoted from the 1s to the 3s. As electrons fall back to the ground state, excess energy is emitted as light energy. Each line in the spectrum corresponds to some energy-level transition.

analyzing the emitted light of its excited state in a spectroscope (see Figure 5.12).

The concentration of sodium ion in blood plasma and in various foodstuffs (after being made water-soluble) can be readily determined by flame emission spectroscopy. A solution containing the sodium ion is sprayed into a high-intensity flame, which excites the sodium. The ion then emits its light spectrum. The emitted light passes through a filter which is selective for the yellow light in the sodium spectrum. The intensity of the light passing through the filter is proportional to the concentration of the sodium ion in the original solution.

The concentrations of many other ions in physiological fluids can be measured by a procedure known as atomic absorption spectroscopy. Here a selected amount of energy, which can be varied for different elements, is pumped into a sample; electrons are promoted to higher levels by this energy. The amount of energy absorbed by the sample can be quantitatively measured and related to the concentration of the metallic ion in question.

Using these two analytical methods, medical and food testing

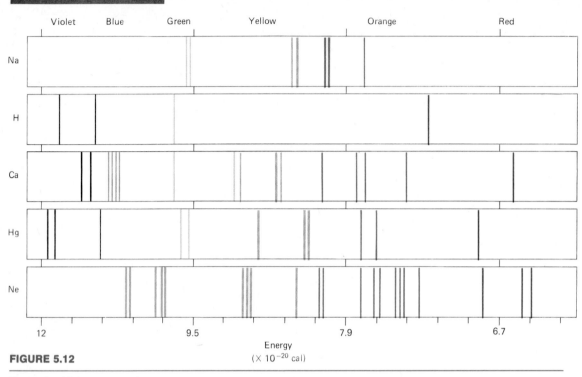

FIGURE 5.12

Energy
($\times 10^{-20}$ cal)

The emitted light spectrum is different and unique for every element.

laboratories can quickly and precisely obtain concentrations of almost every metal ion in physiological fluids, foodstuffs, and other samples of biomass. Forensic specialists and ecologists also employ these techniques extensively.

PERIODIC TABLE

5.8 Can the knowledge of electron configuration provide a deeper understanding of the periodic table and periodic trends? Let us write the electron configurations of lithium, sodium, potassium, and rubidium, elements from Group IA of the periodic table, and see.

Atomic Numbers	Element	Electron Configuration
3	Li	$1s^2 2s^1$
11	Na	$1s^2 2s^2 2p^6 3s^1$
19	K	$1s^2 2s^2 2p^6 3s^2 3p^6 4s^1$
37	Rb	$1s^2 2s^2 2p^6 3s^2 3p^6 4s^2 3d^{10} 4p^6 5s^1$

These four elements exhibit very similar chemical and physical properties (see Section 4.11). Can you detect the similarity in their electron configurations?

In each of these elements there is one electron in the outermost energy level. The outermost levels are the $n = 2, 3, 4,$ and 5 for Li, Na, K, and Rb, respectively. Could it be that the properties of an element are determined by the number of electrons in the outermost energy level? Let us examine the electron configurations of some elements in Group VIIA of the periodic table. Notice that in this case the sublevels are not arranged in the order in which they fill, but rather the sublevels within the major energy level are grouped together in order of increasing distance from the nucleus.

Atomic Numbers	Element	Electron Configuration
9	F	$1s^2 2s^2 2p^5$
17	Cl	$1s^2 2s^2 2p^6 3s^2 3p^5$
35	Br	$1s^2 2s^2 2p^6 3s^2 3p^6 3d^{10} 4s^2 4p^5$
53	I	$1s^2 2s^2 2p^6 3s^2 3p^6 3d^{10} 4s^2 4p^6 4d^{10} 5s^2 5p^5$

If chemical tests are performed on these elements, it is found that fluorine, chlorine, bromine, and iodine all have very similar chemical properties. Each of these elements has seven electrons in the outermost level. The outermost levels are the $n = 2, 3, 4,$ and 5 for F, Cl, Br, and I, respectively. The idea that the chemistry of an element is determined by the number of electrons in the outermost level seems to be correct, and moreover, the group number in the periodic table is simply the number of electrons in the outermost level for elements within that group. Electrons in the outermost level are called **valence electrons.**

Stating the principle somewhat differently, lithium, sodium, potassium, and rubidium are all in the same group in the periodic table because they all have the same number of valence electrons (one). *In general, for the A Group elements, all the elements within a group have the same number of valence electrons. The number of valence electrons is the group number.*

SAMPLE EXERCISE

5.16

How many electrons are there in the outermost level of Group VA elements?

SOLUTION

Five, the group number tells you. Check any member of the group, for example, N, $1s^2 2s^2 2p^3$. There are five electrons in the outermost second level.

Problem 5.10

How many electrons are there in the outermost level of Group VIIIA elements?

Let us check this generalization further by looking at a modified periodic table (Figure 5.13). We find that all the elements within a particular A Group have the same outer-level electron configuration. In Groups IA and IIA (*s*-block elements) the *s* sublevel is being filled, whereas in Groups IIIA through VIIIA (*p*-block elements) the *p* sublevel is being filled. In Chapter 4 the discussion of the periodic table pointed out that the properties of the elements repeat themselves at certain fixed intervals. In this chapter it appears that the underlying basis for this is that a given number of valence electrons is repeated at certain fixed intervals.

By noting the period number and the group number of a particular A Group element, one can immediately write down the outer-level electron configuration. The period number tells the main-level number. The group number tells the total number of electrons that must be distributed between the *s* and *p* sublevels, with two going into the *s* and the rest into the *p* sublevel. For example, for nitrogen the period number is 2 and the group number is VA:

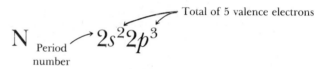

$$\underset{\underset{\text{number}}{\text{Period}}}{N} \longrightarrow 2s^2 2p^3 \quad \longleftarrow \text{Total of 5 valence electrons}$$

Outerlevel electron configuration is repeated within each group.

FIGURE 5.13

Group number→	IA	IIA									IIIA	IVA	VA	VIA	VIIA	VIIIA
1	H s^1															He s^2
2	Li s^1	Be s^2									B s^2p^1	C s^2p^2	N s^2p^3	O s^2p^4	F s^2p^5	Ne s^2p^6
3	Na s^1	Mg s^2	IIIB	IVB	VB	VIB	VIIB	VIIIB	IB	IIB	Al s^2p^1	Si s^2p^2	P s^2p^3	S s^2p^4	Cl s^2p^5	Ar s^2p^6
4	K s^1	Ca s^2				Nonrepresentative elements					Ga s^2p^1	Ge s^2p^2	As s^2p^3	Se s^2p^4	Br s^2p^5	Kr s^2p^6
5	Rb s^1	Sr s^2				*d* block					In s^2p^1	Sn s^2p^2	Sb s^2p^3	Te s^2p^4	I s^2p^5	Xe s^2p^6
6	Cs s^1	Ba s^2									Tl s^2p^1	Pb s^2p^2	Bi s^2p^3	Po s^2p^4	At s^2p^5	Rn s^2p^6
7	Fr s^1	Ra s^2	*f* block													

Period

s block

p block

In the next chapter the key role that the valence electrons play in the bonding properties of an atom will be seen. You should be able to write the outer-level electron configuration for any A Group element by simply noting its position in the periodic table.

SAMPLE EXERCISE

5.17

Write the electron configuration for the valence electrons of sulfur.

SOLUTION

From Figure 5.13 note that sulfur is in the third period, which means that the valence electrons must be in the $n = 3$ level. Sulfur is in Group VIA, which means that there are six valence electrons. Filling the six electrons into the third level produces the arrangement $3s^2 3p^4$ for the valence electrons of sulfur.

Problem 5.11

Using the periodic table in the inside cover of your text, write out the electron configuration for the valence electrons of oxygen.

Notice that these rules are restricted to the groups in the periodic table labeled A. The elements in these groups are known as the **representative,** or **main-group, elements.** For these A Group elements, all the sublevels in the levels below the outermost level are completely filled.

For the **nonrepresentative** B Group **elements**, the inner levels are not always completely filled. For example, the electron configuration of vanadium (atomic number 23) is

$$V: \qquad 1s^2 2s^2 2p^6 3s^2 3p^6 3d^3 4s^2$$

LEWIS ELECTRON DOT STRUCTURES

5.9 Because of the importance of the valence electrons, chemical symbols which represent them are often used. These symbols are called **Lewis electron dot structures** after the theoretical chemist G. N. Lewis, who developed them.

A Lewis structure for an element consists of the element's symbol and surrounding dots to represent the number of valence electrons. To write a Lewis dot structure, write down the symbol of the element and surround the symbol with a number of dots corresponding to the number of valence electrons. Do not put more than two dots around any edge of the symbol. Whether the dots are paired, for example, :Ḃ, or spread out, for example, ·Ḃ·, is not important. The best method depends on the ultimate use to be made of the symbol. Remember that for an A Group element the group number is equal to the number of valence electrons, the number of electrons in the outermost level.

$$\text{H} \cdot \quad \cdot \ddot{\underset{..}{\text{O}}}{:} \quad \text{K} \cdot \quad :\ddot{\underset{..}{\text{F}}}{:} \quad \cdot \dot{\text{Al}}$$

Lewis structures will be a very helpful notation in your study of chemical bonding in Chapter 6.

Problem 5.12

Write Lewis dot structures for the following elements: Na, N, Br, C, Ne, and Mg.

PERIODIC TRENDS

The arrangement of electrons with which you are now familiar not only explains the existence of chemical groups, it also accounts for other observable trends in properties. Look at the special periodic table in Figure 5.14. This periodic table tells the atomic size and ionization energy for every representative group element. The top and side of this periodic table are marked with the trends in these properties among the elements.

Ionization energy is defined as the amount of energy needed to remove one electron from each of the atoms in a gaseous sample, the mass of which in grams is numerically equal to the atomic weight of the element. For example, the ionization energy of sodium is 118 kcal for 23 g. Notice in Figure 5.14 that the ionization energy increases as one proceeds across a period from left to right; that is, it becomes more difficult to remove an electron as one moves across the period from metals to nonmetals. In addition, the atomic size of the elements shrinks from left to right across a period.

Both effects are based on the same feature of electron structure: the nuclear charge (atomic number) increases from left to right, but the added electrons are being placed in the same outermost or valence level. For example, the buildup of electrons in period 3 can be represented as follows:

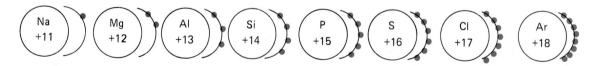

Because the valence electrons are all in the same main level, the distance between the positively charged nucleus and the valence electrons is not significantly increased going across a period. But the positive charge on the nucleus becomes larger, and the bigger the charge, the

Ionization energy **increases**
Atomic radius **decreases**

→

IA	IIA	IIIA	IVA	VA	VIA	VIIA	VIIIA
312 H 0.37							
124 Li 1.23	214 Be 0.89	190 B 0.80	259 C 0.77	334 N 0.74	312 O 0.74	400 F 0.72	494 Ne 0.71
118 Na 1.57	175 Mg 1.36	137 Al 1.25	187 Si 1.17	251 P 1.10	237 S 1.04	299 Cl 0.99	314 Ar 0.98
100 K 2.03	140 Ca 1.74	138 Ga 1.25	187 Ge 1.22	242 As 1.21	224 Se 1.17	272 Br 1.14	321 Kr 1.12
96 Rb 2.16	131 Sr 1.91	133 In 1.50	168 Sn 1.41	138 Sb 1.41	207 Te 1.37	244 I 1.33	279 Xe 1.30
89 Cs 2.35	120 Ba 1.98	140 Tl 1.55	170 Pb 1.54	184 Bi 1.52	193 Po 1.67		

Ionization energy decreases
Atomic radius increases
↓

First ionization
energy (kcal)

Atomic radius (Å)

FIGURE 5.14

First ionization energies and
atomic radii for the
representative elements
(Å is the abbreviation for
angstrom, a unit of length
equivalent to 1×10^{-8} cm).

stronger the attractive pull of the nucleus on the electrons. The size of the atoms decreases because the nucleus pulls electrons closer. Ionization energy increases because more energy must be applied to remove electrons from the stronger nuclear pull. Summarizing horizontal trends, we find:

Decreasing atomic size
→
Increasing ionization energy

Let us now examine vertical trends, i.e., trends within a group. Figure 5.14 shows you that atomic size increases and ionization energy decreases moving down a group.

Increasing | Decreasing
atomic | ionization
size | energy
↓

These trends arise because, proceeding down a group, the valence electrons of each new element are found in a new main level that is farther away from the nucleus. For example, in Group IA, for Na the outer level is 3; for K it is 4; for Rb, 5; and for Cs, 6. The atomic size increases because electrons are found at larger distances from the nucleus. Because it is easier to remove electrons from an atom when they are far away from the nucleus and therefore feel the nuclear pull less strongly, ionization energy decreases as atoms become larger.

SAMPLE EXERCISE

5.18

Using only the periodic table on the inside cover of this text, predict which atom in each of the following sets is larger and which has the higher ionization energy:

a. As and N
b. Ca and Br
c. Cl and Rb

SOLUTION

Refer to the preceding summarization of horizontal ($\rightarrow$) and vertical ($\downarrow$) trends.

a. These elements are in the same group. Arsenic is larger because it is below nitrogen in the periodic table. The ionization energy of arsenic is smaller for the same reason.
b. These elements are in the same period. Bromine is smaller than calcium because it is farther to the right. Bromine has a larger ionization energy for the same reason.
c. Rubidium is larger because it is both below and to the left of chlorine. The ionization energy of rubidium is smaller for the same reason.

Problem 5.13

a. Which of the A Group elements have the lowest ionization energy?

b. Are these metals or nonmetals?

Problem 5.14

In general, across a period, which are the larger atoms, metals or nonmetals?

Problem 5.15

Consulting *only* the periodic table on the inside cover

a. Which atom do you predict has the highest ionization energy?

b. Which do you predict has the lowest ionization energy?

As you saw in Section 5.8, nonrepresentative B Group elements contain an unfilled inner energy level. In general, for the nonrepresentative elements, a *d* sublevel in a main energy level below the outermost is unfilled even though there are two electrons in the *s* sublevel of the outermost energy level. The consequence of this is that these elements do not exhibit the dramatic periodic trends seen for A Group elements. For example, there is little variation in ionization energy among the elements with atomic numbers 21 to 30 (Sc to Zn). The inner 3*d* sublevel is filled as you go from $_{21}$Sc to $_{30}$Zn, but there is no change in outer electron configuration and thus essentially no change in ionization energy. We see a similar pattern as the 4*d* orbitals fill in $_{39}$Y through $_{48}$Cd (see Figure 5.15).

The B Group elements, which are all metals, are also called **transition elements** or **transition metals.** Many of them have important industrial applications and are required in body chemistry. For example, iron is processed into steel and also occurs in hemoglobin, a component of red blood cells essential to the oxygen-carrying capacity of blood.

As you progress in your study of chemistry, you will find numerous other uses for this unique compilation known as the **periodic**

FIGURE 5.15

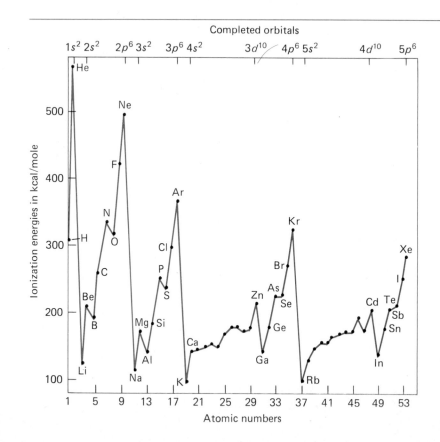

Variation in first ionization energy with change in atomic number.

table, and it should be as helpful to you in studying chemical properties as your electronic calculator is in performing numerical calculations.

SUMMARY

All matter is constructed on the basis of achieving a low-energy, high-stability condition (Section 5.3). For atoms, this low-energy condition is achieved by arranging electrons in orbitals, which are volumes of varying size and shape surrounding the nucleus. The orbitals are constructed on the basis of keeping electrons close to the nucleus but separated from one another (Section 5.4).

We express the exact arrangement of electrons in atoms by using a shorthand notation for electron configuration (Table 5.1) or diagrams employing boxes and arrows (Section 5.5).

Quantum numbers are a numerical code that contains information about the orbital and spin of an electron. Every electron in an atom has a different set of quantum numbers from any other electron because it differs in the orbital it occupies and/or in its spin (Section 5.6).

Practical use is made of quantized electron arrangements in emission and absorption spectroscopy. Concentrations of many common metal ions are measured by methods based on the principles discussed in this chapter (Section 5.7).

The number of valence electrons, i.e., the number of electrons in the outermost level of an atom, determines the group in the periodic table to which the element belongs (Section 5.8). Lewis electron dot structures are convenient representations of the valence electrons of atoms (Section 5.9). Relative atomic sizes and ionization energies can be related to the concept that electrons are arranged in energy levels (Section 5.10).

CHAPTER ACCOMPLISHMENTS

After completing this chapter you should be able to

Introduction

Energy revisited

1. Define energy.

Concept of minimum energy

2. Explain the relationship between minimum energy and maximum stability.

Minimum energy in the atom

3. Describe the electrostatic factors that determine how electrons are arranged in atoms.

4. Define orbital.
5. State the relationships between orbitals, sublevels, and main energy levels.
6. State the spin relationship between electrons in the same orbital.

Electron-configuration notation

7. List the order in which atomic orbitals are filled (Aufbau principle).
8. Write the electron configuration of any A Group element.
9. State and apply Hund's rule for writing electron configuration and determining the number of unpaired electrons in an atom.

Quantum numbers

10. List the names and symbols of the four quantum numbers.
11. State the Pauli exclusion principle.
12. Relate quantum numbers to their physical meaning.
13. Write the four quantum numbers for any electron in an atom.
14. Use the $n + l$ rule to order by energy any set of sublevels.

Evidence and uses of the electron energy levels

15. Describe the result of putting energy into an atom.
16. Explain why the light spectrum emitted by excited atoms is unique for each element.

Periodic table

17. Explain the relationship between electronic arrangement and the periodic table.
18. Define valence electrons.
19. Use the periodic table to obtain the number of valence electrons for any A Group element.
20. Use the periodic table to write the valence electron configuration of any A Group element.

Lewis electron dot structures

21. Write the Lewis dot structure for any element, given the number of valence electrons.

Periodic trends

22. Define ionization energy.
23. Explain the trends in ionization energy and atomic size across a period and within a group of the periodic table.
24. Given a periodic table recognize the transition metals.
25. State the distinguishing difference in electron arrangement between the A and B Group elements.

PROBLEMS

5.2 Energy revisited

5.16 **a.** Give an example of a chemical change that releases energy.
b. Where did this released energy come from?

5.3 Concept of minimum energy

5.17 Explain in terms of an energy change why:
    a. Apples fall from a tree to the ground.
    b. Water runs down a hill not up a hill.
    c. Electrons try to remain close to the nucleus.

5.18 Compare the stabilities of two states, one of which has a high energy content and the other a low energy content.

5.4 Minimum energy in the atom

5.19 **a.** What forces must be overcome to remove an electron from an atom?
b. What forces would be relieved by removing an electron from a neutral atom?
c. Is energy required or is energy released in removing an electron from a neutral atom?
d. Is the type of force in part (a) greater or lesser than that in part (b)? Explain how you arrived at your answer.

5.20 What forces must be balanced in an electrostatically neutral atom?

5.21 **a.** Describe the attractive force in a hydrogen atom.
b. Describe the attractive forces that could exist within and between two hydrogen atoms.
c. Describe the repulsive forces that could exist between two hydrogen atoms.
d. What happens to the attractive and repulsive forces as we change the distance between the two hydrogen atoms? Could a minimum energy structure exist for two hydrogen atoms? What would you call this structure?

5.22 Explain why there is only one s-type orbital, but three p-type orbitals at any given level (above $n = 1$).

5.23 Describe the geometry of an s orbital, a p orbital.

5.24 What is the angular relationship between any two of the three 2p orbitals?

5.25 What requirement is there on the spin relationship between any two electrons in the same orbital?

5.26 **a.** Which generally contains a greater number of electrons: an orbital, a sublevel, or an energy level?
b. Is there any main energy level which can only contain two electrons?

5.27 Within a particular energy level which sublevel is of lowest energy?

5.5 Electron configuration notation

5.28 Without looking at Table 5.1, write out the electron configurations of:
    a. Li   **d.** O
    b. F   **e.** P
    c. B

5.29 Using the notation of Table 5.1, write out the electron configurations of:
    a. Calcium   **d.** Bromine
    b. Arsenic   **e.** Indium
    c. Potassium

5.30 Correct any errors in the following electron configurations:
    a. $1s^2 1p^6 2s^2$
    b. $1s^2 2s^2 2p^6 3s^1 3p^2$
    c. $1s^2 2s^2 2p^6 3s^2 3d^7$
    d. $1s^3 2s^2 2p^6 3s^2 3p^5 4s^1$

5.31 Use the boxes-and-arrows notation to show the electron configurations of:
    a. Sodium   **d.** Carbon
    b. Silicon   **e.** Fluorine
    c. Beryllium   **f.** Cobalt

5.32 How many unpaired electrons are there in each of the atoms in Problem 5.31?

5.33 Why does nitrogen have three unpaired electrons in its p orbitals?

5.6 Quantum numbers

5.34 Give a possible set of quantum numbers for an electron in a 3s orbital.

5.35 **a.** Determine the maximum number of orbitals in the $n = 4$ main level.
b. With what sublevels are these orbitals associated?
c. What is the maximum number of electrons in the $n = 4$ level?

5.36 Explain why the maximum number of electrons in any level is always twice the number of orbitals allowed in that level.

5.37 The following sets of four quantum numbers each describe an electron in an atom. What is the energy level and sublevel in each case?
a. $n = 2, l = 0, m = 0, S = -\frac{1}{2}$
b. $n = 4, l = 2, m = 0, S = +\frac{1}{2}$
c. $n = 3, l = 1, m = -1, S = -\frac{1}{2}$
d. $n = 2, l = 1, m = -1, S = +\frac{1}{2}$
e. $n = 2, l = 1, m = +1, S = +\frac{1}{2}$

5.38 According to the $n + l$ rule, which sublevel has lower energy, 6f or 7p?

5.39 According to the $n + l$ rule, which sublevel has lower energy, 6f or 7d?

5.40 Is there any difference in energy among the $2p_x$, $2p_y$, and $2p_z$ orbitals?

5.41 Is there any difference in energy among the five 3d orbitals?

5.7 Evidence and uses of the electron energy levels

5.42 Fill in the word "emitted" or "absorbed":
In order for an electron to be promoted to an excited state, energy must be _____ . When an electron falls back to the ground state, energy is

_____ .

5.43 Explain in your own words the difference between absorption and emission spectra.

5.44 When we see a yellow flame, what are we seeing on an atomic level?

5.45 After being heated by a bunsen burner, a crucible may glow cherry red. What is the red color due to?

5.8 Periodic table

5.46 What similarity exists in the electron configurations of elements in Group VIA? Group VIIA?

5.47 The atomic numbers of the elements within a certain set are 7, 15, 33, 51, and 83. What similarity exists in the electron configuration of these elements?

5.48 Write out the electron configuration of
a. Strontium
b. Iodine
c. Bismuth
d. A synthetic element with atomic number 104

5.49 Give the symbol of the element with the following electron configuration:
a. $1s^2 2s^2 2p^1$
b. $1s^2 2s^2 2p^6 3s^2 3p^6 4s^1$
c. $1s^2 2s^2 2p^6 3s^2 3p^6 4s^2 3d^7$
d. $1s^2 2s^2 2p^6 3s^2 3p^6 4s^2 3d^{10} 4p^1$
e. $1s^2 2s^2 2p^6 3s^2 3p^6 4s^2 3d^{10} 4p^6$

5.50 Consulting only the periodic table on the inside cover, predict the orbital type for the last electron added to the following elements: Al, Cl, Ca, C, and S.

5.51 Consulting only the periodic table on the inside cover, write out the electron configuration for the valence electrons of:
a. Sulfur
b. Phosphorus
c. Iodine
d. Krypton

5.52 Give the symbol of the element with the following valence electron configurations:
a. $2s^2$ **d.** $1s^1$
b. $4s^1$ **e.** $3s^2 3p^4$
c. $2s^2 2p^5$ **f.** $5s^2$

5.53 Consulting only the periodic table on the inside cover, give the symbol of an element which has chemical properties similar to one with valence electron configuration
a. $3s^2 3p^4$ **d.** $3s^1$
b. $5s^2$ **e.** $1s^2$
c. $2s^2 2p^5$

5.54 **a.** Which groups in the periodic table are the representative elements?

b. Give the symbol of a nonrepresentative element.

5.55 Write Lewis dot structures for the following elements: Li, Si, I, Ar, P, and Ba.

5.56 **a.** Explain why atomic size decreases proceeding across a period from left to right.
b. Why does the ionization energy increase as atomic size decreases?

5.57 **a.** Where in the periodic table are elements with low ionization energies found?
b. Where in the periodic table does one find elements with high ionization energies? What type of elements are these?
c. Which group of elements in the periodic table have the highest ionization energies? What does this say about the stability of the electron arrangement for this group?

5.58 Arrange each set of elements in order of increasing ionization energy:
a. Na, Li, Rb, Cs, K
b. Br, Ge, As, Ga, K, Ca
c. Se, Br, K

5.59 Circle the correct choice in each set:
a. Highest ionization energy: Na, K, C, Si
b. Smallest size: C, Li, Ge, F
c. Largest size: K, Cs, Se
d. Smallest ionization energy: K, Cs, Se

5.60 What is the main difference in electron configuration between a representative and a transition element?

5.61 Define the term transition metal.

5.62 Which elements tend to have higher ionization energies, metals or nonmetals?

5.63 State at least five characteristics of atoms of an element that you can glean from a periodic table.

CHEMICAL
BONDING
CHAPTER

6
C
Carbon
12.01

6.1 Elements are rarely found in uncombined form. Rather, they are combined with other elements to form compounds. There are two major types of compounds: (1) **ionic compounds,** in which the ionic bond holds the elements together, and (2) **molecular compounds,** in which the covalent bond is the holding force. Note that a chemical **bond** is the force that holds elements together in compounds. In this chapter we will explore the nature of the ionic bond and the formulas of ionic compounds. We will answer the question of why there are two types of compounds and then explore the nature of the covalent bond and molecular formulas.

As in Chapter 5, the overriding factor in compound formation is the atom's striving to achieve the lowest-energy (most stable) electron configuration.

HOW CAN ATOMS ACHIEVE LOWER-ENERGY STATES?

6.2 In Chapter 5 you learned that there is a natural tendency for matter to attain a lower-energy state. Therefore, the electrons in an atom are arranged to give the most stable (lowest-energy) electron configuration for that atom. An atom left to itself does the best it can, given the number of electrons that it has, to arrange those electrons in such a way as to produce the lowest possible energy state for that atom. However, the atoms of most elements interact with atoms of other elements; this enables them to form electron arrangements of lower energy than they could achieve as individuals, and this inter-action leads to compound formation. Compound formation can be explained by an extension of the same principles of low-energy ar-rangements that you have seen for elements.

There is a family of elements in the periodic table, Group VIIIA, the **inert** or **noble gases,** that has essentially no tendency to interact with other elements to form compounds.[1] The electron configurations of the noble gases are apparently the most stable (possessing the lowest energy) of all the elements, since these gases have so little tendency to undergo change in their electron configuration. They are stable just as they are. Because the noble gases are a periodic *group*, we know that their electron configurations are similar.

Element	Symbol	Electron Configuration
Helium	He	$1s^2$
Neon	Ne	$1s^2 2s^2 2p^6$
Argon	Ar	$1s^2 2s^2 2p^6 3s^2 3p^6$
Krypton	Kr	$1s^2 2s^2 2p^6 3s^2 3p^6 3d^{10} 4s^2 4p^6$
Xenon	Xe	$1s^2 2s^2 2p^6 3s^2 3p^6 3d^{10} 4s^2 4p^6 4d^{10} 5s^2 5p^6$
Radon	Rn	$1s^2 2s^2 2p^6 3s^2 3p^6 3d^{10} 4s^2 4p^6 4d^{10} 4f^{14} 5s^2 5p^6 5d^{10} 6s^2 6p^6$

The consistent lowest-energy pattern in these configurations is that the outermost *s* and *p* orbitals are *completely filled* with paired electrons. For helium, this corresponds to two valence (outer-level) electrons. For all the other noble gases, completely filled outermost *s* and *p* orbitals require eight valence electrons. Eight valence electrons are often referred to as an **octet.**

Most elements react to form compounds through a process whereby their atoms gain, lose, or share valence electrons in order to achieve the highly stable (octet) electron configuration of a noble gas element. Atoms can gain, lose, or share electrons and form compounds because by so doing they acquire lower-energy electron arrangements.

By gaining, losing, or sharing electrons, an element will acquire the electron configuration of the noble gas to which it is closest in the periodic table. For H, Li, and Be, this means that they strive to attain the He valence electron configuration of $1s^2$, two valence electrons, a **duet.** Other elements strive to achieve eight valence electrons. This is often called the rule of eight or the **octet rule.**

Elements do not always form a noble-gas electron configuration, but that is far and away the general rule. An exception of minor importance is boron, which strives to attain six valence electrons.

METALS LOSE ELECTRONS

6.3   The trends in ionization energy in Section 5.10 reveal that electrons can be removed from metals more readily than from nonmetals.

---

[1] Since 1962 several compounds of the noble gases xenon, radon, and krypton have been prepared. However, these compounds are exceptions to the rule. In this book, we will regard the noble gases as inert.

In this section you will see that the number of electrons that a metal readily loses is the number that gives the metal the stable electron configuration of a noble gas. Electron loss always produces a positively charged metal ion. This is so because upon electron loss from an atom, the positively charged protons in the nucleus outnumber the negatively charged electrons. A positive ion is called a **cation.**

Ionization energy is the amount of energy needed to remove an electron from a gaseous atom or ion completely. In Section 5.10 we discussed only the "first" ionization energy, the energy needed to remove the outermost electron from a neutral atom. This first electron removed is the one that is the least tightly held by the nucleus, because it is the farthest away. Considering aluminum as an example, 138 kcal is required to remove one electron from each atom in 27 grams[2] of gaseous aluminum, and form Al^+ ions.

If we put enough energy in, we can remove an electron from the Al^+ ion. This is called the **second ionization energy** of aluminum, and turns out to be 433 kcal for the same 27-g sample originally described.

We can continue the process and measure the third and the fourth ionization energies, and so on. The second ionization energy for a given element is always greater than the first, since in the case of the second an electron is being removed from an already positively charged species. Remember that separating positively and negatively charged ions requires energy (Section 5.4). For similar reasons, the third ionization energy is greater than the second (Figure 6.1).

Figure 6.2 shows a plot of ionization energies from the first through one past the group number for a few A Group metallic elements. Notice that for a given element there is a gradual increase in ionization energy as one goes from first to second, and so on. Then when we try to remove an electron one past the group number, there is a very large increase in energy. This jump in energy gives strong support to the idea that the metal tends to lose electrons only until it achieves a noble-gas configuration. *For an A Group metal, losing a number of electrons that is equal to the group number produces the noble-gas configuration.* The loss of yet another electron will then destroy the noble-gas configuration. Table 6.1 offers examples. The A Group metals generally form cations with positive charges equal to the group number.

When two particles (atoms or ions) have the same electron configuration, we say they are **isoelectronic.** In the examples in Table 6.1, Li^+ is isoelectronic with He; Mg^{2+}, Al^{3+}, and Ne are all isoelectronic because they all have the same electron configuration. However, isoelectronic species are not identical. They differ from one another

---

[2] The significance of this sample size will unfold in Chapter 8.

in the numbers of protons and neutrons in their nuclei. For example, compare Mg^{2+} with Ne:

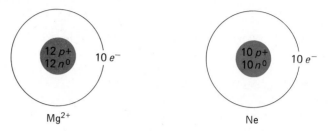

FIGURE 6.1

More and more energy is required as successive electrons are removed from an atom (See also Figure 6.2).

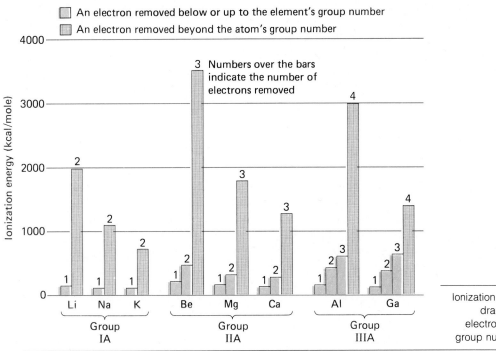

☐ An electron removed below or up to the element's group number
☐ An electron removed beyond the atom's group number

Ionization energy (kcal/mole)

FIGURE 6.2

Ionization energy increases dramatically when an electron one beyond the group number is removed.

NOBLE-GAS CONFIGURATIONS ATTAINED BY ELECTRON LOSS BY METALS   **TABLE 6.1**

Metal and Group Number	Metal Configuration	Noble-Gas Configuration Achieved	Number of Electrons Lost
Li IA	$1s^2 2s^①$	He: $1s^2$	1
Mg IIA	$1s^2 2s^2 2p^6 3s^②$	Ne: $1s^2 2s^2 2p^6$	2
Al IIIA	$1s^2 2s^2 2p^6 3s^② 3p^①$	Ne: $1s^2 2s^2 2p^6$	3

Notice how the number of electrons lost always equals the group number.

Each particle has 10 electrons; therefore, they are isoelectronic, but their nuclei contain different numbers of protons and neutrons. Because the number of protons characterizes an element, the Mg^{2+} ion has different properties from the Ne atom. However, because it is isoelectronic with the inert gas neon, the Mg^{2+} ion is stable and does not tend to change its electron arrangement further.

SAMPLE EXERCISE

6.1

How many electrons must Ca lose to achieve a noble-gas configuration? With which noble gas will the ions be isoelectronic?

SOLUTION

Calcium is in Group IIA, which tells us that it will lose two electrons and become Ca^{2+}. A loss of two electrons results in the argon configuration.

$$Ca\ (20\ e^-) \qquad\qquad Ar\ (18\ e^-) \quad and \quad Ca^{2+}\ (18\ e^-)$$

$$1s^2 2s^2 2p^6 3s^2 3p^6 4s^2 \xrightarrow[2e^-]{remove} \qquad 1s^2 2s^2 2p^6 3s^2 3p^6$$

NONMETALS GAIN ELECTRONS

6.4 Nonmetals do not readily lose electrons. Remember that ionization energy increases going from left to right across the periodic table, so that more energy is required to remove electrons from nonmetals than from metals. Also, nonmetals can more easily achieve noble-gas configurations by gaining electrons than by losing them. For example, consider chlorine, which has 17 electrons:

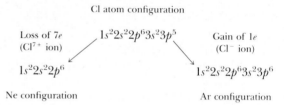

It is much easier to gain one electron than to lose seven. **Electron affinity** is the name given to the energy change that occurs when an electron is added to a gaseous atom or ion.

The halogens (Group VIIA) which need only one electron to reach a stable inert gas electron configuration, show the highest electron affinity or release of energy when an electron is added to a gaseous atom. For example, when 35 g of gaseous chlorine atoms gain electrons to become Cl^- ions, 84 kcal is liberated.

Nonmetals gain electrons to form anions with a noble-gas configuration, that is, to gain an octet (eight electrons). For an A Group nonmetal, the number of electrons gained is equal to the difference between eight and the group number (8 − group number). Chlorine is in Group VIIA, so it gains 8 − 7 = 1 electron. Table 6.2 offers other examples.

Electron gain always produces negatively charged nonmetal ions. A negative ion is called an **anion.** The charge is negative because the gain causes the electrons to outnumber the protons in the nucleus. The difference between the number 8 and the nonmetal anion's group number also tells you the magnitude of the negative charge on a nonmetal anion. A gain of one electron produces an ion with a − 1 charge, as exemplified by chlorine:

NOBLE-GAS CONFIGURATIONS ATTAINED BY ELECTRON GAIN BY NONMETALS   **TABLE 6.2**

Nonmetal	Group Number	8 − Group Number	Nonmetal Configuration	Noble-Gas Configuration Achieved	Number of Electrons Gained*
F	VIIA	8 − 7 = 1	$1s^2 2s^2 2p^5$	Ne: $1s^2 2s^2 2p^6$	1
O	VIA	8 − 6 = 2	$1s^2 2s^2 2p^4$	Ne: $1s^2 2s^2 2p^6$	2
P	VA	8 − 5 = 3	$1s^2 2s^2 2p^6 3s^2 3p^3$	Ar: $1s^2 2s^2 2p^6 3s^2 3p^6$	3

* Notice how the number of electrons gained always equals 8 − group number.

$$\left(\begin{array}{c} 17p+ \\ 18n \end{array} \right) \; 17e^- \; \xrightarrow[1e^-]{\text{add}} \; \left(\begin{array}{c} 17p+ \\ 18n \end{array} \right) \; 18e^-$$

Cl atom   Cl⁻ ion

SAMPLE EXERCISE

6.2

How many electrons must N gain to achieve a noble-gas configuration? With which noble gas will the ions be isoelectronic?

SOLUTION

Nitrogen is in Group VA, which tells us that it will gain 8 − 5 = 3 electrons and become N^{3-}. A gain of three electrons results in the Ne configuration.

$$N\,(7e^-) \qquad Ne\,(10e^-) \text{ and } N^{3-}\,(10e^-)$$
$$1s^2 2s^2 2p^3 \xrightarrow[3e^-]{\text{gain}} \qquad\qquad 1s^2 2s^2 2p^6$$

Problem 6.1

a. How many electrons must aluminum lose to achieve a noble-gas configuration? With which noble gas will the aluminum ions be isoelectronic?

b. How many electrons must sulfur gain to achieve a noble-gas configuration? With which noble gas will the sulfur ions be isoelectronic?

c. Identify the ions in parts *a* and *b* as cations or anions and give the magnitude of the ionic charge.

You will find it necessary to know the ionic charges which common metals and nonmetals acquire. Table 6.3 shows that with a periodic table in hand there is no need to memorize the charges of the A Group ions. Simply apply the relationships among group number, number of electrons gained or lost, and ionic charge acquired.

TABLE 6.3   COMMON IONS OF THE MAIN GROUP ELEMENTS*

IA	IIA		IIIA	IVA	VA	VIA	VIIA
H^+							
Li^+	Be^{2+}				N^{3-}	O^{2-}	F^-
Na^+	Mg^{2+}	Transition	Al^{3+}		P^{3-}	S^{2-}	Cl^-
K^+	Ca^{2+}	elements	Ga^{3+}				Br^-
Rb^+	Sr^{2+}		In^{3+}	(Sn^{4+})			I^-
Cs^+	Ba^{2+}		Tl^{3+}	(Pb^{4+})			

* Atoms such as carbon which do not commonly form ions are omitted from the table. Tin and lead
are shown in parentheses because they more commonly form +2 ions (see Table 6.4).

ELECTRON TRANSFER

6.5   Because metals tend to achieve a noble-gas configuration by
losing electrons and nonmetals tend to do this by gaining electrons,
there is a perfect setup for **electron transfer.** The combination of
metals and nonmetals leads to the ionic bond and ionic compound
formation through electron transfer—metals give electrons to nonmetals, thus forming ions. Ionic compounds are so named because
they are combinations of ions. In the formation of any ionic compound, the total number of electrons lost by the metal *must* equal the
total number gained by the nonmetal so that electrical neutrality is
maintained.

When the compound sodium chloride forms, each sodium atom
(Group IA) gives up *one* electron to one chlorine atom (Group VIIA).

$^{23}_{11}Na$ atom $^{23}_{11}Na^+$ ion

$\left(\begin{array}{c} 11p+ \\ 12n \end{array}\right)$ $1s^2 2s^2 2p^6 3s^1$ $\left(\begin{array}{c} 11p+ \\ 12n \end{array}\right)$ $1s^2 2s^2 2p^6$

electron
transfer

$\left(\begin{array}{c} 17p+ \\ 18n \end{array}\right)$ $1s^2 2s^2 2p^6 3s^2 3p^5$ $\left(\begin{array}{c} 17p+ \\ 18n \end{array}\right)$ $1s^2 2s^2 2p^6 3s^2 3p^6$

$^{35}_{17}Cl$ atom $^{35}_{17}Cl^-$ ion

In the formation of magnesium chloride, each magnesium atom (Group
IIA) loses two electrons. But one chlorine atom can gain only one
electron. Therefore, two chlorine atoms, each of which gains one
electron, are needed to receive the two electrons lost by the magnesium
atom. This transfer is seen pictorially in Figure 6.3. Notice that after

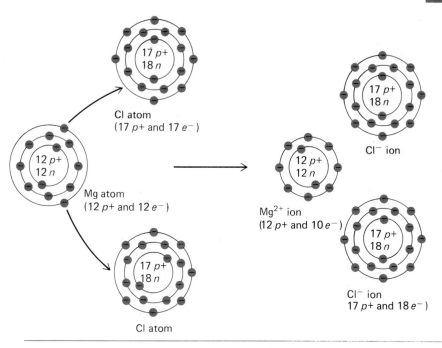

Cl atom
(17 p+ and 17 e⁻)

Mg atom
(12 p+ and 12 e⁻)

Cl atom

Mg^{2+} ion
(12 p+ and 10 e⁻)

Cl^- ion

Cl^- ion
17 p+ and 18 e⁻)

Magnesium atoms give up two electrons to two chlorine atoms. The Mg atom acquires a +2 charge; that is, it becomes the Mg^{2+} ion. Each Cl atoms has acquired an extra electron and so acquires a −1 charge; that is, it becomes the Cl^- ion. The attraction between cations and anions holds the compound $MgCl_2$ together.

FIGURE 6.3

the transfer each ion has a noble-gas configuration; Mg^{2+} has the same electron configuration as Ne, and Cl^- has the Ar electron configuration. It is the attraction between the cations and anions that is the **ionic bond** that holds ionic compounds together.

1. **Metals lose** electrons to attain a noble-gas configuration. For A Group metals, the number of electrons lost equals the group number.

2. **Nonmetals gain** electrons to attain a noble-gas configuration. For A Group nonmetals, the number of electrons gained equals the difference between eight and the group number.

3. Total number of electrons **lost** always **equals** total number of electrons **gained**.

General rules for ionic compound formation

SAMPLE EXERCISE

6.3

Describe how the ionic compound composed of potassium and oxygen forms.

SOLUTION

Follow the preceding general rules:

1. Potassium is in Group IA. Therefore, each atom tends to lose 1e⁻ and becomes the K^+ ion.

2. Oxygen is in Group VIA. Therefore, each atom tends to gain 8 − 6 = 2e^- and becomes the O^{2-} ion.

3. Electrons lost must equal electrons gained, so two K atoms must each lose 1e^- to satisfy one O atom.

This explains why the formula for the compound is K_2O. As mentioned in Section 1.11, subscripts indicate the relative numbers of atoms (or ions) in compounds. Formula writing will be discussed in Section 6.8.

Problem 6.2

Develop a pictorial representation similar to Figure 6.3 for the formation of ionic compounds made from:

a. Calcium and oxygen

b. Aluminum and chlorine

IONIC CHARGES

6.6 Because you now know that electron transfer occurs so that ions achieve noble-gas electron configurations, you could predict the correct combining ratio of metals and nonmetals in ionic compounds. Thus, for example, the analysis of compound formation in Sample Exercise 6.3 led to the formula K_2O because it was found that two K combine with one O. Another method of predicting combining ratios and writing correct formulas is based simply on the idea that compounds must be electrically neutral and does not require an analysis of electron transfer. Your ability to use this method depends on your knowledge of the ionic charges which the common elements acquire.

For the A Group elements, you learned in Sections 6.3 and 6.4 how to use the periodic table to predict the ionic charge acquired by an element. Table 6.3 summarized the ionic charges for the A groups.

Problem 6.3

What is the charge on the ion formed from phosphorus?

It is not possible to predict the magnitude of the positive charges on B Group metals from the periodic table. Fortunately, there are only a few you need to know, and these are listed in Table 6.4. Notice in this table that many so-called transition-metal elements (B Group) and some A Group elements form ions with more than one charge.

POLYATOMIC IONS

6.7 The ions we have discussed so far are *simple* ions, formed from one atom of one element. Simple ions are also called **monatomic ions,**

COMMON TRANSITION METAL IONS AND A GROUP IONS WITH MORE THAN ONE CHARGE* **TABLE 6.4**

+1 Charge	+2 Charge	+3 Charge	+4 Charge
Copper(I) Cu^+ (cuprous)	Copper(II) Cu^{2+} (cupric)	Chromium Cr^{3+}	Tin(IV) Sn^{4+} (stannic)
Silver Ag^+	Iron(II) Fe^{2+} (ferrous)	Iron(III) Fe^{3+} (ferric)	Lead(IV) Pb^{4+} (plumbic)
	Lead Pb^{2+} (plumbous)		
	Mercury(II) Hg^{2+} (mercuric)		
	Nickel Ni^{2+}		
	Tin(II) Sn^{2+} (stannous)		
	Zinc Zn^{2+}		

* Roman numerals are used to distinguish between different ions from the same metal. The roman numeral is the same as the charge on the ion. This is discussed in Section 7.3. The words in parentheses are archaic names used to distinguish metal ions.

because the prefix "mono-" means "one" (the terminal "o" in "mono-" is dropped before a vowel).

There are several **polyatomic ions,** so called because the prefix *poly-* means "many." Thus a polyatomic ion is a charged species made from many atoms. Table 6.5 lists the names and formulas of the more common polyatomic ions. Let us look at the formula for sulfate, SO_4^{2-}. This formula tells us that four oxygen atoms and one sulfur atom are combined in such a way that the group of atoms as a whole has a -2 charge. Subscripts always refer to number of atoms in a unit; when no subscript appears, a 1 is understood. The whole sulfate group is the ion; the component sulfur and oxygen atoms are inseparable. In Section 6.13 you will see that covalent bonding holds the sulfur and oxygens together in the sulfate ion. Polyatomic ions are groups of covalently bonded atoms that have a charge.

Many beginning chemistry students have difficulty recognizing polyatomic ions. Please study Table 6.5 and get acquainted with the idea that a total group of atoms has a charge associated with it. For example, one nitrogen and three oxygens have the charge -1 (NO_3^-), and the combination of one hydrogen, one phosphorus, and four oxygens has a -2 charge (HPO_4^{2-}).

TABLE 6.5   POLYATOMIC IONS

+1 Charge	−1 Charge	−2 Charge	−3 Charge
Ammonium, NH_4^+	Acetate, $C_2H_3O_2^-$	Carbonate, CO_3^{2-}	Phosphate, PO_4^{3-}
	Chlorate, ClO_3^-	Chromate, CrO_4^{2-}	
	Cyanide, CN^-	Dichromate, $Cr_2O_7^{2-}$	
	Dihydrogen phosphate, $H_2PO_4^-$	Hydrogen phosphate, HPO_4^{2-}	
	Hydrogen carbonate*, HCO_3^-	Sulfate, SO_4^{2-}	
	Hydrogen sulfate†, HSO_4^-	Sulfite, SO_3^{2-}	
	Hydroxide, OH^-		
	Nitrate, NO_3^-		
	Nitrite, NO_2^-		
	Perchlorate, ClO_4^-		
	Permanganate, MnO_4^-		

* Commonly called bicarbonate.
† Commonly called bisulfate.

FORMULAS FOR IONIC COMPOUNDS

6.8   Once you know the ionic charges, you can readily write correct formulas for ionic compounds using the principle that cations (+) and anions (−) combine in such a way that the magnitude of total positive charge just equals the magnitude of total negative charge and the compound is thereby electrically neutral. For example, the proper formula for the combination of the elements potassium and bromine in the sedative potassium bromide is KBr because a one-to-one combination of K^+ and Br^- results in electrical neutrality (see Table 6.3).

In writing a formula, we write the positive ion first and then the negative ion. Ionic charges are not shown in the formula. Even though the compound KBr is made up of K^+ and Br^- ions, the formula does not show the charges explicitly. We indicate with subscripts, after each ion, the number of that ion present in the compound. For example, in Sample Exercise 6.3, two potassium ions were indicated in the formula, K_2O. If no subscript is written, then it is understood to be 1, as in KBr or for oxygen in K_2O.

SAMPLE EXERCISE 6.4

What is the formula for the compound formed from Ca^{2+} and Br^-?

SOLUTION

The combination of one Ca^{2+} and one Br^- ion would leave the compound with a +1 charge: $(+2) + (-1) = +1$. Instead, one Ca^{2+} ion and two Br^- ions must combine: $(+2) + 2(-1) = 0$. The formula thus is $CaBr_2$. Note the use of subscript 2 to indicate two bromide ions. $CaBr_2$ is made up of one Ca^{2+} and two Br^- ions.

When polyatomic ions are involved, the rules work very much the same way. What is the formula for the compound formed from Mg^{2+} and OH^-? In order to balance the charges, we will need one magnesium ion combining with two hydroxide ions. But where will we write our subscript to indicate two hydroxide ions clearly? The formula is written $Mg(OH)_2$. Whenever we have more than one of a polyatomic ion, we surround that ion with parentheses and indicate with a subscript after the parentheses the number of polyatomic ions that are present. The compound $Mg(OH)_2$ is made up of one Mg^{2+} and two OH^- ions.

SAMPLE EXERCISE

6.5

What is the formula for the compound formed from Na^+ and OH^-?

SOLUTION

To balance the charges, we simply combine one Na^+ and one OH^-. The formula will then be $NaOH$. We do *not* have to enclose the hydroxide ion within parentheses because there is only one of this polyatomic ion present in $NaOH$. However, you must be able to recognize the unit OH^- in the absence of parentheses.

SAMPLE EXERCISE

6.6

What is the formula for the compound formed from Al^{3+} and SO_4^{2-}?

SOLUTION

How are we to balance the charges $+3$ and -2? Begin by remembering that the common denominator of 3 and 2 is 6. If we take two $+3$ and three -2 ions, the total positive charge is $+6$ and the total negative is -6; the total positive and negative charges are now balanced. Therefore, the correct formula is $Al_2(SO_4)_3$. This formula tells us that the ratio of ions in this ionic compound is two aluminum ions to three sulfate ions.

Problem 6.4

Write formulas for the ionic compounds formed from:

a. Na^+ and I^-

b. NH_4^+ and S^{2-}

c. Al^{3+} and NO_3^-

d. Al^{3+} and CO_3^{2-}

e. Al^{3+} and PO_4^{3-}

f. NH_4^+ and PO_4^{3-}

Problem 6.5

What is the ratio of cations to anions present in each of the following ionic compounds?

a. CaO

b. MgF_2

c. Na_3PO_4

d. $KC_2H_3O_2$

e. $Ca_3(PO_4)_2$

f. $KClO_3$

Suppose we ask you to write a formula for the combination of calcium and sulfur. This time we have not indicated the charges as in the previous examples, so you must first ask yourself what ions these elements form and then proceed as usual. Calcium is in Group IIA; therefore, it forms Ca^{2+}. Sulfur is in Group VIA; therefore, it is S^{2-}. The correct formula is CaS.

SAMPLE EXERCISE

6.7

Write the correct formula for the compound formed by combining zinc and bromine.

SOLUTION

The compound formed from zinc and bromine must be an ionic compound since zinc is a metal and bromine a nonmetal. Remember or look up the charge of the zinc ion, which is Zn^{2+}. Bromine is in Group VIIA; therefore, its charge is -1. To form a neutral compound, the correct combination is one cation and two anions. Thus we get $ZnBr_2$.

If an element can form an ion with more than one charge (Table 6.4), we must be given additional information as to which ion is actually forming in the particular problem at hand.

SAMPLE EXERCISE

6.8

Write a correct formula for the compound formed from iron and chlorine. Assume that iron will form the $+3$ ion.

SOLUTION

We are told that in this problem iron forms the Fe^{3+} ion, and we already know that in all ionic compounds chlorine forms the Cl^- ion. The correct combination to form a neutral compound is one cation and three anions. Thus the formula is $FeCl_3$.

General rules for writing formulas for ionic compounds

1. Determine the ionic charges of the elements which are combining.

2. Choose a combination of the ions such that the sum of the ionic charges equals zero.

3. Write the symbol for the positive ion to the left and that for the negative ion to the right. Do not include the charges.

4. Use subscripts after each ion to indicate the numbers present in the compound. Polyatomic ions require parentheses where there is more than one. The number 1 is understood.

Problem 6.6

Write a correct formula for the compounds formed from the following:

a. Magnesium and sulfur

b. Chromium and chlorine

c. Calcium and nitrogen

d. Lead (assume it forms a +4 ion) and oxygen

There is a shortcut method for writing ionic formulas which you should feel free to use. However, it is a good idea to always remember that the basis for ion combination is the formation of a neutral (uncharged) compound. Let us see how this method works for the compound formed from Al^{3+} and S^{2-}.

Shortcut for writing ionic formulas

Step 1 Write the two ions next to each other with their charges as superscripts:

$$Al^{3+} \qquad S^{2-}$$

Step 2 Then bring the 2 of the sulfur down as the subscript for Al and the 3 from the aluminum as the subscript for the S; that is, "crisscross" the charges as subscripts without signs. Always remember to omit the signs in the subscripts.

$$Al^{3+}\!\!\!\times\!\!\!S^{2-} \qquad \text{gives} \qquad Al_2S_3$$

Step 3 Now check to see that the subscripts are the smallest possible whole-number ratio. If the subscript ratio can be converted into a smaller whole-number ratio, this must be done. In this example the subscript ratio $2:3$ cannot be converted into a smaller whole-number ratio.

If the first two rules are applied to the combination of Ca^{2+} and O^{2-},

$$Ca^{2+}\!\!\!\times\!\!\!O^{2-} \qquad \text{gives} \qquad Ca_2O_2$$

But the subscript ratio $2:2$ can be converted into the smaller whole-number ratio $1:1$ by dividing both subscripts by 2. This gives the correct formula, CaO. Similarly, for Pb^{4+} combining with O^{2-},

$$Pb^{4+}\!\!\!\times\!\!\!O^{2-} \qquad \text{gives} \qquad Pb_2O_4$$

which should be written PbO_2.

Do not forget the parentheses around polyatomic ions. For the combination of Ba^{2+} and NO_3^-,

$$Ba^{2+} \diagdown\!\!\!\!\!\diagup NO_3^- \qquad \text{gives} \qquad Ba(NO_3)_2$$

The subscript ratio of $1:2$ cannot be converted into a smaller whole-number ratio.

Problem 6.7

Redo Problem 6.4 using the shortcut method for formula writing.

THE NATURE OF THE IONIC BOND

6.9 What is called the **ionic bond** is the *electrostatic attraction between cations* $(+)$ *and anions* $(-)$ in an ionic compound. We know from Chapter 5 that bringing together plus and minus charges results in a lower energy state. It is the fact that the cation-anion attraction results in a lowering of energy that enables the ionization process to occur.

Remember that cation formation requires an input of energy. For example, the ionization of $K \rightarrow K^+$ requires 100 kcal/39 g. If we consider forming KCl from 39 g of K and 35 g of Cl[1], then as Cl^- forms, 84 kcal of energy is released. But that means we are 16 kcal short; that is, we need 100 kcal to form K^+, but can get only 84 kcal as Cl^- forms. However, the attractive force between the formed K^+ and Cl^- is 165 kcal, which is more than enough energy to enable the process to occur.

To form an ionic compound from the atoms in the gas phase, there must be a balance among the ionization energy, the electron affinity, and the energy released from the net electrostatic force between ions in the solid state. One expects ionic compounds to form when the ionization energy is relatively low (energy input), the electron affinity relatively high (energy release), and the net attractive force relatively large.

In the formation of magnesium chloride, more energy is required to remove two electrons from the magnesium atom than would be required to remove one electron from potassium. But this energy will be more than returned in the increased net attractive force in $MgCl_2$, which arises from the higher positive charge on the Mg^{2+} ion.

The strong ionic bond, i.e., the attraction between cations and anions, gives ionic compounds their characteristic properties. For example, ionic compounds typically are high-melting solids. In an ionic

---

[1] You will see in Chapter 10 that these are the required amounts to have no leftover K or Cl.

compound the ions are arranged systematically in an alternating cation-anion pattern (Figure 6.4). This gives the material the definite shape characteristics of a solid. Because the cation-anion attractions are so strong, a great deal of energy must be supplied to move them apart. This means that to change a solid to a liquid, a great deal of heat energy must be supplied in order to pull the ions out of their rigid shape and into the shapeless liquid state. Thus ionic compounds have very high melting temperatures.

Each cation in an ionic compound is attracted to all the anions and each anion is attracted to all the cations. That is, no cation has a fixed anion as a separate independent partner. The smallest grouping of ions corresponding to the correct fixed ratio of ions in the compound is called a **formula unit.**

In later chapters some of the other properties of ionic compounds that arise from the nature of the ionic bond will be discussed. Molecular compounds have very different properties because of the nature of the covalent bond.

WHY ARE THERE TWO TYPES OF COMPOUNDS?

6.10   So far in this chapter we have explored the nature of the ionic bond and the formulas of ionic compounds. The combination of metals and nonmetals leads to the ionic bond and ionic compound formation through electron transfer. Combining two or more nonmetals will not lead to electron transfer and the formation of an ionic bond

FIGURE 6.4

(a) Schematic representation of a NaCl crystal. The alternating positive and negative charges maximize positive-negative attraction and minimize positive-positive repulsion. The result is a definite crystalline shape. Also, it is very difficult to pull the ions apart. (b) Actual photograph of a sodium chloride crystal. Note the regular cubic appearance. Magnification = 100×. (*Courtesy of G. W. Luther, Kean College.*)

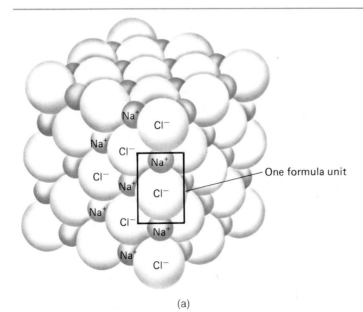

One formula unit

(a)

(b)

because all nonmetals tend to gain electrons. Yet nonmetals do bond to one another. Nonmetals combine to form molecular compounds. Nonmetallic atoms become linked together by forces known as covalent bonds in units called **molecules.**

The atoms in molecules are held together by shared pairs of electrons. A shared pair of electrons is called a **covalent bond.** As in the case of the formation of ionic bonds, covalent bonds form so that atoms can achieve a noble-gas configuration.

Thus compound formation in general occurs in order that atoms can achieve lower-energy, noble-gas configurations. There are two types of compounds because there are two ways to achieve the noble-gas configuration:

1. **Transfer of electrons** from metals to nonmetals leads to ionic bonds and ionic compounds.

2. **Sharing of electrons** between nonmetals leads to molecules that are held together by covalent bonds. Molecular compounds are also called covalent compounds.

DIATOMIC MOLECULES

6.11   You have known since Section 4.13 that certain elements (H_2, O_2, N_2, halogens) exist as diatomic molecules. These elements exist as diatomic molecules rather than individual atoms because the atoms share their electrons in order to achieve a noble-gas configuration.

For example, consider individual chlorine atoms (Group VIIA), for which we would write the Lewis electron dot symbols (Section 5.9):

$$: \overset{..}{\underset{..}{Cl}} \cdot \qquad \cdot \overset{..}{\underset{..}{Cl}} :$$

Seven dots represent the seven valence electrons in the third main level of $_{17}Cl$, $1s^2 2s^2 2p^6 3s^2 3p^5$. Each atom can acquire eight valence electrons (and consequently greater stability and lower energy) if the two come together to share a pair of electrons. As the two atoms come close together, each atom contributes one unpaired electron to form a shared pair between the atoms. The shared pair belongs equally to both atoms, so each now has a complete octet for its valence electrons, i.e., a noble-gas configuration.

Shared pair of electrons, a covalent bond

Each Cl has $8e^-$ around it.

The shared pair can also be indicated by a dash (—).

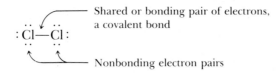

Shared or bonding pair of electrons, a covalent bond

Nonbonding electron pairs

The valence electrons not involved in bonding are called **nonbonding electrons.** They are also called **lone pairs.**

All the other halogens (Group VIIA) form covalent bonds in exactly the same way:

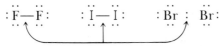

Remember, the dash or shared pair means exactly the same thing.

H_2 molecules are also held together by a shared pair. The two H atoms come together to form H:H, or H—H. Each H atom now has the helium duet, two electrons.

When we try to account for the bonding in O_2 with a single pair of shared electrons, the oxygen atoms do not obtain a complete octet. Oxygen is in Group VIA:

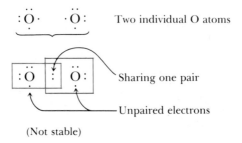

Two individual O atoms

Sharing one pair

Unpaired electrons

(Not stable)

Each O has only seven electrons (count within the boxes around the atoms), but if the oxygens share their remaining unpaired electrons, octets will be attained.[1]

   or $: \ddot{O} = \ddot{O} :$

_____

[1] This electronic structure for oxygen does not account for some of its observed properties and consequently represents a failure of Lewis theory to predict structure. However, it offers a simple example of the concept of a double bond.

Each O has eight electrons. Two shared pairs of electrons are called a double covalent bond, or simply a **double bond.**

In order for each nitrogen in N_2 to have a noble-gas configuration, three shared pairs are required. This is known as a **triple bond.**

$$: N \boxed{\; ::: \;} N : \qquad \text{or} \qquad : N \equiv N :$$

There are no examples of quadruple bonds.

THE NATURE OF THE COVALENT BOND

6.12 You have seen that it is the electrostatic attraction between cations and anions that lowers energy in the formation of ionic bonds. The formation of covalent bonds leads to a lowering of energy because of the attraction between the electrons ($-$) of one atom and the nucleus ($+$) of another atom. Let us see how energy is lowered in the H_2 molecule (relative to two separate H atoms) through covalent bond formation.

In the two separate H atoms the lone valence electron of each isolated atom is attracted to its respective nucleus.

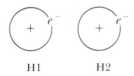

As the atoms come closer together, there are two new attractive forces (electron$_1$ to nucleus$_2$, electron$_2$ to nucleus$_1$) and two repulsive forces (between electron$_1$ and electron$_2$, between nucleus$_1$ and nucleus$_2$). Up to a certain distance apart, called the **bond length,** the new attractive forces are larger than the repulsive forces. This leads to a lowering of energy for the H_2 molecule, i.e., two H atoms held together at a fixed distance compared with two separate H atoms (see Figure 6.5).

The use of a dash (—) to represent the bond or shared pair of electrons better portrays the idea of bond length. The H atoms are held together at some optimum fixed distance:

Optimum bond length

H—H

Lowest energy

Greater repulsion at shorter distances   Less attraction at longer distances

H–H   H——H

Higher energy   Higher energy

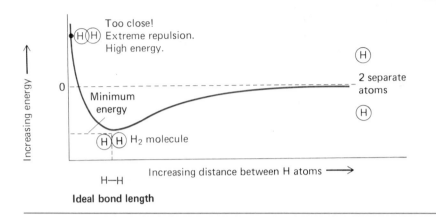

Minimum energy is achieved when the distance between H atoms is the bond length. Separate atoms have higher energy. Atoms closer together have higher energy.

FIGURE 6.5

The atoms cannot move apart to a greater distance than the bond length because this would raise the energy by reducing the attractive forces. They cannot move closer than the bond length because this would also raise the energy by increasing the repulsive forces.

Experimentally the stability of the covalent bond in hydrogen is demonstrated by the fact that it takes 104 kcal to convert 2 g of hydrogen molecules into hydrogen atoms.

$$H_2(g) + 104 \text{ kcal} \rightarrow 2 \text{ H}(g)$$

This energy is known as the **bond-dissociation energy.** The larger the bond-dissociation energy, the stronger the bond. Table 6.6 lists bond-dissociation energies for the diatomic molecules. The numbers are measured in kilocalories per gram amount, which would contain the same number of molecules as there are hydrogen molecules in 2 g of hydrogen (Section 8.5). That is, each number represents breaking the same number of bonds. Clearly, covalent bonds are not all of equal strength. In general, triple bonds are stronger (N_2) than double bonds (O_2), which in turn are stronger than single bonds.

BOND-DISSOCIATION ENERGIES FOR DIATOMIC MOLECULES   **TABLE 6.6**

$X_2(g) \rightarrow 2 X(g)$	Bond-Dissociation Energy, kcal
N_2	226
O_2	118
H_2	104
Cl_2	58
Br_2	46
F_2	38
I_2	36

Problem 6.8

a. Which of the halogens has the strongest bond?

b. Which of the halogens has the weakest bond?

c. What kinds of substances are formed when the bond in a diatomic molecule is broken by the input of the bond-dissociation energy?

LEWIS ELECTRON DOT FORMULAS

6.13 It is important to know *why* covalent bonds form. It is even more necessary to be able to tell *how* electrons are shared in a molecule for which the molecular formula is available. The sharing pattern is shown by writing Lewis electron dot formulas, just as we have done previously for the diatomic molecules.

Another simple example of how to write a Lewis formula can be seen by considering the bonding in HCl. First write the Lewis dot symbols for the atoms:

$$\text{H} \cdot \qquad \cdot \ddot{\ddot{\text{Cl}}} :$$

Group IA   Group VIIA

Now see how the unpaired electrons can be shared so that both H and Cl attain a noble-gas configuration. This is achieved easily for this small molecule:

Shared pair, or covalent bond

$$\text{H} \odot \ddot{\ddot{\text{Cl}}} : \qquad \text{or} \qquad \text{H} \ominus \ddot{\ddot{\text{Cl}}} :$$

Whether we write the formula horizontally, vertically, or at an angle makes no difference:

$$\text{H}-\ddot{\ddot{\text{Cl}}} : \qquad \begin{array}{c} \text{H} \\ | \\ :\ddot{\text{Cl}}: \end{array} \qquad \begin{array}{c} \nearrow \ddot{\ddot{\text{Cl}}}: \\ \text{H} \end{array}$$

H has a duet, and Cl has an octet.

Lewis electron dot formulas are also called **Lewis structures** or **structural formulas** because they tell you how the molecule is *constructed* or built.

Problem 6.9

Write the Lewis electron dot structures for HBr, HF, and HI.

By "playing around" with the unpaired electrons of the individual atoms of H_2O, you probably could figure out that the Lewis structure is H—Ö—H.

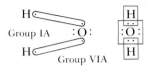

For more complicated molecules, however, particularly those with many atoms, it would be helpful to have a set of guidelines for writing Lewis structures instead of having to just "play around" or flounder.

The guidelines set down here allow you to determine the central atom in the molecule and the number of shared pairs (bonds) that must be present in the Lewis structure. It is easy to determine the difference between the number of electrons *necessary* for each atom to achieve a noble-gas configuration and the number of valence electrons *actually present* in the molecule. This difference reveals how much sharing must be done. After making this determination, you will still need to juggle electrons somewhat, but the job should be greatly simplified. Read through the guidelines carefully and see how they are applied in the Sample Exercises that follow them.

1. Terminology

Guidelines for writing Lewis structures

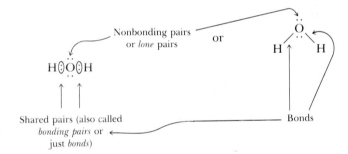

2. Determination of the number of bonding pairs (symbolized b = bonding pairs)

a. Determine the total number of electrons that a molecule would have to have for each hydrogen, H, to have two electrons (each boron, B, to have six electrons) and every other atom to have eight electrons. Give this total the symbol t.

t = 2 (number of H in molecule)
    + 6 (number of B) + 8 (number of all other atoms)

For example, for H_2O

$t = 2(2) + 6(0) + 8(1) = 12$

b. Add up the number of valence electrons for all atoms in the molecule and give this the symbol v. For example, for H_2O

$$\begin{array}{rl} 2\ H \times 1 \text{ valence } e^- \text{ each} & = 2\ e^- \\ 1\ O \times 6 \text{ valence } e^- & = 6\ e^- \\ \hline & v = 8 \end{array}$$

For polyatomic ions, a negative charge indicates extra valence electrons and a positive charge indicates fewer valence electrons. For example, for OH^-

$$\begin{array}{rl} 1\ H \times 1 \text{ valence } e^- & = 1\ e^- \\ 1\ O \times 6 \text{ valence } e^- & = 6\ e^- \\ {}^- \text{ means 1 extra } e^- & = 1\ e^- \\ \hline & v = 8 \end{array}$$

For NH_4^+

$$\begin{array}{rl} 4\ H \times 1 & = 4 \\ 1\ N \times 5 & = 5 \\ {}^+ \text{ means 1 less electron} & = -1 \\ \hline v = & 8 \end{array}$$

c. Bonding pairs, $b = (t - v)/2$. For example, H_2O

From part a   From part b

$$b = \frac{12 - 8}{2} = \frac{4}{2} = 2$$

Two bonds predicted.

3. Determination of the central atom in the molecule or ion.
   a. For covalent compounds containing only two elements and for polyatomic ions, the central atom is the one that appears only once in the formula. (H is *never* a central atom.)

Formula	Central Atom
CH_4	C
SO_3	S
PCl_3	P
SF_4	S
NH_3	N
PO_4^{3-}	P
SO_4^{2-}	S
NH_4^+	N
CO_3^{2-}	C
H_2O	O
HSO_4^-	S

b. For covalent compounds containing H, O, and a third element, the atom other than H or O is the central atom.

c. Carbon is the central atom in any carbon-containing compound.

Formula	Central Atom
H_2SO_4	S
HNO_3	N
$HClO_3$	Cl
HCN	C
$COCl_2$	C

4. Write down the central atom and arrange the other atoms around the central one using up the number of bonded pairs you calculated in step 2. Not all atoms will necessarily be connected to the central atom. In compounds containing H, O, and a third element, distribute the oxygens around the central atom (i.e., avoid O—O bonds), and for most cases connect hydrogens to the oxygens.

Continuing with H_2O as our example, O is central and we must have two bonds—thus H:O:H.

5. Fill in the rest of the structure with nonbonding pairs until each hydrogen has two electrons, each boron has six electrons, and every other atom has eight electrons. For H_2O, the O requires two nonbonding pairs to complete its octet, H:Ö:H.

6. Recheck to be sure that the total number of electrons shown in your structure equals v. The structure we have written for H_2O shows eight electrons, and $v = 8$.

SAMPLE EXERCISE

6.9

Write a Lewis electron dot structure for NH_3.

SOLUTION

Step 1 Determine t, the total number of electrons needed for noble-gas configuration.

$$
\begin{array}{lll}
3\ H & 3 \times 2 & = \ \ 6 \text{ electrons} \\
1\ N & 1 \times 8 & = \ \ 8 \text{ electrons} \\
\hline
& t & = 14 \text{ electrons}
\end{array}
$$

Step 2 Determine v, the number of valence electrons.

$$
\begin{array}{lll}
3\ H & 3 \times 1 & = 3 \text{ electrons} \\
1\ N & 1 \times 5 & = 5 \text{ electrons} \\
\hline
& v & = 8 \text{ electrons}
\end{array}
$$

Step 3 Determine b, the number of bonding pairs.

$$b = \frac{t - v}{2}$$

$$b = \frac{14 \text{ electrons} - 8 \text{ electrons}}{2} = 3 \text{ bonding pairs}$$

Step 4 Determine the central atom. In NH_3, the central atom is N (Guideline 3a).

Step 5 Write down the central atom, arrange the other atoms, and fill in the bonding pairs.

$$H : N : H$$
$$H$$

Step 6 Fill in the nonbonded pairs to complete the nitrogen octet.

$$H : \overset{..}{\underset{..}{N}} : H$$
$$H$$

Step 7 Check that the number of electrons shown equals v, which in this case is 8.

SAMPLE EXERCISE

6.10

Write a Lewis dot structure for HCN.

SOLUTION

Step 1 Determine t.

1 H	1×2	=	2 electrons
1 C	1×8	=	8 electrons
1 N	1×8	=	8 electrons
		$t =$	18 electrons

Step 2 Determine v.

1 H	1×1	=	1 electron
1 C	1×4	=	4 electrons
1 N	1×5	=	5 electrons
		$v =$	10 electrons

Step 3 Determine $b = (t - v)/2$.

$$b = \frac{18 \text{ electrons} - 10 \text{ electrons}}{2} = 4 \text{ bonding pairs}$$

Step 4 The central atom in HCN is C (Guideline 3c).

Step 5 Write down H C N. We must use four bonding pairs. We should recognize that hydrogen can participate in only one bonding pair because one pair gives H its noble-gas configuration of two electrons. Therefore, the other three bonding pairs must be bonding from C to N.

$$H:C:::N$$

Step 6   N needs a nonbonding pair to complete its octet.

$$H:C:::N:$$

Step 7   Ten electrons are showing, which equals v.

SAMPLE EXERCISE

6.11

Write a Lewis electron dot structure for H_2SO_4.

SOLUTION

Step 1   Determine t.

$$
\begin{array}{llll}
2\,H & 2 \times 2 & = & 4 \text{ electrons} \\
1\,S & 1 \times 8 & = & 8 \text{ electrons} \\
4\,O & 4 \times 8 & = & 32 \text{ electrons} \\
\hline
 & & t = & 44 \text{ electrons}
\end{array}
$$

Step 2   Determine v.

$$
\begin{array}{llll}
2\,H & 2 \times 1 & = & 2 \text{ electrons} \\
1\,S & 1 \times 6 & = & 6 \text{ electrons} \\
4\,O & 4 \times 6 & = & 24 \text{ electrons} \\
\hline
 & & v = & 32 \text{ electrons}
\end{array}
$$

Step 3 $b = (t - v)/2$.

$$b = \frac{44 \text{ electrons} - 32 \text{ electrons}}{2} = 6 \text{ bonding pairs}$$

Step 4   The central atom is S (Guideline 3b).

Step 5   Write down S, arrange the O's around it, and attach H's to O's (see Guideline 4).

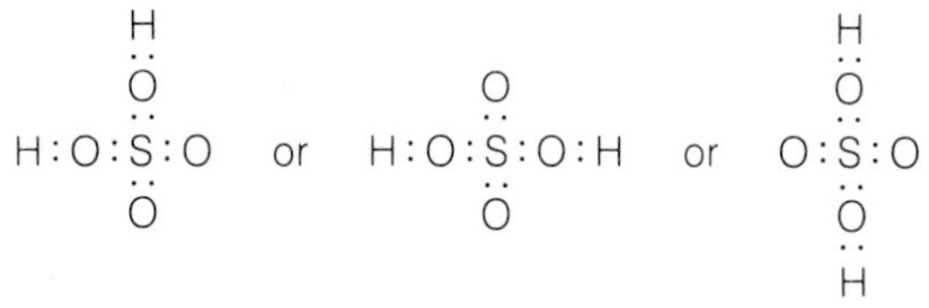

These three forms are equivalent.

Step 6   The oxygens require nonbonding pairs to complete their octets.

This is more easily and more clearly written as

$$
\begin{array}{c}
:\ddot{O}: \\
| \\
H-\ddot{O}-S-\ddot{O}-H \\
| \\
:\ddot{O}:
\end{array}
$$

Step 7 Thirty-two electrons are showing, which equals *v*.

SAMPLE EXERCISE

6.12

Write a Lewis dot structure for CO_3^{2-}.

SOLUTION

Step 1 Determine *t*.

$$
\begin{array}{llll}
1\,C & 1 \times 8 & = & 8 \text{ electrons} \\
3\,O & 3 \times 8 & = & 24 \text{ electrons} \\
\hline
& & t = & 32 \text{ electrons}
\end{array}
$$

Step 2 Determine *v*.

$$
\begin{array}{lll}
1\,C & 1 \times 4 & = 4 \text{ electrons} \\
3\,O & 3 \times 6 & = 18 \text{ electrons} \\
2^- \text{ charge indicates} & & \\
2 \text{ extra electrons} & = 2 \text{ electrons} \\
\hline
& v = 24 \text{ electrons}
\end{array}
$$

Step 3 $b = (t - v)/2$.

$$
b = \frac{32 \text{ electrons} - 24 \text{ electrons}}{2} = 4 \text{ bonding pairs}
$$

Step 4 Central atom is C (Guideline 3c).

Step 5 Write down the central C, arrange the O's around it, and fill in four bonding pairs. There must be two bonding pairs between C and one of the O's, i.e., there must be a double bond.

$$
\begin{array}{ccc}
\overset{\displaystyle O}{\underset{\displaystyle O::C:O}{}} & \text{or} & \overset{\displaystyle O}{\underset{\displaystyle O:C:O}{\ddot{}}} & \text{or} & \overset{\displaystyle O}{\underset{\displaystyle O:C::O}{}}
\end{array}
$$

All three forms are equivalent.

Step 6 The C is now surrounded by eight electrons. We fill in nonbonding pairs so that each oxygen will have an octet. Because the substance in this example is an ion, we write it within brackets with the charge as a superscript.

$$
\left[
\begin{array}{c}
:\ddot{O}: \\
| \\
\ddot{O}=C-\ddot{O}:
\end{array}
\right]^{2-}
$$

Step 7 Twenty-four electrons are showing, which equals *v*.

The negative ion carbonate can bond to cations by an ionic bond. How-ever, the bonding within the ion between the oxygens and the carbon is co-valent. The bonding *within* all polyatomic ions is *covalent*.

Problem 6.10

Write a Lewis electron dot structure for each of the following:

a. CH_4

b. H_3PO_4

c. SO_4^{2-}

d. CH_2O

COORDINATE COVALENT BONDS

6.14 When we first wrote Lewis structures for diatomic molecules, we did so by pairing up unpaired electrons. For example,

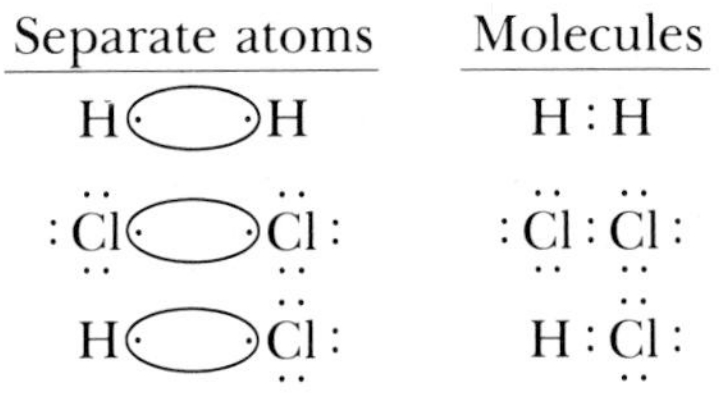

Separate atoms	Molecules
H⟨⟩H	H : H
: Cl⟨⟩Cl :	: Cl : Cl :
H⟨⟩Cl :	H : Cl :

When using the guidelines described in Section 6.13, one does not look for unpaired electrons in writing Lewis structures, but in fact *most covalent bonds are formed by each bonded atom contributing one electron to the pair*. This can be seen in the preceding examples concerning H_2, Cl_2, and HCl. Looking back at NH_3, for which the Lewis structure was determined in Sample Exercise 6.9, we see that the N—H bonds in NH_3 can easily be envisioned as arising from the contribution of one electron from N and one from H.

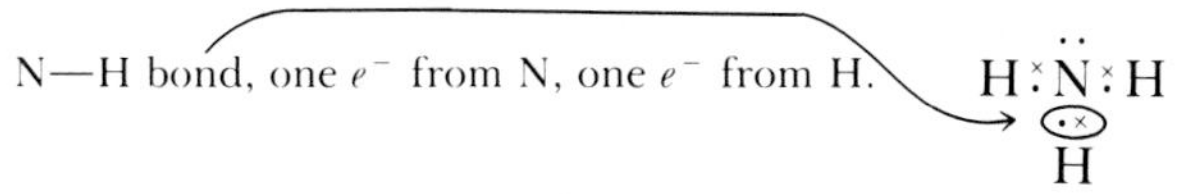

N—H bond, one e^- from N, one e^- from H.

H ⦂ N ⦂ H
H

N has five valence electrons, represented here as ·
Each H has one valence electron, represented here as x.

It is also possible to form a bond in which one of the bonded atoms furnishes both electrons. A bond where one atom furnishes both electrons is called a **coordinate covalent bond.** For example, the ammonium ion NH_4^+ can be viewed as being formed from a coordinate covalent bond in which the two nonbonding electrons on NH_3 form a bond with an H^+ ion which has no electrons to contribute.

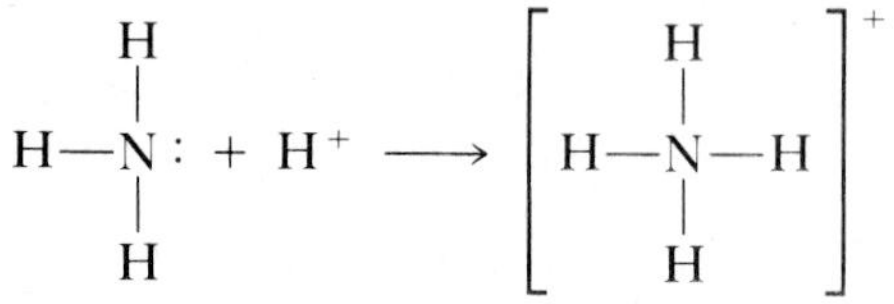

$$
\begin{array}{c}
\text{H} \\
| \\
\text{H—N:} \; + \; \text{H}^+ \\
| \\
\text{H}
\end{array}
\longrightarrow
\left[
\begin{array}{c}
\text{H} \\
| \\
\text{H—N—H} \\
| \\
\text{H}
\end{array}
\right]^+
$$

The four N—H bonds in NH_4^+ are identical. Once formed, a coordinate covalent bond has the same properties as any other covalent bond.

SAMPLE EXERCISE

6.13

Write a Lewis structure for the hydronium ion H_3O^+.

SOLUTION

Step 1 Determine t.

$$
\begin{array}{llll}
3\,\text{H} & 3 \times 2 & = & 6\,e^- \\
1\,\text{O} & 1 \times 8 & = & 8\,e^- \\
\hline
 & & t = & 14\,e^-
\end{array}
$$

Step 2 Determine v.

$$
\begin{array}{llll}
3\,\text{H} & 3 \times 1 & = & 3\,e^- \\
1\,\text{O} & 1 \times 6 & = & 6\,e^- \\
\text{+ charge indicates} & & & \\
\text{1 missing electron} & & = & -1\,e^- \\
\hline
 & & v = & 8\,e^-
\end{array}
$$

Step 3

$$
b = \frac{t - v}{2} = \frac{14 - 8}{2} = 3 \text{ bonds}
$$

Step 4 The central atom is O (Guideline 3a).

Step 5 Write down the central O, surrounded by three bonds to three H's.

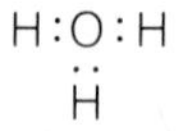

$$
\begin{array}{c}
\text{H:O:H} \\
\cdot\cdot \\
\text{H}
\end{array}
$$

Step 6 Fill in one nonbonding pair to complete the oxygen octet.

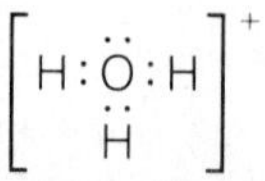

$$
\left[
\begin{array}{c}
\cdot\cdot \\
\text{H:}\ddot{\text{O}}\text{:H} \\
\cdot\cdot \\
\text{H}
\end{array}
\right]^+
$$

Step 7   There are eight electrons shown, and $v = 8$.

Problem 6.11

Show how the ion H_3O^+ can be formed by coordinate covalent bond formation between H_2O and H^+.

A notable example of a coordinate covalent bond is that between oxygen and the iron in the molecule hemoglobin formed in red blood cells. It is this bond that allows hemoglobin to carry oxygen to the cells of the body. Figure 6.6 shows a structure known as **heme,** the iron-containing portion of hemoglobin.

Both bonding electrons for joining oxygen to iron are supplied by oxygen; hence, a coordinate covalent bond is formed.

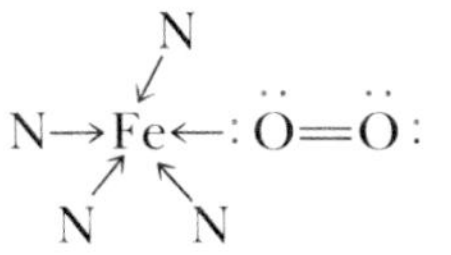

$$N \rightarrow \overset{\overset{\displaystyle N}{\downarrow}}{\underset{\overset{\displaystyle N \quad N}{\nearrow \quad \nwarrow}}{Fe}} \leftarrow : \overset{..}{O} = \overset{..}{O} : \qquad \text{(No geometry is implied.)}$$

Iron is capable of forming such bonds with other electron pairs. For example, a gas found in automobile exhaust, carbon monoxide, can bond to iron in hemoglobin through its lone pairs $(: C \equiv O :)$. Because this bond is stronger than the iron-oxygen bond, oxygen is prevented from linking up with iron and is not carried to the body's cells; this is what makes CO toxic.

ELECTRONEGATIVITY AND POLARITY

6.15   So far, we have discussed the electron pair in a covalent bond as if it were shared equally between the two atoms. This is definitely

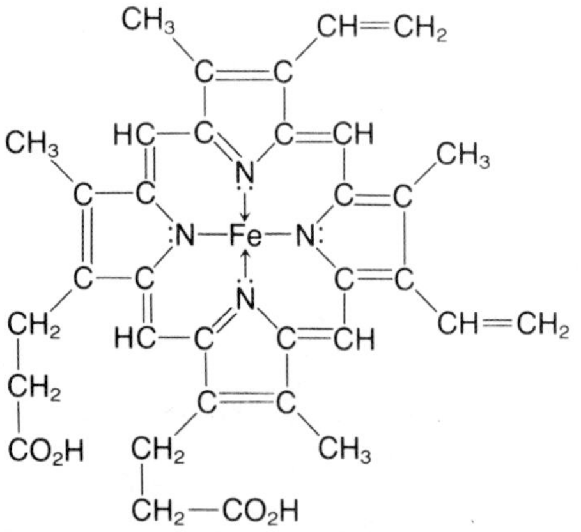

FIGURE 6.6

Heme, an important part of the hemoglobin molecule. The iron ion (Fe^{2+}) is bonded to four nitrogen atoms in a complex molecule known as the protoporphyrin IX. Each five-membered, nitrogen-containing ring is called a porphyrin ring. The Fe^{2+} ion is capable of accepting a pair of electrons from an oxygen molecule to form a coordinate covalent bond.

TABLE 6.7　　ELECTRONEGATIVITIES OF SOME REPRESENTATIVE ELEMENTS

IA	IIA	IIIA	IVA	VA	VIA	VIIA
2.1						
Li	Be	B	C	N	O	F
1.0	1.5	2.0	2.5	3.0	3.5	4.0
Na	Mg	Al	Si	P	S	Cl
0.9	1.2	1.5	1.8	2.1	2.5	3.0
K	Ca	Ga	Ge	As	Se	Br
0.8	1.0	1.6	1.8	2.0	2.4	2.8
Rb	Sr	In	Sn	Sb	Te	I
0.8	0.9	1.7	1.8	1.9	2.1	2.5
Cs	Ba	Tl	Pb	Bi	Po	At
0.7	0.9	1.8	1.8	1.9	2.0	2.2
Fr	Ra					
0.7	0.9					

the case when the two bonded atoms are identical, as with H_2 or Cl_2. However, in most bonds one bonded atom attracts the bonding electrons more strongly. For example, in H—Cl the chlorine has a stronger attraction for electrons than does the hydrogen.

The attractive force that an atom exerts on shared electrons in a chemical bond is known as its **electronegativity.** A scale of relative electronegativities has been devised by the Nobel Laureate Linus Pauling (Nobel prizes: Chemistry 1954, Peace 1962). Values for the representative elements are given in Table 6.7. When you examine this table, you will see that electronegativity is a periodic property. In general it increases from left to right across a period and up a group from bottom to top. This means that nonmetals have high electronegativities and metals have low electronegativities. Because metals have low electronegativities, they are said to be *electropositive.*

The difference in electronegativities between two bonded atoms offers an indication of the **polarity** in the bond. Polarity is a measure of the inequality in the distribution of bonding electrons. A **nonpolar bond** is one in which there is *equal sharing* of bonding electrons. In a **polar bond** there is *unequal sharing;* the more electronegative atom pulls the electrons closer.

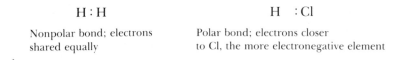

H : H

Nonpolar bond; electrons
shared equally

H　:Cl

Polar bond; electrons closer
to Cl, the more electronegative element

Figure 6.7 shows the sharing pictorially.

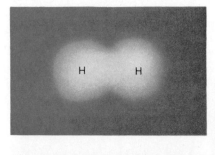

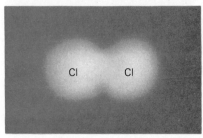

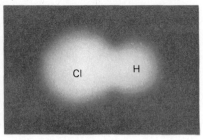

The fuzzy cloud around the atoms is the shared pair of electrons. In H_2 and Cl_2, the cloud is evenly distributed. In HCl the cloud is drawn more toward the more electronegative atom, which is Cl.

FIGURE 6.7

To obtain a rough measure of the polarity of a bond, take the mathematical difference between the electronegativities of the bonded atoms. The larger the difference, the more polar the bond.

For example, in the case of H_2 (H—H) the electronegativity difference (2.1 − 2.1) is zero; thus the bond is a **nonpolar covalent bond,** or simply a **covalent bond.** The two electrons in a (nonpolar) covalent bond are equally distributed between the atoms.

In H—Cl the electronegativity difference between H (2.1) and Cl (3.0) is 0.9 (3.0 − 2.1). Because of the electronegativity difference, the H—Cl bond is called a **polar covalent bond.** If we look more closely at the consequences of unequal sharing of electrons, we will find how this word "polar" originates. In H—Cl, because the bonding electrons are closer to Cl, Cl acquires a partial negative charge (remember that electrons are negative). H acquires a partial positive charge because the electrons neutralizing the nuclear charge are being drawn away. This is written as

$$^{\delta+}H \quad :Cl^{\delta-}$$

where the symbol δ means "partial." The bond has developed a positive pole and a negative pole, just as we speak of the + and − poles of a battery. That is why the bond is said to be polar.

Because there are *two* poles (one + and one −), one can also say that a **dipole** exists in the bond. (The prefix "di-" always means "two.")

A crossed arrow ($\leftrightarrow$) pointing toward the more electronegative element indicates the presence of a dipole.

$$\overset{\longmapsto}{\text{H—Cl}}$$

The greater the electronegativity difference, the more polar the bond and the larger the dipole. As you continue your study of chemistry, bond polarity will play an important role in understanding the properties of molecules.

SAMPLE EXERCISE

6.14

For the bonds indicated below, (1) decide whether they are nonpolar or polar covalent, (2) mark the dipole, and (3) decide which is the most polar:

a. H—I
b. The N—H bond in NH_3
c. The O—H bond in H_2O

SOLUTION

1. Calculate the difference in electronegativities between bonded atoms (see Table 6.7). Subtract the smaller electronegativity from the larger one.

  a. $2.5 - 2.1 = 0.4$
     (I)   (H)
  Although the difference is small, 0.4, we classify this bond as polar based on the discussion so far.
  b. $3.0 - 2.1 = 0.9$
     (N)   (H)
  We classify this bond as polar, because of the electronegativity difference.
  c. $3.5 - 2.1 = 1.4$
     (O)   (H)
  Again, we classify the bond as polar.

2. The dipole is marked by a crossed arrow pointing in the direction of the more electronegative element.

  a. $\overset{\longmapsto}{\text{H—I}}$

  b. $\overset{\longleftarrow}{\text{N—H}}$

  c. $\overset{\longleftarrow}{\text{O—H}}$

3. The O—H bond is the most polar because the electronegativity difference (1.4) is largest.

RECOGNIZING IONIC VERSUS MOLECULAR COMPOUNDS

6.16 A simple generalization that enables us to distinguish between ionic and molecular (covalent) compounds readily is that *most ionic compounds are formed from metals and nonmetals,* whereas *most molecular* *compounds are formed from combinations of nonmetals.* This rule is very convenient for purposes of formula writing, and as you will see (Section 7.2), for naming compounds, because metal plus nonmetal combinations are named according to the rules for ionic compounds, and compounds containing exclusively nonmetals usually are not.

However, in considering the properties of a compound, its ionic or covalent (molecular) nature depends on the electronegativity differences between the bonded atoms. See Table 6.8 for a comparison of the properties of ionic versus covalent compounds. There is a gradual transition from covalent to polar covalent to ionic bonding as the electronegativity differences increase. The properties of the compound depend on the bonding character. The following ranges of electronegativity differences may be used to predict bonding character:

From 0 to 0.5, covalent (to only slightly polar)

From 0.5 to 1.7, polar covalent

Over 1.7, ionic

Because the transition in properties is gradual, the cutoff points of 0.5 and 1.7 are somewhat arbitrary, and bonds with electronegativity differences near ($\pm$ 0.2) these values are expected to be intermediate in character.

Notice that the rule that a combination of a metal and nonmetal yields an ionic compound works very well for most examples we encounter. For example, from Table 6.7 we can obtain the electronegativities for Na and Cl as 0.9 and 3.0, respectively. The difference of 2.1 between these two numbers tells us that the bond between Na and Cl is ionic, and of course, we know from experience that sodium chloride (table salt) exhibits the properties of an ionic compound, as indicated in Table 6.8.

COMPARISON OF PROPERTIES OF IONIC AND COVALENT COMPOUNDS   **TABLE 6.8**

Characteristic	Ionic Compounds	Covalent Compounds
Smallest particles present	Ions	Molecules
Physical state at room temperature	Solids	Gases, liquids, or low-melting-point solids
Electrical conductivity of the molten state	Conductors	Nonconductors
Water solubility	Many are soluble	Most are not soluble
Electrical conductivity of water solution	Conductors	Most are nonconductors (unless they react with water)

In contrast, gasoline exhibits the properties of a covalent compound. Its formula C_8H_{18}, which shows only nonmetals in the compound, and the small (0.4) electronegativity difference between C and H both predict that this should be so.

LAW OF DEFINITE COMPOSITION REVISITED

6.17   You first encountered the Law of Definite Composition in Section 1.9 as part of the first discussion of the properties of compounds. This law and the idea of fixed combining weight ratios are hard to appreciate before you know about the existence of atoms and how they combine. Now that you have explored bonding, you know that elements always show some definite combining ratio of atoms or ions, and this inevitably leads to some definite combining weight ratio. For water (H_2O) the combining atomic ratio is 2 hydrogens to 1 oxygen because this ratio allows all atoms to have octets (O) or duets (H) of electrons. For sodium chloride the combining ionic ratio is $Na:Cl = 1:1$.

Because atoms (and ions) have definite atomic weights, a definite combining ratio of atoms (or ions) means a definite combining ratio by weight. Thus the weight ratio or fixed proportion of elements in H_2O is always $2.0:16 = 1.0:8.0$ because two hydrogens weigh 2.0 amu and one oxygen weighs 16 amu. Similarly the weight ratio for NaCl is always $23.1:35.5$. Notice that the weight of an ion is the same as the weight of an atom. This is so because ions arise by gains or losses of electrons, which we take to have zero mass.

SAMPLE EXERCISE

6.15

Bunsen burner gas, methane, has the combining ratio of one carbon atom with four hydrogen atoms. What is the weight ratio for methane?

SOLUTION

Carbon has a relative mass of 12. Hydrogen has a relative mass of 1. Therefore, one atom weighing 12 combined with four atoms each weighing 1 leads to a weight ratio of $12:4$, or $3:1$.

Problem 6.12

Acetylene gas, a compound used in welding, has the combining ratio of two carbon atoms with two hydrogen atoms. What is the weight ratio for acetylene?

Problem 6.13

Hydrogen sulfide, a noxious gas, has the combining ratio of two hydrogen atoms with one sulfur atom. What is the weight ratio for hydrogen sulfide?

In fluoridated "hard" water, calcium ions (Ca^{2+}) combine with fluoride ions (F^-) in the ratio 1:2. What is the weight ratio in this ionic compound?

SOLUTION

Calcium has a relative mass of 40. Fluorine has a relative mass of 19. Therefore, one ion weighing 40 combined with two ions each weighing 19 (2 × 19 = 38) leads to a weight ratio 40:38.

Problem 6.14

Magnesium sulfide has the combining ratio of one magnesium ion to one sulfide ion. What is the weight ratio in this ionic compound?

SUMMARY

Bonds are the forces that hold elements together in compounds. Ionic bonds hold ions together; covalent bonds hold atoms together in molecules (Section 6.1). Compound formation occurs because atoms can achieve lower-energy states by gaining, losing, or sharing electrons to attain a noble-gas electron configuration, a duet, or more often an octet of valence electrons (Section 6.2).

Metals lose electrons and form cations. Nonmetals gain electrons and form anions. For the A Group elements, the numbers of electrons gained or lost and the ionic charge attained can be predicted from the periodic table (Sections 6.3 and 6.4). A transfer of electrons in which the number lost by the metal equals the number gained by the nonmetal results in an ionic bond (Section 6.5).

It is necessary to know the charges on simple monatomic cations and anions (Section 6.6) and also on the polyatomic ions (Section 6.7). Formulas for ionic compounds can then be established on the principle of positive and negative ionic charges being balanced in electrically neutral compounds. A shortcut to formula writing is the crisscross method (Section 6.8).

An ionic bond is the electrostatic attraction between (+) and (−) ions in an ionic compound, and the strength of the ionic bond gives ionic compounds their characteristic properties (Section 6.9). There are two major categories of compounds: ionic compounds and molecular compounds. Molecular compounds are formed when nonmetallic atoms become linked together in units called molecules by sharing their valence electrons. Shared pairs of electrons are called covalent bonds (Section 6.10).

The elements that exist as diatomic molecules offer the simplest examples of the concept of covalent bond formation, the sharing of electrons to attain a noble gas configuration (Section 6.11). As atoms approach each other, there is an attraction between the electrons of one atom and the nucleus of the other which provides the basis for sharing (Section 6.12).

The electron-sharing pattern in molecules is shown by the use of Lewis electron dot formulas, which can be constructed following a set of guidelines (Section 6.13). When one atom contributes both electrons to the shared pair, the bond is called a coordinate covalent bond (Section 6.14).

Electronegativity is the ability of bonded atoms to attract bonding electron pairs. Unequal sharing of electron pairs because of electronegativity differences leads to different degrees of bond polarity and bond dipoles (Section 6.15). Compounds can be tentatively classified as ionic or covalent on the basis of the metallic or nonmetallic nature of the elements of which they are composed. A more rigorous classification requires an examination of electronegativity differences and bonding character (Section 6.16).

A knowledge of bonding clarifies the relationship between atomic theory and the experimental Law of Definite Composition, or fixed proportions by weight within compounds (Section 6.17).

CHAPTER ACCOMPLISHMENTS

After completing this chapter you should be able to

6.1   Introduction
1.   Give the general names for the two major types of compounds.
2.   Define bond.

6.2   How can atoms achieve lower-energy states?
3.   State why most elements interact to form compounds.
4.   State the common pattern in the electron configuration of noble gases which is responsible for their stability.
5.   State the type of electron configuration atoms strive to obtain when elements combine to form compounds.

6.3   Metals lose electrons
6.   Define ionization energy.
7.   Determine the number of electrons an A Group metal will lose to obtain a noble-gas configuration.
8.   Define the term isoelectronic.

6.4   Nonmetals gain electrons
9.   Define electron affinity.

10.   Determine the number of electrons an element in Group VA, VIA, or VIIA will gain to obtain a noble-gas configuration.

11.   Identify the electron configuration of an A Group ion with that of a noble gas.

12.   Using a periodic table, predict the charge on an ion formed from a given A Group element.

6.5   Electron transfer

13.   Develop a pictorial representation for the formation of ions from atoms.

14.   State the relationship between the number of electrons lost by a metal and the number of electrons gained by a nonmetal in the formation of an ionic compound.

6.6   Ionic charges

15.   Write the symbol and charge of each A Group ion, given a periodic table.

16.   Given the name or the symbol, state the charge(s) of each ion listed in Table 6.4.

6.7   Polyatomic ions

17.   Distinguish between the terms monatomic ion and polyatomic ion.

18.   Given the name, write the formula and charge of a polyatomic ion, or given the formula, name each polyatomic ion in Table 6.5.

6.8   Formulas for ionic compounds

19.   Write a correct formula for an ionic compound, given the combining elements and a periodic table.

20.   Write the formula for an ionic compound, given two combining polyatomic ions or a polyatomic ion and a combining element and a periodic table.

6.9   The nature of the ionic bond

21.   Explain why an ionic bond lowers the energy of a compound relative to the individual ions or atoms.

6.10   Why are there two types of compounds?

22.   Explain why some elements combine to form ionic bonds and others combine to form covalent bonds.

6.11   Diatomic molecules

23.   Define nonbonding electrons.

24.   Distinguish between single, double, and triple covalent bonds.

6.12 The nature of the covalent bond

25. State why there is a lowering of energy relative to the separate atoms when a covalent bond forms.

6.13 Lewis electron dot formulas

26. Write Lewis electron dot structures for molecular compounds.
27. Write Lewis electron dot structures for polyatomic ions.

6.14 Coordinate covalent bonds

28. Define coordinate covalent bond.

6.15 Electronegativity and polarity

29. Define electronegativity.
30. Describe the trends in electronegativity across a row and down a column of the periodic table.
31. State the difference between polar and nonpolar bonds.
32. Given a table of electronegativities, indicate whether a bond formed between two atoms will be nonpolar or polar.
33. Use the crossed arrow ($\leftrightarrow$) to label the dipole in a polar covalent bond.

6.16 Recognizing ionic versus molecular compounds

34. Distinguish between ionic and molecular compounds for purposes of nomenclature and formula writing.
35. Given a table of electronegativities, distinguish between covalent, polar covalent, and ionic bonds.
36. Compare the properties of ionic and molecular compounds.

6.17 Law of definite composition revisited

37. Predict the weight ratio of elements in a compound from the combining ratio of atoms or ions in the compound.

PROBLEMS

6.1 Introduction

6.15 What is a chemical bond?

6.16 List two types of chemical bonds.

6.2 How can atoms achieve lower energy states?

6.17 Describe the characteristic pattern(s) in a noble-gas electron configuration.

6.18 What feature of the structure of an atom is altered when atoms combine to form compounds?

6.19 Describe the two ways by which atoms can obtain a noble-gas electron configuration.

6.3 Metals lose electrons

6.20 Explain why Group IA metals form a 1^+ cation.

6.21 Explain why Ca does not form a Ca^{3+} ion.

6.22 **a.** Write the electron configuration of the K^+ ion.

   b. With which inert gas is this ion isoelectronic?

6.23 a. How many electrons must a barium atom lose to form an ion with a noble-gas configuration?
b. With which noble gas will the barium ion be isoelectronic?

6.24 Give the symbols of two cations that will be isoelectronic with xenon.

6.25 State the noble gas that is isoelectronic with each ion below:
    a. Sr^{2+}
    b. Li^+
    c. Rb^+

6.4 Nonmetals gain electrons

6.26 Explain why nonmetals gain rather than lose electrons in forming ionic bonds.

6.27 Explain why Group VIIA nonmetals form a 1^- anion.

6.28 a. How many electrons must N gain to form a noble-gas configuration?
b. With which noble gas will the N ion be isoelectronic?

6.29 a. Write the electron configuration of the S^{2-} ion.
b. With which inert gas is this ion isoelectronic?

6.30 Give the symbols of three anions that would be isoelectronic with krypton.

6.31 State the noble gas that is isoelectronic with each ion below:
    a. Se^{2-}
    b. P^{3-}
    c. I^-

6.5 Electron transfer

6.32 Explain why two sodium atoms are needed to react with one sulfur atom in the formation of the ionic compound sodium sulfide.

6.33 a. Develop a pictorial representation similar to Figure 6.3 for the formation of an ionic compound from sodium and fluorine.
b. Explain how your diagram accounts for the formula NaF.

6.34 a. Develop a pictorial representation simi-

lar to Figure 6.3 for the formation of an ionic compound from potassium and nitrogen.
b. Explain how your diagram accounts for the formula K_3N.

6.35 a. Develop a pictorial representation similar to that in Figure 6.3 for the formation of an ionic compound from aluminum and oxygen.
b. Explain how your diagram accounts for the formula Al_2O_3.

6.6 Ionic charges

6.36 Write the symbol and charge of the ion that could form from each of the following elements:
    a. Bromine   **d.** Phosphorus
    b. Strontium   **e.** Aluminum
    c. Cesium   **f.** Lithium

6.7 Polyatomic ions

6.37 Give the formula and charge of the following polyatomic ions:
    a. Carbonate
    b. Hydrogen carbonate
    c. Phosphate
    d. Ammonium
    e. Sulfite

6.8 Formulas for ionic compounds

6.38 Write formulas for the compounds formed from the following ions:
    a. Cs^+ and F^- **h.** K^+ and $C_2H_3O_2^-$
    b. Ba^{2+} and I^- **i.** Mg^{2+} and N^{3-}
    c. Ba^{2+} and O^{2-} **j.** Zn^{2+} and
    d. NH_4^+ and CO_3^{2-} HPO_4^{2-}
    e. Al^{3+} and SO_4^{2-} **k.** Pb^{4+} and SO_4^{2-}
    f. Na^+ and MnO_4^- **l.** Fe^{3+} and O^{2-}
    g. Li^+ and PO_4^{3-} **m.** Ca^{2+} and
                                       HCO_3^-

6.39 Write a correct formula for the compound formed by combining:
    a. Lithium and oxygen
    b. Barium and chlorine
    c. Sodium and sulfur
    d. Nickel and nitrogen
    e. Silver and bromine
    f. Iron (the +2 ion) and oxygen

g. Iron (the +3 ion) and oxygen
h. Tin (the +2 ion) and chlorine
i. Tin (the +4 ion) and chlorine
j. Gallium and sulfur
k. Calcium and iodine

6.40 Write the formula of the ionic compound that could form from elements X and Y if X has two valence electrons and Y five valence electrons.

6.41 Write the formula of the ionic compound that could form from elements X and Y if X has one valence electron and Y six valence electrons.

6.42 What is the ratio of cations to anions in each of the following ionic compounds?

a. CaO
b. MgF_2
c. Na_3PO_4
d. $KC_2H_3O_2$
e. $Ca_3(PO_4)_2$

6.9 The nature of the ionic bond

6.43 Which step in ionic bond formation releases the greatest amount of energy?

6.44 Which compound, CaO or KF, will have the strongest electrostatic attractions between the ions? Why?

6.45 Determine the total number of ions in one formula unit of each of the following:

a. $CaCl_2$
b. $Ca_3(PO_4)_2$
c. $Al(HCO_3)_3$
d. $(NH_4)_2SO_4$

6.46 Determine the total number of atoms in one formula unit of each of the compounds in Problem 6.45.

6.10 Why are there two types of compounds?

6.47 Distinguish between ionic and covalent bonds based on the way that they form.

6.11 Diatomic molecules

6.48 Write Lewis electron dot structures for the following diatomic molecules:

a. ICl
b. BrCl
c. ClF
d. HF

6.49 Explain why helium exists as a monatomic gas but hydrogen exists as diatomic molecules.

6.50 Explain why chlorine molecules have a single

covalent bond whereas nitrogen molecules have a triple covalent bond.

6.51 **a.** How many nonbonding electrons surround each chlorine in a chlorine molecule?
   b. How many nonbonding electrons surround each nitrogen in a nitrogen molecule?

6.12 The nature of the covalent bond

6.52 **a.** Write the Lewis symbols for a hydrogen and fluorine atom.
   b. Indicate the new attractive and repulsive forces that arise as the two atoms are brought together to form a covalent bond.

6.53 According to Figure 6.5, what is true about the energy content of a covalent bond at:
   a. Separation distances greater than the bond length?
   b. Separation distances less than the bond length?
   c. A separation distance equal to the bond length?

6.54 The bond-dissociation energies in kilocalories for the hydrogen halides are HF = 135, HCl = 103, HBr = 88, HI = 71.
   a. Which compound has the strongest bond?
   b. Which compound is the least stable?

6.55 The bond-dissociation energy of a chlorine molecule is 58 kcal/71 g of chlorine. Is a chlorine molecule more or less stable than two separated chlorine atoms? Explain your answer.

6.13 Lewis electron dot formulas

6.56 Determine the number of bonding pairs of electrons in each of the following molecules or ions:
   a. H_2S
   b. $CHCl_3$
   c. H_2SO_3
   d. NO_2^-

6.57 Determine the central atom for each of the molecules or ions in Problem 6.56.

6.58 Write the Lewis structures showing all bonding and nonbonding valence electrons for each of the molecules or ions in Problem 6.56.

6.59 Write Lewis structures for the following molecules:

a. CO_2
b. $HClO_3$
c. CO
d. C_2H_6
e. OF_2
f. H_2O_2
g. CH_4S
h. H_2CO_3
i. SO_3

6.60 Write Lewis structures for the following polyatomic ions:

a. OH^-
b. HS^-
c. SO_3^{2-}
d. PO_4^{3-}
e. ClO^-
f. NH_4^+
g. HCO_3^-
h. $C_2H_3O_2^-$

6.61 a. Write Lewis structures for SO_2 and O_3 (ozone).
b. Indicate the similarities in their structure.
c. Experiments show that the bonds in SO_2 are exactly the same. A similar result is found for O_3. Do your structures agree with these results?

6.14 Coordinate covalent bonds

6.62 a. Write Lewis structures for H^+, H, and H^-.
b. Which two could form a coordinate covalent bond?

6.63 a. Write a Lewis structure for BF_3.
b. Indicate a substance with which BF_3 could form a coordinate covalent bond.
c. Write a Lewis dot formula for the complex formed when BF_3 is joined by a coordinate covalent bond to the substance you indicated in part (b).

6.64 In general, what kinds of substances could act as the electron-pair donor in the formation of a coordinate covalent bond?

6.65 In general, what kinds of substances could act as the electron-pair acceptor in the formation of a coordinate covalent bond?

6.66 Which type of electrons, bonding or nonbonding, can participate in forming a coordinate covalent bond?

6.15 Electronegativity and polarity

6.67 Define electronegativity.

6.68 Describe the trends in electronegativity found in the periodic table.

6.69 a. Where in the periodic table do we find elements with the highest electronegativity?
b. Where do we find elements with the lowest electronegativity?

6.70 Using Table 6.7 and the guidelines in Section 6.16, classify the following bonds as nonpolar or polar:

a. $P—Br$
b. $H—O$
c. $I—Br$
d. $N—H$

6.71 Using Table 6.7, arrange the bonds given in Problem 6.70 in order of increasing polarity.

6.72 For each polar bond that you found in Problem 6.70, use the crossed-arrow symbol ($\leftrightarrow$) to indicate the dipole in the bond.

6.73 a. Write a Lewis structure for CF_4.
b. Use the symbol $\leftrightarrow$ to indicate the polarity in each of the bonds in CF_4.
c. Where is the center of the $\delta-$ polarity in CF_4?
d. Where is the center of the $\delta+$ polarity in CF_4?

6.74 For Problem 6.56, use $\leftrightarrow$ to indicate any bond polarities in the Lewis structures.

6.16 Recognizing ionic versus molecular compounds

6.75 Using Table 6.7 and the guidelines in Section 6.16, indicate whether the following bonds are nonpolar, polar, or ionic:

a. $H—Br$
b. $H—F$
c. $Li—O$
d. $Mg—Cl$
e. $Al—Cl$
f. $C—Br$

6.76 Indicate whether the following compounds are ionic or molecular, and state whether the smallest basic grouping is a molecule or a formula unit:

a. H_2O
b. $NaBr$
c. Al_2O_3
d. CH_4O
e. $PbClO_4$
f. HBr
g. K_2CO_3
h. CS_2
i. $CaCl_2$

6.77 Compound A is a high-melting-point, water-soluble solid. Compound B is a liquid which does not conduct electricity. Indicate the likely ionic or covalent nature of compounds A and B.

6.78 Using the guidelines in Section 6.16, predict the type of bond (ionic, polar covalent, or covalent) that would form between atoms of the following elements:

a. C and H   **d.** O and S
b. K and Br   **e.** Ca and Cl
c. N and H   **f.** Al and F

6.79 Predict the order of increasing polarity for the bonds in the following molecules: H_2S, H_2O, H_2Se, and H_2Te.

6.17 Law of definite composition revisited

6.80 Propane, a fuel used in mobile homes, has a combining ratio of three carbon atoms with eight hydrogen atoms. What is the weight ratio for propane?

6.81 Describe how Dalton's atomic model provides an explanation for the Law of Definite Composition.

6.82 When 64 g of sulfur dioxide, SO_2, is decomposed, 32.0 g of sulfur and 32.0 g of oxygen are formed. When 15.0 g of sulfur dioxide is decomposed, 7.50 g of oxygen is formed. How much sulfur will be formed in this decomposition?

6.83 Could a compound have a combining ratio of 1.5 atoms to 2 atoms? Explain your answer.

6.84 When 25.0 g of ammonia NH_3 is decomposed, 20.6 g of nitrogen and 4.41 g of hydrogen are formed. What is the weight ratio of nitrogen to hydrogen in ammonia?

COMPOUNDS:
NOMENCLATURE, WEIGHT, AND PERCENT COMPOSITION
CHAPTER

7

N

Nitrogen

14.01

INTRODUCTION

7.1 Having learned the basic principles of compound formation and mastered the skills of formula writing for both ionic and molecular compounds, you will find it necessary to be able to name compounds. Also, your new familiarity with formulas will allow the extension of atomic weight concepts (Section 4.10) to the weight of formula units and molecules known respectively as **formula weight** and **molecular weight.** Other skills related to formulas and weights, such as computing the percentage composition of the elements within a compound, will unfold. These skills are essential to subsequent chemical calculations.

NOMENCLATURE

7.2 There are over 10 million known compounds today. If each one had a separate arbitrarily assigned name, called a **common name,** we would have a totally impossible situation. To alleviate this, sets of rules for systematically naming compounds have been developed. These naming rules are what are known as **nomenclature.** Despite the advantages of using systematic names, some compounds have been in use so long and are so widely used that they are often called by their common names even by chemists. Table 7.1 lists some common-name compounds.

To some extent, success in learning nomenclature depends on recognizing ionic versus molecular compounds. As you saw in Section 6.16, although there is no sharp dividing line between ionic and covalent bonds and thus between ionic and molecular compounds, for purposes of nomenclature we use the simple generalization that metal and nonmetal combinations are ionic and combinations of nonmetals are molecular. Although the ammonium cation (NH_4^+) is not a metallic ion, compounds containing it are ionic and follow the rules of ionic compound nomenclature.

TABLE 7.1 SOME COMMON HOUSEHOLD CHEMICALS AND THEIR SYSTEMATIC NAMES

Common Name	Formula	Systematic Name	Use
Lye	NaOH	Sodium hydroxide	Unclog drains
Baking soda	$NaHCO_3$	Sodium hydrogen carbonate	Makes cakes and breads rise by giving off CO_2 when heated
Washing soda	$Na_2CO_3 \cdot 10H_2O$	Sodium carbonate decahydrate	General cleanser
Milk of magnesia	$Mg(OH)_2$	Magnesium hydroxide	Antacid and laxative
Borax	$Na_2B_4O_7 \cdot 10H_2O$	Sodium tetraborate decahydrate	Cleanser
Cream of tartar	$KHC_4H_4O_6$	Potassium hydrogen tartrate	Mixed with baking soda in cake making
Lime	CaO	Calcium oxide	Making plaster and mortar
Slaked lime	$Ca(OH)_2$	Calcium hydroxide	Making plaster and mortar
Table salt	NaCl	Sodium chloride	Salty taste in foods

NAMING IONIC COMPOUNDS

7.3 In naming any ionic compound, simply name the cation first, followed by a separate word for the anion:

Cation Anion
First name Last name

Cations are named exactly the same as the metals from which they come. Simple monatomic anions are named by using the root names given in Table 7.2 and adding to the root the suffix "*-ide*." Thus to name NaI, we start with the cation name "*sodium*" and follow it with the word "*iodide*" [(root = *iod*) + *ide*]. Thus NaI is sodium iodide.

TABLE 7.2 NAMING SIMPLE ANIONS

Element, root given in italic type	Root	Anion name
		Root + ide
Oxygen	Ox	Oxide
Chlorine	Chlor	Chloride
Bromine	Brom	Bromide
Iodine	Iod	Iodide
Fluorine	Fluor	Fluoride
Phosphorus	Phosph	Phosphide
Nitrogen	Nitr	Nitride
Sulfur	Sulf	Sulfide

Name the following ionic compounds: MgO, K_2S, and $AlCl_3$.

SOLUTION

MgO   Magnesium oxide.

K_2S Potassium sulfide. Note that no mention is made in the name of the subscript 2 after potassium. This is always true for *ionic* compounds. *The name contains no mention of subscript numbers.*

$AlCl_3$ Aluminum chloride.

Problem 7.1

Name the following compounds: NaCl, $CaBr_2$, and BaO.

If the cation can have a variable charge (see Table 6.4), we follow the name of the cation by a roman numeral, inside parentheses, indicating the magnitude of the charge. For example, Fe^{2+} combines with Cl^- to give the compound $FeCl_2$. This is named iron(II) chloride to distinguish it from the compound that results from the combination of Fe^{3+} and Cl^-, $FeCl_3$, iron(III) chloride.

Name the following ionic compounds: CuBr, $CuBr_2$, and $SnCl_4$.

SOLUTION

CuBr   Copper(I) bromide. Note here the need to indicate the magnitude of the charge on copper. We realize that the charge on Cu is $+1$, since Br is always -1 and there is only one copper ion to balance the -1 charge. Hence the charge on Cu in this compound must be $+1$.

$CuBr_2$ Copper(II) bromide. There are two Br^- ions, and therefore the one copper ion must have a $+2$ charge.

$SnCl_4$ Tin(IV) chloride. There are four Cl^- ions, and therefore the one tin ion, must have a $+4$ charge.

Problem 7.2

Name the following compounds: FeO, Fe_2O_3, and CuO.

So far in this section all the compounds named have been binary ionic compounds. A **binary** compound is one that contains only two kinds of elements. It may contain more than two atoms, but it must contain only two elements. Look back and convince yourself that this is so. For example, aluminum chloride, $AlCl_3$, has four atoms in its formula unit but only two kinds of elements, aluminum and chlorine.

We point this out to help you recognize polyatomic ions in compounds. If an ionic compound contains more than two kinds of elements, i.e., is *not* binary, you should recognize the presence of a polyatomic ion (review Section 6.7 and Table 6.5). For example, Na_2SO_4 is not binary because it contains three elements, Na, S, and O. To name the compound, you must recognize that the cation is sodium and the anion is the polyatomic sulfate ion. Thus the compound is named sodium sulfate; NH_4Cl is ammonium chloride; NH_4OH is ammonium hydroxide.

SAMPLE EXERCISE

7.3

Name the following ionic compounds: Na_3PO_4, K_2SO_3, $(NH_4)_3PO_4$, and Cu_2CO_3.

SOLUTION

Na_3PO_4 Sodium phosphate. Again there is no mention of the subscript number.

K_2SO_3 Potassium sulfite.

$(NH_4)_3PO_4$ Ammonium phosphate.

Cu_2CO_3 Copper(I) carbonate. Carbonate has a -2 charge. The two positive ions therefore have a total positive charge of $+2$, and since there are two copper ions in the formula, each must be charged $+1$.

Problem 7.3

Name the following compounds: K_2CO_3, NH_4NO_3, and $Pb_3(PO_4)_2$.

The number of actual polyatomic ions is considerably larger than that given in Table 6.5; however, in learning the formulas and names you do not have to rely on memory completely. There are many examples of the same two elements combining in different ratios to form different polyatomic anions. Examples of such series are:

$$
\begin{array}{lll}
SO_4^{2-} & PO_4^{3-} & ClO_4^{-} \\
SO_3^{2-} & PO_3^{3-} & ClO_3^{-} \\
 & & ClO_2^{-} \\
 & & ClO^{-}
\end{array}
$$

In each series of the same two elements, the number of oxygens varies but the number of the other nonmetal is fixed at one. In addition, the charge of the anion remains the same throughout the series. If there are only two ions in the series, then the name of the one with the greater number of oxygens ends in -*ate* while the name of the one with the lesser number of oxygens ends in -*ite*.

SAMPLE EXERCISE

7.4

Name SO_4^{2-} and SO_3^{2-}.

SOLUTION

As SO_4^{2-} is sul*fate* (see Table 6.5), from the preceding discussion it follows that SO_3^{2-} is sul*fite*.

In the series of four anions, the ion with the greatest number of oxygens is given the prefix *per-* and the suffix -*ate*. The ion with the least number of oxygens is given the prefix *hypo-* and the suffix -*ite*.

SAMPLE EXERCISE

7.5

Name the ions in the series ClO_4^-, ClO_3^-, ClO_2^-, and ClO^-.

SOLUTION

ClO_4^- *perchlorate* (see Table 6.5)

ClO_3^- *chlorate*

ClO_2^- *chlorite*

ClO^- *hypochlorite*

Writing correct formulas from names

Writing correct formulas from names requires the skills presented in Section 6.8. Review especially Sample Exercises 6.7 and 6.8. From this review you should recognize that the compound calcium chloride, for example, is made up of calcium ions and chloride ions. Write down the formula for the ions with their charges, and achieve a neutral combination by the crisscross method or otherwise:

$$Ca^{2+} \diagdown\diagup Cl^- \quad \text{gives} \quad CaCl_2$$

Given the name copper(I) bromide, notice that the roman numeral tells you that it is Cu^+ that is combined in this compound. Bromide is always Br^-. Thus the correct formula is

$$Cu^+ \diagdown\diagup Br^- \quad \text{gives} \quad CuBr$$

SAMPLE EXERCISE

7.6

Write the formula for the compound ammonium sulfate.

SOLUTION

The name ammonium sulfate is associated with a compound made up of the ammonium ion $NH_4{}^+$ and the sulfate ion $SO_4{}^{2-}$. We write down the two ions:

$$NH_4{}^+ \qquad SO_4{}^{2-}$$

and balance the charges by taking two ammonium ions and one sulfate ion, giving

$$(NH_4)_2SO_4$$

Since the ammonium ion is a polyatomic ion we must enclose it within parentheses when the subscript is greater than 1. Once again we see that the name of the compound does not directly tell us the numbers of each ion involved. It does tell us what ions are involved, and if we know the charges on these ions (as we should), then we can immediately write a correct formula with the proper subscripts.

NAMING BINARY MOLECULAR COMPOUNDS

7.4 Binary molecular compounds, like their ionic counterparts, contain only two elements. Whereas binary ionic compounds contain one metal and one nonmetal, in binary molecular compounds both elements are nonmetallic.

Problem 7.4

Pick out the binary compounds from the following choices:

a. H_2O

b. P_2O_5

c. NH_3

d. $C_6H_6O_6$

e. CH_2O

f. $CHCl_3$

g. CCl_4

Naming binary molecular compounds differs from naming ionic compounds in that the numbers of atoms of each element combining to form molecular compounds are always indicated by the prefixes listed in Table 7.3.

The complete name for a binary molecular compound is composed of the name of the first element (on the left) preceded by the prefix indicating the number of atoms of that element and followed by the name of the second element, which is constructed of a prefix for the number of atoms plus the element root name (Table 7.2) plus the suffix *-ide*. For example, P_2O_5 is diphosphorus pentoxide.

NUMERICAL PREFIXES		TABLE 7.3
Number of Atoms	Prefix	
1	mono-	
2	di-	
3	tri-	
4	tetra-	
5	penta-	
6	hexa-	
7	hepta-	

Prefix + **element name**   *Prefix* + **root name** + *-ide*
    First name   Last name

Where there is only one atom of the first element, it is common to omit the prefix mono- for that element. For example, NO_2 is nitrogen dioxide.

Binary molecular compounds containing hydrogen as the first element symbolized in the formula are an exception to the prefix rule, and like ionic compounds are named without the use of prefixes for either element. The reason for the exception is that whereas most pairs of nonmetals, for example, P and Cl, can form more than one compound (PCl_3 and PCl_5), H and another nonmetal generally yield only one compound. Thus PCl_3 and PCl_5 must be distinguished as phosphorus trichloride and phosphorus pentachloride to indicate combining ratios, but no such distinction is usually necessary for hydrogen compounds. In the case of the combination of hydrogen with oxygen which can form the two compounds, H_2O and H_2O_2, they are distinguished by the common names "water" and "hydrogen peroxide."

SAMPLE EXERCISE

7.7

Write names for the following binary molecular compounds: N_2O_5, SO_2, CO, NCl_3, HCl, and H_2S.

SOLUTION

The names are constructed of a first name and last name as just described.

N_2O_5 Dinitrogen   pentoxide
          Prefix   Element   Prefix   Root   Suffix
          for 2   name   for 5   name

The *a* of the prefix penta- is dropped when the element root begins with a vowel.

SO_2　Sulfur dioxide. Note the absence of the prefix mono- before sulfur.

CO　Carbon monoxide. However, one of the second element in a binary molecular compound must be indicated by mono-.

NCl_3　Nitrogen trichloride.

HCl　Hydrogen chloride. Named without prefixes.

H_2S　Hydrogen sulfide. Named without prefixes.

Problem 7.5

Write names for the following binary molecular compounds: CO_2, CCl_4, PCl_3, N_2O_4, OF_2, and HF.

Writing formulas from names of binary molecular compounds

Because the prefixes in the names of molecular compounds tell the numbers of atoms, it is particularly easy to write formulas from names. For example, carbon dioxide must have the formula CO_2 (one carbon atom because no prefix indicates that *mono-* has been omitted, and two oxygen atoms because of the prefix *di-*).

Problem 7.6

Write formulas for sulfur trioxide, hydrogen bromide, dinitrogen tetroxide, and sulfur tetrafluoride.

Ternary molecular compounds

A **ternary** compound contains three elements. There may be more than three atoms, but the number of elements is limited to three. The only ternary molecular compounds that you will encounter are the ternary oxyacids discussed in the next section.

NAMING ACIDS

7.5　Certain molecular compounds made up of hydrogen and non-metals are classified as acids because of the properties of their water solutions, which we will discuss in Chapter 15. At this point you should become familiar with the naming system used for acids so that a given name will suggest a formula and vice versa. It is easy to recognize acids because usually the formula of an acid is written with H to the left of other symbols. However, although it has the formula H_2O, water is not usually thought of as an acid. For purposes of naming, it is convenient to classify acids into two categories based on whether the compound contains oxygen.

Nonoxyacids

Nonoxyacids are solutions of molecular compounds composed of hydrogen and some nonmetal other than oxygen (or carbon). A listing of some nonoxyacids is given in Table 7.4.

Notice that all the binary compounds listed in Table 7.4 are gases. These gases themselves are named as binary molecular compounds without prefixes. For example, the formula HF is read as hydrogen fluoride. The naming of the acid, i.e., the solution formed when the gas is dissolved in water, is quite different. To name a nonoxyacid, begin with the prefix *hydro-* (for hydrogen), then use the root name for the other element (Table 7.2), attach the suffix "-*ic*," and follow this word with *acid*. In summary:

$$hydro\text{-} + \textbf{root name} + \text{-}ic \text{ acid}$$

SAMPLE EXERCISE

7.8

Name the acid HCl.

SOLUTION

We begin with *hydro-*, then attach the root *chlor-*, then add on the suffix *-ic*, then add the word *acid*, and the complete name is hydrochloric acid.

An exception to these rules is that for the element S the root used in acid nomenclature is *sulfur*, not *sulf-*, as shown in Table 7.2. Notice the name of H_2S in Table 7.4.

The compounds listed in Table 7.4 are usually named as acids (i.e., it is assumed they are dissolved in water), *not* by their pure-gas names.

Oxyacids

An oxyacid is the water solution of a molecular compound made up of hydrogen, some nonmetal, and oxygen. Although oxyacids are truly molecular and therefore not composed of ions, when they are dissolved in water, they break apart into ions. The ions are recognizable as the polyatomic ions you have previously encountered.

NONOXYACIDS **TABLE 7.4**

Formula	Name of the Pure Gas	Name of the Solution Formed by Dissolving the Gas in Water
H_2S	Hydrogen sulfide	Hydrosulfuric acid
HF	Hydrogen fluoride	Hydrofluoric acid
HCl	Hydrogen chloride	Hydrochloric acid
HBr	Hydrogen bromide	Hydrobromic acid
HI	Hydrogen iodide	Hydroiodic acid

SAMPLE EXERCISE

7.9

What polyatomic anion can be found in the following acid formulas: H_2SO_4, HNO_3, H_3PO_4, and $HClO_4$?

SOLUTION

H_2SO_4 yields the sulfate anion (SO_4^{2-}).

HNO_3 yields the nitrate anion (NO_3^-).

H_3PO_4 yields the phosphate anion (PO_4^{3-}).

$HClO_4$ yields the perchlorate anion (ClO_4^-)

Oxyacids are named according to the polyatomic ion which they yield in solution. The nonmetal other than oxygen in the polyatomic ion determines the oxyacid root name. For example, HNO_3 yields NO_3^- in solution. Therefore, the name is based on the root *nitr-* for N. For N, Cl, Br, and I, the roots are as in Table 7.2. Other roots are *sulfur-* for S, *phosphor-* for P, *carbon-* for C, and *acet-* for the ion $C_2H_3O_2^-$.

The polyatomic ion also determines the oxyacid suffix. If the anion ends in *-ate*, the ending of the acid is *-ic,* followed by the word *acid.* If the anion ends in *-ite,* the acid ends in *-ous,* followed by the word *acid.*

$$\textit{-ate} \text{ ions} \longrightarrow \textit{-ic} \text{ acids}$$
$$\textit{-ite} \text{ ions} \longrightarrow \textit{-ous} \text{ acids}$$

If the anion contains the prefix *per-* or *hypo-,* then these prefixes are also used in the names of the acid. The prefix *hydro-* is *never* found in an *oxy*acid.

In summary, oxyacids are named as follows:

Polyatomic anion root + *-ic* or *-ous* acid

SAMPLE EXERCISE

7.10

Name the following acids: H_2SO_4, HNO_3, H_3PO_4, $HClO_2$, and $HClO$.

SOLUTION

H_2SO_4 *Sulfuric* acid. The sulfate ion determines the root *sulfur* and the suffix *-ic* because of the anion ending *-ate.*

HNO_3 *Nitric* acid. From the root *nitr* and suffix *-ic.*

H_3PO_4 *Phosphoric* acid. From the root *phosphor* and the suffix *-ic.*

$HClO_2$ *Chlorous* acid. From the root *chlor* and the suffix *-ous.* It is *-ous* here because the anion ClO_2^-, chlorite, ends in *-ite.*

HClO   *Hypochlorous* acid. The hypochlorite anion gives us the root *hypochlor*, to which we add *-ous* because of the ending *-ite*.

Problem 7.7

Name the following oxyacids: H_2CO_3, $HC_2H_3O_2$, H_2SO_3, $HClO_3$, and $HClO_4$.

Writing formulas from an acid name

To write an acid's formula from its name, begin by deciding whether the name is that of an oxyacid or a nonoxyacid. The prefix *hydro* tells you an acid is a nonoxyacid, i.e., a binary compound of hydrogen and another nonmetal. The root name indicates the nonmetal combined with hydrogen. Thus hydroiodic acid is a combination of hydrogen and iodine. In the correct formula the number of hydrogens (H^+) depends on the number needed to neutralize the negative charge on the anion. For hydroiodic acid, I^- requires one H^+. Therefore, the formula is HI. Hydrogen is always written to the left.

Oxyacids are recognized by the word *acid* and the absence of the prefix *hydro*. The root name tells you the nonmetal other than oxygen in the polyatomic ion combined with an appropriate number of H^+ in the acid formula. If the acid suffix is *ic*, the polyatomic must end in *-ate;* whereas if the acid suffix is *ous*, a polyatomic ion ending in *-ite* must be present.

SAMPLE EXERCISE

7.11

Give formulas for the following acids:

a. Sulfurous acid
b. Hydrobromic acid
c. Chloric acid

SOLUTION

a. H_2SO_3 The absence of *hydro* indicates an oxyacid. The root *sulfur* indicates that the polyatomic anion must be a combination of S and O. The *-ous* suffix then tells us that the polyatomic anion is SO_3^{2-}. Combining H^+ and SO_3^{2-} leads to the formula H_2SO_3.
b. HBr   The prefix *hydro* tells us the acid is a nonoxy binary acid. The root *brom* indicates the anion Br^-. Combining H^+ and Br^- yields HBr.
c. $HClO_3$ An oxyacid formed from the polyatomic ion ClO_3^-.

Problem 7.8

Give formulas for the following acids:

a. Chlorous acid   **b.** Hydrofluoric acid   **c.** Nitric acid

A NOMENCLATURE SUMMARY

7.6 As you have seen in this chapter, there are a variety of rules for naming ionic and molecular compounds, including acids. It is useful to follow a classification scheme to decide on the appropriate rule to use for naming a given compound. One such scheme with references to specific sections appears here.

General rules for nomenclature

1. Classify the compounds as ionic (metal + nonmetal) or molecular (nonmetals only).

2. If the compound is ionic, consult Section 7.3.

3. If the compound is molecular, classify as an acid (H *begins* the formula) or nonacid (no H).

4. If a nonacid, consult Section 7.4.

5. If an acid, classify as nonoxyacid or oxyacid (Section 7.5)

See Figure 7.1.

SAMPLE EXERCISE

7.12

Name the following compounds: HI, Na_2CO_3, HNO_3, KBr, $Ca(OH)_2$, and N_2O_4.

SOLUTION

HI   Molecular because nonmetals only, an acid because of H, nonoxy because no oxygen. Therefore, *hydroiodic acid* according to the rules in Section 7.5 (or hydrogen iodide according to Section 7.4).

Na_2CO_3 Ionic because metal and nonmetals. Therefore, *sodium carbonate* according to Section 7.3.

HNO_3 Molecular, an oxyacid. *Nitric acid* according to Section 7.5.

KBr   Ionic. *Potassium bromide* according to Section 7.3.

$Ca(OH)_2$ Ionic. *Calcium hydroxide* according to Section 7.3.

FIGURE 7.1

A logical approach to classifying compounds for purposes of naming them.

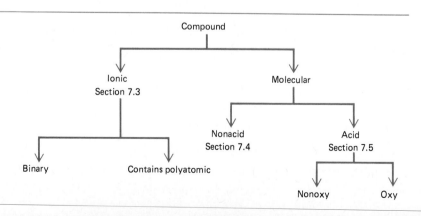

N_2O_4 Molecular, nonacid. *Dinitrogen tetroxide* according to Section 7.4.

Problem 7.9

Name the following compounds: $MgCl_2$, HBr, H_2SO_4, H_2SO_3, $Ca_3(PO_4)_2$, and OF_2.

MOLECULAR WEIGHT AND FORMULA WEIGHT

7.7   At this point you should understand what is meant by a chemical formula. For the molecular compound water, the formula H_2O indicates two hydrogen atoms and one oxygen atom held together by covalent bonds. For sodium sulfate, Na_2SO_4 indicates the ratio of two sodium ions (Na^+) to one sulfate ion (SO_4^{2-}) in this ionic compound. The sulfate ion itself is made up of one sulfur atom and four oxygen atoms hooked together by covalent bonds. In Chapter 4 we discussed the atomic weights of atoms. In Section 6.17 we pointed out that because ions are formed from atoms by the loss or gain of electrons, and because electrons have essentially no mass, the masses of ions are the same as those of the atoms from which they form. It should come as no surprise that the weights assigned to compounds are the sums of the weights of the atoms or ions that make them up.

Molecular weight

Because a water molecule is made up of hydrogen atoms and oxygen atoms and we know the weight of these atoms in atomic mass units, we can assign a weight to one H_2O molecule by simply adding the weights of the atoms involved. We call the weight of one H_2O molecule a **molecular weight.** For H_2O, this weight is 2×1.01 amu for hydrogen plus 16.0 amu for oxygen, to give 18.0 amu. If we could weigh one H_2O molecule, it would weigh $18.0/12.0 = 1.50$ times as heavy as one carbon-12 atom. Molecular weights are relative weights compared with the weight of one carbon-12 atom because they are the sums of relative atomic weights based on that scale.

SAMPLE EXERCISE

7.13

Determine the molecular weight of HNO_3.

SOLUTION

HNO_3 Contains	Mass of Each Atom	$\times$	Number of That Atom Present	=	Total Mass for Each Element
1 H atom	1.01 amu	$\times$	1	=	1.01 amu
1 N atom	14.0 amu	$\times$	1	=	14.0 amu
3 O atoms	16.0 amu	$\times$	3	=	48.0 amu
			Molecular weight	=	63.0 amu

One HNO_3 molecule has a mass of 63.0 amu. One HNO_3 molecule is 63.0/12.0 = 5.25 times as heavy as one carbon-12 atom.

Formula weight

As discussed earlier (Section 6.9), for ionic compounds there are no molecules. Rather, the formula unit tells the smallest grouping of ions corresponding to the correct fixed ratio of ions in that compound. For this reason we call the mass of a formula unit of an ionic compound a **formula weight.** Because ions have the same masses as the atoms from which they are made, mechanically the calculation of formula weight is identical to the calculation of molecular weight. That is, we sum the weights of all ions (atoms) present.

SAMPLE EXERCISE

7.14

What is the formula weight of Na_2SO_4?

SOLUTION

Sodium ions are formed from sodium atoms with no mass change. Sulfate ions are formed from sulfur atoms and oxygen atoms with no mass change.

Na_2SO_4 Contains		Mass of Each Atom (Ion)	×	Number of That Atom (Ion) Present	=	Total Mass of Each Element
2 Na$^+$ ions		23.0 amu	×	2	=	46.0 amu
1 SO$_4^{2-}$ ion {	1 S atom	32.1 amu	×	1	=	32.1 amu
	4 O atoms	16.0 amu	×	4	=	64.0 amu
			Total mass in Na_2SO_4		=	142.1 amu

The formula weight of Na_2SO_4 reported to 3 sig figs is 142. If we could weigh one formula unit of Na_2SO_4, it would weigh 142/12.0 = 11.8 times as heavy as one carbon-12 atom.

When an ion is enclosed in parentheses in a formula, remember that the number outside the parentheses multiplies all the atoms inside the parentheses.

SAMPLE EXERCISE

7.15

What is the formula weight of $Mg(OH)_2$?

SOLUTION

The method is the same as in Sample Exercise 7.14 but points out the multiplication of both atoms O and H within the parentheses.

$Mg(OH)_2$ Contains	Mass of Each Atom (Ion)	× Number of That Atom (Ion) Present	= Total Mass for Each Element
1 Mg^{2+} ion	24.3 amu ×	1	= 24.3 amu
2 OH^- ions {2 O atoms	16.0 amu ×	2	= 32.0 amu
{2 H atoms	1.01 amu ×	2	= 2.02 amu
		Total mass in $Mg(OH)_2$ =	58.3 amu

The formula weight of $Mg(OH)_2$ is 58.3 amu.

Problem 7.10

Determine the molecular weights of SO_2, N_2O_4, and H_2SO_4.

Problem 7.11

Determine the formula weights of NaOH, $CaCl_2$, and $Al_2(CO_3)_3$.

PERCENTAGE COMPOSITION

7.8   When something has more than one component, if we know the size of the whole and the size of each part, we can calculate the percentage that each part contributes to the whole. It is often necessary to know the percentage by weight of each element in a compound. This is the **percentage composition,** the percent by weight of each element in the compound. Before proceeding further you should review percentage calculations in Section 2.6.

Consider the compound CO_2. The relative mass of carbon is 12.0 amu, and that of oxygen is 16.0; therefore the molecular weight of CO_2 is $(1 \times 12.0) + (2 \times 16.0) = 12.0 + 32.0 = 44.0$ amu. The fraction of the molecular weight contributed by carbon is $12.0/44.0 = 0.273$. To convert this fraction to a percent, multiply by 100. Thus percent C $= (12.0/44.0) \times 100 = 0.273 \times 100 = 27.3$ percent. Similarly, for oxygen the fraction is $32.0/44.0 = 0.727$; therefore percent O $= (32.0/44.0) \times 100 = 72.7$ percent.

In general,

$$\begin{array}{l}\text{Percentage} \\ \text{of an element} \\ \text{in a compound}\end{array} = \dfrac{\begin{array}{c}\text{Total mass of the element} \\ \text{in one molecule } or \text{ formula unit}\end{array}}{\begin{array}{ccc}\text{Molecular} & & \text{Formula} \\ \text{weight} & or & \text{weight}\end{array}} \times 100$$

SAMPLE EXERCISE

7.16

Calculate the percentage of sodium, sulfur, and oxygen in Na_2SO_4.

SOLUTION

The formula weight of Na_2SO_4 is 142 amu (see Sample Exercise 7.14).

$$\text{Percent Na} = \frac{\text{total mass Na in one formula unit}}{\text{formula weight}} \times 100$$

$$= \frac{46.0 \text{ amu}}{142 \text{ amu}} \times 100$$

$$= 32.4 \text{ percent}$$

(Note that the unit amu in the numerator cancels with the unit amu in the denominator, "percent" does not have a unit associated with it.)

$$\text{Percent S} = \frac{\text{total mass S in one formula unit}}{\text{formula weight}} \times 100$$

$$= \frac{32.1 \text{ amu}}{142 \text{ amu}} \times 100$$

$$= 22.6 \text{ percent}$$

$$\text{Percent O} = \frac{\text{total mass O in one formula unit}}{\text{formula weight}}$$

$$= \frac{64.0 \text{ amu}}{142 \text{ amu}} \times 100$$

$$= 45.1 \text{ percent}$$

To check, we can add up the percentages of all the parts. The sum should be 100 percent to 3 sig figs.

$$
\begin{aligned}
\text{Percent Na} &= \ \ 32.4 \\
\text{Percent S} &= \ \ 22.6 \\
\text{Percent O} &= \ \underline{45.1} \\
&\ \ \ 100.1 \text{ percent}
\end{aligned}
$$

Rounding to 3 sig figs gives 100 percent.

Problem 7.12

Calculate the percentage of each element in H_2O.

　　Formula weight determinations and an understanding of percentage composition are concepts which you will put to immediate use in Chapter 8, "The Mole Concept."

SUMMARY

Compounds are named in different ways depending on their classification as ionic (Section 7.3), molecular nonacid (Section 7.4), nonoxyacid or oxyacid (Section 7.5).

A molecular or formula weight is the sum of the weights of the atoms or ions of which a compound is composed (Section 7.7). The percentage composition tells the percentage by weight of each element present in a compound (Section 7.8).

CHAPTER ACCOMPLISHMENTS

After completing this chapter you should be able to:

Introduction

Nomenclature
1. Recognize household chemicals by their common names.

Naming ionic compounds
2. Define and recognize a binary compound.
3. Name a binary ionic compound given a chemical formula.
4. Name an ionic compound containing a polyatomic ion, given a chemical formula.
5. Write a correct formula for an ionic compound, given the name of that compound.

Naming binary molecular compounds
6. Name binary molecular compounds given a chemical formula.
7. Write the formula of a binary molecular compound, given the name.

Naming acids
8. Distinguish acids from other compounds.
9. Recognize nonoxyacids.
10. Name nonoxyacids, given the formulas.
11. Write the formulas of nonoxyacids, given the names.
12. Recognize oxyacids.
13. Name oxyacids, given the formulas.
14. Write the formulas of oxyacids, given the names.

A nomenclature summary
15. Name any ionic, binary molecular compound, or acid, given a chemical formula.

Molecular weight and formula weight
16. Given a periodic table and a correct molecular formula, calculate the molecular weight of a molecular compound.
17. Given a periodic table and a correct ionic formula, calculate the formula weight of an ionic compound.

7.8 Percentage composition

18. Calculate the percentage composition of any component in a molecular or ionic compound, given the formula of that compound and a periodic table.

PROBLEMS

7.2 Nomenclature

7.13 Give the systematic name for the following common chemicals:

   a. Table salt
   b. Lye
   c. Baking soda

7.3 Naming ionic compounds

7.14 a. Indicate which of the following metals can form cations with different charges:

   (i) Ba   (ii) Ca   (iii) Sn   (iv) Zn
   b. State the charges that can form on the cation of each variable-charge metal in part (a).

7.15 Indicate which of the following compounds are binary:

   a. $AlCl_3$ **d.** Fe_2O_3
   b. $AlPO_4$ **e.** NaOH
   c. NH_4Cl **f.** Li_3N

7.16 Name the following binary ionic compounds:

   a. NaBr   **e.** Mg_3N_2
   b. BaI_2 **f.** Fe_2O_3
   c. CaO   **g.** PbO_2
   d. $PbCl_2$ **h.** CsF

7.17 Name the binary ionic compound formed by combining:

   a. Potassium and oxygen
   b. Aluminum and chlorine
   c. Calcium and bromine
   d. Silver and oxygen
   e. Copper ($+2$ ion) and iodine
   f. Copper ($+1$ ion) and chlorine
   g. Tin ($+2$ ion) and bromine
   h. Tin ($+4$ ion) and bromine
   i. Gallium and sulfur
   j. Barium and iodine
   k. Nickel and nitrogen

7.18 Name the following ionic compounds:

   a. $CaCO_3$ **f.** $NaC_2H_3O_2$
   b. K_2SO_4 **g.** $ZnHPO_4$
   c. NH_4Cl **h.** $Zn_3(PO_4)_2$
   d. $(NH_4)_2CO_3$ **i.** $KMnO_4$
   e. $Mg(HCO_3)_2$ **j.** $NaClO_4$

7.19 Name the ionic compounds formed by combining the following ions:

   a. Ca^{2+} and HCO_3^- **d.** Fe^{3+} and OH^-
   b. Li^+ and PO_4^{3-} **e.** Mg^{2+} and ClO_3^-
   c. Sn^{2+} and Cl^-

7.20 Write formulas for the following compounds:

   a. Silver oxide
   b. Potassium chloride
   c. Magnesium hydrogen carbonate
   d. Aluminum dihydrogen phosphate
   e. Sodium phosphate
   f. Barium sulfate
   g. Copper(II) iodide
   h. Lead(IV) sulfide
   i. Copper(I) oxide
   j. Tin(II) nitride

7.21 The polyatomic anion IO_3^- is called iodate. Name the polyatomic anions in the series IO_4^-, IO_3^-, IO_2^-, and IO^-.

7.22 What is the meaning of the roman numeral found in parentheses in the name of some ionic compounds?

7.23 Name the compound formed by combining each cation in the left vertical column with each anion in the top horizontal row.

	OH^-	SO_4^{2-}	PO_4^{3-}
NH_4^+			
Fe^{2+}			
Al^{3+}			

7.4 Naming molecular compounds

7.24 Classify the following compounds as binary or ternary:

a. H_2O
b. $HClO$
c. NH_3
d. CBr_4
e. H_2SO_4
f. H_2S
g. CO_2
h. CCl_2Br_2

7.25 Name the following binary molecular compounds:

a. PCl_5
b. SO_2
c. P_2O_5
d. HF (as pure gas)
e. CBr_4
f. Cl_2O_7

7.26 What kinds of elements make up molecular compounds?

7.27 Name each compound in the series: N_2O, NO, N_2O_3, NO_2, N_2O_4, and N_2O_5

7.28 Write the formula of the following molecular compounds:

a. Nitrogen monoxide
b. Carbon tetriodide
c. Phosphorus trichloride
d. Hydrogen chloride
e. Chlorine monobromide
f. Dinitrogen oxide
g. Sulfur dioxide
h. Diphosphorus pentoxide

7.29 Complete the following table of molecular compounds:

Name	Formula
a. Sulfur tetrafluoride	_____
b. _____	NCl_3
c. Boron trichloride	_____
d. Dinitrogen trioxide	_____
e. _____	CBr_4

7.5 Naming acids

7.30 Classify the following as acids or nonacids:

a. H_2SO_4
b. CH_4
c. $NaOH$
d. $HC_2H_3O_2$

7.31 Classify the following as oxy- or nonoxy acids:

a. H_3PO_4
b. HF
c. H_2SO_3
d. H_2S
e. HBr
f. HNO_3

7.32 Name the following as acids:

a. HBr
b. HF
c. H_2S

7.33 Name the following as acids:

a. H_2SO_4
b. H_2SO_3
c. HNO_3
d. H_3PO_4
e. $HClO_3$
f. $HClO_4$
g. H_2CO_3
h. $HC_2H_3O_2$

7.34 Write the formulas of the following acids:

a. Sulfuric acid
b. Hydrosulfuric acid
c. Acetic acid
d. Hydrobromic acid
e. Chlorous acid
f. Hypochlorous acid

7.6 A nomenclature summary

7.35 Name each of the following substances:

a. K_2CrO_4
b. NaF
c. FeO
d. Cu_2CO_3
e. $LiOH$
f. CO
g. HCl
h. $HgCl_2$
i. $HClO_2$
j. Cl_2O

7.36 Write a chemical formula for each of the following substances:

a. Ammonium nitrate
b. Silver chloride
c. Sulfur trioxide
d. Copper(I) sulfate
e. Hydroiodic acid
f. Nitric acid
g. Phosphoric acid
h. Carbon tetrachloride
i. Magnesium hydrogen carbonate
j. Potassium permanganate

7.7 Molecular weight and formula weight

7.37 Calculate the molecular or formula weight of the following compounds:

a. I_2
b. NH_3
c. $(NH_4)_2CO_3$
d. K_2SO_4
e. Cl_2O_7
f. $(NH_4)_3PO_4$
g. H_2SO_3
h. $HClO_4$
i. $KMnO_4$
j. $Ca(HCO_3)_2$

7.8 Percentage composition

7.38 Calculate the percentage of each element in:

a. H_3PO_4
b. $Ca(HCO_3)_2$
c. Potassium chromate

d. Barium nitrate
e. Na_2CO_3

7.39 What is the percentage of silver in silver nitrate?

7.40 When 100. g of tin(IV) oxide is decomposed, 78.8 g of tin and 21.2 g of oxygen are formed. Calculate the percentage of tin and oxygen in tin(IV) oxide.

7.41 When 100. g of tin(II) oxide is decomposed, 88.1 g of tin and 11.9 g of oxygen are formed. Calculate the percentage of tin and oxygen in tin(II) oxide.

7.42 When 5.000 g of vitamin C is decomposed, 2.045 g carbon, 0.2290 g hydrogen, and 2.725 g of oxygen are formed. Calculate the percentage of carbon, hydrogen, and oxygen in vitamin C.

7.43 When 25.00 g of an unknown compound is decomposed, 19.97 g of copper and 5.030 g of oxygen are formed. Calculate the percentage of copper and oxygen in this unknown compound.

THE MOLE CONCEPT

CHAPTER 8

8

O

Oxygen

16.00

INDIVIDUALS VERSUS "PACKAGES"

8.1 By the time you read this chapter you should have a mental picture of what individual atoms, ions, and molecules are like. You should also be aware that these particles are extremely tiny—we cannot see, let alone hold, an individual atom, ion, or molecule. Samples of matter that we handle are collections of enormous numbers of these particles. Aluminum foil is a huge collection of Al atoms, a glass of water contains a gigantic number of H_2O molecules, and table salt is a huge collection of Na^+ ions and Cl^- ions.

In this chapter we will see how one can keep track of the numbers of particles in a sample, and, furthermore, how numbers of particles are related to the mass of a sample. Knowledge of this relationship is absolutely essential if one is to develop successful "recipes" for doing chemical reactions in the laboratory.

Historically, the relationship between numbers of particles and mass was worked out in the early nineteenth century, long before there was any detailed knowledge of the nature of particles of matter. The strategy that chemists employed to relate numbers of particles to mass involved defining a unit of matter containing a definite number of particles. This unit of matter is called a **mole** (abbreviated mol). Think of a mole as a "package" containing a definite number of particles (atoms, molecules, etc.) just as a dozen is a package containing 12 eggs or 12 doughnuts or 12 anything (see Figure 8.1). Besides the dozen, we daily encounter other "package" concepts:

Package	Number of Items within Package
Dozen	12
Pair	2
Gross	144
Mole	6.022×10^{23}

THE MOLE CONCEPT

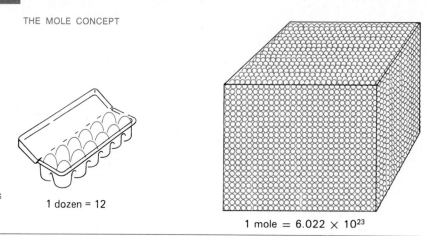

Eggs come in dozens. Atoms and molecules come in moles.

1 dozen = 12

1 mole = 6.022 × 10²³

FIGURE 8.1

The number of items in a mole is so large (6.022×10^{23}) that it is hard to imagine, but the concept is the same as that of the dozen. Moles are packages of particles; dozens are packages of various items, such as eggs.

The mole concept bridges the gap between the **microscopic** world of atoms, molecules, and formula units which we cannot see and handle and the **macroscopic** world of elements and compounds which we can see and hold. Before seeing how the mole package concept can be related to the mass of a sample, it might be useful to review Sections 4.9, 4.10, and 7.7 before proceeding. These sections review atomic, molecular, and formula weight.

Microscopic
atoms, molecules
formula units

amu

invisible

not weighable

The mole
bridge

Macroscopic
"chunks" of matter

grams

visible

weighable

RELATIVE WEIGHTS

8.2 The concept of relative weights (Section 4.9) leads directly to the relationship between numbers of particles, packages of particles, and mass in a sample. Let us choose the elements carbon and oxygen to demonstrate the relationship between numbers of atoms and mass of a sample.

From the periodic table we see that to 4 sig figs the average atomic weights of carbon and oxygen are 12.01 and 16.00 amu, respectively.

As a matter of practicality[1] we can accept these numbers as the relative masses of individual atoms:

Atomic Ratio	Mass Ratio
$\dfrac{1 \text{ atom O}}{1 \text{ atom C}}$	$\dfrac{16.00 \text{ amu}}{12.01 \text{ amu}} = 1.332$

Table 8.1 shows that the mass ratio remains the same so long as the number of oxygen atoms equals the number of carbon atoms. Notice also in this table that the units of mass cancel out because they are identical and therefore equal to 1. The *mass ratio* has no special units associated with it. What Table 8.1 establishes is that *for equal numbers of atoms the mass ratio is always the same as the ratio of the atomic weights given in the periodic table.*

The reverse of this statement is also true and is of great importance: **If the mass ratio of two samples corresponds to the ratio of atomic weights, then the two samples must contain equal numbers of atoms.** This is true of all the examples in Table 8.1 and will always be true, regardless of the units of mass, because identical mass units cancel out.

MASS RATIOS FOR VARYING NUMBERS OF OXYGEN AND CARBON ATOMS   **TABLE 8.1**

Atomic Ratio	Mass Ratio
$\dfrac{1 \text{ atom O}}{1 \text{ atom C}}$	$\dfrac{1 \times 16.00 \text{ amu}}{1 \times 12.01 \text{ amu}} = \dfrac{16.00}{12.01} = 1.332$
$\dfrac{2 \text{ atoms O}}{2 \text{ atoms C}}$	$\dfrac{2 \times 16.00 \text{ amu}}{2 \times 12.01 \text{ amu}} = \dfrac{32.00}{24.02} = 1.332$
$\dfrac{10 \text{ atoms O}}{10 \text{ atoms C}}$	$\dfrac{10 \times 16.00 \text{ amu}}{10 \times 12.01 \text{ amu}} = \dfrac{160.0}{120.1} = 1.332$
$\dfrac{10^6 \text{ atoms O}}{10^6 \text{ atoms C}}$	$\dfrac{10^6 \times 16.00 \text{ amu}}{10^6 \times 12.01 \text{ amu}} = \dfrac{1.600 \times 10^7}{1.201 \times 10^7} = 1.332$
$\dfrac{6.022 \times 10^{23} \text{ atoms O}}{6.022 \times 10^{23} \text{ atoms C}}$	$\dfrac{6.022 \times 10^{23} \times 16.00 \text{ amu}}{6.022 \times 10^{23} \times 12.01 \text{ amu}} = \dfrac{9.635 \times 10^{24}}{7.232 \times 10^{24}} = 1.332$

SAMPLE EXERCISE

8.1

Will 98.07 g of nitrogen atoms and 7.056 g of hydrogen atoms have equal numbers of nitrogen and hydrogen atoms?

---

[1] Section 4.10 carefully explains that average atomic weight is a weighted average of the masses of all isotopes of an element, and therefore no single atom actually has that weight. However, because we must deal with atoms "on the average," we routinely use the periodic table of atomic weights for the weights of single atoms.

SOLUTION

The ratio of the given masses is

$$\frac{\text{Nitrogen}}{\text{Hydrogen}} = \frac{98.07 \cancel{g}}{7.056 \cancel{g}} = 13.90$$

From the periodic table the ratio of the average atomic weights is

$$\frac{14.01 \cancel{\text{amu}}}{1.008 \cancel{\text{amu}}} = 13.90$$

Because the ratio of the masses equals the ratio of the average atomic weights (to 4 sig figs), the numbers of nitrogen and hydrogen atoms will be equal. We have no information as to how many nitrogen or hydrogen atoms are present, but we can be sure that however many nitrogen atoms are present, there must be an equal number of hydrogen atoms.

HOW MANY PARTICLES ARE IN A MOLE?

8.3 Because maintaining equal numbers of atoms in samples of two elements depends on their atomic-weight ratios, and because the unit of mass in the metric system is the gram, it is convenient to consider packages of material corresponding to the atomic weight of the material in grams, or the **gram atomic weight** (GAW). The GAW for the element carbon is 12.01 g. For the element oxygen, GAW = 16.00 g. We know from Table 8.1 that $\frac{16.00 \cancel{g}}{12.01 \cancel{g}} = 1.332$; therefore, the number of atoms in 12.01 g of carbon must be equal to the number of atoms in 16.00 g of oxygen. Or, in general, *the number of atoms in the GAW of one element is always equal to the number of atoms in the GAW of any other element.*

It turns out that the number of atoms in a GAW of any element is 6.022×10^{23}. Unfortunately, the experiments by which this actual number is determined are too complicated to be described in this text. We ask you to accept this number without proof. Thus, the most important concept is

$$\begin{array}{l} \text{GAW} \\ \text{for any} \\ \text{element} \end{array} = \begin{array}{l} \text{the same number of atoms} \\ \text{as for any other element,} \\ \text{namely, } 6.022 \times 10^{23} \text{ atoms} \end{array}$$

This number, 6.022×10^{23}, is called **Avogadro's number** to honor the Italian physicist Amedeo Avogadro, who in the early nineteenth century recognized the importance of being able to relate numbers of particles to measurable quantities of matter. Avogadro never knew his number, but he did know the principle. You must know both the principle and the number.

Now you know what the unit or package called the **mole** contains. Because of the reasoning just detailed, a mole is defined as Avogadro's number of constituent units, and this corresponds to easily assignable weights. For atoms,

$$1 \text{ mole} = 6.022 \times 10^{23} \text{ atoms} = \text{GAW}$$

This gives us equalities and thus conversion factors for relating moles, grams, and numbers of atoms.

SAMPLE EXERCISE

8.2

a. How many grams are there in 1 mole of Mg?
b. How many grams are there in 2 moles of S?

SOLUTION

a. Consult the periodic table and find that the atomic weight of magnesium is 24.31 amu. Therefore, the GAW is 24.31 g.

$$1 \text{ mole Mg} = \text{GAW} = 24.31 \text{ g Mg}$$

b. The GAW of S is 32.06 g. Therefore,

$$1 \text{ mole S} = 32.06 \text{ g S}$$
$$2 \text{ moles S} = 64.12 \text{ g S}$$

Using the Unit Conversion Method,

Given quantity × conversion factor = new

$$2 \text{ moles S} \times \frac{32.06 \text{ g S}}{1 \text{ mole S}} = 64.12 \text{ g S}$$

Problem 8.1

What are the gram atomic weights (GAW) of Al, F, and Br?

Problem 8.2

How many grams are in 3 moles of Al?

CHEMICAL FORMULAS REVISITED

8.4 Before proceeding you should be absolutely sure that you understand the meaning of a chemical formula.

Throughout Chapters 6 and 7, and especially in Sections 6.8, 6.13, and 7.7, we discussed the meaning of chemical formulas and subscripts. For example, the subscripts in its formula tell us that the molecule HNO_3 is made up of one H atom, one N atom, and three O atoms.

For an ionic compound such as Al_2S_3, the subscripts of the formula tell you that there are two Al^{3+} ions and three S^{2-} ions in one formula unit. For $Mg(OH)_2$, the subscripts indicate that one formula unit contains one Mg^{2+} ion and two OH^- ions. We also know that the 2 outside the parentheses is a multiplier for both O and H inside the parentheses.

It is sometimes useful to construct conversion factors which relate the numbers of atoms or ions within a molecule or formula unit to the whole molecule or formula unit. We use our knowledge of subscripts to do this. For example, for HNO_3 we can write the conversion factors

$$\frac{1 \text{ H atom}}{1 \text{ molecule } HNO_3} \qquad \frac{1 \text{ N atom}}{1 \text{ molecule } HNO_3} \qquad \frac{3 \text{ O atoms}}{1 \text{ molecule } HNO_3}$$

For the formula unit $Mg(OH)_2$ we can write

$$\frac{1 \text{ } Mg^{2+} \text{ ion}}{1 \text{ formula unit } Mg(OH)_2} \qquad \frac{2 \text{ } OH^- \text{ ions}}{1 \text{ formula unit } Mg(OH)_2}$$

Thus we see an extension of the conversion-factor concept: the numerator and the denominator of a fraction are considered equivalent if one expresses the numbers of subunits contained in the larger unit.

SAMPLE EXERCISE

8.3

Write the conversion factors which relate atoms within molecules or ions within formula units for the following:

a. N_2O_5 **c.** Na_2SO_4

b. $CaCl_2$ **d.** $Mg_3(PO_4)_2$

SOLUTION

a. This is a molecular compound because N and O are both nonmetals; therefore,

$$\frac{2 \text{ N atoms}}{1 \text{ } N_2O_5 \text{ molecule}} \qquad \frac{5 \text{ O atoms}}{1 \text{ } N_2O_5 \text{ molecule}}$$

The compounds in (b), (c), and (d) are all ionic because they are combinations of metals (Ca, Na, Mg) and nonmetals:

b. $\dfrac{1 \text{ } Ca^{2+} \text{ ion}}{1 \text{ } CaCl_2 \text{ formula unit}} \qquad \dfrac{2 \text{ } Cl^- \text{ ions}}{1 \text{ } CaCl_2 \text{ formula unit}}$

c. $\dfrac{2 \text{ } Na^+ \text{ ions}}{1 \text{ } Na_2SO_4 \text{ formula unit}} \qquad \dfrac{1 \text{ } SO_4^{2-} \text{ ion}}{1 \text{ } Na_2SO_4 \text{ formula unit}}$

d. $\dfrac{3 \text{ } Mg^{2+} \text{ ions}}{1 \text{ } Mg_3(PO_4)_2 \text{ formula unit}} \qquad \dfrac{2 \text{ } PO_4^{3-} \text{ ions}}{1 \text{ } Mg_3(PO_4)_2 \text{ formula unit}}$

When a *coefficient* is placed before a chemical formula, it *multiplies the entire formula.* For example, $2 N_2O_5$ means two molecules of N_2O_5; therefore, four N atoms and ten O atoms must be present:

$$2 \text{ molecules } N_2O_5 \times \frac{2 \text{ N atoms}}{1 \text{ molecule } N_2O_5} = 4 \text{ N atoms}$$

$$2 \text{ molecules } N_2O_5 \times \frac{5 \text{ O atoms}}{1 \text{ molecule } N_2O_5} = 10 \text{ O atoms}$$

Similarly, $3 Na_2SO_4$ means three formula units of Na_2SO_4, and therefore a total of six Na^+ ions and three SO_4^{2-} ions are present:

$$3 \text{ formula units } Na_2SO_4 \times \frac{2 \text{ Na}^+ \text{ ions}}{1 \text{ formula unit } Na_2SO_4} = 6 \text{ Na}^+ \text{ ions}$$

$$3 \text{ formula units } Na_2SO_4 \times \frac{1 \text{ SO}_4^{2-} \text{ ions}}{1 \text{ formula unit } Na_2SO_4} = 3 \text{ SO}_4^{2-} \text{ ions}$$

As we proceed, the new ideas introduced in this section—namely, the construction of conversion factors relating the number of constituent atoms or ions to their molecules or formula units and the meaning of coefficients—will be used repeatedly.

Problem 8.3

How many PO_4^{3-} ions are represented by $4 Mg_3(PO_4)_2$?

MOLES OF COMPOUNDS

8.5 Because the mole is defined as Avogadro's number of constituent units, this means that for compounds,

1 **mole** of a molecular compound $= 6.022 \times 10^{23}$ molecules

1 **mole** of an ionic compound $= 6.022 \times 10^{23}$ formula units

Just as 1 mole = Avogadro's number = GAW for elements, it can be seen that for molecular compounds,

$$1 \text{ mole} = 6.022 \times 10^{23} \text{ molecules} = GMW$$

or for ionic compounds,

$$1 \text{ mole} = 6.022 \times 10^{23} \text{ formula units} = GFW$$

where GMW is the molecular weight expressed in grams, and the GFW is the formula weight expressed in grams.

Therefore, to relate moles and grams for any material, look up or calculate the atomic weight, molecular weight, or formula weight (Section 7.7), and that number of grams is 1 mole.

Collections of atoms: 1 mole = GAW
Molecular compounds: 1 mole = GMW
Ionic compounds: 1 mole = GFW

SAMPLE EXERCISE

8.4

How many grams are in 1 mole of each of the following?
a.   Al   b. N_2O_5 c. Na_2SO_4 d.   O   e. O_2

SOLUTION

a.   The atomic weight of aluminum is 26.98 amu. The atomic weight in grams, or gram atomic weight (GAW), is 26.98 g.

$$1 \text{ mole Al} = 26.98 \text{ g Al}$$

b.   To find the molecular weight (Section 7.7), we sum up the atomic weights of all the atoms present. For N_2O_5 this is

$$
\begin{array}{rcll}
2 \times N = 2 \times 14.01 = & 28.02 \text{ amu} \\
5 \times O = 5 \times 16.00 = & \underline{80.00 \text{ amu}} \\
& 108.02 \text{ amu}
\end{array}
$$

The molecular weight of N_2O_5 is 108.02 amu. The molecular weight in grams, or gram molecular weight (GMW), is 108.02 g.

$$1 \text{ mole } N_2O_5 = 108.02 \text{ g } N_2O_5.$$

c.   To find the formula weight (Section 7.7), we sum up the weights of all the ions present in one formula unit. For Na_2SO_4 this is

$$
\begin{array}{rcll}
2 \times Na^+ \ \ = 2 \times 22.99 \text{ amu} = & 45.98 \text{ amu} \\
1 \times SO_4^{2-} = 1 \times 96.06 \text{ amu} = & \underline{96.06 \text{ amu}} \\
& 142.04 \text{ amu}
\end{array}
$$

The formula weight of Na_2SO_4 is 142.04 amu. The formula weight in grams is 142.04 g.

$$1 \text{ mole } Na_2SO_4 = 142.04 \text{ g } Na_2SO_4.$$

d.   The atomic weight of oxygen is 16.00 amu. The atomic weight in grams, or gram atomic weight (GAW), is 16.00 g.

$$1 \text{ mole O} = 16.00 \text{ g O}$$

e.   The molecular weight of oxygen is

$$2 \times O = 2 \times 16.00 = 32.00 \text{ amu}$$

The molecular weight in grams, or gram molecular weight (GMW), is 32.00 g.

$$1 \text{ mole } O_2 = 32.00 \text{ g}$$

Let us show that for a molecular compound if 1 mole is defined as 6.022×10^{23} molecules, then 1 mole must also equal the gram molecular weight (GMW). As an example we will examine the molecular compound carbon disulfide, CS_2. One molecule of CS_2 contains one carbon atom and two sulfur atoms.

$S \; C \; S$ 1 molecule

One mole of CS_2 is 6.022×10^{23} molecules of CS_2. In 6.022×10^{23} molecules of CS_2 there must be 6.022×10^{23} carbon atoms and $2 \times (6.022 \times 10^{23})$ sulfur atoms.

$$6.022 \times 10^{23} \; \text{molecules } CS_2 \times \frac{1 \; C \; \text{atom}}{1 \; \text{molecule } CS_2}$$

$$= 6.022 \times 10^{23} \; C \; \text{atoms}$$

$$= 1 \; \text{mole of } C$$

$$6.022 \times 10^{23} \; \text{molecules } CS_2 \times \frac{2 \; S \; \text{atoms}}{1 \; \text{molecule } CS_2}$$

$$= 2 \times 6.022 \times 10^{23} \; S \; \text{atoms}$$

$$= 2 \; \text{moles of } S$$

or 1 mole of C and 2 moles of S.

The weight of 1 mole of CS_2 must then be the sum of the weights of 1 mole C and 2 moles S (see Sample Exercise 8.2).

$$
\begin{aligned}
1 \; \text{mole C} &= 12.01 \; \text{g} \\
2 \; \text{moles S} &= \underline{64.12 \; \text{g}} \\
1 \; \text{mole } CS_2 &= 76.13 \; \text{g } CS_2
\end{aligned}
$$

This is the GMW of CS_2 (see Figure 8.2). By following similar reasoning you now can show that 1 mole is equal to the gram formula weight (GFW) of an ionic compound.

Problem 8.4

How many grams are there in 1 mole of $CaCl_2$?

FIGURE 8.2

A mole of material weighs exactly the GAW, GMW, or GFW of that material.

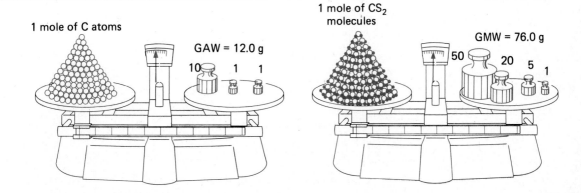

1 mole of C atoms

GAW = 12.0 g
10   1   1

1 mole of CS_2 molecules

GMW = 76.0 g
50   20   5   1

GRAM-MOLE-PARTICLE CONVERSIONS

8.6 Probably the conversion factors that you will use most often in chemistry will be those which relate moles to grams through the GAW, GMW, or GFW. Occasionally you will also need to remember the relationship of moles to numbers of particles (atoms, molecules, formula units) as well.

Because the GAW, GMW, and GFW all refer to the weight of 1 mole of material, we can use the more general term **"molar weight"** in referring to any of the three. The molar weight of a monatomic element is the element's GAW; the molar weight of a molecular compound is its GMW; and the molar weight of an ionic compound is its GFW.

Molar weight	is another name for	GAW, GMW, or GFW.

From now on in this text we will use the term "molar weight" instead of GAW, GMW, or GFW. The molar weight is the weight of 1 mole of anything. Figure 8.3 displays molar quantities of some common chemical substances.

SAMPLE EXERCISE

8.5

Perform the conversions indicated.

a. How many grams are contained in 1.45 moles H_2O?
b. How many moles correspond to 50.5 g of H_2S?
c. 14.9 g Na_2SO_4 is equivalent to how many moles of Na_2SO_4?

FIGURE 8.3

Molar quantities of some common chemical substances. (*Photo by Bob Rogers.*)

d. How many molecules of NH_3 are in 7.2 moles NH_3?

e. How many P atoms weigh 15.91 g?

SOLUTION

In all the parts of this exercise we will be using one or more of the conversion factors resulting from the equalities

$$1 \text{ mole} = 6.022 \times 10^{23} \text{ particles} = \textbf{molar weight}$$

and the Unit Conversion Method.

a. *Step 1* The given quantity and unit is 1.45 moles H_2O.

Step 2 The new unit is grams H_2O.

Step 3 The equality relating the given and new is

$$1 \text{ mole} = \text{molar weight}$$

$$1 \text{ mole } H_2O = 18.0 \text{ g } H_2O$$

Step 4

Given × conversion factor = new quantity and unit

$$\frac{\text{(new units)}}{\text{(given units)}}$$

Therefore,

$$1.45 \;\cancel{\text{moles } H_2O} \times \frac{18.0 \text{ g } H_2O}{1 \;\cancel{\text{mole } H_2O}} = 26.1 \text{ g } H_2O$$

b. *Step 1* Given is 50.5 g H_2S.

Step 2 New unit is moles H_2S.

Step 3 Equality is

$$1 \text{ mole} = \text{molar weight}$$

$$1 \text{ mole } H_2S = 34.1 \text{ g } H_2S$$

Step 4

$$\textbf{Given} \times \frac{\textbf{new units}}{\textbf{old units}} = \textbf{answer}$$

$$50.5 \;\cancel{\text{g } H_2S} \times \frac{1 \text{ mole } H_2S}{34.1 \;\cancel{\text{g } H_2S}} = 1.48 \text{ moles } H_2S$$

c. *Step 1* Given is 14.9 g Na_2SO_4.

Step 2 New unit is moles Na_2SO_4.

Step 3 Equality is

$$1 \text{ mole} = \text{molar weight}$$

$$1 \text{ mole } Na_2SO_4 = 142 \text{ g } Na_2SO_4$$

Step 4

$$\text{Given} \times \frac{\text{new units}}{\text{old units}} = \text{answer}$$

$$14.9 \text{ g Na}_2\text{SO}_4 \times \frac{1 \text{ mole Na}_2\text{SO}_4}{142 \text{ g Na}_2\text{SO}_4} = 0.105 \text{ mole Na}_2\text{SO}_4$$

d. *Step 1* Given is 7.2 moles NH_3.

Step 2 New unit is molecules of NH_3.

Step 3 Equality is 1 mole = 6.022×10^{23} molecules.

Step 4

$$\text{Given} \times \text{conversion factor} = \text{answer}$$

$$7.2 \text{ moles NH}_3 \times \frac{6.022 \times 10^{23} \text{ molecules NH}_3}{1 \text{ mole NH}_3}$$

$$= 4.3 \times 10^{24} \text{ molecules NH}_3$$

e. *Step 1* Given is 15.91 g P.

Step 2 New unit is atoms of P.

Step 3 In this case, two equalities and two conversion factors are necessary

$$1 \text{ mole P} = 30.97 \text{ g P}$$

$$1 \text{ mole P} = 6.022 \times 10^{23} \text{ atoms P}$$

Step 4 Set up the conversion factors so that given units cancel out and new units arise.

$$\text{Given} \times \text{conversion factors} = \text{answer}$$

$$15.91 \text{ g P} \times \frac{1 \text{ mole P}}{30.97 \text{ g P}} \times \frac{6.022 \times 10^{23} \text{ atoms P}}{1 \text{ mole P}} = 3.094 \times 10^{23} \text{ atoms P}$$

Problem 8.5

How many moles of lithium sulfide correspond to 75.8 g of lithium sulfide?

MOLES WITHIN MOLES

8.7 In Section 8.4 we reviewed the meaning of chemical formulas and you saw how to deal with coefficients when they precede formulas. Therefore, you know that, for example, 2 NaCl indicates 2 formula units of sodium chloride, hence 2 Na^+ ions and 2 Cl^- ions.

When we write 1 mole of NaCl, this means 6.022×10^{23} formula units of NaCl, and hence 6.022×10^{23} Na^+ ions and 6.022×10^{23} Cl^- ions.

$$6.022 \times 10^{23} \text{ NaCl} \quad \text{contains} \quad \begin{cases} 6.022 \times 10^{23} \text{ Na}^+ \text{ ions} \\ \text{and} \\ 6.022 \times 10^{23} \text{ Cl}^- \text{ ions} \end{cases}$$

So we see that 1 mole of NaCl contains 1 mole of Na^+ ions (6.022×10^{23} ions) and 1 mole of Cl^- ions (6.022×10^{23} ions).

$$1 \text{ mole NaCl} \quad \text{contains} \quad \begin{cases} 1 \text{ mole Na}^+ \text{ ions} \\ 1 \text{ mole Cl}^- \text{ ions} \\ 2 \text{ moles of ions} \end{cases}$$

Some students have difficulty visualizing 2 moles of particles coming from 1 mole. Remembering that the mole concept is like the dozen concept might be helpful in this case. In one dozen eggs, there are one dozen yolks and one dozen whites. So we see two dozen subunits (one dozen yolks and one dozen whites) within one dozen of the whole eggs. As another example, in 1 mole of married couples, there are 1 mole of males and 1 mole of females, that is, 2 moles of human beings within 1 mole of couples.

Just as the subscripts in a chemical formula tell us the number of atoms or ions within a molecule or formula unit, the *subscripts also tell us the numbers of moles of atoms or ions contained within 1 mole of the compound.* For NaCl, the unexpressed subscripts are ones, so 1 mole NaCl holds within it 1 mole Na^+ and 1 mole Cl^-. In 1 mole of Na_2SO_4 there are 2 moles Na^+ (because the subscript is 2) and 1 mole $SO_4{}^{2-}$. In 1 mole N_2O_5 there are 2 moles of N and 5 moles of O.

Conversion factors can be written that relate the moles of the components of a compound to the moles of the compound itself. For example, for $Mg(OH)_2$,

$$\frac{1 \text{ mole Mg}^{2+}}{1 \text{ mole Mg(OH)}_2} \quad \text{and} \quad \frac{2 \text{ moles OH}^-}{1 \text{ mole Mg(OH)}_2}$$

are two possible conversion factors.

SAMPLE EXERCISE

8.6

How many moles of Li^+ ions and S^{2-} ions are there in 3.75 moles of Li_2S?

SOLUTION

Given $\times$ conversion factor = number of moles of each ion

$$3.75 \text{ moles Li}_2\text{S} \times \frac{2 \text{ moles Li}^+}{1 \text{ mole Li}_2\text{S}} \qquad = 7.50 \text{ moles Li}^+$$

$$3.75 \text{ moles } Li_2S \times \frac{1 \text{ mole } S^{2-}}{1 \text{ mole } Li_2S} \qquad = 3.75 \text{ moles } S^{2-}$$

SAMPLE EXERCISE

8.7

How many moles of nitrogen atoms and oxygen atoms are contained in 3.00 moles of N_2O_4?

SOLUTION

Given × conversion factor = answer

$$3.00 \text{ moles } N_2O_4 \times \frac{2 \text{ moles N atoms}}{1 \text{ mole } N_2O_4} = 6.00 \text{ moles N atoms}$$

$$3.00 \text{ moles } N_2O_4 \times \frac{4 \text{ moles O atoms}}{1 \text{ mole } N_2O_4} = 12.0 \text{ moles O atoms}$$

Of course, as you already saw in Chapter 6 (especially Section 6.13), there are no *free* N or O atoms in N_2O_4. In this example all 6.00 moles of N atoms and 12.0 moles of O atoms are bonded together to form 3.00 moles of N_2O_4 molecules.

Problem 8.6

How many moles of K^+ ions and PO_4^{3-} ions are there in 4.25 moles of K_3PO_4?

EMPIRICAL FORMULAS DEFINED

8.8   There are other kinds of chemical formulas besides the molecular or ionic formulas that we have been discussing. The simplest formula of all, and the one that is obtained directly from experimental data, is the **empirical** formula. In science, "empirical" is just another word for "experimental."

   The empirical formula of a compound is one in which the subscripts are in the form of the simplest whole-number ratio. For example, for the compound nitrogen tetroxide, for which the molecular formula is N_2O_4, the empirical formula is NO_2 because the ratio 2:4 can be reduced to the simpler ratio 1:2. Nitrogen tetroxide only contains molecules in which two nitrogen and four oxygen atoms are bonded together. However, the simplest ratio of the bonding atoms, called the **empirical formula,** is 1:2. Similarly, the compound benzene has the molecular formula C_6H_6, but the empirical formula would be CH because the ratio 6:6 can be reduced to 1:1. Often the molecular and empirical formulas are identical. For example, for the compound carbon dioxide, for which the molecular formula is CO_2, the empirical formula is also CO_2, since the ratio 1:2 cannot be reduced further. For ionic compounds, the ionic formula is almost always identical with the empirical formula. For example, consider Na_2SO_4, Li_2S, and Al_2O_3.

SAMPLE EXERCISE

8.8

Give the empirical formulas for each of the following compounds, which are represented by their molecular formulas:

a. Glucose, $C_6H_{12}O_6$
b. Chloroform, $CHCl_3$
c. Ethylene glycol, $C_2H_6O_2$

SOLUTION

a. $6:12:6$ can be reduced to $1:2:1$ by dividing through by 6. Therefore, the empirical formula is CH_2O.
b. $CHCl_3$ cannot be further simplified. $CHCl_3$ is both the molecular and empirical formula.
c. $2:6:2$ can be reduced to $1:3:1$ by dividing through by 2. Therefore, the empirical formula is CH_3O.

Remember that the subscripts in a formula tell you not only the ratio of atoms in a molecule (or ions in a formula unit), but also the ratio of the moles of the constituent components of a compound. For example, the ratio $1:2$ in CO_2 represents the ratio one carbon atom to two oxygen atoms in one CO_2 molecule, and also 1 mole of carbon to 2 moles of oxygen in 1 mole of CO_2.

SAMPLE EXERCISE

8.9

Give the molar ratio of the elements in the following compounds:

a. NH_3
b. $CaBr_2$
c. Al_2S_3

SOLUTION

a. The subscripts are in the ratio $1:3$; therefore, the molar ratio is 1 mole of N to 3 moles of H.
b. 1 mole of Ca to 2 moles of Br
c. 2 moles of Al to 3 moles of S

CALCULATION OF EMPIRICAL FORMULAS

8.9 In Section 7.8 we discussed percentage composition of elements in a compound. Now that we have also discussed moles and defined empirical formulas, it can be understood how chemists use experimental data obtained in the laboratory to determine correct chemical formulas for compounds. Percentage-composition data is obtained directly from experiment. Such data can then be translated into empirical formulas, as you shall see shortly. Then in Section 8.10 you will see how empirical formulas are translated into molecular formulas.

Calculation of empirical formulas from percentage composition by weight

1. Calculate the number of grams of each element present in some number of grams (usually choose 100 g) of compound.

Percentage of element × grams of compound = grams of element
    (In decimal form)

(If 100 g of compound is chosen, then percent of element equals grams of element.)

2. Calculate the number of moles of each element in the compound.

$$\text{Grams of element} \times \frac{1 \text{ mole element}}{\text{molar weight}} = \text{moles element}$$

This will establish a molar ratio of the elements in the compound.

3. Simplify the molar ratio by dividing through by the smallest number.

4. Round off the simplified ratio to 2 sig figs.

5. If the ratio turns out to be all whole numbers to 2 sig figs, you have the answer. If not, skip to step 6.

6. If the ratio is not integral to 2 significant figures, find a number which when multiplied times the simplified experimental ratio, converts that ratio to a ratio of whole numbers to 2 sig figs?

Carefully follow through Sample Exercises 8.10 and 8.11 because they apply this stepwise procedure.

SAMPLE EXERCISE

8.10

A compound is found to contain 32.38 percent Na, 22.58 percent S, and 45.07 percent O. What is the empirical formula of the compound?

SOLUTION

Follow the procedure just described for the calculation of empirical formulas.

Step 1 Choose a 100-g sample of the compound in order to simplify the calculation:

Percent of element × grams of compound = grams of element
(In decimal form)

Na:	0.3238	× 100	= 32.38 g Na
S:	0.2258	× 100	= 22.58 g S
O:	0.4507	× 100	= 45.07 g O

Notice that the gram amounts = the percentages given. This is only because we chose 100 g of compound.

Step 2 Convert grams of the element to moles of the element.

$$Na: 32.38 \text{ g Na} \times \frac{1 \text{ mole Na}}{22.99 \text{ g Na}} = 1.408 \text{ moles Na}$$

$$S: 22.58 \text{ g S} \times \frac{1 \text{ mole S}}{32.06 \text{ g S}} = 0.7043 \text{ moles S}$$

$$O: 45.07 \text{ g O} \times \frac{1 \text{ mole O}}{16.00 \text{ g O}} = 2.817 \text{ moles O}$$

Step 3 The molar ratio is

$$Na:S:O = 1.408:0.7043:2.817$$

Simplify by dividing through by the smallest number, which in this case is 0.7043.

$$\frac{1.408}{0.7043} = 1.999 \qquad \frac{0.7043}{0.7043} = 1.000 \qquad \frac{2.817}{0.7043} = 4.000$$

Now

$$Na:S:O = 1.999:1.000:4.000$$

Step 4 Rounding off the ratio to 2 sig figs, we get

$$Na:S:O = 2.0:1.0:4.0$$

Step 5 The numbers of the ratio are all whole numbers, so we have the subscripts for the empirical formula: Na_2SO_4.

SAMPLE EXERCISE

8.11

A compound is found to contain 69.94 percent Fe and 30.06 percent O. Determine the empirical formula.

SOLUTION

Step 1 In 100 g of compound there are

$$0.6994 \times 100 = 69.94 \text{ g Fe}$$

$$0.3006 \times 100 = 30.06 \text{ g O}$$

Step 2   The numbers of moles of elements in 100 g of compound are:

$$69.94 \text{ g Fe} \times \frac{1 \text{ mole Fe}}{55.85 \text{ g Fe}} = 1.252 \text{ moles Fe}$$

$$30.06 \text{ g O} \times \frac{1 \text{ mole O}}{16.00 \text{ g O}} = 1.879 \text{ moles O}$$

Step 3   The molar ratio is Fe:O = 1.252:1.879, which is simplified by division by 1.252 to give Fe:O = 1.000:1.501.

Step 4   Rounding off to 2 sig figs, we get Fe:O = 1.0:1.5.

Step 5   The numbers of the ratio are not whole numbers, so we must skip to step 6.

Step 6   If we multiply the simplified experimental ratio by 2, then we will have achieved a whole-number ratio to 2 significant figures:

$$2(1.000:1.501) = 2.000:3.002 = 2.0:3.0$$

Therefore, the subscripts are 2 and 3, and the formula is Fe_2O_3.

Very often the molar ratio is integral and we can stop at step 5 (e.g., Sample Exercise 8.10). When step 6 is necessary (e.g., Sample Exercise 8.11), the multiplier is usually 2, 3, or 4. Just keep multiplying until whole numbers are achieved.

SAMPLE EXERCISE

8.12

The molar ratio of elements in a compound is found to be 1.00:1.33. Convert this to a whole-number ratio.

SOLUTION

Apply steps 4, 5, and 6.

Step 4   Round to 2 sig figs: 1.0:1.3.

Step 5   Because the ratio is not composed of whole numbers, you must proceed to step 6.

Step 6   Try 2 as a multiplier.

$$2(1.00:1.33) = 2.000:2.666 = 2.0:2.7$$

Obviously multiplication by 2 does not achieve a whole-number ratio to 2 significant figures. Try 3 as a multiplier.

$$3(1.00:1.33) = 3.00:3.99 = 3.0:4.0$$

This time we have achieved a whole-number ratio to 2 sig figs.

Problem 8.7

Calculate the empirical formulas for compounds with the following percentage compositions:

a. 52.4 percent K and 47.6 percent Cl

b. 37.2 percent C, 7.82 percent H, and 55.0 percent Cl

Sometimes experimental data about the elemental makeup of a compound is given in a form other than percentage composition. For example, we might be told that a compound is made up of exactly 17.9 g of C and 4.50 g of H. This type of data simplifies the calculation of empirical formula because in this case we are given the grams of each element in the compound and can proceed directly to the calculation of the molar ratio.

$$17.9 \ \cancel{g \ C} \times \frac{1 \ \text{mole C}}{12.0 \ \cancel{g \ C}} = 1.49 \ \text{moles C}$$

$$4.50 \ \cancel{g \ H} \times \frac{1 \ \text{mole H}}{1.01 \ \cancel{g \ H}} = 4.46 \ \text{moles H}$$

$$C:H = 1.49:4.46 = 1.00:2.99$$

To 2 sig figs this is $1.0:3.0$, and the empirical formula is CH_3.

Problem 8.8

In the laboratory, 56.00 g of oxygen is found to combine with 3.528 g of hydrogen to form a compound. The compound is found to have properties that are distinct from those of water. What is the empirical formula of this unknown compound?

MOLECULAR FORMULAS

8.10 We have seen (Section 8.8) that the difference between a molecular formula and an empirical formula is that the molecular formula represents the actual number of atoms in one molecule whereas the empirical formula gives only the smallest whole-number ratio of the atoms in one molecule. Sample Exercise 8.8 demonstrated that the relationship between the subscripts of a molecular formula and the subscripts of an empirical formula was given by some whole-number (integral) multiplier.

Molecular formula = whole number × empirical formula

Experimental percentage-composition data enables us to calculate an empirical formula. In order to determine the appropriate whole-number multiplier and hence the molecular formula from the empirical formula, we also need to know the experimental molecular weight of the compound.

The method of calculating the molecular formula from the empirical formula is therefore:

1. Calculate the **empirical weight,** the weight of the empirical formula.

TABLE 8.2   EMPIRICAL AND MOLECULAR FORMULAS

Name of Substance	Empirical Formula	Empirical Weight	Molecular Weight	$\dfrac{\text{Molecular Weight}}{\text{Empirical Weight}} = $ Integral Multiple	Molecular Formula
Benzene	CH	13	78	6	C_6H_6
Hydrogen peroxide	HO	17	34	2	H_2O_2
Dinitrogen tetroxide	NO_2	46	92	2	N_2O_4
Propylene	CH_2	14	42	3	C_3H_6
Sulfur dioxide	SO_2	64	64	1	SO_2

2. Divide the molecular weight by the empirical weight to find the integral multiple relating them.

3. Multiply the subscripts of the empirical formula by the multiplier determined in step 2.

SAMPLE EXERCISE

8.13

A compound of hydrogen and oxygen has an empirical formula HO and a molecular weight of 34. What is the molecular formula?

SOLUTION

1. Since the empirical formula is HO, the empirical weight is 17 (1.0 for H + 16 for O).

2. $\dfrac{\text{Molecular weight}}{\text{Empirical weight}} = \dfrac{34}{17} = 2$

3. We multiply each subscript in the empirical formula by the integral multiple to obtain the molecular formula.

$$\text{Molecular formula} = H_{1\times2}O_{1\times2} = H_2O_2$$

Sometimes a problem asks you to find both the empirical and the molecular formulas. Remember to always find the empirical formula first.

SAMPLE EXERCISE

8.14

Acetylene, a compound used in welding, has a percentage composition of 92.24 percent C and 7.742 percent H and a molecular weight of 26.04. What is the molecular formula of acetylene?

SOLUTION

First determine the empirical formula:

Step 1 In 100.0 g of acetylene there are 92.24 g C and 7.742 g H.

Step 2 Determine the number of moles of C and H:

$$92.24\ \cancel{g\ C} \times \frac{1\ \text{mole C}}{12.01\ \cancel{g\ C}} = 7.680\ \text{moles C}$$

$$7.742\ \cancel{g\ H} \times \frac{1\ \text{mole H}}{1.008\ \cancel{g\ H}} = 7.681\ \text{moles H}$$

Step 3 The molar ratio is C:H = 7.680:7.681, which is simplified to C:H = 1.0:1.0. Thus the empirical formula is CH.

Then use the empirical formula to find the molecular formula:

Step 1 Calculate the empirical weight for CH:

$$12.01 + 1.008 = 13.02$$

Step 2

$$\frac{\text{Molecular weight}}{\text{Empirical weight}} = \frac{26.04}{13.02} = 2$$

Step 3

$$\text{Molecular formula} = C_{1 \times 2}H_{1 \times 2} = C_2H_2$$

SUMMARY

Chemists think of matter as coming in packages of large numbers of particles (Section 8.1). These packages are called **moles.** One mole always contains 6.022×10^{23} constituent units of a sample of matter, and the mole always corresponds to the relative weight of the constituent unit in grams, that is, GAW, GMW, or GFW (Sections 8.3 and 8.5). The term "molar weight" is used in place of any of the more specific terms GAW, GMW, or GFW. The concept of relative weights (Section 8.2) leads us directly to these conclusions about the contents of the packages called moles. Gram-mole-particle conversions (Section 8.6) are done by the Unit Conversion Method, using conversion factors derived from the equalities

1 mole = 6.022×10^{23} constituent units = molar weight

Coefficients that appear before chemical formulas act as multipliers for the entire formula (Section 8.4). Subscripts of chemical formulas can be interpreted as describing the numbers of constituent atoms or ions within a molecule or formula unit, *and* they also tell us the numbers of moles of atoms or ions within 1 mole of a compound (Section 8.7).

An empirical formula is one in which the subscripts are in the simplest possible whole-number ratio (Section 8.8). Empirical formulas are obtained directly from experimental percentage-composition data (Section 8.9). If the molecular weight of a compound is already known, the molecular formula can be calculated from the empirical formula (Section 8.10).

CHAPTER ACCOMPLISHMENTS

After completing this chapter you should be able to

8.1 Individuals versus "packages"

1. Compare the use of the mole as a package unit to the use of the dozen as a package unit.

8.2 Relative weights

2. Recognize that equal numbers of atoms are present in any two samples of elements if the mass ratio of the samples corresponds to the atomic-weight ratio of the elements.

8.3 How many particles are in a mole?

3. Recognize that the abbreviation GAW stands for gram atomic weight.
4. State the relationship between the number of atoms in a GAW of one element and the number of atoms in a GAW of any other element.
5. Define mole.
6. State Avogadro's number.
7. Relate 1 mole of atoms to the number of atoms in a mole.
8. Relate 1 mole of atoms in an element to the mass of the element.

8.4 Chemical formulas revisited

9. Construct conversion factors that relate the numbers of atoms in a molecule to the molecule that contains them.
10. Construct conversion factors that relate the numbers of ions in a formula unit to the formula unit that contains them.
11. Correctly interpret coefficients that are placed before chemical formulas.

8.5 Moles of compounds

12. Relate 1 mole of a molecular compound to a number of molecules.
13. Relate 1 mole of an ionic compound to a number of formula units.
14. Recognize that the abbreviation GMW stands for gram molecular weight.
15. Recognize that the abbreviation GFW stands for gram formula weight.
16. Relate 1 mole of a compound to the mass of the compound.

8.6 Gram-mole-particle conversions

17. Define molar weight.
18. Given the formula of a substance, convert
    a. A given number of moles of that substance to grams of that substance.
    b. A given number of grams of that substance to moles of that substance.

c. A given number of moles of that substance to particles of that substance.

d. A given number of grams of that substance to particles of that substance.

8.7 Moles within moles

19.   Interpret subscripts of chemical formulas in terms of moles of constituent particles.

20.   Construct conversion factors that relate moles of atoms to moles of molecular compounds.

21.   Construct conversion factors that relate moles of ions to moles of ionic compounds.

8.8 Empirical formulas defined

22.   Define empirical formula.

23.   Given a molecular formula, determine the empirical formula.

8.9 Calculation of empirical formulas

24.   Calculate an empirical formula from percentage composition by weight.

25.   Calculate an empirical formula from information about combining weights.

8.10 Molecular formulas

26.   Calculate a molecular formula from the empirical formula and the molecular weight.

PROBLEMS

8.1 Individual versus packages

8.9 What kinds of packages, other than a dozen, are you familiar with?

8.10 **a.** Give an example of an object found in the microscopic world.
b. Give an example of an object found in the macroscopic world.
c. Describe one difference between the microscopic and macroscopic worlds.

8.2 Relative weights

8.11 Calculate the mass ratio of a neon atom to a hydrogen atom.

8.12 Refer to Problem 8.11. Which atom is heavier, Ne or H? How many times heavier?

8.13 **a.** Will the number of sodium atoms in 22.99 g of sodium be equal to the number of sulfur atoms in 22.99 g of sulfur?
b. Which mass, the sodium or the sulfur, will contain the greater number of atoms?

8.14 Do 64.00 tons of molecular oxygen and 48.04 tons of carbon contain equal numbers of atoms (4 significant figures)?

8.3 How many particles in a mole?

8.15 Give the gram atomic weight, to 4 sig figs, for the following elements:
   a. He
   b. Se
   c. F
   d. Au
   e. Li
   f. Ge

8.16 a. Give the gram atomic weight of helium to 4 sig figs.
b. How many helium atoms are present in the mass of helium described in part (a)?
c. Calculate the mass in grams of one helium atom.

8.17 What would be the mass of Avogadro's number of magnesium atoms?

8.18 a. There are approximately 5 billion (5 × 10^9) people on the earth. The budget of the U.S. government is approximately a trillion dollars ($1 × 10^{12}). How many dollars would be distributed if every person on earth were to receive an amount of money equal to the entire budget of the U.S. government?
b. How many dollars would be distributed if a mole of people each received one penny?
c. In which case [(a) or (b)] would the greater number of dollars be distributed?

8.19 a. An atom of "starwarsane" (a new element) has a mass of 4.367 × 10^{-22} g. If a robot requires Avogadro's number of starwarsane atoms for smooth running, how many grams does the robot require?
b. What is the GAW of starwarsane?

8.20 a. You just inherited Avogadro's number of dollars. Assuming that you can spend or give money away at a rate of one million dollars per second, how many years would it take to spend your inheritance?
b. The estimated age of the earth is 4.6 billion years. Is your spending time longer or shorter than the earth's age?

8.21 The average mass of an adult male is 70. kg. What is the mass of Avogadro's number of average adult males? Compare this mass to the mass of the earth (6.0 × 10^{24} kg).

8.22 How many grams are there in 2.00 moles of potassium?

8.23 How many grams are there in 3.95 moles of sulfur?

8.24 How many grams are there in 5.99 moles of copper?

8.25 How many moles are there in 5.19 g of magnesium?

8.26 How many moles are there in 87.9 g of aluminum?

8.27 How many moles are there in 2.19 g of iron?

8.28 How many moles are there in 211 mg of lithium?

8.29 How many moles of chromium do we need in order to have as many chromium atoms as there are iron atoms in 1.75 moles of iron?

8.30 You have 7.5 moles of silicon. How many moles of carbon should be measured out to ensure that you have an equal number of Si atoms and C atoms?

8.31 You already possess 0.250 moles of krypton. How many grams of argon would ensure a number of argon atoms that equals the number of atoms in the krypton sample?

8.32 a. A student has weighed out 14.5 g of zinc; how many grams of nickel should the student weigh out to ensure equal numbers of zinc and nickel atoms?
b. Suppose the student needs twice as many nickel atoms as zinc atoms; how many grams of nickel should be weighed out?

8.33 Complete the following table:

Element	Number of moles	Number of atoms	Mass, g
Lithium	1.50	_____	_____
Calcium	_____	6.022 × 10^{23}	_____
Phosphorus	_____	2.50 × 10^{23}	_____
Silicon	_____	_____	88.8g
Zinc	3.90	_____	_____

8.34 Arrange the following in order from highest mass to lowest mass:
a. 1.00 moles of magnesium
b. 1.00 moles of iron
c. 1.00 moles of neon

8.4 Chemical formulas revisited

8.35 Construct conversion factors which relate the numbers of hydrogen, nitrogen, and oxygen atoms to one molecule of HNO_3.

8.36 Construct conversion factors which relate the number of calcium ions and the number of phosphate ions to one formula unit of $Ca_3(PO_4)_2$.

8.37 Construct conversion factors which relate the

numbers of atoms of each element in C_3H_7NO to one molecule of that compound.

8.38 What is the total number of ions that can be found in one formula unit of $(NH_4)_2CO_3$? What are the ions?

8.39 How many aluminum ions are present in 6 $AlCl_3$? How many chloride ions?

8.40 Construct a conversion factor which relates the number of oxygen atoms to one molecule of oxygen.

8.5 Moles of compounds

8.41 Compute the molar weight of each of the following:

a. KCl	d. $(NH_4)_2S$
b. SO_3	e. $NaHCO_3$
c. H_2SO_4	f. $Mg_3(PO_4)_2$

8.42 How many grams are there in 1.00 mole of each of the following:

a. N_2	d. Na^+
b. I_2	e. CO_3^{2-}
c. $C_6H_6O_6$	f. $Al(HCO_3)_3$

8.43 **a.** Suppose you wish to combine equal numbers of H_2 molecules and I_2 molecules. If you have 1.50 moles of H_2, how many moles of I_2 should you measure out?
b. If you had 4.04 g of H_2, how many moles of I_2 would be required to ensure equal numbers of H_2 and I_2 molecules? How many grams of I_2 is this?

8.44 What is the mass of Avogadro's number of MgO formula units?

8.45 What is the mass of Avogadro's number of P_2O_5 molecules?

8.6 Gram-mole-particle conversions

8.46 Calculate the number of moles in
a. 64.0 g O_2
b. 58.9 g Cl
c. 0.205 g NH_3
d. 1.91×10^2 g $Ba_3(PO_4)_2$
e. 90.5 g $(NH_4)_2SO_4$
f. 24.3 g Mg^{2+}

8.47 Calculate the number of grams contained in each of the following:

a. 0.500 moles F_2
b. 9.75 moles H_2
c. 2.50×10^{-2} moles NH_3
d. 11.3 moles $BaSO_4$
e. 3.011×10^{23} formula units LiF
f. 3.90 moles S^{2-}

8.48 Calculate the number of molecules present in each of the following samples:
a. 18.0 g NH_3
b. 180. g H_2O
c. 9.80×10^{-3} g CO_2
d. 1.09×10^{-7} moles SO_2

8.49 Calculate the mass of each of the following samples:
a. 2.50×10^{23} molecules H_2O
b. 6.022×10^{22} formula units MgF_2
c. 3.01×10^{25} molecules NH_3
d. 1.51×10^{24} formula units $(NH_4)_2CO_3$

8.50 Arrange the following in order of largest *mass* to smallest *mass*:
a. 20.0 g H_2O
b. 1.55 moles H_2O
c. 9.05×10^{23} molecules H_2O

8.51 The density of carbon tetrachloride is 1.595 g/mL. What is the volume of 1.00 mole of CCl_4?

8.7 Moles within moles

8.52 State the meaning of each expression in terms of either moles of molecules, moles of formula units, or moles of atoms that are present:

a. 5 H_2	d. 8 H_2O
b. 4 Mg	e. 4 O
c. 16 P_2O_5	f. 3 MgO

8.53 State the number of moles of each ion contained in each of the following:
a. 5.25 moles KI
b. 4.00 moles $AlCl_3$
c. 1.10×10^{-3} moles $Ca(NO_3)_2$
d. 0.250 moles $Ba_3(PO_4)_2$

8.54 Calculate the number of moles of *each atom* contained in the given amount of each of the following molecules:
a. 1.50 moles Br_2
b. 3.50 moles H_2
c. 3.00 moles H_2O
d. 1.25×10^{-1} moles $C_6H_6O_6$
e. 0.450 moles H_2SO_4

8.55 **a.** What is the molar weight of SO_2 (4 sig figs)?
   b. How many SO_2 molecules are present in the mass of SO_2 from part (a)?
   c. Calculate the mass in grams of one SO_2 molecule.

8.56 **a.** What is the molar weight of Na_2SO_4 (4 sig figs)?
   b. How many Na^+ ions are present in the mass of Na_2SO_4 from part (a)?

8.8 Empirical formulas defined

8.57 Distinguish between empirical and molecular formulas.

8.58 Each of the following compounds is represented by its molecular formula; state its empirical formula.
   a. Ethanol, C_2H_6O
   b. Methylene chloride, CH_2Cl_2
   c. Lindane, $C_6H_6Cl_6$
   d. Sucrose, $C_{12}H_{22}O_{11}$

8.9 Calculation of empirical formulas

8.59 Calculate the empirical formula of each compound from the mass-composition data given below:
   a. 46.7 percent N and 53.3 percent O
   b. 92.3 percent C and 7.70 percent H
   c. 75.0 percent C and 25.0 percent H
   d. 11.6 percent N and 88.4 percent Cl
   e. 60.0 percent C, 13.4 percent H, and 26.6 percent O
   f. 68.3 percent Pb, 10.6 percent S, and 21.1 percent O
   g. 45.9 percent K, 16.5 percent N, and 37.6 percent O
   h. 42.1 percent Na, 18.9 percent P, and 39.0 percent O

8.60 When 9.16 g of copper and 2.31 g of oxygen combine, they form a copper oxide. What is the empirical formula of this compound?

8.61 When 7.56 g of iron and 4.34 g of sulfur combine, they form an iron sulfide. What is the empirical formula of this compound?

8.62 Carbon and oxygen can combine to form more than one compound.

	Weight Carbon	Weight Oxygen
Compound I	0.168 g	0.448 g
Compound II	0.515 g	0.686 g

Calculate the empirical formula of each compound.

8.10 Molecular formulas

8.63 Analysis of a compound indicates that its mass composition is 80.0 percent C and 20.0 percent H. Its molecular weight is found to be 30.0. What is its molecular formula?

8.64 Acetone, a liquid often used as a nail polish remover, is found to obtain 62.0 percent carbon, 10.4 percent hydrogen, and 27.5 percent oxygen. Its molecular weight is found to be 58.1. What is the molecular formula of acetone?

8.65 A liquid of molecular weight 60.0 was found to contain 40.0 percent C, 6.7 percent H, and 53.3 percent O by weight. What is the molecular formula of the compound?

8.66 Vitamin C is a compound which upon analysis is found to contain 40.92 percent C, 4.58 percent H, and 54.51 percent O. The molecular weight of vitamin C is 176.1. What is the molecular formula of vitamin C?

8.67 Upon analysis, histidine, an amino acid found in proteins, yielded the following mass composition: 46.38 percent C, 5.90 percent H, 27.01 percent N, and 20.71 percent O. The molecular weight of histidine was found to be 155.1. What is its molecular formula?

8.68 When 125.0 g of a compound is decomposed, it is found to yield 50.00 g of C, 8.25 g of H, and 66.75 g of O. The molecular weight is found to be 90.0. What is the molecular formula of this compound?

8.69 A compound forms by combining 22.65 g of carbon, 1.90 g of hydrogen, and 8.73 g of nitrogen. The molecular weight of that compound is found to be 106. What is its molecular formula?

CHEMICAL REACTIONS
CHAPTER

| 9 |
| F |
| Fluorine |
| 19.00 |

WHAT IS A REACTION?

9.1   So far this book has dealt with chemicals only one compound or element at a time. The excitement of chemistry builds when you combine materials and watch chemical reactions occur. A **chemical reaction** is the process by which one or more chemical substances are converted into one or more *different* chemical substances.

Within your body chemical reactions occur constantly. Your car hums as a consequence of the chemical reaction between gasoline and air. Batteries energize because of chemical reactions. Nails rust and silver tarnishes as a result of chemical reactions. In all these cases, **reactants** (or starting materials) are converted into **products** (new materials) (see Figure 9.1).

In this chapter you will learn to recognize or predict the products of a few simple reactions. The first task involves learning how chemists represent chemical reactions through chemical equations.

CHEMICAL EQUATIONS

9.2   In Section 1.11 you first encountered the idea of chemistry as a new language to be mastered. By now you are thoroughly familiar with chemical **symbols** (letters) and have a large vocabulary of chemical **formulas** (words). To describe chemical reactions, we will use chemical **equations** (sentences), which are shorter descriptions than descriptions by English sentences.

For example, if you swallow an antacid tablet like Tums for indigestion, the reaction that occurs would be described in English as follows: calcium carbonate (Tums) reacts with hydrochloric acid (stomach acid) to yield calcium chloride, carbon dioxide, and water. Clearly the chemists' way of saying this is shorter:

$$CaCO_3 + 2HCl \longrightarrow CaCl_2 + CO_2 + H_2O$$

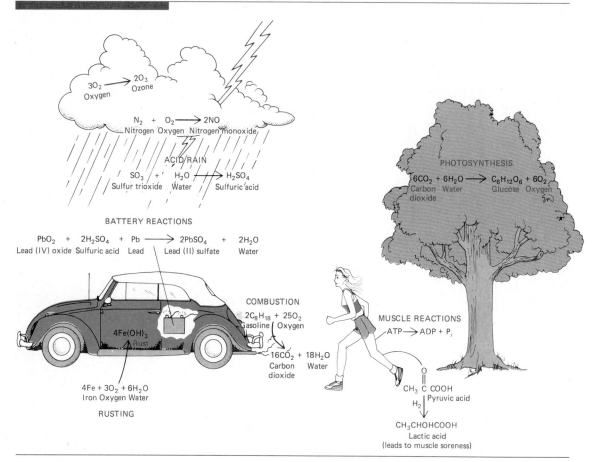

FIGURE 9.1

Chemical reactions abound in everyday life.

In addition to being shorter, the chemical equation contains a great deal more useful information than the English sentence does.

Chemical equations always have the form:

$$\text{\textbf{Reactants}} \longrightarrow \textbf{products}$$

| **Starting** | **Final** |
| **materials** | **materials** |

The arrow means "yields" or "gives." It is important to remember that the arrow can also be regarded as an equals (=) sign. We can translate an English sentence that describes a chemical reaction into a chemical equation that describes the same reaction by:

1. Writing correct formulas that correspond to the names of the chemical substance mentioned

2. Deciding which formulas are reactants and which are products

3. Following the preceding format, writing reactant formulas to the left of the arrow and product formulas to the right of the arrow

For example, let us translate the English sentence that describes what happens when you burn charcoal for your barbecue. Burning charcoal involves carbon (charcoal) reacting with oxygen to yield carbon dioxide. Following the preceding steps we write:

1. C for carbon, O_2 for oxygen (the element oxygen occurs in air as a diatomic molecule), and CO_2 for carbon dioxide. *Establishing correct formulas is the most important step.*

2. "Carbon reacting with oxygen" tells us that C and O_2 are the reactants; "to yield carbon dioxide" tells us that CO_2 is the product.

3. Thus

$$C + O_2 \longrightarrow CO_2$$

To repeat, the most important step in writing chemical equations is to write *correct chemical formulas.*

SAMPLE EXERCISE

9.1

Translate the following English sentence into a chemical equation: sodium chloride reacts with silver nitrate to yield silver chloride and sodium nitrate.

SOLUTION

1. Establish correct formulas
NaCl (Na^+, Cl^-)
$AgNO_3$ (Ag^+, NO_3^-)
AgCl
$NaNO_3$
2. NaCl and $AgNO_3$ are reactants
AgCl and $NaNO_3$ are products
3. $NaCl + AgNO_3 \longrightarrow AgCl + NaNO_3$

Problem 9.1

Translate the following English sentence into a chemical equation: Lead reacts with oxygen to yield lead(IV) oxide.

THE MEANING OF BALANCED EQUATIONS

9.3 There is usually another step required in order to write a correct chemical equation. After following the three-step procedure noted in Section 9.2, one must be certain the equation is *balanced,* that is, that there are the same numbers of atoms of each kind on both sides of the arrow. An equation must be balanced in order to satisfy the Law of Conservation of Matter, which states that "matter can neither be created nor destroyed in a chemical reaction." Chemical reactions are "rearrangements," or "reshufflings," of atoms, but no new atoms are formed, nor old ones lost. Atoms are conserved during a chemical reaction.

Let us look at the preceding examples to see what is meant by "balanced":

$$C \quad + \quad O_2 \quad \longrightarrow \quad CO_2$$

| 1 carbon atom | 2 oxygen atoms bonded together in a diatomic molecule | 1 carbon and 2 oxygen atoms bonded together in 1 carbon dioxide molecule |

This equation is balanced because on each side of the arrow there are one carbon and two oxygen atoms. What has changed is the bonding; the electrons are shared differently in the products than the reactants (Figure 9.2).

The example in Sample Exercise 9.1 is also balanced:

$$NaCl \quad + \quad AgNO_3 \longrightarrow AgCl \quad + \quad NaNO_3$$

| 1 Na^+ ion | 1 Ag^+ ion | 1 Ag^+ ion | 1 Na^+ ion |
| 1 Cl^- ion | 1 NO_3^- ion | 1 Cl^- ion | 1 NO_3^- ion |

Each side of the equation contains one of each ion, but the cations and anions have switched partners in moving from reactants to products.

Notice that in the first example of an equation, the Tums equation, there is a coefficient of 2 before the HCl. This coefficient is necessary in order to have a balanced equation, i.e., in order to have equal numbers of each kind of atom on both sides of the equation.

Balanced: $CaCO_3 + 2HCl \longrightarrow CaCl_2 + CO_2 + H_2O$

1 Ca   1 C   3 O   2 H   2 Cl   1 Ca   2 Cl   1 C   2 H   3 O

Unbalanced: $CaCO_3 + HCl \longrightarrow CaCl_2 + CO_2 + H_2O$

1 H   1 Cl   2 Cl   2 H

When the translation of an English sentence into a chemical equation by the three steps outlined in Section 9.2 does not lead automatically to a balanced equation, balancing is accomplished by inserting proper coefficients. The coefficients should always be whole numbers.

BALANCING EQUATIONS

9.4 Hydrogen reacts with oxygen to yield water. Let us write a *balanced* chemical equation for this reaction:

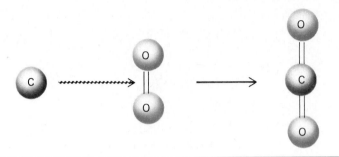

FIGURE 9.2

Chemical reactions are rearrangements of atoms. No atoms are lost or created. In this example, a separate C atom inserts itself between the two atoms of the O_2 molecule. The result is a new molecule, CO_2.

1. The correct formulas are H_2, O_2, and H_2O.

2. The reactants are H_2 and O_2. The product is H_2O.

3. Thus

$$H_2 + O_2 \longrightarrow H_2O$$

The next step is to determine whether the equation is balanced. This is done by counting the numbers of each kind of atom on each side of the equation and comparing. We determine the number of atoms of a kind in each formula by multiplying for each formula

Coefficient $\times$ subscript = number of atoms

Remember that when there is no coefficient, the coefficient is understood to be 1.

$$H_2 \quad + \quad O_2 \quad \longrightarrow \quad H_2O$$

$1 \times 2 = 2\,H$ $1 \times 2 = 2\,O$ $1 \times 2 = 2\,H$ $1 \times 1 = 1\,O$

Subscript   Subscript   Subscript   Subscript
Coefficient   Coefficient   Coefficient   Coefficient

We see that this equation is *not* balanced, because whereas there are two oxygen atoms on the left, there is only one oxygen atom on the right. To balance an equation one inserts coefficients. *Never* alter subscripts because that would produce incorrect formulas.

For this equation we begin the balancing procedure by placing the coefficient 2 before H_2O:

$$H_2 + O_2 \longrightarrow 2\,H_2O$$

$2\,H \quad 2\,O$ $2 \times 2 = 4\,H$ $2 \times 1 = 2\,O$

Subscript   Subscript
Coefficient   Coefficient

Now there are two oxygens on each side of the equation, but the hydrogen balance is upset. To correct this we put a 2 before H_2:

$$2H_2 + O_2 \longrightarrow 2H_2O$$

$$\underset{(2 \times 2)}{4\,H} \quad \underset{(1 \times 2)}{2\,O} \qquad \underset{(2 \times 2)}{4\,H} \quad \underset{(2 \times 1)}{2\,O}$$

Now the equation is balanced.

To repeat, balance an unbalanced equation by inserting coefficients. *Never* alter subscripts. Notice in the preceding equation that $H_2 + O_2 \rightarrow H_2O$ could be "balanced" by writing $H_2 + O_2 \rightarrow H_2O_2$. But this would be very wrong because H_2O_2 is not water.

Problem 9.2

Decide whether the following are balanced:

a. $KOH + H_2SO_4 \longrightarrow K_2SO_4 + H_2O$

b. $2Na + H_2O \longrightarrow 2NaOH + H_2$

c. $H_2 + Cl_2 \longrightarrow 2HCl$

Let us summarize the total method for equation writing and balancing:

Step 1   Write the equation using the stepwise procedure described in Section 9.2.

Step 2   Determine whether the equation is balanced by multiplying the coefficient times the subscript for each atom in each formula on each side and comparing.

Step 3   If the equation is *not* balanced, balance it by inserting the correct coefficients. When an equation contains unbalanced atoms other than H and O, it is most readily balanced by first inserting coefficients to balance these other atoms and then balance H and O.

Step 4   When you think the equation is balanced, check it by again multiplying the coefficient times the subscript for each atom on either side.

After you have gained some familiarity with balancing equations, we will introduce some hints in Section 9.6 that may provide additional help in accomplishing the balancing process.

SAMPLE EXERCISE

9.2

Write a balanced equation for the reaction between nitrogen and hydrogen which yields ammonia (NH_3).

SOLUTION

Follow Steps 1 through 4 as outlined immediately preceding this exercise.

Step 1 $N_2 + H_2 \quad\quad \longrightarrow \quad\quad NH_3$

Step 2

Coefficient × subscript	Coefficient × subscript
$1 \times 2 = 2N$	$1 \times 1 = 1N$
$1 \times 2 = 2H$	$1 \times 3 = 3H$

Both N and H are not balanced.

Step 3 Begin by balancing N, which can be accomplished by placing a coefficient of 2 before NH_3:

$$N_2 + H_2 \longrightarrow 2NH_3$$

This coefficient multiplies the entire NH_3 formula, so that we now have $2 \times 3 = 6$ H on the right. We can get 6 H on the left side by using the coefficient 3 for H_2:

$$N_2 + 3H_2 \longrightarrow 2NH_3$$

Step 4

Coefficient × subscript	Coefficient × subscript
$1 \times 2 = 2N$	$2 \times 1 = 2N$
$3 \times 2 = 6H$	$2 \times 3 = 6H$

The equation is balanced because we have exactly 2N and 6H on either side.

Problem 9.3

Write a balanced equation for the reaction between hydrogen and bromine which yields hydrogen bromide.

When polyatomic ions appear on *both* sides of an equation, they can be counted and balanced as a whole. In the following example we can balance $SO_4{}^{2-}$ rather than balancing S and O individually.

$$Al \quad + \quad H_2SO_4 \quad \longrightarrow \quad Al_2(SO_4)_3 \quad + \quad H_2$$

Count:

$$1 \times 1 = 1Al \quad 1 \times 2 = 2H \quad 1 \times 2 = 2\,Al \quad 1 \times 2 = 2H$$
$$1 \times 1 = 1SO_4 \quad 1 \times 3 = 3SO_4$$

Balance:

$$2Al \quad + \quad 3H_2SO_4 \quad \longrightarrow \quad Al_2(SO_4)_3 \quad + \quad 3H_2$$

Count:

$$2 \times 1 = 2Al \quad 3 \times 2 = 6H \quad 1 \times 2 = 2Al \quad 3 \times 2 = 6H$$
$$3 \times 1 = 3SO_4 \quad 1 \times 3 = 3SO_4$$

SAMPLE EXERCISE

9.3

Balance the equation:

$$MgCl_2 + K_3PO_4 \longrightarrow Mg_3(PO_4)_2 + KCl$$

SOLUTION

Begin with Step 2 of the balancing procedure, since the equation is given.

Step 2

Reactants	Products
Coefficient × subscript ·	Coefficient × subscript
$1 \times 1 = 1$ Mg	$1 \times 3 = 3$ Mg
$1 \times 2 = 2$ Cl	$1 \times 1 = 1$ Cl
$1 \times 3 = 3$ K	$1 \times 1 = 1$ K
$1 \times 1 = 1$ PO$_4$	$1 \times 2 = 2$ PO$_4$

All atoms are unbalanced.

Step 3 Proceed to balance in an orderly fashion one atom or ion at a time going from left to right.

(*a*) To balance Mg, use the coefficient 3 before MgCl$_2$:

$$3MgCl_2 + K_3PO_4 \longrightarrow Mg_3(PO_4)_2 + KCl$$

(*b*) There are now $3 \times 2 = 6$ Cl on the left, so the coefficient 6 is required before KCl

$$3MgCl_2 + K_3PO_4 \longrightarrow Mg_3(PO_4)_2 + 6 KCl$$

(*c*) There are now 3 K on the left and 6 on the right, so use the coefficient 2 in front of K$_3$PO$_4$

$$3MgCl_2 + 2K_3PO_4 \longrightarrow Mg_3(PO_4)_2 + 6 KCl$$

(*d*) There are now two (PO$_4{}^{3-}$) ions on each side of the equation.

Step 4 Check:

Coefficient × subscript	Coefficient × subscript
$3 \times 1 = 3$ Mg	$1 \times 3 = 3$ Mg
$3 \times 2 = 6$ Cl	$6 \times 1 = 6$ Cl
$2 \times 3 = 6$ K	$6 \times 1 = 6$ K
$2 \times 1 = 2$ PO$_4$	$1 \times 2 = 2$ PO$_4$

SAMPLE EXERCISE

9.4

Balance the equation:

$$KClO_3 \longrightarrow KCl + O_2$$

SOLUTION

Begin with Step 2.

Step 2

$$1 \times 1 = 1\ K \qquad 1 \times 1 = 1\ K$$
$$1 \times 1 = 1\ Cl \qquad 1 \times 1 = 1\ Cl$$
$$1 \times 3 = 3\ O \qquad 1 \times 2 = 2\ O$$

Clearly, oxygen requires balancing.

Step 3 When we see three of a kind on one side and two on the other, we must use the idea of the lowest common denominator (6 in this case) in order to balance. Thus

$$2KClO_3 \longrightarrow KCl + 3O_2$$
$$2 \times 3\,O = 6\,O \qquad 3 \times 2\,O = 6\,O$$

The 2 in front of $KClO_3$ requires us to put a 2 before KCl. Thus

$$2KClO_3 \longrightarrow 2KCl + 3O_2$$

Step 4   Check:

$$2 \times 1 = 2\,K \qquad 2 \times 1 = 2\,K$$
$$2 \times 1 = 2\,Cl \qquad 2 \times 1 = 2\,Cl$$
$$2 \times 3 = 6\,O \qquad 3 \times 2 = 6\,O$$

The only way to learn to balance equations comfortably is through practice. Proceed slowly atom by atom, and keep track by writing down the numbers of atoms on either side of the equation at each step. Always make a final check.

Problem 9.4

Balance the following equations:

a. $Zn + Pb(NO_3)_2 \longrightarrow Zn(NO_3)_2 + Pb$

b. $Al + Cl_2 \longrightarrow AlCl_3$

c. $Ag_2O \longrightarrow O_2 + Ag$

USE CORRECT FORMULAS

9.5 Always remember that *correct equations require correct formulas*. You will find that most equations you encounter can be balanced quite easily, but if you experience great difficulty in balancing an equation, it is probably because you have not used the correct formulas. For example, suppose you were asked to write an equation to show that aluminum chloride reacts with potassium carbonate to yield aluminum carbonate and potassium chloride. Begin by writing correct formulas according to the rules in Sections 6.8 and 6.13.

Aluminum chloride	+	Potassium carbonate		Aluminum carbonate	+	Potassium chloride
$Al^{3+} \diagdown\!\!\!\!\!\diagup Cl^-$		$K^+ \diagdown\!\!\!\!\!\diagup CO_3{}^{2-}$		$Al^{3+} \diagdown\!\!\!\!\!\diagup CO_3{}^{2-}$		$K^+ \diagdown\!\!\!\!\!\diagup Cl^-$
$AlCl_3$	+	K_2CO_3	$\longrightarrow$	$Al_2(CO_3)_3$	+	KCl

This equation is balanced quite readily:

1. *2* $AlCl_3$ balances Al

2. *6 KCl* then balances Cl

3. *3* K_2CO_3 then balances K and CO_3

$$2AlCl_3 + 3K_2CO_3 \longrightarrow Al_2(CO_3)_3 + 6KCl$$

But suppose that an incorrect formula had been used accidentally. For example,

$$AlCl_3 + K_2CO_3 \longrightarrow \underset{\text{Wrong}}{AlCO_3} + KCl$$

It is *impossible* to balance this. A few attempts should convince you. Try

$$2\ AlCl_3 + 3\ K_2CO_3 \longrightarrow 2\ AlCO_3 + 6\ KCl$$

Al, Cl, and K are balanced, but not CO_3. Try

$$2\ AlCl_3 + 3\ K_2CO_3 \longrightarrow 3\ AlCO_3 + 6\ KCl$$

Cl, K, and CO_3 are balanced, but not Al. Write to us if you find a way to balance this with coefficients only. Remember, to balance you use coefficients only. The subscripts are supposed to be those of a correct formula to begin with.

Problem 9.5

Write a balanced equation for the reaction between calcium bromide and silver nitrate that yields calcium nitrate and silver bromide.

HELPFUL HINTS

9.6   As you practice balancing, you will develop some tricks of your own. In this section we will mention some simplifying devices that others have noticed.

Hint 1   Balance polyatomic ions as single entities if they appear on both sides of an equation. We have seen this idea used in Section 9.4.

Hint 2   If hydroxide (OH^-) appears on *only one* side of an equation and water on the other, it is convenient to write water (H_2O) as HOH and regard it as being composed of H^+ and OH^-. For example,

$$Mg(OH)_2 + HCl \longrightarrow MgCl_2 + H_2O$$

can be balanced more easily when it is written

	$Mg(OH)_2$	+	HCl $\longrightarrow$	$MgCl_2$	+	HOH
Count:	1 Mg			1 Mg		
	2 OH			1 OH		
	1 H			1 H		
	1 Cl			2 Cl		

Balance:

OH	by	2 HOH
Cl	by	2 HCl

$$Mg(OH)_2 \ + \ 2HCl \longrightarrow MgCl_2 \ + \ 2HOH$$

Count:

1 Mg	1 Mg
2 OH	2 OH
2 H	2 H
2 Cl	2 Cl

Hint 3 Begin balancing an equation with an unbalanced atom that has an ionic charge greater than 1. For example, if you begin to balance the equation

$$K_2S + AlCl_3 \longrightarrow KCl + Al_2S_3$$

by starting with K (K^+), then you would put a 2 in front of KCl.

$$K_2S + AlCl_3 \longrightarrow 2 \ KCl + Al_2S_3$$

Then when you consider S and find that 3 K_2S is required, you would have to change 2 KCl to 6 KCl. If you had begun with S (S^{2-}) or Al (Al^{3+}), no change of coefficients would have been required. Beginning with S, you would proceed:

$3 \ K_2S + AlCl_3 \longrightarrow KCl + Al_2S_3$ then fix K:
$3 \ K_2S + AlCl_3 \longrightarrow 6 \ KCl + Al_2S_3$ then Al:
$3 \ K_2S + 2 \ AlCl_3 \longrightarrow 6 \ KCl + Al_2S_3$ The job is done!

Hint 4 When there is an *even* number of one atom on one side of an equation and an *odd* number of that same atom on the other side, begin balancing by multiplying the formula with the odd number by 2. For example, if you are asked to balance

$$KClO_3 \longrightarrow KCl + O_2,$$

notice that 3 O (odd) on the left and the 2 O (even) on the right. Begin by multiplying the odd by 2.

$$2 \ KClO_3 \longrightarrow KCl + O_2$$

Now count:

2 K	1 K
2 Cl	1 Cl
6 O	2 O

Balance: $2 \ KClO_3 \longrightarrow 2 \ KCl + 3 \ O_2$

This is an alternative to the lowest common denominator idea discussed previously.

Another example of the application of Hint 4 is seen in balancing

$$CO + O_2 \longrightarrow CO_2$$

Count: 1 C   1 + 2 = 3 O   1 C
2 O

Noticing that there is an odd number of O on the left, you begin by "evening up" by multiplying CO by 2.

$$2\,CO + O_2 \longrightarrow CO_2$$

Now balance C: $2\,CO + O_2 \qquad\qquad \longrightarrow 2\,CO_2$
This job is done.
Count:
2 C $\qquad\qquad\qquad$ 2 C
2 O $+ 2\,O = 4\,O \qquad$ 4 O

Problem 9.6

Balance the following equations by using the appropriate helpful hints.

a. $K + H_2O \longrightarrow KOH + H_2$

b. $NO + O_2 \longrightarrow NO_2$

c. $Na_2CO_3 + Fe(NO_3)_2 \longrightarrow NaNO_3 + FeCO_3$

SPECIAL SYMBOLS

9.7 By using certain symbols, chemists are able to pack even more information into a chemical equation than we have seen thus far. For example, the following equation indicates the physical state of each material as it is observed when the reaction occurs in the laboratory.

$$2Na(s) + 2H_2O(l) \longrightarrow H_2(g) + 2NaOH(aq)$$
$\qquad\uparrow\qquad\qquad\uparrow\qquad\qquad\uparrow\qquad\qquad\uparrow$
$\quad$ Solid $\qquad$ Liquid $\qquad$ Gas $\qquad$ Aqueous

Reactants and products may be either pure solids, liquids, or gases. More often reactions are done in water (aqueous) solution so that some of the reactants or products are dissolved rather than being in some pure physical state. A reactant or product that is in water solution is designated by (*aq*), short for aqueous, a word which derives from *aqua*, the Latin word for water.

The formation of a gas or solid can be indicated in other ways. The previous example could have been written

$$2Na(s) + 2H_2O(l) \longrightarrow H_2\uparrow + 2NaOH(aq)$$

The arrow pointing upward (↑) indicates that a gas is escaping. The equation for the reaction of silver nitrate solution with sodium chloride solution might be

$$AgNO_3(aq) + NaCl(aq) \longrightarrow AgCl(s) + NaNO_3(aq)$$

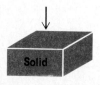 Thud! or

$$AgNO_3(aq) + NaCl(aq) \longrightarrow AgCl\downarrow + NaNO_3(aq)$$

Whee!

Gas

Solid

The arrow pointing downward (↓) indicates the formation of a solid which is *not* water soluble and which therefore "falls out" of solution. An insoluble solid which falls out of solution is called a **precipitate.** Notice that ↑ for gas and ↓ for a solid are used only for *products,* i.e., only on the right-hand side of the equation.

Symbols written above the yield arrow (⟶) indicate important reaction conditions. For example,

$$CaCO_3(s) \xrightarrow{\Delta} CaO(s) + CO_2 \uparrow$$

Δ is the symbol for *heat.* Heat must be applied for this reaction to occur.

A chemical formula written above the yield arrow indicates a **catalyst.** In the following reaction, MnO_2 is a catalyst.

$$2KClO_3(s) \xrightarrow{MnO_2} 2KCl(s) + 3O_2 \uparrow$$

A catalyst is a substance which affects the speed of a chemical reaction without itself being changed in the course of the reaction. Since it does not change, it is neither a reactant nor a product and so appears above the arrow rather than on either side (see Table 9.1).

TYPES OF REACTIONS

9.8 In all the previous examples you have been shown the reactants and the products of a chemical reaction. We would like you to be able to predict the products of a chemical reaction. That is, when given the reactants that are mixed together, you should be able to predict

SUMMARY OF SPECIAL SYMBOLS USED IN CHEMICAL EQUATIONS   **TABLE 9.1**

Symbol	Meaning
(s)	Solid reactant or product
(l)	Liquid reactant or product
(g)	Gaseous reactant or product
(aq)	Reactant or product in water (aqueous) solution
↑	Gaseous product
↓	Solid product (precipitate)
Δ	Heat
$\xrightarrow{\text{Formula}}$	Catalyst

the products. This ability to predict products takes a great deal of experience, and realistically, at the end of this chapter your ability will be very limited. However, you will be prepared to develop this skill as you proceed in the chapters that follow.

The study of chemical reactions can be greatly simplified by realizing that almost all reactions can be classified into four major types:

1. Combination

2. Decomposition

3. Single replacement

4. Double replacement

Learning to recognize reaction types organizes and simplifies the study of reactions.

Combination reactions   As the name implies, in combination reactions two reactants *combine* or join together to form one product.

Combination model equation

$A + B \longrightarrow AB$

Two reactants   One product

All reactions in which a compound is formed from its elements are combination reactions. Notice how the following examples fit the model equation.

$$C + O_2 \longrightarrow CO_2$$
$$Mg + Cl_2 \longrightarrow MgCl_2$$
$$2Na + Br_2 \longrightarrow 2NaBr$$

In each example, *two elements* react to give *one product* (a compound). These are the combination reactions you will encounter most frequently. Now, whenever you see two elements reacting, you can predict that the product will be the compound that they form. For the combination of metallic and nonmetallic elements, you can complete the equation by writing a correct formula for the ionic compound formed according to the rules of Section 6.8. For example, given the uncompleted equation,

$$Al + Cl_2 \longrightarrow$$

you should recognize the reactants as the metallic element aluminum and the nonmetallic element chlorine, and realize that the product will be the ionic compound that they form, namely, aluminum chloride. Complete the equation by writing the correct formula for aluminum chloride, and then balance the equation.

Complete:

$$Al + Cl_2 \longrightarrow AlCl_3 \qquad (Al^{3+} \diagdown\!\!\!\diagup Cl^{1-})$$

Balance:

$$2Al + 3Cl_2 \longrightarrow 2AlCl_3$$

If the metallic element can form more than one cation (Section 6.6) then you need to be told which cation is forming in the given reaction.

Complete the following equation:

$$Ca + O_2 \longrightarrow$$

SOLUTION

Step 1 Recognize that the reactants are both elements, which signals that this is a combination reaction.

Step 2 Recognize Ca as a metal and oxygen as a nonmetal, which tells you that they will form an ionic compound.

Step 3 Write a correct formula for calcium oxide, and complete the equation.

$$Ca^{2+} \diagdown\!\!\!\diagup O^{2+} \quad Ca_2O_2 \quad CaO$$

$$Ca + O_2 \longrightarrow CaO$$

Step 4 Balance the equation according to the procedure outlined earlier.

$$2Ca + O_2 \longrightarrow 2CaO$$

For the reaction of *non*metallic elements, we can predict that the product will be a molecular compound, but we cannot complete the equation unless we are told the formula for the molecular compound. In many cases the same nonmetallic elements can combine to form more than one compound:

$$2P + 3Cl_2 \longrightarrow 2PCl_3$$
$$2P + 5Cl_2 \longrightarrow 2PCl_5$$

Another example of a combination reaction is the reaction between the compound water and an oxide compound. Notice how the following examples fit the model equation.

$$\text{H}_2\text{O} \; + \quad \text{SO}_3 \qquad \longrightarrow \qquad \text{H}_2\text{SO}_4$$
Water   Nonmetallic oxide   Oxyacid

$$\text{H}_2\text{O} \; + \qquad \text{MgO} \qquad \longrightarrow \qquad \text{Mg(OH)}_2$$
Water   Metallic oxide   Metal hydroxide

In each example, two compounds react to give one product (a new compound). If the reacting oxide is nonmetallic, the product is always an oxyacid. If the reacting oxide is *metallic*, the product is always a metal hydroxide. Metal hydroxides are also called **bases.**

Decomposition reactions   As the name implies, in decomposition reactions one reactant *decomposes*, or breaks apart, into two or more products.

Decomposition model equation

$$XY \qquad \longrightarrow \qquad X + Y$$

One reactant   Two (or more) products

EXAMPLES

$$\text{CaCO}_3 \xrightarrow{\Delta} \text{CaO} + \text{CO}_2$$ Metal carbonates decompose when heated to yield CO_2.

$$2\text{HgO} \xrightarrow{\Delta} 2\text{Hg} + \text{O}_2$$ Some metal oxides decompose when heated to yield oxygen.

$$2\text{KClO}_3 \xrightarrow{\Delta} 2\text{KCl} + 3\text{O}_2$$ Chlorates decompose when heated to yield oxygen.

Smelling salts, $(\text{NH}_4)_2\text{CO}_3$, work because of their decomposition to the products shown:

$$(\text{NH}_4)_2\text{CO}_3 \longrightarrow 2\text{NH}_3\uparrow \; + \; \text{CO}_2\uparrow \; + \; \text{H}_2\text{O}$$

The pungency of the liberated NH_3 gas is what revives the semiconscious person. Decomposition reactions are very easy to recognize because they are the only type of reaction in which there is *only one* reactant. But it is difficult to predict the products of decomposition because there are no general rules. There are only specific rules for specific sets of compounds such as those stated in the examples given. You must be able to recognize decomposition reactions.

Single-replacement reactions   In single-replacement reactions an element reacts with a compound in such a way that the element replaces one of the existing elements in the compound. An analogy can be made to a single person cutting in on a dance team producing a new couple and a single person.

Single-replacement model equation

$$E + AB \longrightarrow EB + A$$

| Two reactants | Two products |
| (One element + One compound) | (One compound + One element) |

Some metallic elements replace other metal cations in ionic compounds. Remember that ionic compounds are made up of ions even though the chemical formulas do not show the ionic charges.

$$Zn + CuSO_4 \longrightarrow ZnSO_4 + Cu$$

Metallic element   Ionic compound $(Cu^{2+}SO_4^{2-})$ Ionic compound $(Zn^{2+}SO_4^{2-})$ Metallic element

Notice how this example fits the model equation. Notice also that zinc atoms (Zn) are becoming zinc ions (Zn^{2+}) and the replaced cations (Cu^{2+}) are becoming copper atoms (Cu). Other examples all show these same features, namely, one metallic element is changed from atoms to ions, while the other metallic element undergoes the change from ions to atoms.

$$2K + Pb(NO_3)_2 \longrightarrow 2KNO_3 + Pb$$
$$(Pb^{2+}) \qquad (K^+)$$

$$Sn + 2AgNO_3 \longrightarrow Sn(NO_3)_2 + 2Ag$$
$$(Ag^+) \qquad (Sn^{2+})$$

Whereas zinc will replace copper in a compound, as shown earlier, copper will not replace zinc; i.e., mixing together Cu and $ZnSO_4$ results in no reaction (see Figure 9.3). We say that Zn is more reactive than Cu. *More reactive metals replace less reactive metals in compounds.* Table 9.2, the Activity Series of metals, tells us which are the more reactive metals.

Look back at the other preceding examples and notice that K can replace Pb because K is higher than Pb in the Activity Series. Similarly, Sn is above Ag in the Activities Series and so Sn can replace Ag^+ from $AgNO_3$.

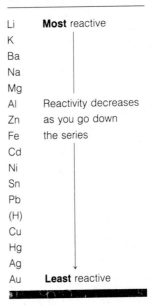

TABLE 9.2 ACTIVITY SERIES OF METALS

Li	**Most** reactive
K	
Ba	
Na	
Mg	
Al	Reactivity decreases
Zn	as you go down
Fe	the series
Cd	
Ni	
Sn	
Pb	
(H)	
Cu	
Hg	
Ag	
Au	**Least** reactive

SAMPLE EXERCISE

9.6

Predict whether the following replacement reactions will occur.

a. Na + $AgNO_3$ $\longrightarrow$
b. Al + KCl $\longrightarrow$
c. Pb + $ZnSO_4$ $\longrightarrow$

SOLUTION

Consult the Activity Series (Table 9.2). Replacement occurs if the metallic

FIGURE 9.3

The beaker on the left contains a strip of copper metal in a clear solution of zinc sulfate. The shiny, untarnished appearance of the metal and the clarity of the solution indicate that there is no reaction between the materials. The beaker on the right contains a strip of zinc metal in Cu(II) sulfate solution. The portion of the metal submerged has darkened as it interacts with the solution. The particles at the bottom of the beaker are Cu particles that come out of solution through the reaction Zn(s) + CuSO₄(aq) → ZnSO₄(aq) + Cu(s). (*Photograph by Bryan Lees.*)

element is higher in the series than the metallic cation in the compound.

a. Na is above Ag. The reaction will occur.
b. No reaction. Al is below K.
c. No reaction. Pb is below Zn.

Hydrogen appears in the Activity Series even though it is not a metal because it turns out that active metals can replace hydrogen in acids.

Metallic Element	Acid	Ionic Compound	Element (Hydrogen Gas)
Zn	+ H_2SO_4 ⟶	$ZnSO_4$	+ H_2
Mg	+ 2HCl ⟶	$MgCl_2$	+ H_2

Notice how these examples fit the model equation. Metals above (H) in the series replace it; metals below do not.

Problem 9.7

Predict whether the following replacement reactions will occur:

a. Au + HCl ⟶

b. Al + HNO_3 ⟶

The order of metals in the Activity Series depends on the ease with which the metal atoms become metal cations. The more easily they form cations, the more reactive they are. An important factor is the ionization energy of the metal (Section 6.3). We will discuss this further in Chapter 16 (Section 16.9).

Whenever you see a metal reacting with either an ionic compound or an acid, you should be able to predict the products.

SAMPLE EXERCISE

9.7

Complete the following equation:

$$Ca + AgNO_3 \longrightarrow$$

SOLUTION

Step 1 Recognize that there is a metal (Ca) reacting with an ionic compound. This signals a single-replacement reaction.

Step 2 Check the Activity Series (Table 9.2). In this case, Ca is above Ag and so a reaction will occur. (If Ca were below Ag, you would write "no reaction" and proceed no further.)

Step 3 Write correct formulas for the products, which will be Ag metal and the ionic compound that forms from Ca^{2+} and NO_3^-, namely, $Ca(NO_3)_2$.

$$Ca + AgNO_3 \longrightarrow Ca(NO_3)_2 + Ag$$

Step 4 Balance the equation according to the procedure in Section 9.4

$$Ca + 2AgNO_3 \longrightarrow Ca(NO_3)_2 + 2Ag$$

SAMPLE EXERCISE

9.8

Complete the following equation:

$$Al + HCl \longrightarrow$$

SOLUTION

Step 1 Recognize that this is a metal (Al) reacting with an acid. This signals a single-replacement reaction.

Step 2 Check the Activity Series. In this case, Al is above (H), so a reaction will occur. (If Al were below (H), you would write "no reaction" and proceed no further.)

Step 3 Write correct formulas for the products. One product of the reaction of an active metal and acid is always H_2. (Remember, the element hydrogen exists as diatomic molecules.) The other product is the ionic compound that forms from Al^{3+} and Cl^-, namely, $AlCl_3$.

$$Al + HCl \longrightarrow AlCl_3 + H_2$$

Step 4   Balance the equation.

$$2\,Al + 6\,HCl \longrightarrow 2\,AlCl_3 + 3\,H_2$$

We will discuss these reactions again in Chapter 13 when we discuss ionic equations.

Double-replacement reactions   In double-replacement reactions two compounds react with each other to form two different compounds.

Double-replacement model equation

$$AB + XY \quad \longrightarrow \quad AY + XB$$

Two reactants	Two products
(Two compounds)	(Two compounds)

There is a double replacement (two replacements) in the sense that *A* replaces *X* in *XY* and *X* replaces *A* in *AB*. It is perhaps easier to view the reaction as a "switching of partners." The *A-B* and *X-Y* partnerships are dissolved and the *A-Y* and *X-B* partnerships are formed in their place.

Ionic compounds can react in double-replacement reactions.

$$NaCl \quad + AgNO_3 \quad \longrightarrow \quad NaNO_3 \quad + AgCl \downarrow$$
$$(Na^+, Cl^-) + (Ag^+, NO_3^-) \qquad (Na^+, NO_3^-) \quad (Ag^+, Cl^-)$$

Notice how this example fits the model equation. The cations and anions switch partners to form new ionic compounds. Look at some more examples:

$$K_2S \quad + MgSO_4 \quad \longrightarrow K_2SO_4 \quad + MgS \downarrow$$
$$(K^+, S^{2-}) \quad (Mg^{2+}, SO_4^{2-}) \qquad (K^+, SO_4^{2-}) \quad (Mg^{2+}, S^{2-})$$

$$Na_2CO_3 \quad + CaCl_2 \longrightarrow 2NaCl \quad + CaCO_3 \downarrow$$
$$(Na^+, CO_3^{2-}) \quad (Ca^{2+}, Cl^-) \qquad (Na^+, Cl^-) \quad (Ca^{2+}, CO_3^{2-})$$

Not all ionic compounds actually react to yield the products shown exclusively or in an easily isolated form. One driving force that ensures that the products form as written is the occurrence that one of the products is a precipitate ($\downarrow$). Please notice that one product in each of the above double-replacement examples was a precipitate. We will not consider how to predict the appearance of a precipitate until Chapter 13 (Section 13.10). In this chapter we will give you examples of ionic compounds which do react in double-replacement reactions and ask you to complete the equations correctly for the reactions by switching cation-anion partners. To do this you must write correct formulas for ionic compounds according to the principles in Section 6.8.

Complete the following equation:

$$MgCl_2 + K_2CO_3 \longrightarrow$$

SOLUTION

Step 1   Recognize that the reactants are two ionic compounds. This signals a double-replacement reaction.

Step 2   Identify the cation-anion pairs in the reactants. $MgCl_2$ is made up of Mg^{2+} cations and Cl^- anions, and K_2CO_3 is made up of K^+ cations and CO_3^{2-} anions.

Step 3   Switch the cation-anion partners, and write correct formulas for the new cation-anion pairs. Mg^{2+} teams up with CO_3^{2-} to give $MgCO_3$, and K^+ teams up with Cl^- to give KCl:

$$MgCl_2 + K_2CO_3 \longrightarrow MgCO_3 + KCl$$

Step 4   Balance the equation.

$$MgCl_2 + K_2CO_3 \longrightarrow MgCO_3 + 2KCl$$

Complete the following equation:

$$Na_3PO_4 + Fe_2(SO_4)_3 \longrightarrow$$

SOLUTION

Step 1   Recognize the signal for a double-replacement reaction: the reactants are two ionic compounds.

Step 2   Identify the cation-anion pairs in the reactants. Na_3PO_4 is made up of Na^+ cations and PO_4^{3-} anions, and $Fe_2(SO_4)_3$ is made up of Fe^{3+} cations and SO_4^{2-} anions. Note: You can ascertain the ionic charges by using a reverse crisscross procedure:

$$Fe_2 \diagdown\!\!\!\!\diagup (SO_4)_3$$

Step 3   Switch the cation-anion partners, and write correct formulas for the new cation-anion pairs. Na^+ teams up with SO_4^{2-} to give Na_2SO_4, and Fe^{3+} teams up with PO_4^{3-} to give $FePO_4$:

$$Na_3PO_4 + Fe_2(SO_4)_3 \longrightarrow Na_2SO_4 + FePO_4$$

Step 4   Balance the equation:

$$2Na_3PO_4 + Fe_2(SO_4)_3 \longrightarrow 3Na_2SO_4 + 2FePO_4$$

We will resume a consideration of these reactions in Chapter 13.

Another commonly encountered double-replacement reaction is the reaction between an acid and a base (metal hydroxide) to produce water and an ionic compound. For example,

$$\overbrace{HCl + NaOH}^{} \longrightarrow NaCl + HOH$$

HCl	+	NaOH	NaCl	+	HOH
Acid		Base	Ionic compound		Water

We will explore this reaction in more detail in Chapter 15 when we discuss acids and bases. For now, Table 9.3 summarizes the features of the four types of reactions with which you should be familiar.

SAMPLE EXERCISE

9.11

Identify the type of reaction represented by each equation below:

a. $2Li + Ni(NO_3)_2 \longrightarrow 2LiNO_3 + Ni$
b. $2HgO \longrightarrow 2Hg + O_2$
c. $N_2 + 3H_2 \longrightarrow 2NH_3$
d. $2KCl + Pb(NO_3)_2 \longrightarrow 2KNO_3 + PbCl_2$

SOLUTION

Compare the equation to the model equations to see which one fits. Start by counting the number of reactants and products.

a. Two reactants $\longrightarrow$ two products; therefore, this must be a replacement reaction. It is a single-replacement reaction. Li replaces Ni.
b. One reactant $\longrightarrow$ two products; therefore, it is a decomposition reaction.
c. Two reactants $\longrightarrow$ one product; therefore, it is a combination reaction.
d. Two reactants $\longrightarrow$ two products; therefore, this must be a replacement reaction. It is a double-replacement reaction in which there is a switch of cation-anion pairs.

SAMPLE EXERCISE

9.12

Complete and balance the following equations:

a. $Zn + HNO_3 \longrightarrow$
b. $Zn + O_2 \longrightarrow$
c. $ZnCl_2 + K_2CO_3 \longrightarrow$

SOLUTION

Decide on the reaction type and then proceed as previously described for that type.

a. Fits   E   +   AB $\longrightarrow$
              Element   Compound

SUMMARY OF THE FOUR REACTION TYPES

TABLE 9.3

Reaction Type	Model Equation	Number of Reactants	Number of Products	Important Examples*
Combination	$A + B \longrightarrow AB$	2	1	Metal + nonmetal $\longrightarrow$ ionic compound
Decomposition	$XY \longrightarrow X + Y$	1	2 (or more)	
Single replacement	$E + AB \longrightarrow EB + A$	2	2	Metal + ionic compound $\longrightarrow$ ionic compound + metal Metal + acid $\longrightarrow$ ionic compound + H_2
Double replacement	$AB + XY \longrightarrow AY + XB$	2	2	2 ionic compounds $\longrightarrow$ 2 ionic compounds

* You must be able to recognize and distinguish among all four reaction types. In addition, you should be able to complete equations for the reactions designated as "Important examples."

Therefore, it is a single-replacement reaction. This is the reaction of an active metal plus an acid. See Sample Exercise 9.8

$$Zn + 2HNO_3 \longrightarrow Zn(NO_3)_2 + H_2 \uparrow$$

b. This is the combination of two elements to give a compound. See Sample Exercise 9.5.

$$2Zn + O_2 \longrightarrow 2ZnO$$

c. Two ionic compounds as reactants signal a double-replacement reaction. See Sample Exercises 9.9 and 9.10.

$$ZnCl_2 + K_2CO_3 \longrightarrow ZnCO_3 \downarrow + 2KCl$$

SUMMARY

A chemical reaction is the process by which one or more chemical substances (reactants) are converted into one or more different chemical substances (products) (Section 9.1). Chemists represent chemical reactions through the use of correct chemical formulas that are arranged in chemical equations (Section 9.2). In order to satisfy the Law of Conservation of Matter, a chemical equation must be balanced, that is, the same numbers of each kind of atom should appear on both sides of the yield sign (Section 9.3).

The technique of balancing equations is to insert coefficients and count up atoms (or ions) on either side of the equation until both sides have the same numbers of every kind of atom (or ion) (Section 9.4). The importance of using correct chemical formulas in writing balanced chemical equations cannot be overemphasized (Section 9.5). There are certain "tricks of the trade" that help in balancing (Section

9.6). Chemists use certain special symbols in order to include more information in chemical equations (Section 9.7).

There are four major reaction types: combination, decomposition, single replacement, and double replacement. The four types have characteristic features which are summarized in Table 9.3. The Activity Series enables us to predict the occurrence or nonoccurrence of single-replacement reactions (Section 9.8).

CHAPTER ACCOMPLISHMENTS

After completing this chapter you should be able to

What is a reaction?
1. Define chemical reaction.
2. Define the terms reactant and product.

Chemical equations
3. Translate English sentences that describe chemical reactions into chemical equations.
4. Recognize the importance of correct chemical formulas in chemical equations.

The meaning of balanced equations
5. State the Law of Conservation of Matter.
6. State the meaning of the word *balanced* as it refers to a chemical equation.

Balancing equations
7. Recognize whether an equation is unbalanced or balanced.
8. Balance a chemical equation.

Use correct formulas

Helpful hints

Special symbols
9. State the meaning of the special symbols in Table 9.1.

Types of reactions
10. Name the four major types of reactions.
11. Recognize a combination reaction.
12. Predict the nature of the product that will be formed from the combination of any two elements.

13. Given metal and nonmetal reactants, write a balanced chemical equation for their chemical reaction.
14. Recognize a decomposition reaction.
15. Recognize a single-replacement reaction.
16. Given an Activity Series, predict whether or not a given single-replacement reaction will occur.
17. Given an Activity Series and metal and ionic compound reactants, write a balanced chemical equation for their chemical reaction.
18. Given an Activity Series and metal and acid reactants, write a balanced equation for their reaction.
19. Recognize a double-replacement reaction.
20. Given two ionic compound reactants, write a balanced equation for their reaction.

PROBLEMS

9.1 What is a reaction?

9.8 Define or explain the following terms:
  a. Chemical reaction
  b. Chemical equation
  c. Reactant
  d. Product

9.2 Chemical equations

9.9 What symbol do chemists use to represent the word yields?

9.10 Associate the terms *reactant* and *product* with the words right and left.

9.11 $Zn + 2HCl \longrightarrow ZnCl_2 + H_2$
  a. Write the names of the reactants in the above equation.
  b. Write the names of the products for the above reaction.

9.12 Translate the following English sentences into chemical equations:
  a. Iron reacts with sulfur to yield iron(II) sulfide.
  b. Hydrochloric acid reacts with sodium hydroxide to yield sodium chloride and water.
  c. Tin reacts with oxygen to form tin(IV) oxide.
  d. Nitric acid reacts with lithium carbonate to yield lithium nitrate, water, and carbon dioxide.

9.13 Translate these equations into English sentences:

  a. $BaCl_2 + Na_2SO_4 \longrightarrow 2NaCl + BaSO_4$
  b. $Si + O_2 \longrightarrow SiO_2$
  c. $2K + I_2 \longrightarrow 2KI$
  d. $H_2SO_4 + Ca(OH)_2 \longrightarrow CaSO_4 + 2H_2O$

9.3 The meaning of balanced equations

9.14 What is the relationship between the Law of Conservation of Matter and a balanced chemical equation?

9.15 Identify each of the following equations as balanced or unbalanced:
  a. $2KI + Br_2 \longrightarrow 2KBr + I_2$
  b. $2P + 5O_2 \longrightarrow 2P_2O_5$
  c. $Al + 3HBr \longrightarrow AlBr_3 + 3H_2$
  d. $2NH_3 + H_2SO_4 \longrightarrow (NH_4)_2SO_4$
  e. $H_2O_2 \longrightarrow H_2O + O_2$
  f. $H_2 + Cl_2 \longrightarrow HCl$

9.16 A chemist carries out a reaction in which $A + B \longrightarrow C + D$. She finds that 18 g of product form when she uses 8 g of reactant A. What is the minimum amount of reactant B that the chemist must have combined with reactant A?

9.4 Balancing equations

9.17 To balance chemical equations one should use _____ . In balancing equations, _____ in chemical formulas should never be altered.

9.18 Balance the equations that you identified as unbalanced in Problem 9.15.

9.19 Balance the following equations:
  a. $KI + Cl_2 \longrightarrow KCl + I_2$
  b. $Cu + O_2 \longrightarrow CuO$
  c. $Li_2O + H_2O \longrightarrow LiOH$
  d. $SO_2 + O_2 \longrightarrow SO_3$
  e. $H_2 + N_2 \longrightarrow NH_3$
  f. $Na + ZnSO_4 \longrightarrow Na_2SO_4 + Zn$

9.20 Balance the following equations:
  a. $N_2 + O_2 \longrightarrow NO$
  b. $NaBr + Cl_2 \longrightarrow NaCl + Br_2$
  c. $P + Cl_2 \longrightarrow PCl_5$
  d. $BaCl_2 + (NH_4)_2SO_4 \longrightarrow BaSO_4 + NH_4Cl$
  e. $K_2O + H_2O \longrightarrow KOH$
  f. $Fe + O_2 \longrightarrow Fe_2O_3$
  g. $CaC_2 + H_2O \longrightarrow C_2H_2 + Ca(OH)_2$
  h. $Zn + HNO_3 \longrightarrow Zn(NO_3)_2 + H_2$
  i. $NH_4NO_2 \longrightarrow N_2 + H_2O$
  j. $PbO + O_2 \longrightarrow PbO_2$

9.21 Balance the following equations:
  a. $CH_4 + O_2 \longrightarrow CO_2 + H_2O$
  b. $FeS_2 + O_2 \longrightarrow FeO + SO_2$
  c. $P_4O_{10} + H_2O \longrightarrow H_3PO_4$
  d. $Cl_2 + H_2O \longrightarrow HCl + HClO$
  e. $(NH_4)_2Cr_2O_7 \longrightarrow Cr_2O_3 + N_2 + H_2O$
  f. $Al + CuSO_4 \longrightarrow Al_2(SO_4)_3 + Cu$

9.22 Balance the following equations:
  a. $C_2H_6 + O_2 \longrightarrow CO_2 + H_2O$
  b. $C_4H_{10} + O_2 \longrightarrow CO_2 + H_2O$
  c. $IBr + NH_3 \longrightarrow NH_4Br + NI_3$
  d. $Fe(OH)_3 + H_2SO_4 \longrightarrow Fe_2(SO_4)_3 + H_2O$
  e. $Al + Sn(NO_3)_2 \longrightarrow Al(NO_3)_3 + Sn$
  f. $NaOH + H_3PO_4 \longrightarrow Na_3PO_4 + H_2O$

9.23 Translate the following English sentences into balanced chemical equations:
  a. Carbon reacts with chlorine to form carbon tetrachloride.
  b. Potassium reacts with nitrogen to form potassium nitride.
  c. Barium nitrate reacts with sulfuric acid to form barium sulfate and nitric acid.
  d. Calcium hydroxide decomposes to calcium oxide and water.
  e. Phosphorus and oxygen combine to yield diphosphorus pentoxide.

9.24 Translate the following English sentences into balanced chemical equations:
  a. Sodium chloride reacts with silver nitrate to form sodium nitrate and silver chloride.

  b. Iron reacts with hydrochloric acid to yield iron(III) chloride and hydrogen.
  c. Sodium bicarbonate reacts with acetic acid to form sodium acetate, carbon dioxide, and water.
  d. Copper reacts with sulfur to form copper(II) sulfide.

9.5 Use correct formulas

9.25 Find and correct the translation error in each of the following.
  a. Sentence: Sodium reacts with iodine to form sodium iodide.
  Equation: $Na + I_2 \longrightarrow NaI_2$
  b. Sentence: Silver oxide decomposes to silver and oxygen.
  Equation: $Ag_2O \longrightarrow 2Ag + O$
  c. Sentence: Lead(II) nitrate reacts with potassium chloride to form lead(II) chloride and potassium nitrate.
  Equation: $Pb(NO_3)_2 + KCl \longrightarrow PbCl + K(NO_3)_2$
  d. Sentence: Silicon reacts with oxygen to form silicon dioxide.
  Equation: $S + O_2 \longrightarrow SO_2$

9.26 The following equations are *incorrect*, although they are balanced. Identify what is wrong with each one, and write a correct equation.
  a. $K_2CO_3 + CaCl_2 \longrightarrow CaCO_3 + K_2Cl_2$
  b. $N + 3H \longrightarrow NH_3$
  c. $C + Cl_4 \longrightarrow CCl_4$
  d. $K_2O + H_2O \longrightarrow K_2(OH)_2$

9.27 The following equations are *incorrect*. Identify what is wrong with each one and write a correct equation.
  a. $Al + H_2SO_4 \longrightarrow AlSO_4 + H_2$
  b. $N_2 + H_2 \longrightarrow NH_4$
  c. $BaCl_2 + Na_2SO_4 \longrightarrow Na_2Cl_2 + BaSO_4$
  d. $Mg + O_2 \longrightarrow MgO_2$

9.6 Helpful hints

9.28 Balance the following equations:
  a. $Fe(OH)_3 + H_2SO_4 \longrightarrow Fe_2(SO_4)_3 + H_2O$
  b. $C_2H_6 + O_2 \longrightarrow CO_2 + H_2O$
  c. $NaOH + Al(OH)_3 \longrightarrow NaAlO_2 + H_2O$
  d. $P_4O_{10} + H_2O \longrightarrow H_3PO_4$
  e. $K_2O + P_4O_{10} \longrightarrow K_3PO_4$
  f. $MgI_2 + H_2SO_4 \longrightarrow HI + MgSO_4$

g. $PCl_5 + H_2O \longrightarrow HCl + H_3PO_4$

h. $Al + Sn(NO_3)_2 \longrightarrow Al(NO_3)_3 + Sn$

9.7 Special symbols

9.29 What are the meanings of the symbols (s), (l), (g), (aq), Δ, $\downarrow$, and $\uparrow$?

9.30 Balance the following equations:

a. $Cl_2O_7(l) + H_2O(l) \longrightarrow HClO_4(aq)$

b. $NH_4NO_3(s) \overset{\Delta}{\longrightarrow} N_2O(g) + H_2O(l)$

c. $BaCl_2(aq) + (NH_4)_2CO_3(aq) \longrightarrow$
$$BaCO_3 \downarrow + NH_4Cl(aq)$$

d. $Ca(s) + HCl(g) \longrightarrow CaCl_2(s) + H_2 \uparrow$

9.31 Translate into balanced equations:

a. Solid lithium reacts with liquid water to yield aqueous lithium hydroxide and hydrogen gas.

b. Heating solid tin(II) carbonate produces solid tin(II) oxide and carbon dioxide gas.

c. Aqueous potassium chloride reacts with aqueous silver nitrate, producing a precipitate of silver chloride and aqueous potassium nitrate.

9.32 Translate into balanced equations:

a. Iodine crystals react with chlorine gas to form solid iodine trichloride.

b. Solid zinc reacts with hydrochloric acid to yield aqueous zinc chloride and hydrogen gas.

c. Heating solid silver oxide produces metallic silver and oxygen gas.

d. Aluminum metal reacts with aqueous copper(II) sulfate to form aqueous aluminum sulfate and copper metal.

9.8 Types of reactions

9.33 What are the four types of chemical reactions discussed in this chapter?

9.34 Write a model equation using A's, B's, X's, and Y's for each of the four reaction types.

9.35 Look through the equations in Problems 9.11 through 9.32, and find one example of each reaction type.

9.36 Identify each of the following reactions as a combination, decomposition, single replacement, or double replacement.

a. $3Na + Al(NO_3)_3 \longrightarrow 3NaNO_3 + Al$

b. $Na_3PO_4 + Al(NO_3)_3 \longrightarrow$
$$AlPO_4 \downarrow + 3NaNO_3$$

c. $2H_2O_2 \longrightarrow 2H_2O + O_2 \uparrow$

d. $BaO + SO_3 \longrightarrow BaSO_4$

e. $2NaClO \longrightarrow 2NaCl + O_2 \uparrow$

f. $Cl_2 + 2 KI \longrightarrow 2KCl + I_2$

g. $Ba(OH)_2 + 2HNO_3 \longrightarrow Ba(NO_3)_2 + 2H_2O$

9.37 Elements react in combination reactions to form _____.

9.38 Complete and balance the following:

a. $Mg + I_2 \longrightarrow$

b. $Li + O_2 \longrightarrow$

c. $Al + S \longrightarrow$

d. $K + P \longrightarrow$

e. $Ca + N_2 \longrightarrow$

9.39 Complete and balance the following equations:

a. Calcium reacts with bromine to yield _____ .

b. Sulfur dioxide combines with water to form _____ .

c. Magnesium reacts with nitrogen to yield _____ .

d. Calcium oxide combines with water to form _____ .

9.40 Give the formula of a reactant that could decompose to yield the following products:

a. $CO_2 + H_2O$

b. $MgO + CO_2$

c. $Ag + O_2$

d. $NaCl + O_2$

9.41 Consult the Activity Series in Table 9.2 and predict whether the following single-replacement reactions will occur:

a. $Mg + CuSO_4 \longrightarrow$

b. $Mg + CaSO_4 \longrightarrow$

c. $Ba + HCl \longrightarrow$

d. $Ag + HNO_3 \longrightarrow$

e. $Al + Ni(NO_3)_2 \longrightarrow$

f. $Ca + H_3PO_4 \longrightarrow$

g. $Al + SnCl_2 \longrightarrow$

h. $Au + H_2SO_4 \longrightarrow$

9.42 Complete and balance those equations in Problem 9.41 for which reactions do occur.

9.43 (a) Write an example of a balanced chemical equation illustrating the replacement of a metal by a more active metal. (b) Write an example of a balanced chemical equation illustrating a metal replacing H_2 from an acid.

9.44 Complete and balance the following equations (write NR for no reaction where appropriate):

  a. $Cd + AgNO_3 \longrightarrow$

  b. $Hg + HCl \longrightarrow$

  c. $Li + H_3PO_4 \longrightarrow$

9.45 Complete and balance the following equations representing double-replacement reactions:

  a. $K_2SO_4 + Ba(NO_3)_2 \longrightarrow$

  b. $(NH_4)_2CO_3 + MgCl_2 \longrightarrow$

  c. $(NH_4)_3PO_4 + Ca(NO_3)_2 \longrightarrow$

  d. $FeCl_2 + K_3PO_4 \longrightarrow$

  e. $Na_2S + Ni(NO_3)_2 \longrightarrow$

9.46 Complete and balance the following double-replacement reactions:

  a. $Ba(OH)_2 + HNO_3 \longrightarrow$

  b. $Fe(NO_3)_3 + NaOH \longrightarrow$

  c. $(NH_4)_2S + BaI_2 \longrightarrow$

  d. $H_2SO_3 + Al(OH)_3 \longrightarrow$

  e. $(NH_4)_3PO_4 + Ni(NO_3)_2 \longrightarrow$

9.47 Classify each of the following reactions as one of the four basic types; then complete and balance:

  a. $Li + AuCl_3 \longrightarrow$

  b. $Al(NO_3)_3 + K_2CO_3 \longrightarrow$

  c. $Ba + F_2 \longrightarrow$

  d. $Ba + SnF_2 \longrightarrow$

  e. $Mg + P \longrightarrow$

  f. $SnCl_2 + Na_3PO_4 \longrightarrow$

9.48 Classify each of the following reactions as one of the four basic types; then complete and balance:

  a. $BaCO_3 \longrightarrow$

  b. $Zn + H_2SO_4 \longrightarrow$

  c. $Pb(NO_3)_2 + NaI \longrightarrow$

  d. $HClO_3 + KOH \longrightarrow$

  e. $Li + O_2 \longrightarrow$

  f. $Al + Br_2 \longrightarrow$

9.49 Baking soda (sodium bicarbonate) is used in cakes and breads to make them rise. Its action results from the fact that solid sodium bicarbonate decomposes when heated to form gaseous carbon dioxide, gaseous water, and solid sodium carbonate. Write a balanced equation for this reaction.

9.50 Rusting involves the chemical reaction of iron, water, and oxygen to form iron(III) hydroxide. Write a balanced equation for the rusting process.

9.51 Zinc metal can be obtained by the reaction of zinc oxide with hydrogen gas. The other product is a familiar substance. Complete and balance the equation for the reaction of zinc oxide and hydrogen.

9.52 The chemical reaction that supplies current in an automobile battery is the reaction of lead plus lead(IV) oxide plus sulfuric acid to give lead(II) sulfate and water. Write a balanced equation for this reaction.

9.53 The combustion of gasoline is the reaction between octane (C_8H_{18}) and oxygen to form carbon dioxide and water. Write a balanced equation for the combustion.

9.54 Photosynthesis is the combination of carbon dioxide and water in the presence of light to form glucose ($C_6H_{12}O_6$) and oxygen. Write a balanced equation for photosynthesis.

9.55 Sulfur dioxide, a gaseous pollutant of air, readily combines with oxygen to form sulfur trioxide, which in turn dissolves in rain water to form sulfuric acid. The sulfuric acid reacts with calcium carbonate, the chemical composition of marble structures, to form calcium sulfate, carbon dioxide and water. Write balanced chemical equations for each of these three reactions.

STOICHIOMETRY
CHAPTER

10

Ne

Neon

20.18

WHAT IS STOICHIOMETRY?

10.1   The rather forbidding title of this chapter, **Stoichiometry,** is pronounced "stoy-key-ah′-meh-tree," and is defined as calculations relating the amounts of reactants and products in chemical reactions. If you have mastered Chapters 8 and 9, then you should find this chapter fairly easy. Stoichiometry applies the mole concept (Chapter 8) to the balanced chemical equation (Chapter 9). When you finish this chapter you will be fully prepared to go into a laboratory, measure out the proper amounts of reactants required for a particular reaction, and predict how much product should be produced.

MOLAR INTERPRETATION OF THE BALANCED EQUATION

10.2   The meaning of a balanced chemical equation was discussed in Section 9.3 in terms of interacting particles, that is, atoms, molecules, or ions. This is certainly one useful interpretation of a balanced equation. For example,

$$C + O_2 \longrightarrow CO_2$$

tells us that one C atom plus one molecule of O_2 yields one molecule of CO_2. Similarly,

$$3H_2 + N_2 \longrightarrow 2NH_3$$

means that three molecules of H_2 plus one molecule of N_2 yields two molecules of NH_3.

SAMPLE EXERCISE

10.1

Interpret the following equation in terms of the interacting particles:

$$Ca + 2H_2O \longrightarrow Ca(OH)_2 + H_2 \uparrow$$

SOLUTION

Identify the nature of each reactant and product. That is, ask yourself of what kinds of units or particles are they constructed. The coefficients tell you the relative numbers of the particles.

Ca　A symbol alone (no coefficient or subscript) denotes a single *atom*.

H_2O　Combinations of nonmetallic elements are molecular compounds; therefore, this is a *molecule*.

$Ca(OH)_2$　Combinations of metals and nonmetals are ionic compounds; therefore, this is a *formula unit* made up of a Ca^{2+} ion and two OH^- ions.

H_2　The subscript 2 reminds us that hydrogen is an element that exists as a *diatomic molecule*.

Thus the interpretation of the equation is one Ca atom plus two H_2O molecules yields one $Ca(OH)_2$ formula unit plus one H_2 molecule.

If we multiply all the coefficients of a balanced equation by the same number, we still have a balanced equation. For example, the coefficients of the equation $C + O_2 \rightarrow CO_2$ can be multiplied by 2 to give

$$2C + 2O_2 \longrightarrow 2CO_2$$

This is still a balanced equation. We now have two carbon atoms on each side and four oxygen atoms on each side. Similarly, multiplying $3H_2 + N_2 \longrightarrow 2NH_3$ by 3 gives

$$9H_2 + 3N_2 \longrightarrow 6NH_3$$
$$\text{18 H} \quad \text{6 N} \qquad \text{6 N} \quad \text{18 H}$$

In balancing equations we always try to use the smallest set of numbers, just as we try to use the simplest ratio in any mathematical relationship. However, an exact multiple of the simplest set of coefficients still gives a balanced equation.

Let us multiply through by Avogadro's number.

C	+	O_2	$\longrightarrow$	CO_2
One C atom		One molecule of O_2		One molecule of CO_2

$$6.022 \times 10^{23}\ C \quad + \quad 6.022 \times 10^{23}\ O_2 \longrightarrow 6.022 \times 10^{23}\ CO_2$$

6.022×10^{23} C atoms	6.022×10^{23} molecules of O_2	6.022×10^{23} molecules of CO_2
1 mole C	1 mole O_2	1 mole CO_2

We see by doing this that the coefficients of a balanced equation not only tell us the ratio of reacting particles, but also tell us the reacting **mole ratio.** We can read the coefficients of balanced equations as the numbers of moles of reactant or product.

Prove that the coefficients of the equation $3H_2 + N_2 \rightarrow 2NH_3$ tell you the molar ratio of reactants and products.

SOLUTION

1. The coefficients tell the ratio of reacting particles:

3 molecules H_2 + 1 molecule $N_2 \longrightarrow$ 2 molecules NH_3

2. Multiplication of all the coefficients by Avogadro's number will not disrupt the balancing.

$3(6.022 \times 10^{23})$ molecules H_2 + $1(6.022 \times 10^{23})$ molecules $N_2 \longrightarrow$
$2(6.022 \times 10^{23})$ molecules NH_3

3. Because 1 mole = 6.022×10^{23} molecules, we can also read the equation as

3 moles H_2 + 1 mole $N_2 \longrightarrow$ 2 moles NH_3

Molar amounts are easily related to weighable gram amounts through the molar weight (Section 8.6), and in practice in the laboratory, chemists almost always interpret the coefficients of balanced chemical equations as numbers of moles of reactants or products. Thus we can, and should, read chemical equations in terms of moles:

$Ca + 2H_2O \longrightarrow Ca(OH)_2 + H_2$

1 mole Ca + 2 moles $H_2O \longrightarrow$ 1 mole $Ca(OH)_2$ + 1 mole H_2

$2KClO_3 \longrightarrow 2KCl + 3O_2$

2 moles $KClO_3 \longrightarrow$ 2 moles KCl + 3 moles O_2

Problem 10.1

Interpret the following equations in terms of the numbers of moles of reactants and products.

a. $2Al(OH)_3 + 3H_2SO_4 \longrightarrow Al_2(SO_4)_3 + 6H_2O$

b. $4Li + O_2 \longrightarrow 2Li_2O$

"EQUALITIES" AND CONVERSION FACTORS FROM CHEMICAL EQUATIONS

10.3 Given any balanced chemical equation, you can write "equalities" that relate molar amounts of one reactant to molar amounts of another reactant, or molar amounts of reactant to molar amounts of product. For example, for the balanced equation $N_2 + 3H_2 \rightarrow 2NH_3$

we can write the following equalities by using the molar interpretation of coefficients:

$$1 \text{ mole } N_2 \simeq 3 \text{ moles } H_2$$

$$1 \text{ mole } N_2 \simeq 2 \text{ moles } NH_3$$

$$3 \text{ moles } H_2 \simeq 2 \text{ moles } NH_3$$

Or in summary

$$1 \text{ mole } N_2 \simeq 3 \text{ moles } H_2 \simeq 2 \text{ moles } NH_3$$

Notice that we have put "equalities" in quotation marks and use the symbol $\simeq$ rather than a true equals sign ($=$). This is because these equalities between molar amounts apply only to the particular reaction described by the balanced chemical equation. For different reactions involving these materials, we obtain different equalities. For example, N_2 and H_2 can also react to form hydrazine, N_2H_4, a rocket fuel. This reaction is represented by the equation

$$N_2 + 2 H_2 \longrightarrow N_2H_4$$

For this reaction the equality between N_2 and H_2 is 1 mole $N_2 \simeq$ 2 moles H_2.

These equalities can be made into conversion factors just as we have done previously for other equalities. For example, for the reaction between N_2 and H_2 to form NH_3 ($N_2 + 3 H_2 \rightarrow 2 NH_3$) for which the equalities (such as 1 mole $N_2 \simeq 3$ moles H_2) were given, we can write the conversion factors as follows:

$$\frac{1 \text{ mole } N_2}{3 \text{ moles } H_2} \qquad \frac{1 \text{ mole } N_2}{2 \text{ moles } NH_3} \qquad \frac{3 \text{ moles } H_2}{2 \text{ moles } NH_3}$$

Of course, we can also write the reciprocals of these fractions. These conversion factors are the **mole ratios** that relate molar amounts of one reactant to molar amounts of another reactant, or molar amounts of reactant to molar amounts of product.

SAMPLE EXERCISE

10.3

What is the mole ratio of H_2O to NaOH in the following reaction?

$$2Na + 2H_2O \longrightarrow 2NaOH + H_2 \uparrow$$

SOLUTION

The coefficients give you equalities between molar amounts. In this case, 2 moles $H_2O \simeq 2$ moles NaOH. The conversion factor written from this equality is the mole ratio

$$\frac{2 \text{ moles } H_2O}{2 \text{ moles } NaOH}$$

SAMPLE EXERCISE 10.4

What is the mole ratio of HF to SnF_2 in the following reaction, which is a means of making stannous fluoride, the "fluoride" in toothpaste?

$$Sn + 2HF \longrightarrow SnF_2 + H_2$$

SOLUTION

The coefficients give you the ratio of the reacting molar amounts. In this case, 2 mole of HF produces 1 mole of SnF_2. The mole ratio, which can be written and used as a conversion factor is 2 mole of HF/1 mole of SnF_2.

To do stoichiometry one must *always* use an appropriate *mole ratio* that has been obtained from a balanced chemical equation. Mole ratios are used just as any other conversion factor in the Unit Conversion Method.

Problem 10.2

What is the mole ratio of Na to H_2 in the reaction given in Sample Exercise 10.3?

Problem 10.3

What is the mole ratio of Sn to H_2 in the reaction given in Sample Exercise 10.4?

MOLE-MOLE CONVERSIONS

10.4 The simplest stoichiometry problems are those in which we are given a balanced chemical equation and a molar amount of one reactant or product and asked to calculate the molar amounts of the other reactants and products. These problems are handled by the Unit Conversion Method in the usual manner (Section 3.3). We identify the given quantity and extract a conversion factor from the balanced equation.

SAMPLE EXERCISE 10.5

The decomposition of $MgCO_3$ is represented by the following equation:

$$MgCO_3 \longrightarrow MgO + CO_2$$

If 6 moles of $MgCO_3$ decompose, how many moles of CO_2 are produced?

SOLUTION

Follow the steps of the Unit Conversion Method.

Step 1 The given quantity is 6 moles $MgCO_3$.

Step 2   The new quantity is moles CO_2.

Step 3   The equality relating the given and new quantities is obtained from the coefficients of the balanced chemical equation:

$$1 \text{ mole } MgCO_3 \simeq 1 \text{ mole } CO_2$$

We choose the conversion factor with "new" in the numerator and "given" in the denominator. Thus

$$\frac{1 \text{ mole } CO_2}{1 \text{ mole } MgCO_3}$$

Step 4   Set up in the proper format:

Given × conversion factor = new

$$6 \text{ moles } \cancel{MgCO_3} \times \frac{1 \text{ mole } CO_2}{1 \text{ mole } \cancel{MgCO_3}} = 6 \text{ moles } CO_2$$

Mole-mole conversions are simple, one-step unit conversions, and you are undoubtedly ready to apply the method even to very complex and unfamiliar reactions.

SAMPLE EXERCISE

10.6

How many moles of $H_2C_2O_4$ are required to react completely with 1.50 moles of $KMnO_4$ according to the following equation:

$$2KMnO_4 + 6HCl + 5H_2C_2O_4 \longrightarrow 2MnCl_2 + 10CO_2 + 2KCl + 8H_2O$$

SOLUTION

Step 1   Given is 1.50 moles of $KMnO_4$.

Step 2   New is moles of $H_2C_2O_4$.

Step 3   The equality from the balanced equation is

$$2 \text{ moles } KMnO_4 \simeq 5 \text{ moles } H_2C_2O_4$$

and the conversion factor $\dfrac{\textbf{new}}{\textbf{given}}$ is $\dfrac{5 \text{ moles } H_2C_2O_4}{2 \text{ moles } KMnO_4}$.

Step 4   The format is

$$1.50 \text{ moles } \cancel{KMnO_4} \times \frac{5 \text{ moles } H_2C_2O_4}{2 \text{ moles } \cancel{KMnO_4}}$$

$$= \frac{1.50 \times 5 \text{ moles } H_2C_2O_4}{2} = 3.75 \text{ moles } H_2C_2O_4$$

Notice that the coefficients from the equation (5 and 2) are exact numbers, so they do not limit the number of sig figs in the answer.

Problem 10.4

Write a balanced equation for the reaction between K and Br_2 that forms the sedative KBr and calculate the number of moles of KBr produced by the reaction of 7.50 moles of Br_2.

GRAM-MOLE, MOLE-GRAM CONVERSIONS

10.5   In Section 8.6 we told you that the conversion factors most often used in chemistry are those relating moles and grams through the molar weight. We are going to use these factors extensively in stoichiometry. Using both gram-mole or mole-gram conversion factors and mole-ratio conversion factors from balanced equations enables us to relate gram amounts of one reactant or product to molar amounts of other reactants or products, or vice versa.

Central to all stoichiometry problems is the concept that the mole ratio relates the "new" quantity to the "given" quantity. Identify the mole ratio first. Then identify the gram-mole conversions necessary to use this mole ratio as a factor. Notice how this order of identifying equalities and conversion factors is used in Sample Exercises 10.7 and 10.8.

SAMPLE EXERCISE

10.7

According to the equation

$$Fe_2O_3 + 2Al \longrightarrow Al_2O_3 + 2Fe$$

how many moles of Al_2O_3 are produced by the reaction of 81 g of Al?

SOLUTION

Use the Unit Conversion Method. Two conversion factors will be required.

Step 1   The given quantity and unit is 81 g Al.

Step 2   The new unit is moles Al_2O_3.

Step 3   The mole ratio necessary to convert moles Al to moles Al_2O_3 is

$$\frac{1 \text{ mole } Al_2O_3}{2 \text{ moles Al}} \qquad \frac{\textbf{new} \text{ material}}{\textbf{given} \text{ material}}$$

To use this factor, grams Al must be changed to moles Al through the molar weight.

$$\frac{1 \text{ mole Al}}{27 \text{ g Al}} \qquad \frac{\textbf{new} \text{ unit}}{\textbf{given} \text{ unit}}$$

Step 4   Use the format beginning with the *given* and using conversion factors of the form **new/given** to cancel out units.

$$81 \text{ g Al} \times \frac{1 \text{ mole Al}}{27 \text{ g Al}} \times \frac{1 \text{ mole } Al_2O_3}{2 \text{ moles Al}} = \frac{81 \times 1 \times 1 \text{ mole } Al_2O_3}{27 \times 2}$$
$$= 1.5 \text{ moles } Al_2O_3$$

SAMPLE EXERCISE

10.8

According to the equation

$$2KClO_3 \longrightarrow 2KCl + 3O_2$$

how many grams of $KClO_3$ are required in order to produce 4.50 moles of O_2?

SOLUTION

Use the Unit Conversion Method. Two conversion factors will be required.

Step 1   The given quantity and unit is 4.50 moles of O_2.

Step 2   The new unit is grams of $KClO_3$.

Step 3   The mole ratio from the balanced equation necessary to convert moles O_2 to moles $KClO_3$ is

$$\frac{2 \text{ moles } KClO_3}{3 \text{ moles } O_2} \qquad \frac{\textbf{new} \text{ material}}{\textbf{given} \text{ material}}$$

To use this factor and get the answer in grams $KClO_3$, moles $KClO_3$ must be converted to grams through the molar weight.

$$K\ 1 \times 39.1 = \quad 39.1$$
$$Cl\ 1 \times 35.5 = \quad 35.5$$
$$O\ 3 \times 16.0 = \quad \underline{48.0}$$
$$122.6$$

$$1 \text{ mole} = 122.6 \text{ g } KClO_3$$

Thus the appropriate conversion factor is

$$\frac{122.6 \text{ g } KClO_3}{1 \text{ mole } KClO_3} \qquad \frac{\textbf{new} \text{ units}}{\textbf{given} \text{ units}}$$

Step 4   Use the format beginning with the given and using conversion factors of the form **new/given** to cancel out units.

$$\textbf{Given} \times \textbf{conversion factors} = \textbf{new}$$

$$4.50 \text{ moles } O_2 \times \frac{2 \text{ moles } KClO_3}{3 \text{ moles } O_2} \times \frac{122.6 \text{ g } KClO_3}{1 \text{ mole } KClO_3}$$

$$= \frac{4.50 \times 2 \times 122.6 \text{ g } KClO_3}{3 \times 1} = 368 \text{ g } KClO_3$$

(Three sig figs allowed)

Problem 10.5

Refer to the balanced equation in Sample Exercise 10.8 to do this problem. How many grams of KCl are produced when 3.00 moles of $KClO_3$ react?

GRAM-GRAM CONVERSIONS

10.6 When chemists do chemical reactions in the laboratory, they measure out reactants and products in grams. It is essential to be able to relate gram amounts of reactants to gram amounts of products.

As in previous stoichiometry problems, to do gram-gram stoichiometry problems we need a balanced chemical equation to give us a mole-ratio conversion factor that relates the materials involved in the conversion. Once again the mole ratio will be the central focus and should be determined first. The other conversion factors will be gram-mole or mole-gram factors.

SAMPLE EXERCISE

10.9

Sodium iodide can be made in the laboratory by combining sodium metal and iodine. The balanced equation for the reaction is $2Na + I_2 \rightarrow 2NaI$. How many grams of I_2 are required to produce 225 g of NaI?

SOLUTION

Use the Unit Conversion Method. Three conversion factors will be required. Determine the mole ratio first.

Step 1   The given is 225 g NaI.

Step 2   The new is grams I_2.

Step 3   The mole ratio necessary to convert NaI to I_2 is

$$\frac{1 \text{ mole } I_2}{2 \text{ moles NaI}} \qquad \frac{\textbf{new} \text{ material}}{\textbf{given} \text{ material}}$$

To use this factor, grams of NaI must be changed to moles of NaI through the molar weight.

$$\frac{\textbf{New} \text{ unit}}{\textbf{Given} \text{ unit}} \qquad \frac{1 \text{ mol of NaI}}{150. \text{ g of NaI}} \qquad \left(\begin{array}{l} \text{Atomic} \\ \text{weights} \\ \text{Na} \quad 23.0 \\ \text{I} \quad \underline{127.} \\ \quad\quad 150. \end{array} \right)$$

To obtain the answer in grams of I_2, moles of I_2 must be converted to grams of I_2 through the molar weight.

$$\frac{\textbf{New} \text{ unit}}{\textbf{Given} \text{ unit}} \qquad \frac{254 \text{ g of } I_2}{1 \text{ mol of } I_2} \qquad (2 \times 127 = 254)$$

Step 4   Use the usual format, beginning with the given:

$$225 \text{ g NaI} \times \frac{1 \text{ mole NaI}}{150. \text{ g NaI}} \times \boxed{\frac{1 \text{ mole } I_2}{2 \text{ moles NaI}}} \times \frac{254 \text{ g } I_2}{1 \text{ mole } I_2}$$

$$= \frac{225 \times 1 \times 1 \times 254 \text{ g } I_2}{150. \times 2 \times 1} = 191 \text{ g } I_2$$

Notice that recognizing the mole ratio as the central focus in Sample Exercise 10.9, or any problem, helps in setting up the format. All gram-gram conversions have the following "skeleton":

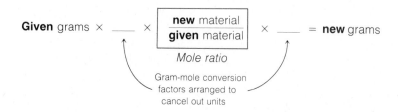

Given grams × ____ × $\dfrac{\textbf{new} \text{ material}}{\textbf{given} \text{ material}}$ × ____ = **new** grams

Mole ratio

Gram-mole conversion factors arranged to cancel out units

SAMPLE EXERCISE

10.10

How many grams of Li are required to react completely with 161 g S to produce Li_2S? The equation for the reaction is

$$2Li + S \longrightarrow Li_2S$$

SOLUTION

Work this through as in Sample Exercise 10.9, recognizing 161 g S as the **given** and grams Li as **new**, *or* try fitting the data into the preceding "skeleton."

We begin this by inserting the quantities into the clearly labeled parts of the "skeleton."

$$161 \text{ g S} \times \underbrace{\text{_____}}_{\textbf{Given}} \times \boxed{\dfrac{2 \text{ moles Li}}{1 \text{ mole S}}} \times \underbrace{\text{_____}}_{\textbf{New}} = \text{grams Li}$$

Mole ratio

Now the first conversion factor must cancel grams S. So it will be

$$\dfrac{1 \text{ mole S}}{32.1 \text{ g S}} \qquad \text{from } 1 \text{ mole S} = 32.1 \text{ g S}$$

The last conversion factor must produce the new unit grams Li. So it will be

$$\dfrac{6.94 \text{ g Li}}{1 \text{ mole Li}} \qquad \text{from } 1 \text{ mole Li} = 6.94 \text{ g Li}$$

Thus the setup is

$$161 \text{ g S} \times \dfrac{1 \text{ mole S}}{32.1 \text{ g S}} \times \boxed{\dfrac{2 \text{ moles Li}}{1 \text{ mole S}}} \times \dfrac{6.94 \text{ g Li}}{1 \text{ mole Li}}$$

$$= \dfrac{161 \times 1 \times 2 \times 6.94 \text{ g Li}}{32.1 \times 1 \times 1} = 69.6 \text{ g Li}$$

Problem 10.6

How many grams of Li_2S are produced by the reaction described in Sample Exercise 10.10?

CONVERSIONS SUMMARIZED

10.7 All stoichiometric conversions require a balanced equation from which one determines an appropriate mole ratio. Because *the mole ratio must be a conversion factor* relating moles of new material to moles of given material, gram amounts of material must be converted into molar amounts (or vice versa) in the course of stoichiometric calculations. A general plan of attack for all stoichiometry problems appears below:

1. Determine the mole ratio necessary to convert the given material into the new material. The mole ratio comes from the coefficients of the balanced equation and will have the form

$$\frac{\text{Number of moles } \mathbf{new} \text{ material}}{\text{Number of moles } \mathbf{given} \text{ material}}$$

2. If both the given and new units are stated in moles, then the problem will be a one-step conversion, as seen in Section 10.4.

3. If either or both the given and new units are stated in grams, then there must be gram-mole or mole-gram conversion factors, as seen in Sections 10.5 and 10.6 (see also Figure 10.1).

LIMITING REACTANT

10.8 Suppose that you set out to bake some loaves of bread, the main ingredients of which are flour and eggs. The recipe for one loaf calls for 3 cups of flour and 1 egg. Now suppose that you have available in your kitchen 3 eggs and 12 cups of flour. With the given recipe and these amounts available you can make a maximum of 3 loaves of

FIGURE 10.1

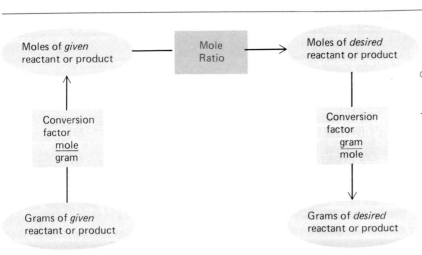

The big picture of stoichiometry! Begin with grams or moles given and end with moles or grams desired by proceeding along the part of this pathway that applies.

bread. The number of eggs *limits* the number of loaves. Twelve cups of flour is sufficient to make 4 loaves [12 ~~cups~~ × (1 loaf/3 ~~cups~~) = 4 loaves], but there are not enough eggs for 4 loaves. Thus the amount of flour available is in excess of what can be used up (see Figure 10.2).

No, this has not turned into a cookbook. We have described the preceding kitchen dilemma because it gives us a familiar example of a similar problem that arises in doing some stoichiometry problems. Let us summarize the preceding flour, egg, and loaf data in the form of a chemical equation:

Reactants ⟶ Products

FIGURE 10.2

"Balanced" equation:

$$3 \text{ cups flour} + 1 \text{ egg} \longrightarrow 1 \text{ loaf}$$

Available reactants and product obtained:

$$12 \text{ cups flour} + 3 \text{ eggs} \longrightarrow 3 \text{ loaves} + 3 \text{ cups flour left over}$$

Excess   Limiting reactant   (Only 9 cups of flour can "react" with 3 eggs)

The 3 eggs limit you to make 3 loaves of bread by the recipe, even though you have enough flour to make 4 loaves. There is flour left over. Likewise, 4 moles of Na limits you to make 4 moles of NaCl, even though there is enough Cl_2 to make 6 moles of NaCl. Some Cl_2 is left over.

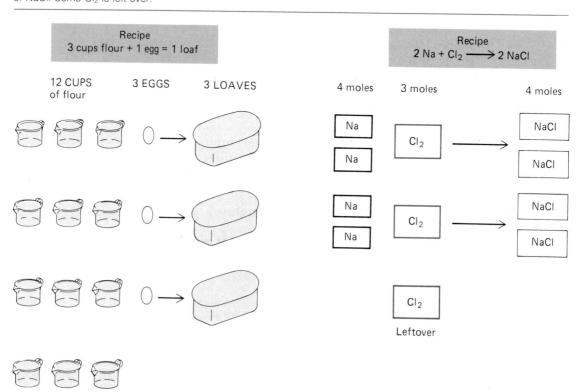

Recipe
3 cups flour + 1 egg = 1 loaf

12 CUPS of flour   3 EGGS   3 LOAVES

Leftover flour

Recipe
$2 Na + Cl_2 \longrightarrow 2 NaCl$

4 moles   3 moles   4 moles

Na Cl_2 NaCl
Na   NaCl

Na Cl_2 NaCl
Na   NaCl

Cl_2

Leftover

Whenever you are told the amounts of two or more reactants in a chemical reaction, you must identify the **limiting reactant**, that is, the one that limits the amount of product that can be obtained. The other reactant or reactants are in excess. When you are told the amount of only one reactant that reactant is the limiting reactant and any other reactants are assumed to be present in excess.

Just as for the bread problem, we determine the limiting reactant (reagent) by deciding which reactant gives the least amount of product. The limiting reactant is totally consumed in the reaction. There is more of the excess reactant (or reactants) than is needed in the reaction, so some of this reactant (or reactants) remains unused.

For example, suppose we have 4 moles of Na and 3 moles of Cl_2 and wish to make NaCl. The "recipe" for this is the balanced equation

$$2Na + Cl_2 \longrightarrow 2NaCl$$

It tells us through mole ratios the amount of product that can be obtained from a given amount of reactant. Thus from 4 moles of Na we can obtain 4 moles of NaCl (Figure 10.2)

$$4 \text{ moles Na} \times \frac{2 \text{ moles NaCl}}{2 \text{ moles Na}} = 4 \text{ moles NaCl}$$

From 3 moles of Cl_2 we can make 6 moles of NaCl.

$$3 \text{ moles Cl}_2 \times \frac{2 \text{ moles NaCl}}{1 \text{ mole Cl}_2} = 6 \text{ moles NaCl}$$

The 4 moles of Na is the limiting reactant because it produces the lesser amount of product. The Cl_2 is in excess. Four moles Na requires only 2 moles of Cl_2 to form 4 moles of NaCl. (Note that if $2Na + Cl_2 \rightarrow 2NaCl$, then $4Na + 2Cl_2 \rightarrow 4NaCl$. Therefore, the 4 moles of Na is totally consumed by reacting with 2 moles of Cl_2 to give 4 moles of NaCl.)

Initially	Reacted or Formed	Finally
4 moles Na	− 4 reacted	= 0 mole Na
3 moles Cl_2	− 2 reacted	= 1 mole Cl_2
0 mole NaCl	+ 4 formed	= 4 moles NaCl

The limiting reactant limits the amount of product formed.

The limiting reactant is always identified by working in moles, as was done just previously for the NaCl formation. This means that if you are told the amounts of reactants in grams, you must convert grams to moles for each reactant before attempting to identify the limiting reactant.

General method   **1.** Recognize the necessity of determining the limiting reactant when
for determining   you are given the amounts of two or more reactants.
the limiting
reactant   **2.** Convert amounts of reactants given to moles.

3. Calculate the molar amount of product obtained based on the
amount of *one* of the reactants as if it were the only reactant amount
given. Then calculate the amount of product obtained based on the
other reactant as if it were the only one given.

4. Choose the reactant which produces the least amount of product
as the limiting reactant. This amount of product is the maximum
amount that can be produced (see Section 10.9 "Theoretical Yield").

SAMPLE EXERCISE

10.11

A 50.-g sample of $Mg(OH)_2$ is mixed with 70. g of H_3PO_4. A neutralization
reaction takes place which is represented by the equation

$$3Mg(OH)_2 + 2H_3PO_4 \longrightarrow Mg_3(PO_4)_2 + 6H_2O$$

How many grams of $Mg_3(PO_4)_2$ are produced? (*Note*: The formula weight of
$Mg(OH)_2 = 58.3$; the molecular weight of $H_3PO_4 = 98.0$; and the formula
weight of $Mg_3(PO_4)_2 = 262.9$.)

SOLUTION

1. Because the amounts of *both reactants* are given, you must decide which
is the limiting reactant and proceed according to the general method.

2. Use the formula weight and molecular weight given to calculate the moles
of each reactant:

$$50. \text{ g Mg(OH)}_2 \times \frac{1 \text{ mole Mg(OH)}_2}{58.3 \text{ g Mg(OH)}_2} = 0.86 \text{ mole Mg(OH)}_2$$

$$70. \text{ g H}_3\text{PO}_4 \times \frac{1 \text{ mole H}_3\text{PO}_4}{98.0 \text{ g H}_3\text{PO}_4} = 0.71 \text{ mole H}_3\text{PO}_4$$

3. Perform a mole-mole conversion on each reactant to determine how much
product could be obtained from that reactant:

$$0.86 \text{ mole Mg(OH)}_2 \times \frac{1 \text{ mole Mg}_3(\text{PO}_4)_2}{3 \text{ moles Mg(OH)}_2} = 0.29 \text{ mole Mg}_3(\text{PO}_4)_2$$

↑

mole ratios from
balanced equation

↓

$$0.71 \text{ mole H}_3\text{PO}_4 \times \frac{1 \text{ mole Mg}_3(\text{PO}_4)_2}{2 \text{ moles H}_3\text{PO}_4} = 0.36 \text{ mole Mg}_3(\text{PO}_4)_2$$

4. $Mg(OH)_2$ is the limiting reactant because it produces the least amount of product, 0.29 mole $Mg_3(PO_4)_2$.

5. We know the molar amount of product. To get the gram amount, we use the given formula weight:

$$0.29 \text{ mole } Mg_3(PO_4)_2 \times \frac{262.9 \text{ g } Mg_3(PO_4)_2}{1 \text{ mole } Mg_3(PO_4)_2} = 76 \text{ g } Mg_3(PO_4)_2$$

Once the limiting reactant is determined, the amount of product obtained is also known, and the amount of excess reactant left over can be calculated.

1. Calculate the number of moles of the excess reactant that must have reacted with the limiting reactant to form the product.

2. Subtract the amount of the excess reactant that reacted from the original molar amount present.
Number of moles originally − number of moles reacted = number of moles left over

3. Use a gram-mole conversion factor to change moles to grams if that is asked for.

Method for calculating excess reactant left over after reaction

SAMPLE EXERCISE

10.12

How much in grams of the excess reactant is left over after the reaction described in Sample Exercise 10.11 is completed?

SOLUTION

The excess reactant was H_3PO_4.

1. $0.86 \text{ mole } Mg(OH)_2 \times \dfrac{2 \text{ moles } H_3PO_4}{3 \text{ moles } Mg(OH)_2}$
$= 0.57 \text{ mole } H_3PO_4 \text{ react with } 0.86 \text{ mole } Mg(OH)_2$

2. Originally there was 0.71 mole H_3PO_4:
0.71 mole originally − 0.57 mole reacted = 0.14 mole left over

3. $0.14 \text{ mole } H_3PO_4 \times \dfrac{98.0 \text{ g } H_3PO_4}{1 \text{ mole } H_3PO_4} = 14 \text{ g } H_3PO_4 \text{ left over}$

SAMPLE EXERCISE

10.13

A mixture of 30.0 g of Al and 80.0 g of Fe_2O_3 is prepared. The mixture is heated and a vigorous single replacement occurs according to the following equation:

$$Fe_2O_3(s) + 2Al(s) \longrightarrow Al_2O_3(s) + 2Fe(l)$$

What is the mass of iron produced? (Atomic weight Al = 27.0, formula weight Fe_2O_3 = 160., and the atomic weight Fe = 55.9.)

SOLUTION

1. The given amounts of both reactants signals a limiting reactant problem.

2. Use the molar weights to calculate the moles of each reactant:

$$30.0 \text{ g Al} \times \frac{1 \text{ mole Al}}{27.0 \text{ g Al}} = 1.11 \text{ moles Al}$$

$$80.0 \text{ g Fe}_2\text{O}_3 \times \frac{1 \text{ mole Fe}_2\text{O}_3}{160. \text{ g Fe}_2\text{O}_3} = 0.500 \text{ mole Fe}_2\text{O}_3$$

3. Determine the molar amount of iron that could be obtained from each reactant:

$$1.11 \text{ moles Al} \times \frac{2 \text{ moles Fe}}{2 \text{ moles Al}} = 1.11 \text{ moles Fe}$$

$$0.500 \text{ mole Fe}_2\text{O}_3 \times \frac{2 \text{ moles Fe}}{1 \text{ mole Fe}_2\text{O}_3} = 1.00 \text{ moles Fe}$$

4. Fe_2O_3 is the limiting reactant because it produces the least amount of product, 1.00 mole of Fe. Notice that Fe_2O_3 is limiting even though we began with a larger mass of Fe_2O_3. The limiting reactant *cannot* be determined by simply looking at the given masses of the two reactants.

5. The mass of iron product is obtained from the molar amount using the atomic weight of iron:

$$1.00 \text{ mole Fe} \times \frac{55.9 \text{ g Fe}}{1 \text{ mole Fe}} = 55.9 \text{ g Fe}$$

SAMPLE EXERCISE

10.14

Calculate the amount of excess Al in Sample Exercise 10.13.

SOLUTION

1. Determine the amount of Al that reacted with the Fe_2O_3:

$$0.500 \text{ mole Fe}_2\text{O}_3 \times \frac{2 \text{ moles Al}}{1 \text{ mole Fe}_2\text{O}_3} = 1.00 \text{ mole Al}$$

2. Initially there was 1.11 moles of Al:
    1.11 moles initially − 1.00 moles reacted = 0.11 mole Al left over.

3. $0.11 \text{ mole Al} \times \dfrac{27.0 \text{ g Al}}{1.00 \text{ mole Al}} = 3.0 \text{ g Al}$

Problem 10.7

What mass of SO_3 is produced when 3.60 g SO_2 is combined with 2.40 g O_2?

$$2SO_2(g) + O_2(g) \longrightarrow 2SO_3(g)$$

Perhaps you are asking yourself why chemists don't simply use correct stoichiometric amounts of all reactants in doing chemical reactions—that is, why not use amounts such that all reactants are consumed and none are in excess. In Sample Exercise 10.11, this would require the reaction of 50. g (0.86 mole) $Mg(OH)_2$ and 56 g (0.57 mole) H_3PO_4 to give exactly 76 g (0.29 mole) $Mg_3(PO_4)_2$. More than stoichiometric amounts are often used because many reactions do not work as well as we would like them to; that is, reactants are not completely converted into products, or products once formed interact to reform reactants. One way to encourage a reaction to go completely to products is to use one reactant in excess. We will further explore this idea in Chapter 14.

A typical way that chemists decide on the amounts of reactants to use in a reaction is to:

1. Decide the amount of product that is desired.

2. Consider the relative availability and cost of the reactants. Usually one has less of the more expensive reactant.

3. Calculate the correct stoichiometric amount of the less available (more expensive) reactant needed for the desired amount of product.

4. Use an excess of the more available (cheaper) reactant.

1. A chemist wants to make 12.0 g AgF by the reaction

$$2 \text{ Ag} + F_2 \longrightarrow 2 \text{ AgF}$$

2. Ag is very expensive and there is not much in the laboratory. F_2 is abundantly available.

3. $12.0 \text{ g AgF} \times \dfrac{1 \text{ mole AgF}}{127 \text{ g AgF}}$

$\times \dfrac{2 \text{ moles Ag}}{2 \text{ moles AgF}} \times \dfrac{108 \text{ g Ag}}{1 \text{ mole Ag}}$

$= \dfrac{12.0 \times 108 \text{ g Ag}}{127} = 10.2 \text{ g Ag}$

4. Use an excess of F_2, that is, use more than 1.80 g F_2, which is the correct stoichiometric amount.

$12.0 \text{ g AgF} \times \dfrac{1 \text{ mole AgF}}{127 \text{ g AgF}}$

$\times \dfrac{1 \text{ mole } F_2}{2 \text{ moles AgF}} \times \dfrac{38.0 \text{ g } F_2}{1 \text{ mole } F_2}$

$= \dfrac{12.0 \times 38.0 \text{ g } F_2}{127 \times 2} = 1.80 \text{ g } F_2$

THEORETICAL YIELD

10.9 The **theoretical yield** is the maximum number of grams of product that can be obtained based on the amount of limiting reactant given and the balanced chemical equation. Whenever we use stoichiometry to calculate the number of grams of product obtained from a specified amount of reactant, we are calculating the theoretical yield. Thus you already know how to do the theoretical yield calculation. This is just an added bit of terminology.

SAMPLE EXERCISE

10.15

Calculate the theoretical yield of AlF_3 obtained from the reaction of 0.56 mole of Al in the reaction $2Al + 3F_2 \longrightarrow 2AlF_3$.

SOLUTION

The mass of AlF_3 is determined from the moles of Al, the only given amount of reactant.

Step 1   The mole ratio is

$$\frac{2 \text{ moles } AlF_3}{2 \text{ moles } Al} \quad \begin{array}{l} \textbf{(new)} \\ \textbf{(given)} \end{array}$$

Step 2   Because the answer must be in grams of AlF_3, a mole-gram conversion will be necessary using the relation

$$1 \text{ mole } AlF_3 = 84 \text{ g } AlF_3 \qquad \begin{array}{l} 1 \times Al \ 27 \\ 3 \times F \ \underline{57} \\ 84 \end{array}$$

Step 3   Set up the unit conversion format:

$$0.56 \text{ mole Al} \times \frac{2 \text{ moles } AlF_3}{2 \text{ moles Al}} \times \frac{84 \text{ g } AlF_3}{1 \text{ mole } AlF_3} = 47 \text{ g } AlF_3$$

The theoretical yield is 47 g AlF_3, the maximum amount possible from 0.56 mole of Al.

ACTUAL YIELD

10.10   When doing chemical reactions in the laboratory, one rarely can collect the entire theoretical yield of product. Some product is generally "lost" as the reaction is done. Apparent loss of product happens because the reaction does not proceed to completion, because some product stays in solution, or because some material is left behind when the chemist transfers it from one flask to another. The Law of Conservation of Mass is not violated; the lost material can be accounted for. But the amount of product we actually collect is *less than* the theoretical yield. What we actually collect is the **actual yield**. Note that the actual yield can only be determined experimentally; it cannot be calculated.

PERCENTAGE YIELD

10.11 The **percentage yield** is a measure of the closeness of the actual yield and the theoretical or maximum possible yield.

$$\frac{\text{Actual yield}}{\text{Theoretical yield}} \times 100 = \text{percentage yield}$$

SAMPLE EXERCISE

10.16

In the reaction described in Sample Exercise 10.15, suppose that the amount of product actually obtained was 40. g. Calculate the percentage yield.

SOLUTION

$$\text{Percentage yield} = \frac{\text{actual yield}}{\text{theoretical yield}} \times 100 = \frac{40.\ \text{g}}{47\ \text{g}} \times 100 = 85\ \text{percent}$$

HEAT AS A REACTANT OR PRODUCT

10.12 Whenever a chemical reaction occurs, there is an energy change. It is convenient to measure these changes in the form of heat energy using the units of kilocalories (kcal). Heat can be released as the reaction proceeds. When heat is released, we can regard heat as a product. A reaction in which heat is a product is called an **exothermic reaction** ("exo" means "out"; "thermic" means "heat"; hence, heat comes out or is released). An example is the reaction of methane, CH_4, the gas in bunsen burners, with O_2.

$$CH_4(g) + 2O_2(g) \longrightarrow CO_2(g) + 2H_2O(l) + 213\ \text{kcal} \quad (10.1)$$
$$\text{Product}$$

Sometimes heat must be supplied in order for a reaction to occur. In this case we can regard heat as a reactant. A reaction in which heat is a reactant is called an **endothermic reaction** ("endo" means "in"; "thermic" means "heat"; thus, heat is put in or is absorbed). The formation of laughing gas, N_2O, from N_2 and O_2 is an example of an endothermic reaction.

$$2N_2(g) + O_2(g) + 39.0\ \text{kcal} \longrightarrow 2N_2O(g) \quad (10.2)$$
$$\text{Reactant}$$

Problem 10.8

Identify the following reactions as *exo*thermic or *endo*thermic:

a. $2PCl_3 + 146\ \text{kcal} \longrightarrow 2P + 3Cl_2$

b. $C + O_2 \longrightarrow CO_2 + 94.1\ \text{kcal}$

c. $2O_3 \longrightarrow 3O_2 + 68.1\ \text{kcal}$

Because heat can be regarded as a reactant or product in a chemical reaction, one can obtain from the balanced equation "equalities" and conversion factors which relate the amount of heat that has been supplied or produced to the amounts of reactants used or the amounts of products formed. The equalities from Equations (10.1) and (10.2) are:

From (10.1)

$$1 \text{ mole } CH_4 \simeq 2 \text{ moles } O_2 \simeq 1 \text{ mole } CO_2 \simeq 2 \text{ moles } H_2O \simeq 213 \text{ kcal}$$

From (10.2)

$$2 \text{ moles } N_2 \simeq 1 \text{ mole } O_2 \simeq 39.0 \text{ kcal} \simeq 2 \text{ moles } N_2O$$

Conversion factors can be constructed from the equalities in the usual way. Using the same general format as for other stoichiometry problems, we can calculate the amount of heat that must be supplied or the amount of heat that has been produced given an amount of some chemical reactant or product.

SAMPLE EXERCISE

10.17

How much heat is released when 8.00 g of CH_4 burns? Burning is the reaction of CH_4 with O_2 shown in Equation (10.1).

SOLUTION

Use the Unit Conversion Method. The equation gives you the equality

$$1 \text{ mole } CH_4 \simeq 213 \text{ kcal}$$

The molecular weight of CH_4 is 16.0; thus

$$1 \text{ mole } CH_4 = 16.0 \text{ g}$$

Given $\times$ **conversion factors** $=$ **new**

$$8.00 \text{ g } \cancel{CH_4} \times \frac{1 \cancel{\text{ mole } CH_4}}{16.0 \text{ g } \cancel{CH_4}} \times \frac{213 \text{ kcal}}{1 \cancel{\text{ mole } CH_4}} = 107 \text{ kcal}$$

Problem 10.9

How much heat is required to produce 132 g of laughing gas, N_2O, according to Equation (10.2)?

A PAGE FROM A LABORATORY NOTEBOOK

10.13 You are now fully prepared to understand a typical page from the laboratory record of a working chemist (see Figure 10.3). A chemist working on insecticide research set out to prepare 500 g of

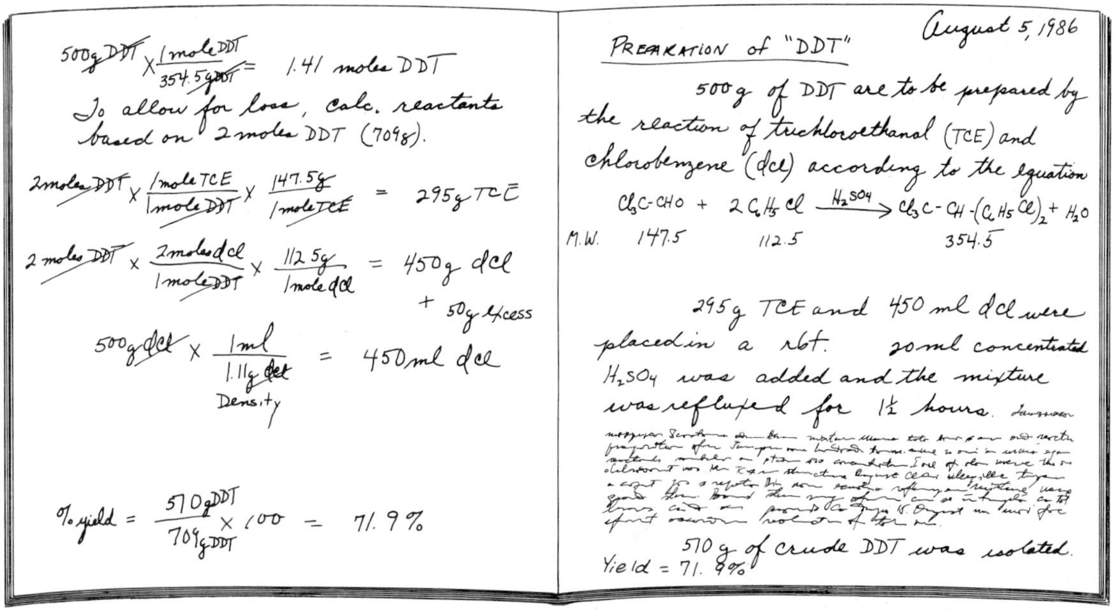
500 g DDT × 1 mole DDT / 354.5 g DDT = 1.41 moles DDT
To allow for loss, calc. reactants based on 2 moles DDT (709 g).
2 moles DDT × 1 mole TCE / 1 mole DDT × 147.5 g / 1 mole TCE = 295 g TCE
2 moles DDT × 2 moles φCl / 1 mole DDT × 112.5 g / 1 mole φCl = 450 g φCl
+ 50 g excess
500 g φCl × 1 ml / 1.11 g φCl = 450 ml φCl
Density
% yield = 510 g DDT / 709 g DDT × 100 = 71.9%
PREPARATION of "DDT"
August 5, 1986
500 g of DDT are to be prepared by the reaction of trichloroethanal (TCE) and chlorobenzene (φCl) according to the equation
Cl3C-CHO + 2 C6H5Cl → Cl3C-CH-(C6H5Cl)2 + H2O
M.W. 147.5 112.5 354.5
295 g TCE and 450 ml φCl were placed in a rbf. 20 ml concentrated H2SO4 was added and the mixture was refluxed for 1½ hours.
510 g of crude DDT was isolated.
Yield = 71.9%

FIGURE 10.3

Facing pages from the notebook of a working chemist. The formulas in the equations are partial structural formulas that organic chemists sometimes use rather than the molecular formulas that you are used to. Also, trichloroethanal is abbreviated TCE and chlorobenzene is abbreviated φCl.

DDT ($C_{14}H_9Cl_5$) from chlorobenzene (C_6H_5Cl) and trichloroethanal (C_2HOCl_3). His first task was to write a balanced chemical equation:

$$C_2HOCl_3 + 2C_6H_5Cl \longrightarrow C_{14}H_9Cl_5 + H_2O$$

From the balanced equation he determined the necessary amounts of reactants through appropriate *mole ratios.* In this case, he used 295 g of trichloroethanal as the *limiting reactant* and 500 g of chlorobenzene (450 g was needed; the *excess* helped the reaction go to completion) to obtain a theoretical yield of 709 g of DDT. He set his theoretical yield higher than the amount he actually wanted because he knew that the actual yield might be considerably less than the theoretical yield.

When the reaction was complete, he found that his *actual yield* was 510. g, i.e., more than the amount desired. However, the actual yield was less than the 709-g *theoretical yield* that was possible from the reactant amounts. He calculated the *percentage yield* as 71.9 percent.

SUMMARY

"Stoichiometry" is the name given to calculations relating amounts of reactants and products in chemical reactions (Section 10.1). To do stoichiometry, one interprets the coefficients of balanced chemical equations as the numbers of moles of reactants and products (Section 10.2). This interpretation leads us to the concept of the mole ratio, the central focus in all stoichiometry problems (Section 10.3).

Mole ratios enable us to determine the amount of reactant reacted or product produced, given an amount of another reactant or product. We can use the mole ratio in one-step conversions between moles of materials (Section 10.4). By the use of conversion factors that relate grams to moles, we can also use the mole ratio to convert between grams of one substance and moles of another or between gram amounts of different substances (Sections 10.5, 10.6, and 10.7).

When the amounts of two or more reactants are given, we must determine which is the limiting reactant and which reactant or reactants are in excess (Section 10.8). The theoretical yield is the maximum limiting amount of product that may be obtained based on the amount of reactant given (Section 10.9). The actual yield is what is actually obtained in a chemical reaction, which is always less than the theoretical yield (Section 10.10). The percentage yield compares the actual and theoretical yields (Section 10.11).

Heat can be regarded as either a reactant or a product in a chemical reaction (Section 10.12).

CHAPTER ACCOMPLISHMENTS

After completing this chapter you should be able to

10.1 What is stoichiometry?

1. Define stoichiometry.

10.2 Molar interpretation of the balanced equation

2. Interpret a chemical equation in terms of numbers of interacting particles.
3. Interpret the coefficients of a chemical equation as numbers of moles of reactants and products.

10.3 "Equalities" and conversion factors from chemical equations

4. Given a balanced chemical equation, write equalities between molar amounts of reactants and/or products.
5. Construct mole-ratio conversion factors from molar equalities.

10.4 Mole-mole conversions

6. Given a balanced equation and a molar amount of one reactant or product, calculate the *molar* amounts of other substances reacting or produced.

10.5   Gram-mole, mole-gram conversions

7. Given a balanced equation and a gram amount of one reactant or product, calculate the molar amounts of other substances reacting or produced.

8. Given a balanced equation and a molar amount of one reactant or product, calculate the *gram* amounts of other substances reacting or produced.

10.6   Gram-gram conversions

9. Given a balanced equation and a gram amount of one reactant or product, calculate the gram amounts of other substances reacting or produced.

10.7   Conversions summarized

10. State the importance of the mole ratio in all stoichiometry calculations.

10.8   Limiting reactant

11. Given the amounts of two or more reactants and a chemical equation, identify the limiting reactant.

12. Determine the maximum amount of product that can be obtained based on the limiting reactant.

13. Determine how much of an excess reactant is left over after the completion of a reaction.

10.9   Theoretical yield

14. Given some amount of reactant or reactants, calculate the theoretical yield of product.

10.10   Actual yield

15. Define actual yield.

10.11   Percentage yield

16. Given the actual and theoretical yields, calculate the percentage yield.

17. Given the actual yield and enough information to calculate the theoretical yield, calculate the percentage yield.

10.12   Heat as a reactant or product

18. Define exothermic reaction.
19. Define endothermic reaction.
20. Identify chemical equations as representing either exothermic or endothermic reactions.
21. Given a balanced chemical equation, write "equalities" between heat and the molar amount of reactants or products.

22. Given a balanced equation and an amount of one reactant or product, calculate the quantity of heat consumed or produced as the reaction proceeds.

23. Given a balanced equation and an amount of heat consumed or produced, calculate the amount of reactant consumed or product produced.

10.13 A page from a laboratory notebook

24. Recognize the stoichiometric calculations in a laboratory note-book.

PROBLEMS

10.1 What is stoichiometry?

10.10 Define in your own words the term stoichiometry.

10.11 What must be true about any equation used in a stoichiometric calculation?

10.2 Molar interpretation of the balanced equation

10.12 Interpret the following balanced chemical equations in terms of numbers of interacting particles:

- **a.** $2P(s) + 3H_2(g) \longrightarrow 2PH_3(g)$
- **b.** $HBr(g) + KOH(aq) \longrightarrow KBr(aq) + H_2O(l)$
- **c.** $2CO(g) + O_2(g) \longrightarrow 2CO_2(g)$
- **d.** $C(s) + 2Cl_2(g) \longrightarrow CCl_4(l)$
- **e.** $Mg(s) + 2HCl(aq) \longrightarrow MgCl_2(aq) + H_2 \uparrow$

10.13 Interpret the equations in Problem 10.12 in terms of the numbers of moles of reactants and products.

10.14 Which interpretation, moles (Problem 10.13) or particles (Problem 10.12), is used in doing stoichiometry problems?

10.15 Could an equation be read directly in units of mass? For example, can $C + O_2 \rightarrow CO_2$ be read 1 g of C + 1 g of O_2 yields 1 g of CO_2? Explain your answer.

10.3 "Equalities" and conversion factors from chemical equations

10.16 Write all the "equalities" between reactants and products which the balanced equations in Problem 10.12, parts (a) and (e) indicate.

10.17 Write the mole ratios corresponding to the equalities you wrote in Problem 10.16.

10.18 Write all possible mole ratio conversion factors that relate the number of moles of each product to the number of moles of each reactant:

$$2C_2H_6 + 7O_2 \longrightarrow 4CO_2 + 6H_2O$$

10.19 Explain how the law of conservation of mass is satisfied by the use of mole ratios in stoichiometry problems.

10.4 Mole-mole conversions

10.20 Given the balanced equation

$$Na_2Cr_2O_7 + 6HI + 4H_2SO_4 \longrightarrow$$
$$3I_2 + Cr_2(SO_4)_3 + Na_2SO_4 + 7H_2O$$

- **a.** How many moles of I_2 form when 3.0 moles of HI react?
- **b.** How many moles of HI react with 3.0 moles of H_2SO_4?
- **c.** How many moles of water form at the same time that 9.0 moles of I_2 form?
- **d.** How many moles of HI are required to produce 0.41 mole of H_2O?

10.21 Given the balanced equation

$$5C + 2SO_2 \longrightarrow CS_2 + 4CO$$

- **a.** How many moles of CS_2 form when 3.00 moles of carbon react?
- **b.** How many moles of SO_2 are needed to react with 3.00 moles of carbon?

c. How many moles of CO form at the same time that 7.00 moles of CS_2 form?

d. How many moles of C are required to produce 1.50 moles of CS_2?

10.22 The balanced equation for the reaction of methane (CH_4) with oxygen is:

$$CH_4 + 2O_2 \longrightarrow CO_2 + 2H_2O$$

a. How many moles of oxygen are required to react with 5.00 moles of CH_4?

b. How many moles of oxygen are required to produce 3.50 moles of CO_2?

c. How many moles of water will be produced from 1.15 moles of CH_4?

10.23 **a.** Write a balanced equation and calculate how many moles of aluminum oxide can be produced from the reaction of 6.0 moles of aluminum with oxygen. *A balanced equation is always necessary to do stoichiometry.*

b. How many moles of oxygen must react with the 6.0 moles of Al?

10.5 Gram-mole, mole-gram conversions

10.24 Given the equation

$$2Al + 6HCl \longrightarrow 2AlCl_3 + 3H_2 \uparrow$$

a. How many grams of Al are needed to release 0.54 mole of H_2?

b. How many grams of $AlCl_3$ are obtained from 12 moles of HCl?

c. How many grams of H_2 are released from 8.0 moles of HCl?

10.25 Given the balanced equation

$$2KOH + H_2SO_4 \longrightarrow K_2SO_4 + 2H_2O$$

a. How many moles of H_2SO_4 are needed to make 78 g of K_2SO_4?

b. How many moles of H_2O are produced from the reaction of 16.9 g of H_2SO_4?

c. How many moles of KOH are needed in order to produce 125 g of water?

10.26 One of the steps in the manufacture of nitric acid is represented by the following reaction:

$$3HNO_2 \longrightarrow 2NO + HNO_3 + H_2O$$

a. How many grams of HNO_2 are required to make 5.0 moles of HNO_3?

b. How many grams of NO are produced along with the 5.0 moles of HNO_3?

10.27 A propellant used in rocket engines is a mixture of hydrazine (N_2H_4) and hydrogen peroxide (H_2O_2). This mixture reacts spontaneously to yield N_2 and H_2O:

$$N_2H_4 + 2H_2O_2 \longrightarrow N_2 + 4H_2O$$

a. How much hydrazine in grams is needed to react completely with 0.50 mole of H_2O_2?

b. How many grams of N_2 are produced by the combination of ingredients described in part (*a*).

10.28 Chlorine is prepared by passing an electric current through a solution of sodium chloride. Sodium hydroxide and hydrogen are important by-products of this reaction:

$$2NaCl(aq) + 2H_2O(l) \xrightarrow{\text{electricity}}$$
$$2NaOH(aq) + Cl_2(g) + H_2(g)$$

a. How many grams of chlorine are produced from 1.50 moles of NaCl?

b. If you wish to prepare 9.00 moles of Cl_2, how many grams of NaCl are needed?

c. How many moles of NaOH and H_2 will be produced if you start with the number of grams of NaCl calculated in part (*b*)?

10.6 Gram-gram conversions

10.29 Given the equation

$$MnO_2 + 4HCl \longrightarrow MnCl_2 + Cl_2 + 2H_2O$$

a. How many grams of chlorine are formed from 12.5 g of MnO_2?

b. How many grams of HCl are needed to make 3.00 g of $MnCl_2$?

c. How many grams of water form when 18.0 g of HCl react?

10.30 The reaction that occurs in Chlorox that leads to whitening is

$$NaClO + NaCl + H_2O \longrightarrow Cl_2 + 2NaOH$$

In a cup of Chlorox there is approximately 12 g of NaClO.

a. How many moles of Cl_2 are liberated by 12 g NaClO?

b. How many grams of NaOH are produced?

10.31 Soda-lime glass is made by the reaction among Na_2CO_3, limestone ($CaCO_3$), and sand (SiO_2):

$$Na_2CO_3 + CaCO_3 + 6SiO_2 \longrightarrow$$
$$Na_2O \cdot CaO \cdot 6SiO_2 + 2CO_2 \uparrow$$
Soda-lime glass

a. How many grams of sand are needed to make one champagne bottle weighing 618 g?
b. How many moles of CO_2 are released as the glass for the champagne bottle forms?
c. How many moles of limestone are needed to make 0.720 mole of glass?

10.32 Write a balanced equation and calculate how many grams of chlorine (Cl_2) are needed to produce 153 g of sodium chloride from sodium metal and chlorine gas.

10.33 When iron reacts with bromine to form iron(III) bromide,
a. How many grams of iron are needed to produce 65.0 g of the compound?
b. How many moles of Br_2 will be needed for this reaction?

10.34 Aluminum displaces copper from copper(II) sulfate. How many grams of Al are required to displace 34.0 g of Cu?

10.35 An important source of energy for human and other living species is the breakdown of energy-rich glucose ($C_6H_{12}O_6$) to carbon dioxide and water. The overall chemical reaction is:

$$C_6H_{12}O_6 + 6O_2 \longrightarrow 6CO_2 + 6H_2O$$

a. Calculate the number of grams of oxygen required to react with 280. g of $C_6H_{12}O_6$.
b. Calculate the number of grams of water produced from the given amount of glucose in part (a).

10.36 Some organisms derive their energy from the breakdown of glucose ($C_6H_{12}O_6$) to carbon dioxide and ethyl alcohol (C_2H_6O), a reaction known as **fermentation**:

$$C_6H_{12}O_6 \longrightarrow 2C_2H_6O + 2CO_2$$

What mass of glucose is needed to form 10.0 g of ethyl alcohol?

10.37 The combination of nitrogen with hydrogen to form ammonia (NH_3) is an important industrial process:

$$N_2(g) + 3H_2(g) \longrightarrow 2NH_3(g)$$

a. How much ammonia can be produced from 2.00 kg of N_2?
b. How much H_2 is necessary to prepare 8.50 kg of NH_3?

10.7 Conversions summarized

10.38 Which conversion factor is used in all stoichiometry problems?

10.8 Limiting reactant

10.39 Chloroform ($CHCl_3$) can be made by the reaction of chlorine and methane (CH_4) according to the equation

$$3Cl_2 + CH_4 \longrightarrow CHCl_3 + 3HCl$$

For each of the following combinations of reactants, decide which is the limiting reactant:
a. 3.00 moles Cl_2 and 3.00 moles CH_4
b. 5.00 moles Cl_2 and 1.50 mole CH_4
c. 0.50 mole Cl_2 and 0.20 mole CH_4

10.40 The "fluoride" in toothpaste is stannous fluoride [tin(II) fluoride], which can be made through the reaction of Sn and HF.

$$Sn + 2HF \longrightarrow SnF_2 + H_2$$

For each of the following combinations of reactants, decide which is the limiting reactant:
a. 50.0 g Sn and 1.00 mole HF
b. 120. g Sn and 60. g HF
c. 120. g Sn and 39 g HF

10.41 How many moles of $CHCl_3$ are produced in the reaction in Problem 10.39, part (a)?

10.42 How many grams of SnF_2 are produced in the reaction in Problem 10.40, part (b)?

10.43 How much of the excess reactant is left over after the reaction is complete in Problem 10.40, part (b)?

10.44 How many grams of magnesium nitride are produced when 63 g Mg and 41 g N_2 react?

$$3Mg + N_2 \longrightarrow Mg_3N_2$$

10.45 How much of the excess reactant is left over after the reaction is complete in Problem 10.44?

10.46 Calcium hydroxide reacts with phosphoric acid according to the equation:

$$3Ca(OH)_2 + 2H_3PO_4 \longrightarrow Ca_3(PO_4)_2 + 6H_2O$$

    a. When 215 g of $Ca(OH)_2$ reacts with 201 g of H_3PO_4, which reactant is limiting according to the above equation?
    b. What mass of $Ca_3(PO_4)_2$ can form?
    c. How much of the excess reactant is left over at the end of the reaction?

10.9 Theoretical yield

10.47 What is the theoretical yield of CO_2 based on 115 g of $CaCO_3$ according to the equation in Problem 10.31?

10.48 In a laboratory 34.0 g of Li metal reacts with an excess of Br_2. What is the theoretical yield of LiBr?

10.49 Old oil paintings are darkened by PbS which forms by the reaction of the Pb in paint with H_2S in air.

$$Pb + H_2S \longrightarrow PbS + H_2$$

Suppose 0.40 g of H_2S comes in contact with 2.0 g of Pb. What is the theoretical yield of PbS?

10.50 Old oil paintings can be cleaned by the reaction of H_2O_2 with PbS.

$$PbS + 4H_2O_2 \longrightarrow PbSO_4 + 4H_2O$$

    a. What is the theoretical yield of $PbSO_4$ based on 0.80 g of H_2O_2?
    b. How much PbS will be removed?

10.10 Actual yield

10.51 **a.** What is meant by the actual yield of a chemical reaction?
    b. Can the actual yield be calculated before the reaction is carried out?

10.11 Percentage yield

10.52 In doing the reaction described in Problem 10.48 a chemist obtains 340.0 g of LiBr. What is the percentage yield?

10.53 An art restorer actually obtains 1.16 g of $PbSO_4$ in the cleaning process described in Problem 10.50. What is the percentage yield?

10.54 Chromium metal can be removed from its oxide ore by reaction with carbon:

$$Cr_2O_3 + 3C \longrightarrow 2Cr + 3CO$$

When 1.00 kg Cr_2O_3 reacts with 0.300 kg C,
    a. What is the theoretical yield of Cr?
    b. How much of which reactant is left over in excess?
    c. The actual yield obtained was 531 g of Cr. What is the percentage yield?

10.55 Elemental iron can be recovered from iron oxide ore by reaction with carbon monoxide:

$$Fe_2O_3 + 3CO \longrightarrow 2Fe + 3CO_2$$

    a. When 145 g of iron(III) oxide reacts with 95 g of carbon monoxide, how much iron will form?
    b. Assume that 93 g of iron is produced in the reaction in part (a). What is the percentage yield of this reaction?

10.56 A chemist sets out to make nitroglycerin from the reaction of glycerine with nitric acid:

$$C_3H_8O_3 + 3HNO_3 \longrightarrow C_3H_5O_9N_3 + 3H_2O$$
Glycerine   Nitroglycerin

She uses 4.6 g glycerine and 15 g HNO_3 and gets 9.8 g nitroglycerin. What is the percentage yield?

10.12 Heat as a reactant or product

10.57 **a.** Give an example of an exothermic process that you are familiar with.
    b. Give an example of an endothermic process.

10.58 Identify the following reactions as exothermic or endothermic:
    a. $CaCO_3 + 42 \text{ kcal} \longrightarrow CaO + CO_2$
    b. $MnO_2 + Mn \longrightarrow 2MnO + 59.5 \text{ kcal}$
    c. $C + H_2O + 31 \text{ kcal} \longrightarrow CO + H_2$

10.59 The propellant reaction given in Problem 10.27 is highly exothermic:

$$N_2H_4(l) + 2H_2O_2(l) \longrightarrow$$
$$N_2(g) + 4H_2O(g) + 153 \text{ kcal}$$

a. How much heat is liberated when 5.00 moles of H_2O_2 react?

b. How much heat is released as 1.50 moles of water form?

10.60 The air pollutant SO_3 combines with water to form H_2SO_4, which is destructive to marble and limestone, and the reaction also contributes to thermal pollution (excessive heat in the atmosphere):

$$SO_3(g) + H_2O(l) \longrightarrow H_2SO_4(aq) + 31 \text{ kcal}$$

How much H_2SO_4 in grams and how much heat in kilocalories is produced by 21 g of SO_3?

10.61 The reaction of H_2SO_4 with marble can be represented

$$H_2SO_4(aq) + CaCO_3(s) \longrightarrow$$
$$CaSO_4(s) + H_2O(l) + CO_2(g) + 27 \text{ kcal}$$

a. When 26 g $CaCO_3$ reacts, how much heat is released?

b. Look back at Problem 10.60. How much $CaCO_3$ is destroyed by 21 g of SO_3?

10.62 How many moles of carbon dioxide can be released from $MgCO_3$ if 11.2 kcal of energy is supplied?

$$MgCO_3 + 28.1 \text{ kcal} \longrightarrow MgO + CO_2$$

10.63 When 5.0 g of nitroglycerin blows up, the reaction can be represented by the equation

$$4C_3H_5O_9N_3 \longrightarrow 12CO_2 \uparrow + 6N_2 \uparrow$$
$$+ O_2 \uparrow + 10H_2O + 1725 \text{ kcal}$$

a. How much heat is evolved?

b. Given that 1 kcal of heat can raise the temperature of 1 kg of water 1°C, what will be the final temperature of 200. g of water initially at 40°C if it is subjected to the heat evolved in part (a)?

GASES
CHAPTER

11

Na
Sodium
22.99

INTRODUCTION

11.1 In the space between you and this book flows a mixture of gases. We call this mixture **air** (see Table 11.1). Let us explore the physical properties of a typical gas such as air.

In Chapter 1, we summarized the characteristics of the three physical states (Table 1.1). Gases were described as having indefinite shapes and indefinite volumes. Another way of saying the same thing appears in statements 1 and 2 in the list of gas characteristics that follows; that is, gases are easily *compressed* and readily *expand*. Other properties that are unique to the gaseous state can also be found in this list.

Characteristics of Gases

1. *Gases can be easily compressed* by applying pressure to a movable piston in a container with rigid walls. (Figure 11.1a). Compression implies a *decrease* in volume. When we try the "pressing" experiment

COMPOSITION OF CLEAN, DRY AIR	TABLE 11.1

Gas	Percent by Volume
N_2	78.09
O_2	20.94
Ar	0.93
CO_2	0.03
Noble gases other than Ar	0.0024
CH_4	0.00015
Trace amounts of H_2, N_2O, CO, O_3, NH_3, and SO_2.	

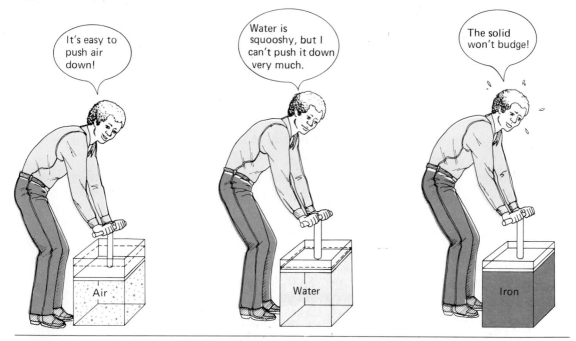

FIGURE 11.1

Gases are easily compressed. Liquids are difficult to compress and solids are the least compressible of the three physical states.

on a container filled with water, we find that liquids are not easily compressed (Figure 11.1*b*). Solids are the least compressible of the three states (Figure 11.1*c*).

2. *Gases expand to fill the entire volume of their container.* Expansion implies an *increase* in volume. If a small container of a gas is opened in a classroom, the gas escapes and soon expands to fill the entire room. The mass of the gas undergoing expansion remains the same, only the volume changes; that is, the same amount of material is spread out within a larger space.

3. *Gases have indefinite densities.* Because density is the ratio of mass to volume and the volume of a gas varies with its container but the mass remains constant, density also varies as the volume of the container does. In statement 2 the expanding gas described has a lower density when it fills the room than when it was in a small container.

Problem 11.1

Originally 0.96 g of a gas at 20°C occupies 0.50 L. The gas is allowed to expand to fill 12 L at 20°C. Calculate the original and final densities.

4. *Gases have a low density* compared with liquids and solids (Table 3.5). This is why gas bubbles within a liquid tend to rise.

5. Assuming that they do not react chemically, *gases can diffuse (mix) rapidly through each other* in all directions. Ammonia gas released at the front of a classroom rapidly diffuses through the air, as can be verified by a student sitting in the back. The odor of ammonia is quite pungent. This student's observation would also be a demonstration of a gas sample expanding to fill an entire room.

6. *Gases exert a pressure on the wall of any container or surface that they touch*. At sea level, air pushes against every square inch of our bodies with a force of 14.7 lb.

7. *Gases expand in volume when they are heated and contract when they are cooled*. Liquids and solids also expand and contract with temperature, but not to the same extent as gases.

These properties of gases can all be explained and accounted for by the Kinetic Theory (Section 11.12). Because this theory has been developed fairly recently, we have saved it for later in the chapter so that it appears in its proper place in the development of the knowledge of gases. First, we will explore the "gas laws," which tell us the quantitative relationships among the gas variables,[1] i.e., among the number of moles, the volume, the pressure, and the temperature of a gas. In order to understand how the gas laws were developed, it is necessary to know how to measure the gas variables, the number of moles n, the volume V, the pressure P, and the temperature T. The designated letters small n and capitals V, P, and T are the standard symbols for the gas variables.

n, T, AND *V* MEASUREMENTS

11.2 The number of moles n can be determined by measuring the mass of a gas sample. The relationship between mass and moles has been well known to you since Chapter 8. Just as we measure the mass of a liquid by difference—that is, we determine the mass of both empty and filled containers and subtract the mass of the empty container from the mass of the filled container to find the liquid's mass—so do we determine the mass of a gas (see Figure 11.2).

The temperature T of a gas can be measured with a thermometer. You will see that the relationships among the temperature, volume, and pressure of a gas can be greatly simplified by the introduction of a new temperature scale known as the **Kelvin scale** (Section 11.6).

The volume V of a gas can be measured by placing the gas in a

---

[1] We use this word because, as you have seen in the previous description of gas characteristics, gas volume "varies" with temperature and pressure and would naturally vary as the amount is increased or decreased.

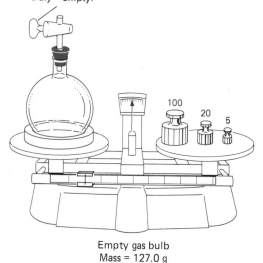

Stopcock allows the flask to be evacuated; i.e., the air is pumped out by a vacuum pump so that the bulb is truly "empty."

Empty gas bulb
Mass = 127.0 g

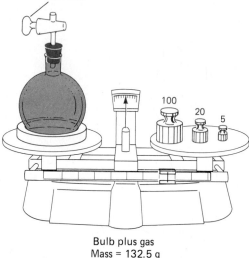

The gas sample to be weighed is put into the flask through the stopcock.

Bulb plus gas
Mass = 132.5 g

FIGURE 11.2

The mass of a gas is determined by difference just as we determine the mass of a liquid. The container in which the gas is to be weighed must be evacuated; that is, air must be pumped out to ensure an empty container.

container of known volume. The unit of volume most often employed in gas measurements is the liter.

Because pressure and its measurement are probably the least familiar to you of all the gas variables, we will devote an entire section (Section 11.3) to pressure.

PRESSURE

11.3 Pressure P is defined as the force exerted per unit area, or

$$P = \frac{\text{force}}{\text{area}}$$

Some typical forces are pushing, pulling, and collisions. Given the same area, the larger the force, the greater the pressure. Given the same force, pressure is greater when the force operates over a smaller area (see Figure 11.3). Notice that force and pressure are not the same thing. An identical force can lead to very different pressures depending on the area over which it is distributed.

You will see in Section 11.12 that particles in a sample of matter are constantly moving. The pressure of a gas comes from the force

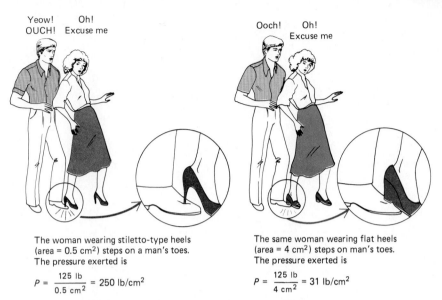

The woman wearing stiletto-type heels (area = 0.5 cm²) steps on a man's toes. The pressure exerted is

$$P = \frac{125 \text{ lb}}{0.5 \text{ cm}^2} = 250 \text{ lb/cm}^2$$

It really HURTS!

The same woman wearing flat heels (area = 4 cm²) steps on man's toes. The pressure exerted is

$$P = \frac{125 \text{ lb}}{4 \text{ cm}^2} = 31 \text{ lb/cm}^2$$

It doesn't hurt nearly as much.

Pressure is eight times greater in the case of the smaller heel, even though the force remained the same.

FIGURE 11.3

Pressure varies inversely with area. The same force over a smaller area results in a greater pressure.

exerted on the inside area of the gas's container by the collisions of moving molecules (see Figure 11.4). The pressure of the air around us, which is known as **atmospheric pressure,** is the pressure exerted when molecules in air collide with our bodies, the ground, or other objects on earth. This pressure is approximately 14.7 pounds per square inch (lb/in²). **P**ounds per **s**quare **i**nch can also be abbreviated **psi.**

$$\frac{14.7 \text{ lb}}{1 \text{ in}^2} \qquad \frac{\text{force}}{\text{area}}$$

FIGURE 11.4

The pressure of a gas comes from the force of collisions of moving molecules exerted on the inside area of the gas's container.

Air pressure, or atmospheric pressure, can be measured with a barometer (see Figure 11.5). A long (about 80 cm), narrow (diameter = 1 cm) tube, open at one end, is filled with mercury. The tube is stoppered and inverted into a dish containing mercury. When the stopper is removed, some of the mercury empties out of the tube into the dish, but a column of mercury approximately 76 cm (760 mm) high remains in the tube. Because the tube was originally completely filled with mercury, the space that is left above the mercury column when some of the mercury flows out contains *no* air. A vacuum, i.e., a totally empty space, is created, and there is no force pushing down on the top of the mercury column. There is a balance between the pressure from the weight of the mercury in the tube and the pressure

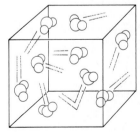

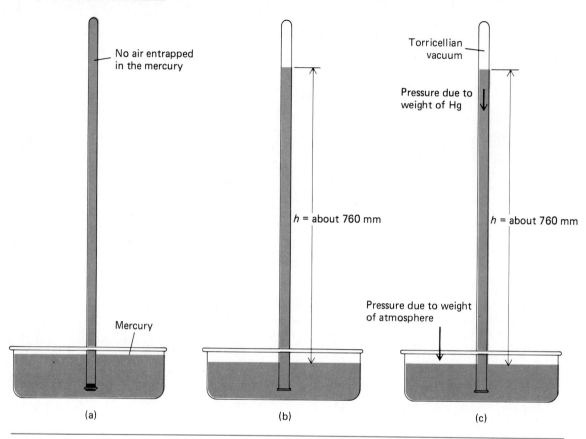

No air entrapped
in the mercury

Mercury

(a)

Torricellian
vacuum

Pressure due to
weight of Hg

h = about 760 mm

Pressure due to weight
of atmosphere

(b)

h = about 760 mm

(c)

FIGURE 11.5

The construction of a barometer. (a) A stoppered tube full of mercury is inverted in a mercury bath. (b) The stopper is removed and some mercury falls out of the tube. The remaining height of mercury above the mercury surface exposed to the atmosphere is 760 mm. (c) The weight of the atmosphere pushes down on the open bath of mercury and thus holds up the column of mercury in the tube. Atmospheric pressure exactly balances the pressure exerted by the mercury column.

from the atmosphere, that is, the pressure of the weight of air on the mercury in the dish. Examine Figure 11.5 carefully.

The height of the mercury in the tube gives us a measure of the atmospheric pressure. If the atmospheric pressure increases, the height of the mercury in the tube will increase; if the atmospheric pressure decreases, the height in the tube will decrease. At sea level on an average day, atmospheric pressure can support a column of mercury 760 mm high. This pressure is called **one atmosphere** (1 atm). In honor of the Italian scientist Torricelli, who first developed the barometer, the unit of pressure, millimeters of mercury, has been given the name torr. So 760 mmHg = 760 torr. Table 11.2 illustrates equalities among common units of pressure. From these equalities we can construct conversion factors in the usual way.

$$1 \text{ atm} = 760 \text{ mmHg} = 760 \text{ torr}$$

1 atm = 760 mmHg	760 mmHg = 760 torr
1 atm = 760 torr	1 mmHg = 1 torr
1 atm = 14.7 lb/in²	760 mmHg = 14.7 lb/in²

Problem 11.2

Weather reporters typically say, "Today's barometric pressure is 30. and falling." The 30. is the day's atmospheric pressure in inches of mercury. What is the pressure in torr?

Now after considering the gas variables and their measurement, it is time to tackle the gas laws.

DALTON'S LAW OF PARTIAL PRESSURES

11.4 Suppose that we have three 1.0-L containers. The first one contains a gas A at pressure P_a, and the second one contains a gas B at a pressure of P_b. Now we empty the gases from containers 1 and 2 into container 3. The volume of the combined gases is 1.0 L, the same as the volume of the original pure gases. The pressure in the third container is found to be equal to the sum of the pressures that each gas exerted by itself.

$$P_{total} = P_a + P_b$$

John Dalton (the same scientist who proposed the Atomic Theory) conducted many experiments of the type just described and found that the total pressure of a mixture of gases was always equal to the sum of the pressures of the pure gases taken separately. Stated another way, each gas in a mixture exerts a pressure as if it were alone in the container. The pressure of each gas in the mixture is known as the **partial pressure** of that gas. **Dalton's Law of Partial Pressures** states that the total pressure of a mixture of gases is equal to the sum of all the partial pressures of each gas in the mixture:

$$P_{total} = p_1 + p_2 + p_3 + p_4 + \cdots \qquad (11.1)$$

Partial pressures are commonly represented by small p's.

Problem 11.3

A container holds three gases, argon, neon, and krypton, with partial pressures of 1.0 atm, 0.5 atm, and 1.5 atm, respectively. What is the total gas pressure in the container?

Container 1

Container 2

Container 3

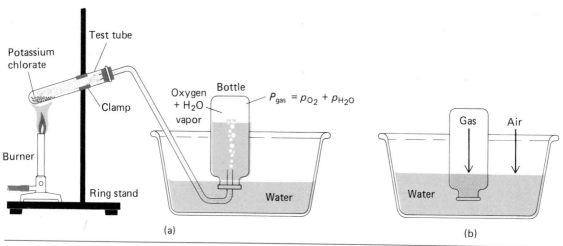

FIGURE 11.6

KClO$_3$ decomposes to form O$_2$, which is collected over water. The total pressure of the gas mixture (O$_2$ + water vapor) in the bottle can be obtained by equalizing the water levels inside and outside the bottle. (a) Collection of O$_2$ gas by displacement of water. (b) Gas pressure can be measured by equalizing the water levels inside and outside the bottle. When the water levels are equal, $P_{inside} = P_{outside}$. $P_{outside}$ is the atmospheric pressure.

Gases such as oxygen which are only very slightly soluble in water can be collected by displacing water from a collection bottle (Figure 11.6). The oxygen obtained is *not* pure, but rather it is mixed with water vapor. If Dalton's law is applied, the total pressure of the mixture must equal the sum of the individual pressures of O$_2$ and H$_2$O vapor.

$$P_{total} = p_{O_2} + p_{H_2O}$$

If the water level inside the bottle is carefully adjusted so that it is the same as in the trough (as it is in Figure 11.6*b*), the total gas pressure inside the bottle must be P_{atm}, which is the pressure on the water in the trough. Then

$$P_{total} = P_{atm} = p_{O_2} + p_{H_2O}$$

or subtracting p_{H_2O} from each side,

$$P_{atm} - p_{H_2O} = p_{O_2}$$

The measurement of P_{atm} by a barometer has already been described. The vapor pressure of water p_{H_2O} depends only on the temperature of the water (Section 12.7) and can be looked up in a table such as the one in App. 3 of this text.

SAMPLE EXERCISE

11.1

A sample of N$_2$ is collected by water displacement in a setup similar to that in Figure 11.6. The water level inside the bottle is equalized with that in the trough. Barometric pressure is found to be 757 mmHg, and the temperature of the water is 22°C. What is the partial pressure of N$_2$?

SOLUTION

The total gas pressure inside the bottle is

$$P_{total} = p_{N_2} + p_{H_2O}$$

Since the water levels inside and outside the bottle were equalized, the total gas pressure inside the bottle must equal P_{atm}.

$$P_{total} = P_{atm} = p_{N_2} + p_{H_2O}$$

$$p_{N_2} = P_{atm} - p_{H_2O}$$

P_{atm} is given:

$$P_{atm} = 757 \text{ mmHg}$$

$$p_{N_2} = 757 \text{ mmHg} - p_{H_2O}$$

Appendix 3 indicates that p_{H_2O} at 22°C is 19.8 mmHg.

$$p_{N_2} = 757 \text{ mmHg} - 19.8 \text{ mmHg}$$

$$p_{N_2} = 737 \text{ mmHg}$$

Problem 11.4

A sample of oxygen is collected by water displacement. What would be the pressure of dry oxygen if the temperature is 18°C and the barometric pressure is 765 mmHg?

The concept of partial pressures plays an important role in understanding respiration in multicellular organisms. Respiration is the physiological process by which oxygen and carbon dioxide are exchanged between the cells, the blood, the lungs, and the outside atmosphere (Figure 11.7).

Gases flow spontaneously from a region of higher partial pressure to a region of lower partial pressure. As Figure 11.7 shows, atmospheric oxygen flows into the alveoli, small sacs in the lungs, because p_{O_2} is greater in the inspired air (158 mmHg) than in alveolar air (101 mmHg). Carbon dioxide flows in the reverse direction from the alveolar sacs to the outside air because p_{CO_2} in the alveoli (40 mmHg) is much greater than in the atmosphere (0.3 mmHg).

In a second exchange, O_2 diffuses from the alveolar sacs (p_{O_2} = 101 mmHg) across the alveolar capillary membrane into the circulating venous blood (p_{O_2} = 45 mmHg). The oxygen in the now arterial blood has a p_{O_2} of about 100 mmHg. Although a small amount is dissolved in blood, oxygen is predominantly transported in chemical combination with hemoglobin, a component of red blood cells. At the same time that oxygen moves from the lungs to the blood, carbon dioxide migrates from the venous blood (p_{CO_2} = 46 mmHg) into alveolar sacs (p_{CO_2} = 40 mmHg).

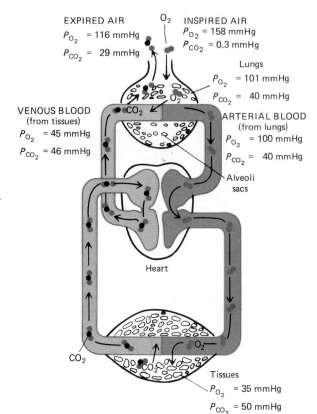

EXPIRED AIR
P_{O_2} = 116 mmHg
P_{CO_2} = 29 mmHg

O_2 INSPIRED AIR
P_{O_2} = 158 mmHg
P_{CO_2} = 0.3 mmHg

Lungs
P_{O_2} = 101 mmHg
P_{CO_2} = 40 mmHg

CO_2

O_2

VENOUS BLOOD
(from tissues)
P_{O_2} = 45 mmHg
P_{CO_2} = 46 mmHg

ARTERIAL BLOOD
(from lungs)
P_{O_2} = 100 mmHg
P_{CO_2} = 40 mmHg

Alveoli
sacs

Heart

CO_2

O_2

CO_2

Tissues
P_{O_2} = 35 mmHg
P_{CO_2} = 50 mmHg

FIGURE 11.7

Inspired air, with its relatively high P_{O_2} enriches the lungs' alveoli with oxygen, and oxygen then enters the bloodstream. This arterial blood then takes oxygen to the tissues. Carbon dioxide from tissues enters the blood. Blood with a depleted oxygen supply and high CO_2 concentration is venous blood, which returns to the lungs, where CO_2 is expired and inspired oxygen "converts" venous blood to arterial blood.

The partial pressure of O_2 in tissue is about 35 mmHg, and therefore O_2 migrates from the red blood cells (p_{O_2} = 100 mmHg) into the tissues. The deoxygenated blood (venous blood) has a p_{O_2} of about 45 mmHg and returns to the lungs, where the cycle is repeated. Meanwhile CO_2 flows from the tissues (p_{CO_2} = 50 mmHg) into arterial blood (p_{CO_2} = 40 mmHg). The venous blood flowing back to the lungs has a p_{CO_2} = 46 mmHg, which will be reduced at the lung capillaries, where CO_2 will diffuse into the alveoli, repeating the CO_2 cycle.

BOYLE'S LAW

11.5   When we increase the pressure on a bubble of gas by squeezing it, its volume gets smaller. When we lessen the pressure, the volume gets larger again. There appears to be an *inverse* relationship between the volume and the pressure of a gas; raising or lowering one has the opposite effect on the other gas variable.

In the seventeenth century Robert Boyle investigated the quan-

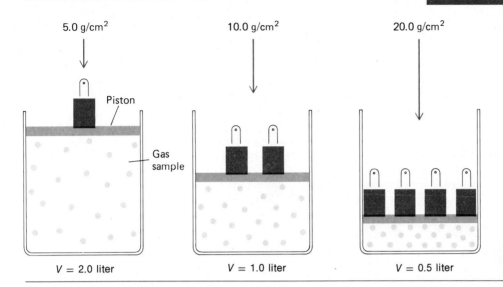

FIGURE 11.8

Pressure increases as more weight is distributed over the same area. Because the piston is not moving, we know that the internal gas pressure is identical to the external applied pressure. The volume decreases as the pressure increases.

titative realtionship between the volume and pressure of a gas. Although this was not his experimental apparatus, Boyle's experiments can be illustrated by a gas confined within a cylinder (Figure 11.8). In his experiments Boyle held the temperature and the number of moles of gas at fixed values and observed the change in volume as he varied the external pressure on the gas. The pressure of the gas inside the cylinder must be the same as the external pressure, because the piston is not moving. That is, the pressure exerted upward from the gas must be exactly equal to the pressure pushing downward from the weights.

Table 11.3 organizes the data from the illustration in Figure 11.8, and Figure 11.9 shows a plot of volume versus pressure from Table 11.3.

Looking at Table 11.3 and the plot in Figure 11.9, notice that when the pressure is doubled, the volume is cut in half. When the

PRESSURE-VOLUME DATA FOR A FIXED MASS OF
GAS AT A CONSTANT TEMPERATURE

TABLE 11.3

Pressure, g/cm^2	Volume, L	$P \times V$, $(g/cm^2) \cdot L$
2.5	4.0	10.
6.0	2.0	10.
10.	1.0	10.
20.	0.5	10.
40.	0.25	10.

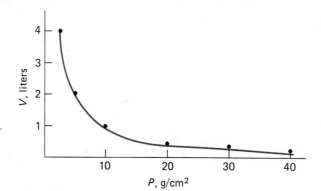

FIGURE 11.9

Plot of the change in volume as pressure changes. Three points of this graph are pictured in Figure 11.8.

pressure is quadrupled, the volume is reduced to one quarter the original volume. When the pressure is cut in half, the volume is doubled. These results can be stated mathematically by Boyle's law:

$$V \propto \frac{1}{P} \qquad \text{at constant } n \text{ and } T \qquad (11.2)$$

where $\propto$ is the symbol which expresses proportionality. This expression is read, "The volume of a gas is *inversely* proportional to the pressure when the temperature and the number of moles are held constant." This statement is **Boyle's law.**

A proportionality sign can always be replaced by an equals sign and a constant, thus converting the relationship into an equation (see Appendix 1F). In this case we get

$$V = \frac{k}{P} \qquad (11.3)$$

where k is a constant, a fixed number. Multiplication of both sides of the equation by P gives us

$$PV = k \qquad \text{at constant } n \text{ and } T \qquad (11.4)$$

This equation states that for a fixed number of moles at a given temperature, the product of the pressure and volume of a gas is equal to a constant. We find confirmation of Equation (11.4) in the third column of Table 11.3.

For a fixed quantity of gas at a particular temperature, if we vary the pressure from some value P_1 to a new value P_2, the volume must change from V_1 to V_2, but the product ($P \times V$) will still be equal to the same constant k.

$$\begin{array}{l} P_1 V_1 = k \\ P_2 V_2 = k \end{array} \qquad \text{at constant } n \text{ and } T$$

or

$$P_2V_2 = P_1V_1 \qquad (11.5)$$

Notice again the variation in P and V but the constancy of $P \times V$ shown in Table 11.3. You will see problems involving Boyle's law in Section 11.7.

   An application of the inverse relationship between volume and pressure can be seen in the physiological process of breathing. Moving your diaphragm downward (Figure 11.10a) expands the rib cage and increases the volume of the thoracic cavity ($V_T\uparrow$). The principles in this section tell us that as V_T increases, the pressure in the thoracic cavity P_T must decrease, that is, fall below atmospheric pressure. The air outside at atmospheric pressure then rushes into the lungs. Moving the diaphragm upward (Figure 11.10b) contracts the rib cage, decreasing the volume of the thoracic cavity ($V_T\downarrow$). The pressure P_T is increased and air is pushed out of the lungs.

CHARLES' LAW

11.6   If we take a balloon filled with air and place it in boiling water, the volume of the balloon visibly increases (see Figure 11.11). If we

FIGURE 11.10

Breathing is a direct demonstration of the inverse relationship between volume and pressure. Increased volume reduces pressure in the lungs. Decreased volume increases pressure in the lungs. The volume of the thoracic cavity is represented by V_T, V_{T_I} for inhalation, and V_{T_E} for exhalation.

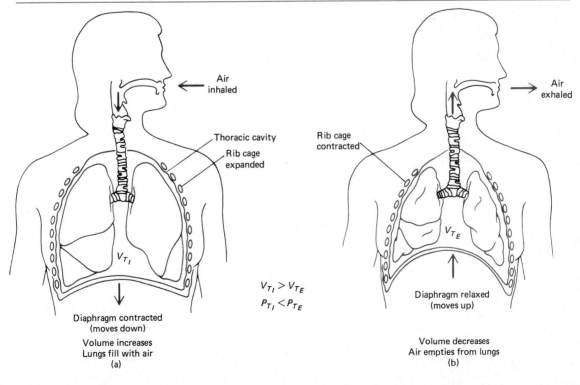

Air inhaled

Thoracic cavity
Rib cage expanded

V_{T_I}

Diaphragm contracted (moves down)

Volume increases
Lungs fill with air
(a)

$V_{T_I} > V_{T_E}$
$P_{T_I} < P_{T_E}$

Rib cage contracted

Air exhaled

V_{T_E}

Diaphragm relaxed (moves up)

Volume decreases
Air empties from lungs
(b)

FIGURE 11.11

The volume of the air in the balloon increases as the temperature increases. The pressure is constant and equal to atmospheric P.

take the balloon and place it in the freezer compartment of our refrigerator, the volume decreases. In both cases, the volume change occurs at a constant pressure (P_{atm}) and with a fixed amount of gas. We can conclude that when the temperature of a gas increases, its volume increases. When the temperature decreases, the volume of the gas decreases. That is, there is a *direct* relationship between the volume and the temperature of a gas. Raising or lowering one has the same effect on the other gas variable.

In 1787 Jacques Charles investigated the quantitative relationship between the volume and the temperature of a fixed quantity of gas

FIGURE 11.12

The volume of air trapped below the mercury plug expands as the temperature increases. The P is constant (P_{atm} + small P from the constant weight of Hg). See Table 11.4 and Figures 11.13 and 11.14. (Volumes are not drawn to scale; increases are exaggerated for visibility.)

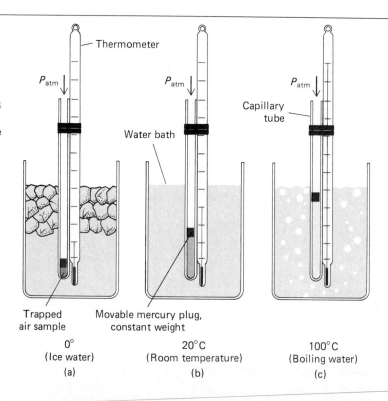

held at a constant pressure. Figure 11.12 shows an experimental setup similar to the one he used. In the experiment a small plug of liquid mercury is placed in a capillary tube that is open at one end. A fixed mass (therefore, a fixed number of moles) of air is trapped below the mercury. Because the only pressure on the air mass is P_{atm} plus the pressure from the fixed weight of mercury, $P_{air\ mass}$ is constant. The capillary tube is attached to a thermometer, which can be used to read the temperature (in °C) and the volume of the trapped air mass.

A typical set of data for the preceding experiment is shown in Table 11.4. We observe that as the temperature increases, the volume increases, and as the temperature decreases, the volume decreases.

A plot of volume versus temperature (in °C) gives the solid straight line shown in Figure 11.13. The line can be extended (broken dashes) to its intercept on the temperature axis. The value is -273°C for all gases subjected to this experiment. The horizontal intercept is the value of the point where the line crosses the horizontal axis. In principle, this intercept represents the temperature at which the volume of any gas should be zero. In practice, however, all gases liquefy before reaching this temperature.

Mathematically, a straight-line relationship between two variables is simpler if the horizontal intercept is zero. We can accomplish this for the volume-temperature relationship by establishing a new temperature scale with the zero of this new scale equal to -273°C. This temperature is called absolute zero because it is not possible to achieve lower temperatures than this. We call this new scale the **Kelvin temperature scale** (after Lord Kelvin, who constructed it) or the **absolute temperature scale.** The units of this scale are represented by a capital

VOLUME-TEMPERATURE DATA FOR A FIXED MASS OF GAS
AT A CONSTANT PRESSURE

TABLE 11.4

Volume, $cm^3 \times 10^2$	Temperature, °C	Temperature, K
9.6	-10.0	263
10.0	0.0	273
10.5	10.0	283
10.7	20.0	293
11.1	30.0	303
11.5	40.0	313
11.9	50.0	323
12.2	60.0	333
12.6	70.0	343
13.0	80.0	353
13.4	90.0	363

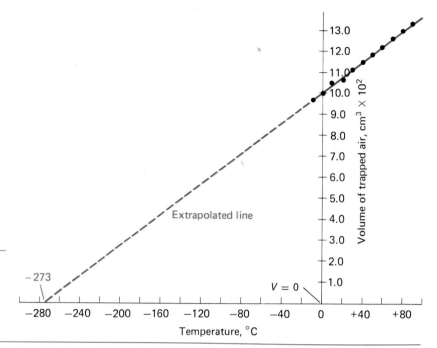

FIGURE 11.13

Plot of volume versus temperature (°C) from Table 11.4. Notice that the line extrapolated to zero volume intercepts the temperature axis at $t = -273°C$.

letter K. One does *not* use the degree symbol as in °C and °F. Any Celsius temperature can be converted to Kelvin by using the formula

$$K = °C + 273$$

In the third column of Table 11.4 we list the Kelvin temperature for each Celsius reading in the second column. By convention a capital T designates a Kelvin temperature and a small t designates a Celsius temperature. In Figure 11.14 we plot volume versus Kelvin temperature.

Problem 11.5

Convert the following temperatures to Kelvin temperatures:

a. 0°C

b. 25°C

c. 212°F

The results of the preceding experiment can be stated mathematically by Charles' law:

$$V \propto T \qquad \textbf{at constant } n \textbf{ and } P \tag{11.6}$$

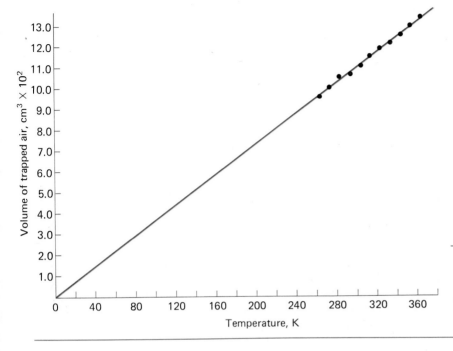

FIGURE 11.14

Plot of volume versus Kelvin temperature from Table 11.4. This graph through the origin ($T = 0$; $V = 0$) clearly shows the direct proportionality between V and Kelvin T.

This expression is read, "The volume of a gas is *directly* proportional to the Kelvin temperature when the pressure and number of moles are held constant." This statement is **Charles' law.**

As we did with Boyle's law, we can convert the proportionality into an equation by using a constant (see Appendix 1F):

$$V = mT \qquad \textbf{at constant } n \textbf{ and } P \qquad (11.7)$$

Dividing each side of Equation (11.7) by T gives us

$$\frac{V}{T} = m \qquad \textbf{at constant } n \textbf{ and } P \qquad (11.8)$$

This equation states that for a fixed number of moles of gas at a given pressure, the quotient of the volume divided by the Kelvin temperature is equal to a constant.

If an experimenter holds the pressure P and quantity of gas n fixed but varies the temperature from T_1 to T_2, the volume will change from V_1 to V_2. However, the quotient (V/T) must still equal m:

$$\frac{V_1}{T_1} = m$$

$$\textbf{at constant } n \textbf{ and } P$$

$$\frac{V_2}{T_2} = m$$

or

$$\frac{V_2}{T_2} = \frac{V_1}{T_1} \tag{11.9}$$

Remember that the volume is directly proportional only to the Kelvin temperature; no direct proportionality exists between volume and Celsius temperature. In gas law problems, Celsius temperatures must always be converted to Kelvin.

Problem 11.6

a. Using the data in Table 11.4, show that the ratio V/T is a constant.

b. Using the data in Table 11.4, show that the ratio V/t (in °C) is *not* a constant.

You will encounter problems that require Charles' law in Section 11.7.

COMBINED GAS LAWS

11.7 Boyle's law and Charles' law can be combined into a single expression:

$$V \propto \frac{1}{P} \qquad \text{at constant } n \text{ and } T \qquad \text{Boyle's law} \qquad (11.2)$$

$$V \propto T \qquad \text{at constant } n \text{ and } P \qquad \text{Charles' law} \qquad (11.6)$$

Therefore,

$$V \propto \frac{T}{P} \qquad \text{at constant } n \qquad \text{Combined law} \qquad (11.10)$$

This expression is read, "For a fixed mass of gas the volume is directly proportional to the Kelvin temperature and inversely proportional to the pressure." We can write this in an equation form by replacing the proportionality sign by an equals sign and a constant (r).

$$V = \frac{rT}{P} \qquad \text{at constant } n \qquad (11.11)$$

Multiplying both sides of the equation by P and dividing by T gives

$$\frac{PV}{T} = r \qquad \text{at constant } n \qquad (11.12)$$

If the pressure, volume, and temperature are changed from P_1, V_1, and T_1 to P_2, T_2, and V_2, then

$$\frac{P_1 V_1}{T_1} = r$$

$$\frac{P_2 V_2}{T_2} = r$$

and so

$$\frac{P_2 V_2}{T_2} = \frac{P_1 V_1}{T_1} \qquad \text{at constant } n \qquad (11.13)$$

Equation (11.13) can be used to solve problems involving a change in pressure, volume, and temperature for a fixed mass of gas (constant number of moles). *The units used on both sides of the equation must be the same.*

SAMPLE EXERCISE

11.2

In the laboratory, 4.00 L of helium gas are trapped in a cylinder at a pressure of 7.00 atm. The pressure is decreased, at a constant temperature, to 2.00 atm. What is the new volume?

SOLUTION

Step 1 Write down all given data in the form of initial and final conditions. If necessary, convert all pressures to the same unit, all volumes to the same unit, and temperature to Kelvin:

Initial	Final
$P_1 = 7.00$ atm	$P_2 = 2.00$ atm
$V_1 = 4.00$ L	$V_2 = ?$

Temperature is constant ($T_1 = T_2$).

Step 2 We recognize this as a problem involving pressure and volume changes at constant temperature and mass. We write down the combined gas law:

$$\frac{P_2 V_2}{T_2} = \frac{P_1 V_1}{T_1} \qquad (11.13)$$

Step 3 Omit terms which are identical on both sides. Since in this problem $T_1 = T_2$, we can omit the temperature term:

$$P_2 V_2 = P_1 V_1$$

Step 4 Isolate the unknown variable to one side. We are asked to solve for the new volume, so we divide both sides by P_2, isolating V_2 on one side:

$$\frac{\cancel{P_2} V_2}{\cancel{P_2}} = \frac{P_1 V_1}{P_2}$$

$$V_2 = \frac{P_1 V_1}{P_2}$$

Step 5 Fill in the given data, and be careful to have all pressures in the same unit and all volumes in the same unit. Do arithmetic operations and cancel the common units.

$$V_2 = \frac{(7.00 \ \cancel{\text{atm}})(4.00 \ \text{liters})}{(2.00 \ \cancel{\text{atm}})}$$

$$V_2 = 14.0 \ \text{liters}$$

ANSWER CHECK

It is a very good idea to check the reasonableness of your answer by thinking through the problem in terms of the qualitative discussions of P, V, and T changes in Sections 11.5 and 11.6. In this problem, because the volume is varying with a change in pressure at constant temperature, a Boyle's law (Section 11.5) relationship should be recognized. Because the pressure is decreasing (from 7 to 2 atm), the volume must increase; i.e., the final volume must be greater than the initial volume of 4 L. The final answer of 14 L is reasonable according to this qualitative analysis.

SAMPLE EXERCISE

11.3

A balloon containing 0.80 L of air at 25.0°C is warmed up to 80°C. What is the new volume of the balloon?

SOLUTION

Step 1

Initial	Final
V_1 = 0.80 L	V_2 = ?
T_1 = 25°C + 273	T_2 = 80°C + 273
= 298 K	= 353 K

Pressure is constant ($P_1 = P_2$).

Step 2   This problem involves volume and temperature changes. The mass of air inside the balloon is fixed, and the pressure (P_{atm}) is constant.

$$\frac{P_2 V_2}{T_2} = \frac{P_1 V_1}{T_1} \qquad (11.13)$$

Step 3   Since $P_1 = P_2$, we have

$$\frac{V_2}{T_2} = \frac{V_1}{T_1}$$

Step 4   Isolate the unknown variable by multiplying both sides by T_2

$$\frac{V_2 \cancel{T_2}}{\cancel{T_2}} = \frac{V_1 T_2}{T_1}$$

$$V_2 = \frac{V_1 T_2}{T_1}$$

Step 5

$$V_2 = \frac{(0.80 \text{ L})(353 \cancel{K})}{298 \cancel{K}}$$

$$V_2 = 0.95 \text{ L}$$

ANSWER CHECK

The volume is varying because of a change in temperature at constant pressure. This is a Charles' law (Section 11.6) relationship. Because the temperature is increasing (from 298 to 353 K), the volume must increase; i.e., the final volume must be greater than the initial volume of 0.80 L. The final answer of 0.95 L is reasonable.

SAMPLE EXERCISE

11.4

An automobile tire is filled to a pressure of 28 lb/in² at a temperature of 20.0°C. The tire is driven hard, and the temperature increases to 46°C. Assuming that the volume of the tire does not change, calculate the new pressure.

SOLUTION

Step 1

Initial	Final
$P_1 = 28$ lb/in²	$P_2 = ?$
$T_1 = 20.0°C + 273$	$T_2 = 46°C + 273$
$= 293$ K	$= 319$ K

Step 2 This problem involves pressure and temperature changes. Mass and volume are fixed.

$$\frac{P_2 V_2}{T_2} = \frac{P_1 V_1}{T_1} \qquad\qquad (11.13)$$

Step 3 Since $V_1 = V_2$,

$$\frac{P_2}{T_2} = \frac{P_1}{T_1}$$

This direct relationship between pressure and temperature at constant volume and mass is known as **Gay-Lussac's law.**

Step 4 Isolate the unknown variable by multiplying both sides by T_2

$$\frac{P_2 \cancel{T_2}}{\cancel{T_2}} = \frac{P_1 T_2}{T_1} \quad \text{or} \quad P_2 = \frac{P_1 T_2}{T_1}$$

Step 5

$$P_2 = \frac{(28 \text{ lb/in}^2)(319 \cancel{K})}{(293 \cancel{K})}$$

$$P_2 = 30. \text{ lb/in}^2$$

SAMPLE EXERCISE

11.5

A sample of helium gas has a volume of 1.25 L at −125°C and 5.00 atm. The gas is compressed at 50.0 atm to a volume of 325 mL. What is the final temperature of the helium gas in °C?

SOLUTION

Step 1

Initial	Final
P_1 = 5.00 atm	P_2 = 50.0 atm
V_1 = 1.25 L	V_2 = 325 mL
T_1 = $-125°C$ + 273	T_2 = ?
= 148 K	

The volumes must be expressed in the same unit. We will convert 325 mL to liters:

$$V_2 = 325 \text{ m\cancel{L}} \left(\frac{1 \text{ L}}{1000 \text{ m\cancel{L}}} \right) = 0.325 \text{ L}$$

Step 2 The problem involves a pressure, volume, and temperature change. Mass is fixed.

Step 3 Since *P*, *V*, and *T* all change, there are no terms to be omitted.

Step 4 Remember, when two fractions are equal, numerator$_a$ × denominator$_b$ = numerator$_b$ × denominator$_a$, so

$$P_1V_1T_2 = P_2V_2T_1$$

Now isolate the unknown variable by dividing both sides by P_1V_1,

$$\frac{\cancel{P_1}\cancel{V_1}T_2}{\cancel{P_1}\cancel{V_1}} = \frac{P_2V_2T_1}{P_1V_1}$$

$$T_2 = \frac{P_2V_2T_1}{P_1V_1}$$

$$T_2 = \frac{(50.0 \text{ \cancel{atm}})(0.325 \text{ \cancel{L}})(148 \text{ K})}{(5.00 \text{ \cancel{atm}})(1.25 \text{ \cancel{L}})}$$

$$T_2 = 385 \text{ K}$$

We must convert this to °C:

$$K = °C + 273$$
$$385 = °C + 273$$
$$385 - 273 = °C$$
$$112° = °C$$

Problem 11.7

A 738-mL sample of a gas at 0°C and 760. mmHg is cooled to $-200°C$ and compressed to 75.0 atm. What will be the new volume of this gas sample?

STANDARD TEMPERATURE AND PRESSURE

11.8 A statement about the volume of a gas is meaningless unless we also know the temperature and pressure of the gas, because as we have seen, V changes as T and P change. Volumes of gases can only be compared at the same temperature and pressure. Scientists have agreed on a convenient set of reference conditions. The temperature 273 K (0°C) and the pressure 1 atm (760 mmHg) have been chosen as **standard temperature** and **standard pressure**. The abbreviation "STP" is often used for standard temperature and pressure.

$$\text{STP} = \textbf{273 K (0°C) and 1 atm (760 mmHg)}$$

Since 760. mmHg is defined as 1 atm, it does not limit the number of sig figs in a calculation.

Sample Exercise 11.6 illustrates the conversion of the volume of a gas from a given set of conditions to STP.

SAMPLE EXERCISE

11.6

A given mass of gas occupies a volume of 435 mL at 25°C and 740. mmHg. What will be the new volume at STP?

SOLUTION

Step 1

Initial	Final
P_1 = 740. mmHg	The statement "at STP" tells you
V_1 = 435 mL	the final T and P conditions.
T_1 = 25°C + 273	P_2 = 760. mmHg
= 298 K	V_2 = ?
	T_2 = 273 K

Step 2 This problem involves pressure, volume, and temperature changes. Mass is fixed.

$$\frac{P_2 V_2}{T_2} = \frac{P_1 V_1}{T_1} \qquad (11.13)$$

Step 3 Since P, V, and T all change, there are no terms to be omitted.

Step 4 To isolate the unknown V_2, multiply both sides by T_2/P_2:

$$\frac{\cancel{T_2}}{\cancel{P_2}} \times \frac{\cancel{P_2} V_2}{\cancel{T_2}} = \frac{P_1 V_1}{T_1} \times \frac{T_2}{P_2}$$

$$V_2 = \frac{P_1 V_1 T_2}{T_1 P_2}$$

$$V_2 = \frac{(740. \cancel{\text{mmHg}})(435 \text{ mL})(273 \cancel{\text{K}})}{(298 \cancel{\text{K}})(760. \cancel{\text{mmHg}})}$$

$$V_2 = 388 \text{ mL}$$

AVOGADRO'S LAW

11.9 In the foregoing sections we have worked with a *fixed* amount of gas and observed quantitative variations in P, V, and T. If we take a balloon (at constant T and P) and simply add more gas to it, the volume of the balloon increases. Of course, it is true for any sample of matter that if we add more mass to the sample, it gets bigger (volume increases). But the remarkable thing about gases is that there is a **direct proportionality** between volume and the amount of material in moles which is the same for all gases. It is remarkable because this proportionality does not hold for the liquid or solid state (see Figure 11.15).

As we have seen previously, a direct proportion is expressed

$$V \propto n \qquad \text{at constant } T \text{ and } P \qquad (11.14)$$

Writing this as an equation yields

$$V = An \qquad (11.15)$$

FIGURE 11.15

As larger numbers of moles are added to the tubes, the volume increases. More matter occupies more space. At constant pressure and temperature, the volume increase is directly proportional to the molar increase.

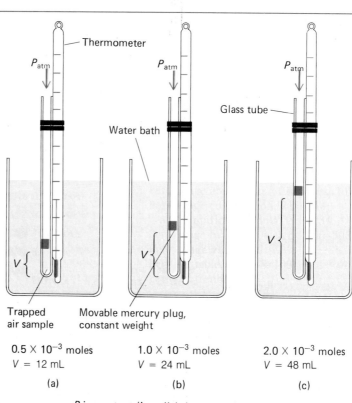

Thermometer

P_{atm}

P_{atm}

P_{atm}

Glass tube

Water bath

V

V

V

Trapped air sample

Movable mercury plug, constant weight

0.5 × 10^{-3} moles
V = 12 mL

(a)

1.0 × 10^{-3} moles
V = 24 mL

(b)

2.0 × 10^{-3} moles
V = 48 mL

(c)

P is constant (just slightly greater than P_{atm})
T is held constant at 293 K (20°C)

where A is the constant of proportionality.

Dividing each side of the equation by n, we get

$$\frac{V}{n} = A \qquad \text{at constant } T \text{ and } P \qquad (11.16)$$

Holding the pressure and temperature fixed and varying V from V_1 to V_2 and n from n_1 to n_2, we have

$$\frac{V_1}{n_1} = A$$

$$\frac{V_2}{n_2} = A$$

so

$$\frac{V_1}{n_1} = \frac{V_2}{n_2} \qquad \text{at constant } T \text{ and } P \qquad (11.17)$$

At the same temperature and pressure, if the volume of one gas V_1 equals the volume of a second gas, V_2, then according to Equation (11.17), the number of moles of the first gas n_1 must equal the number of moles of the second gas n_2:

$$\frac{V_1}{n_1} = \frac{V_2}{n_2} \qquad (11.17)$$

If $V_1 = V_2$, then we can cancel out V on each side. Thus

$$\frac{1}{n_1} = \frac{1}{n_2}$$

Cross-multiply to find

$$n_2 = n_1$$

Equal volumes of gases at the same temperature and pressure contain the same number of moles.

This very significant statement is **Avogadro's law,** which states that *equal volumes of all gases at the same temperature and pressure contain the same number of molecules*. Because the number of molecules is directly proportional to the number of moles, the two statements on the equal volumes of gases are identical.

SAMPLE EXERCISE

11.7

At 738 mmHg and 24°C 0.393 mole of nitrogen gas is contained in 9.87 L. What volume would 0.393 mole of argon gas occupy at 738 mmHg and 24°C?

SOLUTION

The two gases are at the same temperature and pressure. Therefore, 0.393

mole of argon must be contained in the same volume as 0.393 mole of nitrogen. Since the volume of nitrogen is 9.87 liters, the volume of argon is 9.87 L.

Problem 11.8

At STP 0.76 mole of helium occupies 17 L. How many moles of hydrogen gas would occupy this volume at STP?

Avogadro developed his original hypothesis from Gay-Lussac's **Law of Combining Volumes of Gases,** which states that the ratio of the volumes of reacting gases are small whole numbers (at constant T and P). For example, H_2 and Cl_2 combine in the volume ratio $1:1$; for H_2 and O_2, the volume ratio is $2:1$; for H_2 and N_2, $3:1$. Notice that these ratios represent the reactant coefficients in the balanced equations for the reactions.

$$H_2 + Cl_2 \longrightarrow 2HCl$$

$$2H_2 + O_2 \longrightarrow 2H_2O$$

$$3H_2 + N_2 \longrightarrow 2NH_3$$

From the data about combining volumes, Avogadro reasoned that equal gas volumes must contain equal numbers of reacting particles (see Figure 11.16). In light of our knowledge about chemical formulas and the diatomic elements, this conclusion seems obvious. But in 1811 Avogadro did not have this information. For this reason his hypothesis was exceptionally brilliant, especially since it established the mole concept, which we know to be central to so many chemical principles and calculations.

MOLAR GAS VOLUME

11.10   The volume of 1 mole of gas is the **molar gas volume.** *At the special conditions of STP, we find experimentally that 1 mole of any gas occupies a volume of 22.4 L.*

Problems involving gases at STP can often be solved most readily by use of the following equality:

1 mole of a gas at STP = 22.4 liters

SAMPLE EXERCISE

11.8

What is the volume of 11.9 g of nitrogen gas at STP?

SOLUTION

This problem can be solved by the standard Unit Conversion Method.

Step 1   The given quantity and unit is 11.9 g of N_2 at STP.

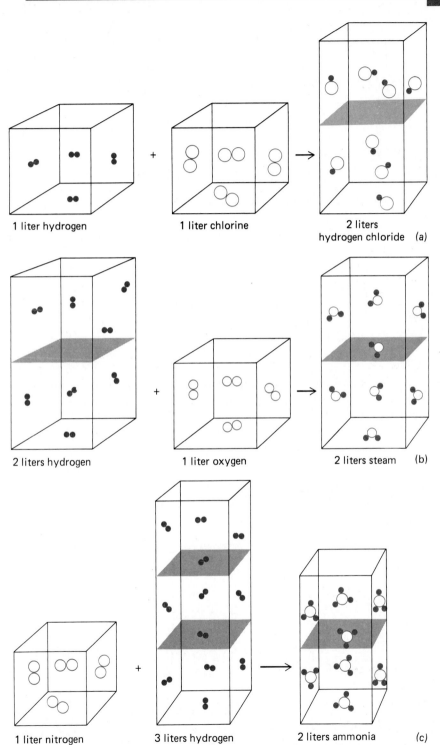

1 liter hydrogen + 1 liter chlorine → 2 liters hydrogen chloride (a)

2 liters hydrogen + 1 liter oxygen → 2 liters steam (b)

1 liter nitrogen + 3 liters hydrogen → 2 liters ammonia (c)

FIGURE 11.16

For gases at constant T and P, the reacting volume ratios are identical to the molar ratios expressed by the coefficients of the balanced equation. This is because equal volumes of gases contain equal numbers of molecules (at constant T and P). These pictures show 4 molecules per liter in each case: (a) H_2 and Cl_2 react on a 1:1 molecular basis $H_2 + Cl_2 \rightarrow 2HCl$ and therefore a 1:1 volume basis because equal volumes contain equal numbers of molecules; (b) H_2 and O_2 react on a 2:1 molecular basis $2H_2 + O_2 \rightarrow 2H_2O$ and therefore a 2:1 volume basis because equal volumes contain equal numbers of molecules; (c) N_2 and H_2 react on a 1:3 molecular basis $N_2 + 3H_2 \rightarrow 2NH_3$ and therefore a 1:3 volume basis because equal volumes contain equal numbers of molecules.

Step 2   The new quantity and unit to be determined is number of liters at STP.

Step 3   We do not have a conversion factor to convert grams to liters directly, but at STP we know that 1 mole N_2 = 22.4 L of N_2 and that the molar weight relates moles to grams. In this case, 1 mole of N_2 = 28.0 g N_2.

Step 4   Set up in the proper format:

Given × **conversion factors** = **answer**

$$11.9 \text{ g N}_2 \times \frac{1 \text{ mole N}_2}{28.0 \text{ g N}_2} \times \frac{22.4 \text{ L N}_2}{1 \text{ mole N}_2} = 9.52 \text{ L N}_2$$

SAMPLE EXERCISE

11.9

What is the density of hydrogen gas at STP?

SOLUTION

The **given** information in this case comes from the statement that the gas is at STP. At STP for any gas,

$$1 \text{ mole} = 22.4 \text{ L}$$

For H_2 gas in particular, 1 mole of H_2 = 2.02 g of H_2.
   Density is mass divided by volume. For gases, the units used are grams per liter. At STP, we know both the mass and volume of a 1-mole sample of gas and can substitute in the equation for density:

$$\text{Density (H}_2) = \frac{2.02 \text{ g}}{22.4 \text{ L}} = 0.0902 \text{ g/L}$$

Problem 11.9

What is the density of helium gas at STP?

Problem 11.10

Calculate the volume of 7.50 g of hydrogen gas at STP.

IDEAL GAS LAW

11.11   The fact that 1 mole of any gas occupies 22.4 L at STP allows us to calculate the volume of any number of moles of a gas at STP. Now it would be convenient to find a relationship which would allow the calculation of the fourth variable given any three of the four variables P, V, T, or n at any condition.
   In Section 11.9 we found that

$$V \propto n \qquad \text{at constant } T \text{ and } P \tag{11.14}$$

Combining this proportionality with Equation (11.10), $V \propto \dfrac{T}{P}$, gives

$$V \propto \frac{nT}{P} \tag{11.18}$$

Replacing the proportionality by the constant R gives us the so-called **Ideal Gas Law,** Equation (11.20):

$$V = \frac{RnT}{P} \tag{11.19}$$

or

$$PV = nRT \tag{11.20}$$

We can rewrite this equation, isolating R, by dividing both sides by nT:

$$\frac{PV}{nT} = \frac{\cancel{n}R\cancel{T}}{\cancel{n}\cancel{T}}$$

$$\frac{PV}{nT} = R \tag{11.21}$$

If we now substitute the STP conditions into Equation (11.21), we can determine the value and units of R for *all* conditions. Remember, R is a constant; it is the same for any proper combination of n, V, T, and P.

$$R = \frac{(1.00\ \text{atm})(22.4\ \text{L})}{(1.00\ \text{mole})(273\ \text{K})}$$

$$R = 0.0821\ \frac{(\text{L})(\text{atm})}{(\text{mole})(\text{K})} \tag{11.22}$$

When P is in atmospheres, use this value of R.

Should it be desirable to work with pressure expressed in millimeters of mercury, then R would have the following value and units:

$$R = \frac{(760\ \text{mmHg})(22.4\ \text{L})}{(1.00\ \text{mole})(273\ \text{K})}$$

$$R = 62.4\ \frac{(\text{L})(\text{mmHg})}{(\text{mole})(\text{K})} \tag{11.23}$$

When P is in millimeters of mercury, use this value of R. Considering the four variables of a gas, if any three are known, then the fourth can be calculated using the Ideal Gas Law and the gas constant R.

Hints on the use of the Ideal Gas Law equation

1. Pay special attention to the units.
    a. As always for gas law problems, T must be in Kelvin.
    b. V must be in liters.
    c. If P is in atmospheres, use

$$R = 0.0821 \frac{\text{L·atm}}{\text{mole·K}} \qquad (11.22)$$

If P is in millimeters of mercury, use

$$R = 62.4 \frac{\text{L·mmHg}}{\text{mole·K}} \qquad (11.23)$$

2. Because determining the number of moles of material, given a number of grams, always involves dividing grams by molar weight (see, for example, Sample Exercise 11.8), we can substitute grams per molar weight for n in the Ideal Gas Law equation:

$$PV = nRT \qquad (11.20)$$

$$PV = \frac{g}{\text{molar weight}} RT \qquad (11.24)$$

This Equation (11.24) allows us to determine molar weights experimentally in a manner which is exemplified in Sample Exercise 11.12.

3. The ideal gas equation is *not* used when there are *changing* conditions of P, V, or T. In problems involving a change in these variables, use the combined gas law equation (11.13):

$$\frac{P_2 V_2}{T_2} = \frac{P_1 V_1}{T_1} \qquad (11.13)$$

Let us look at some applications of the Ideal Gas Law.

SAMPLE EXERCISE

11.10

What volume will be occupied by 3.25 moles of oxygen gas at 73.5 mmHg and 25°C?

SOLUTION

We recognize that this problem can be solved by the Ideal Gas Law because we are given three variables (n, P, and T) and asked to find the fourth gas variable (V). Therefore, we will use

$$PV = nRT \qquad (11.20)$$

Step 1   Isolate the variable we are being asked to find. In this problem, the unknown variable is V, so we write

$$V = \frac{nRT}{P}$$

Step 2 Choose the value of *R* that you will use based on the units of *P* given. In this case, because *P* is given in millimeters of mercury, choose the *R* of Equation (11.23).

Step 3 Perform any necessary conversion of given units to conform to the units of *R*. In this case,

$$25°C = 298 K \quad (25°C + 273)$$

Step 4 Substitute the numerical values with their proper units in the equation shown in Step 1. Cancel out common units and perform the indicated arithmetic.

$$V = \frac{(3.25 \text{ moles}) \left(\frac{62.4 \text{ L·mmHg}}{\text{mole·K}} \right) (298 \text{ K})}{735 \text{ mmHg}}$$

$$V = 82.2 \text{ L}$$

Step 5 Check to be sure that the unit you are left with, after cancellation of units, is the proper unit for the unknown variable. In this case, liters is the proper unit for volume.

SAMPLE EXERCISE

11.11

How many moles of argon gas are present in 785 mL at 100°C and a pressure of 25.0 atm?

SOLUTION

Three variables are given (*V*, *T*, and *P*), and the fourth (*n*) must be determined. Use

$$PV = nRT \qquad (11.20)$$

Step 1 Isolate *n*:

$$\frac{PV}{RT} = n$$

Step 2 Choose *R* based on the *P* units given. In this case, *P* is given in atmospheres, so that the *R* of Equation (11.22) is chosen.

Step 3 Convert units as necessary:

$$100°C = 373 K \quad (100°C + 273)$$

$$785 \text{ mL} \times \frac{1 \text{ L}}{1000 \text{ mL}} = 0.785 \text{ L}$$

Step 4 Substitute, cancel, and do the arithmetic:

$$n = \frac{(25.0 \text{ atm})(0.785 \text{ L})}{\left(\frac{0.0821 \text{ L·atm}}{(\text{mole·K})} \right)(373 \text{ K})}$$

$$n = 0.641 \, \frac{1}{1/\text{mole}} = 0.641 \text{ mole}$$

Note: The unit of moles belongs in the numerator because

$$\frac{1}{1/\text{mole}} = 1 \div \frac{1}{\text{mole}} = 1 \times \frac{\text{mole}}{1} = \text{mole}$$

Step 5   Moles is the proper unit for n.

SAMPLE EXERCISE

11.12

In the laboratory, 10.0 g of an unknown gas is found to occupy a volume of 5.60 liters at 20.0°C and 740. mmHg. What is the molar weight of this unknown gas?

SOLUTION

Whenever we are asked to determine the molar weight of a gas, we use Equation 11.24 (see Hint 2).

$$PV = \frac{g}{\text{molar weight}} RT \tag{11.24}$$

Step 1   Isolate molar weight by multiplying each side by molar weight:

$$PV \text{ (molar weight)} = gRT$$

and dividing each side by PV:

$$\text{Molar weight} = \frac{gRT}{PV}$$

Step 2   Choose the R corresponding to millimeters of mercury.

Step 3   Convert temperature:

$$20.0°C = 293 \text{ K} \quad (20.0°C + 273)$$

Step 4   Substitute, cancel, and do the arithmetic:

$$\text{Molar weight} = \frac{(10.0 \text{ g}) \left(62.4 \, \frac{\text{L·mmHg}}{\text{mole·K}} \right) (293 \text{ K})}{(740. \text{ mmHg})(5.60 \text{ L})} = 44.1 \, \frac{g}{\text{mole}}$$

Step 5   Grams per mole is a proper unit for molar weight.

Problem 11.11

How many moles of gas are there in a 400.-mL aerosol can at a temperature of 20°C and a pressure of 3.00 atm?

Problem 11.12

At STP 700. mL of a gas weigh 1.452 g. What is the molar weight of the gas?

PLATE 1

Elements are classified as metals or nonmetals. See Section 1.9. In copper ore, the shiny metal can be seen embedded in earth and rock. The nonmetal is sulfur, a dull-yellow material. *(Upper photo [trans. no. 486(2)] courtesy Department Library Services, American Museum of Natural History. Lower photo by Yoav Levy/ Phototake.)*

PLATE 2

Many household chemicals are acids or bases (see Chapter 15) and are known by common names rather than by systematic nomenclature. See Section 7.2, especially Table 7.1. *(Photo by Bob Rogers.)*

PLATE 3

One mole of a particular chemical compound corresponds to a unique mass. The liquid is water, H_2O, 1 mole = 18 g. The orange solid is ammonium dichromate, $(NH_4)_2Cr_2O_7$, 1 mole = 252 g. For green nickel nitrate, $Ni(NO_3)_2$, 1 mole = 183 g. One mole of blue copper (II) sulfate, $CuSO_4$, is 160 g. The dark purple material is potassium permanganate, $KMnO_4$, for which 1 mole = 158 g. *(Photo by Bob Rogers.)*

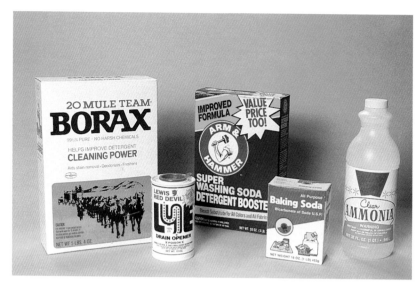

PLATE 4

A yellow precipitate (PbI_2) forms from two clear solutions when $Pb(NO_3)_2$ and NaI are mixed. See Section 9.8 Double displacement reactions and Section 13.9 Ionic equations. *(Photo by Yoav Levy/Phototake.)*

PLATE 5

The halogens are the Periodic Group VII. See Section 4.11. A periodic trend can be seen in that proceeding from the top to bottom of the group, the color darkens. See Section 5.10. Chlorine is light yellow; bromine, orange; and iodine, purple. *(Photo by Yoav Levy/Phototake.)*

PLATE 6

Crystals come in many colors, shapes, and forms. See Section 12.9 Classes of crystalline solids. Green emerald is a precious form of beryl, a mixture of oxides of beryllium, aluminum, and silicon. The yellow color in barite is an impurity trapped in the colorless crystal lattice of $BaSO_4$. Snow is a molecular crystal; two snowflakes are shown. (Upper photos [trans. nos. 2784, K13061] courtesy Department Library Services, American Museum of Natural History. Upper left photo by O. Bauer. Lower photos by Eric V. Grave/Phototake.)

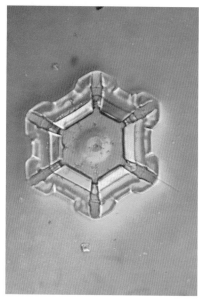

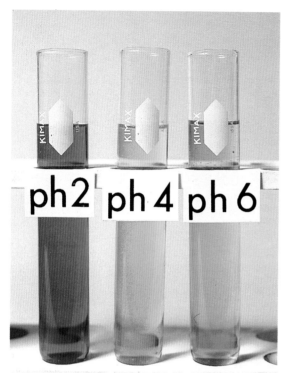

ph 2 ph 4 ph 6   ph 5 ph 7 ph 9

PLATE 9

Certain dyes, called indicators, change color as pH (a measure of acidity) changes. See Section 15.7. Bromthymol blue changes from yellow to blue. Methyl red changes from red to yellow.
(Photos by Bob Rogers.)

(Opposite page) **PLATE 10**

Elements show emission spectra, i.e., a characteristic array of colors. See Section 5.7. A simple demonstration is seen in flame tests. The colors of the flames distinguish Li (red), Na (yellow), K (violet) in the top row and Ca (yellow), Sr (red), and Ba (yellow) in the bottom row. Note the variations in reds and yellows.
(Photos by Yoav Levy/Phototake.)

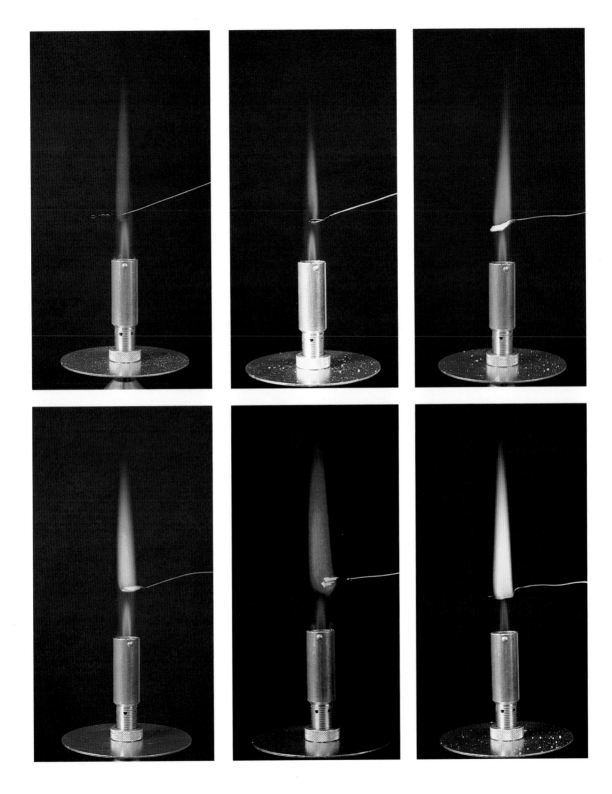

PLATE 11

Density is a characteristic property of matter. See Section 3.4. Colorless water is denser than olive oil. A cork floats because it is less dense than either oil or water. The very dense metal falls to the bottom of the beaker.
(Photo by Bob Rogers.)

PLATE 12

Components of mixtures can be separated by physical methods such as filtration. See Section 1.5. The pink solid is separated from the colorless liquid.
(Photo by Bob Rogers.)

KINETIC THEORY OF GASES

11.12 The Kinetic Theory of Gases was developed to account for the physical properties and gas laws that we have discussed in this chapter. The theory proposes a model for gases that is based on the following assumptions:

1. Gases are made up of small particles (either atoms or molecules) which are constantly moving in random, straight-line motion.

2. The distance between particles is very large compared with the size of the particles. A gas is mostly empty space.

EXPLANATION OF PHYSICAL PROPERTIES AND GAS LAWS BASED ON THE KINETIC THEORY OF GASES

TABLE 11.5

Observation	Explanation
1. Gases can be compressed easily	1. Because gas particles are far apart, they easily can be squeezed closer together by an outside force
2. Gases expand to fill the volume of their container	2. Gas particles are constantly moving with no attractive forces between particles, so they will expand until they meet an outside force, namely, the wall of the container
3. Gases have a low density	3. Because a gas is mostly empty space, there are few particles (low mass) per unit volume
4. Gases can diffuse through each other	4. Gas particles are constantly moving and are separated by large distances; this leads to freedom for particles of one gas to move through the empty space of another gas
5. Gases exert a pressure on container walls	5. Moving gas particles collide with container walls, thus exerting a force on every square inch
6. Boyle's law: $$V \propto \frac{1}{P}$$	6. When the volume of gas is *decreased*, the particles collide with the walls more often, leading to a *greater pressure*, when volume of gas is *increased*, particles collide less often, leading to a *decreased pressure*
7. Charles' law: $$V \propto T$$	7. Increased temperature causes particles to move faster, leading to more and "harder" collisions with walls. Pressure inside the walls is increased until the volume expands to the point where the pressure inside the walls is again equal to the pressure outside
8. Dalton's law of partial pressure: $P_{total} = p_1 + p_2 + p_3 + \cdots$	8. Since the particles move independently of one another, each gas in a mixture will exert a pressure independent of the pressure of the other gases; the total pressure will be the sum of the individual pressures

3. There are no attractive forces between particles. The particles move independently of each other.

4. The particles collide with each other and with the walls of the container without incurring a loss of energy.

5. The average kinetic energy of the particles is directly proportional to the Kelvin temperature of the gas. (When the average kinetic energy increases, the particles move faster.)

Table 11.5 illustrates how the Kinetic Theory of Gases explains some of our observations about gases.

A gas that has no attractive forces between its particles is an **ideal gas**. The Kinetic Theory of Gases gives us a model of an ideal gas. The gas laws described in this chapter hold exactly only for ideal gases. Most "real" gases show approximately ideal behavior except at low temperature and high pressure. At low temperature, the motion of gas particles becomes so slow that attractive forces begin to play an important role. At high pressure, particles are squeezed close enough together for attractive forces to be important.

GAS STOICHIOMETRY

11.13 A. Volume-volume problems   When we have a balanced equation in which substances are present in the gas phase at a constant temperature and pressure, the molar coefficients of the gases also represent combining or reacting volumes. This follows directly from Avogadro's law (Section 11.9). For example, for the equation $3H_2(g) + N_2(g) \longrightarrow 2NH_3(g)$, we can read this as "3 moles of H_2 combine with 1 mole of N_2 to form 2 moles of ammonia," or "3 volumes of hydrogen combine with 1 volume of nitrogen to form 2 volumes of ammonia," because these reactants and products are gases.

SAMPLE EXERCISE

11.13

How many liters of hydrogen would be required to react completely with 6.00 L of oxygen to form liquid water at a constant temperature and pressure?

SOLUTION

Write the balanced equation;

$$2H_2(g) + O_2(g) \longrightarrow 2H_2O(l)$$

This problem can be completed in a manner similar to that of the stoichiometric mole-mole conversions we did in Chapter 10 (Section 10.4).

Step 1   The given quantity is 6.00 L of O_2.

Step 2 The new quantity is liters of H_2.

Step 3 Just as the balanced equation gives us the "equality":

$$2 \text{ moles } H_2 \simeq 1 \text{ mole } O_2$$

and the corresponding conversion factor, so too does the equation relate volumes for gases. In this case;

$$2 \text{ L } H_2 \simeq 1 \text{ L } O_2$$

Step 4 Set up the proper format:

Given × conversion factor = answer

$$6.00 \, \cancel{L \, O_2} \times \frac{2 \text{ L } H_2}{1 \, \cancel{L \, O_2}} = 12.0 \text{ L } H_2$$

This is a **volume ratio.** Note its similarity to the **mole ratio.**

Since H_2O is not present in the gas phase in Sample Exercise 11.13, Avogadro's law does *not* apply to it, and the molar coefficients between either of the gaseous reactants and the liquid water do not represent volume ratios.

B. Weight-volume problems In many stoichiometric problems involving gases you are given a weight of some reactant or product and asked to calculate a volume of a gaseous reactant or product. Apply the techniques developed in Chapter 10 to calculate the number of moles of gas, and then, if the reaction is done at STP, simply multiply the number of moles by the conversion factor:

$$\frac{22.4 \text{ L}}{\text{mole}} \quad (\text{from 1 mole } = 22.4 \text{ L at STP})$$

This is exemplified in Sample Exercise 11.15. Or if the reaction is *not* at STP, use the Ideal Gas Law to calculate the volume corresponding to the number of moles at the *T* and *P* conditions given (see Sample Exercise 11.14). Compare Sample Exercises 11.14 and 11.15.

SAMPLE EXERCISE

11.14

How many liters of oxygen can be produced from the decomposition of 18.5 g of potassium chlorate at 150°C and 750. mmHg according to the balanced equation

$$2KClO_3(s) \longrightarrow 2KCl(s) + 3O_2(g)$$

SOLUTION

Because the given substance, $KClO_3$, is present as a solid, we can employ the usual techniques of stoichiometry involving the mole ratio from the balanced equation and then apply our knowledge of the gas laws.

Stoichiometry
(Review Section)
10.7 if necessary)

Step 1 The mole ratio relating the given and new quantities is

$$\frac{3 \text{ moles } O_2}{2 \text{ moles } KClO_3} \quad \frac{\textbf{new}}{\textbf{given}}$$

Step 2 Because grams of $KClO_3$ is given, we will need the conversion factor we get from the molar weight of $KClO_3$, 122.6 g:

$$\frac{1 \text{ mole } KClO_3}{122.6 \text{ g } KClO_3} \quad \frac{\textbf{new} \text{ unit}}{\textbf{given} \text{ unit}}$$

Step 3 Set up the proper format:

$$18.5 \text{ g } KClO_3 \times \frac{1 \text{ mole } KClO_3}{122.6 \text{ g } KClO_3} \times \frac{3 \text{ moles } O_2}{2 \text{ moles } KClO_3}$$
$$= 0.226 \text{ moles } O_2$$

The conditions given are not STP, so we must use the equation $PV = nRT$ to solve for V.

Ideal Gas Law
(Section 11.11)

Step 1 Isolate V

$$V = \frac{nRT}{P}$$

Step 2 R is chosen as 62.4 (L·mmHg)/(mole·K) because P is given in millimeters of mercury.

Step 3

$$150°C = 423 \text{ K} \quad (150°C + 273)$$

Step 4 Substitute

$$V = \frac{0.226 \text{ moles } O_2 \left(\dfrac{62.4 \text{ L·mmHg}}{\text{mole·K}} \right)(423 \text{ K})}{750. \text{ mmHg}}$$

$$V = 7.95 \text{ L } O_2$$

If you are given a volume of gas and asked to calculate the weight of another substance in a balanced equation, first convert your volume of gas to moles of gas. The problem will then become a typical mole-gram-conversion stoichiometry problem.

How many grams of calcium oxide can be produced from the reaction of 6.00 L of oxygen with excess calcium at 0°C and 1.00 atm, according to the balanced equation

$$2Ca(s) + O_2(g) \longrightarrow 2CaO(s)$$

SOLUTION

If the given 6.00 liters of O_2 is converted to moles of O_2, then the problem can be complete according to the techniques learned in Chapter 10. Because the conditions given are STP, we use the molar gas volume (Section 11.10) for the liter-mole conversion:

$$6.00 \text{ L O}_2 \times \frac{1 \text{ mole O}_2}{22.4 \text{ liters O}_2} = 0.268 \text{ mole O}_2$$

Molar weight

$$0.268 \text{ mole O}_2 \times \frac{2 \text{ moles CaO}}{1 \text{ mole O}_2} \times \frac{56.1 \text{ g CaO}}{1 \text{ mole CaO}} = 30.1 \text{ g CaO}$$

Mole ratio

If the volume had been given at non-STP conditions, you would use the ideal gas equation ($PV = nRT$) to convert the given volume to moles.

How many grams of water can be produced from the reaction of 4.00 L of oxygen with excess hydrogen at 25°C and 1.10 atm, according to the balanced equation $2H_2(g) + O_2(g) \longrightarrow 2H_2O(l)$

SOLUTION

Use the ideal gas equation in the form $n = PV/RT$ to convert 4.00 L of O_2 to moles of O_2 and then proceed as in any typical stoichiometry problem.

Step 1 Establish the values to be used in the ideal gas equation.

$$P = 1.10 \text{ atm} \qquad V = 4.00 \text{ L}$$

$$R = 0.0821 \frac{\text{L·atm}}{\text{mole·K}} \qquad \text{because P is in atmospheres}$$

$$T = 25°C = 298 \text{ K } (25 + 273)$$

Step 2 Substitute in the ideal gas equation and solve for *n*:

$$n = \frac{(1.10 \text{ atm})(4.00 \text{ L})}{0.0821 \frac{\text{L·atm}}{\text{mole·K}} (298 \text{ K})}$$

$$n = 0.180 \text{ mole O}_2$$

Step 3   Use the mole ratio and molar weight to convert moles of O_2 to grams of H_2O:

$$0.180 \;\text{mole } O_2 \times \frac{2 \text{ moles } H_2O}{1 \text{ mole } O_2} \times \frac{18.0 \text{ g } H_2O}{1 \text{ mole } H_2O} = 6.45 \text{ g } H_2O$$

Problem 11.13

How many liters of chlorine gas are needed to react with 15.5 g of magnesium to form magnesium chloride at 25°C and 755 mmHg?

Problem 11.14

How many grams of sulfur tetrafluoride can be formed from the reaction of 4.93 L of fluorine with excess sulfur at 30°C and 1.00 atm?

SUMMARY

Gases have several properties that are unique to that physical state. Whenever a quantitative treatment of gases is desired, all four of the gas variables n, V, T, and P must be considered (Secion 11.1). Measurements of n, V, and T are familiar (Section 11.2). Pressure P is force per unit area. Air pressure is usually measured with a barometer, and that is the origin of the units of pressure commonly used (Section 11.3).

There are several gas laws. Dalton's law tells us that gases in a mixture exert pressures independent of other gases present (Section 11.4). Boyle's law cites the exact inverse relationship between volume and pressure (Section 11.5). Jacques Charles defined the direct relationship between volume and Kelvin temperature (Section 11.6). The mathematical combination of Boyle's and Charles' laws can be used to solve a variety of problems involving changes in gas variables (Section 11.7).

It is convenient to have some standard temperature and pressure in order to compare gas volumes (Section 11.8). Equal volumes of gas at the same T and P contain the same number of moles or molecules (Section 11.9). At STP, 22.4 L is the volume in which is contained 1 mole, or 6.022×10^{23} molecules, of any gas. This fact enables us to relate volume, moles, mass, and density at STP (Section 11.10).

The Ideal Gas Law facilitates the calculation of any of the four gas variables once the other three are given. We can also calculate molar weights from n, V, T, and P data (Section 11.11). The Kinetic Theory of Gases provides a model for the unique behavior of gases and the gas laws (Section 11.12).

Stoichiometry involving gases utilizes previously discussed techniques (Chapter 10) along with the Ideal Gas Law or the molar volume at STP (Section 11.13).

CHAPTER ACCOMPLISHMENTS

After completing this chapter you should be able to

11.1   Introduction
1.   State the characteristic properties of gases.
2.   State the four measurable gas variables.

11.2   *n, T,* and *V* measurements
3.   Describe how the number of moles, volume, and temperature of a gas are measured.

11.3   Pressure
4.   State the definition of pressure and describe how the pressure changes with increasing (or decreasing) force and increasing (or decreasing) area.
5.   Describe how atmospheric pressure is measured.
6.   State the molecular basis of gas pressure.
7.   State the "equalities" among 1 atm and millimeters of mercury, torr, and lb/in².

11.4   Dalton's law of partial pressures
8.   State Dalton's law of partial pressures.
9.   Given the pressure of each gas in a mixture of gases, calculate the total pressure.
10.   Calculate the partial pressure of a dry gas that has been collected by displacement of water, given the atmospheric pressure and temperature.
11.   Explain the relationship between partial pressure and gas exchange between two regions of different partial pressures.

11.5   Boyle's law
12.   State Boyle's law in words and mathematically.
13.   Given an increase or decrease of pressure by a given factor, state the effect on the volume of a fixed amount of gas at constant temperature.

11.6   Charles' law
14.   State Charles' law in words and mathematically.
15.   Convert a Celsius temperature to Kelvin.
16.   Given an increase or decrease of the Kelvin temperature by a given factor, state the effect on the volume of a fixed amount of gas at constant pressure.

11.7   Combined gas laws
17.   Considering the three variables, volume, temperature, and pres-

sure, and holding one variable constant (for example, P), calculate the effect of a change in the second variable (for example, T) on a given amount of the third variable (for example, V).

18.   Considering the three variables, volume, temperature, and pressure, calculate the effect of a change in two variables (for example, P and T) on a given amount of the third variable (for example, V).

11.8   Standard temperature and pressure

19.   Define standard temperature and pressure.

20.   Given the volume of a fixed amount of gas at a given temperature and pressure, calculate the volume at STP.

11.9   Avogadro's law

21.   Given the volumes of two gases at the same temperature and pressure, compare the relative numbers of moles present.

11.10   Molar gas volume

22.   Define molar gas volume.

23.   State the volume of 1 mole of any gas at STP.

24.   Calculate the volume of a given mass of gas at STP.

25.   Calculate the density of a given gas at STP.

26.   Calculate the number of moles that corresponds to a given volume of gas at STP.

11.11   Ideal Gas Law

27.   State the ideal gas equation that relates the four variables of a gas.

28.   Choose the proper value of R, the ideal gas constant, for a given value of the pressure.

29.   Given three (for example, P, T, and n) of the gas variables, calculate the fourth (for example, V).

30.   Solve for the molar weight of a gas, given the mass, volume, temperature, and pressure of the gas.

31.   Solve for the mass of a known gas, given the volume, temperature, and pressure of the gas.

11.12   Kinetic theory of gases

32.   State the assumptions of the Kinetic Theory of Gases.

33.   Explain the characteristic properties of gases, Boyle's law, Charles' law, and Dalton's law of partial pressure in terms of the Kinetic Theory of Gases.

34.   Define ideal gas.

35.   State the conditions under which gas ideality (zero attractive forces) is most nearly approached.

11.13 Gas stoichiometry

36. Given a balanced chemical equation for a reaction occurring at constant T and P and a given volume of one gaseous reactant or product, calculate the volume of a second gaseous reactant or product.

37. Given a balanced chemical equation, a given T and P, and the mass of one reactant or product, calculate the volume of a gaseous reactant or product.

38. Given a balanced chemical equation, a given T and P, and a volume of a gaseous reactant or product, calculate the mass of another reactant or product.

PROBLEMS

11.1 Introduction

11.15 State three physical properties of gases in general.

11.16 Describe at least two ways in which the properties of gases differ from those of liquids and solids.

11.2 Measurements of gas variables n, T, and V.

11.17 Describe why a container must be evacuated before it can be used in determining the mass of a gas.

11.18 Originally 3.57 g of a gas at 20°C occupies 2.5 L. The gas is allowed to expand to fill 15 L at 20°C. Calculate the original and final densities.

11.3 Measurements of gas variables P

11.19 Describe the cause of atmospheric pressure.

11.20 Why might atmospheric pressure be lower on top of a mountain than at sea level?

11.21 Explain why all the mercury does not run out of a barometer.

11.22 Do the following conversions:
- **a.** 735 mmHg to atmospheres
- **b.** 1.75×10^{-2} atm to millimeters of mercury

11.23 The atmospheric pressure in Mexico City is about 580. torr.

- **a.** What is this pressure in millimeters of mercury
- **b.** In atmospheres?

11.4 Dalton's law of partial pressures

11.24 In your own words, state Dalton's law of partial pressures.

11.25 Oxygen and chlorine gas are mixed in a container with partial pressures of 401 mmHg and 0.639 atm, respectively. What is the total pressure inside the container?

11.26 Oxygen gas produced from the decomposition of mercury(II) oxide is collected by water displacement on a day when the barometric pressure and temperature are 765 mmHg and 21°C, respectively. What is the partial pressure of the oxygen gas?

11.27 Two mixtures of gases A and B are separated by a barrier that gases can cross. In the A mixture, gas 1 has a partial pressure of 80 mmHg and gas 2 has 138 mmHg. In the B mixture, $p_1 = 98$ mmHg and $p_2 = 120$ mmHg. Will there be any gas flow between A and B; if so in which direction? Explain.

11.28 If 2.0 L of N_2 at 355 mmHg, 2.0 L of H_2 at 1.9 atm, and 2.0 L of O_2 at 751 torr are all placed in a 2.0-L container, what will be the total pressure inside the container?

11.5   Boyle's law

11.29   a.   State Boyle's law in terms of a proportionality.
b.   State Boyle's law in the form of an equation.
c.   Give a common example that demonstrates Boyle's law.

11.30   Describe how Boyle's law is related to the breathing process.

11.31   Consider the data in the following volume-pressure graph:

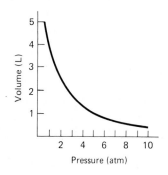

a.   At what pressure will the gas occupy a volume of 3 L?
b.   What is the value of PV for the volume in part a?
c.   What is the value of PV at a volume of 1 liter?
d.   What is the value of PV at a volume of 2 L?

11.6   Charles' law

11.32   a.   State Charles' law in the form of a proportionality.
b.   State Charles' law in the form of an equation.
c.   Give a common example that demonstrates Charles' law.

11.33   How can the thermometer be used to measure the volume of trapped air in Figure 11.12?

11.34   Convert the following:
a.   37°C to Kelvin
b.   −221°C to Kelvin
c.   398 K to degrees Celsius
d.   212°F to Kelvin.

11.35   Why does warm air rise up over cold air?

11.36   a.   A gas is at a temperature of 25°C. Assuming that the pressure and mass are held constant, will the volume of the gas double if the temperature is increased to 50°C?
b.   To what temperature must the gas be heated to double its volume?

11.37   a.   Is the volume of a gas directly proportional to the Celsius temperature?
b.   What is the relationship between the volume of a gas and the Celsius temperature?

11.7   Combined gas laws

11.38   A 1.00-liter balloon is released at sea level where the pressure is 760. mmHg. Assuming that the temperature remains constant, what will be the volume of the balloon at an altitude where the pressure is 580. mmHg?

11.39   At constant pressure, 1.20 L of exhaled gas undergoes a change in temperature from 98°F to 18°C. What is the new volume of the gas?

11.40   In a laboratory, 273 mL of a gas at 0°C is cooled to −1°C at constant pressure.
a.   What is the new volume of the gas?
b.   By what factor of the original volume has the gas decreased?
c.   If the gas were cooled to −2°C, what would be the new volume?
d.   Now by what factor of the original volume has the gas decreased?

11.41   Would your answers in Problem 11.40 be the same if P were not held constant? Explain.

11.42   The pressure on a gas is tripled at constant temperature.
a.   Will the volume increase or decrease?
b.   By what factor will the volume change?

11.43   Gasoline vapor is injected into a cylinder of volume 4.00 mL at a temperature and pressure of 21°C and 753 mmHg. The vapor is compressed under a pressure of 9.50 atm to 1.07 mL. What is the final temperature of the gasoline vapor?

11.44   A 4.00-L balloon of air is contained within a rocket ship at a temperature of 18°C and 1.00 atm. The balloon is released into space at a temperature of $-245°C$ and 2.0 mmHg pressure. What is the new volume of the balloon?

11.45 An aerosol can at 19°C contains a gas at 2.20 atm. What pressure is exerted by the gas at 95°C? Assume that the can is rigid and thus there is no volume change.

11.46 A balloon filled with helium has a volume of 2.50 L at 25°C. The balloon is placed in a freezer at $-12°C$. What is the volume of the balloon in the freezer?

11.47 If 325 L of a gas at 18°C and 1.00 atm is compressed into a cylinder at 135 atm and 21°C, what is the volume of the cylinder?

11.8 Standard temperature and pressure

11.48 A student collects 31.8 mL of argon gas at a temperature of 21°C and a pressure of 755 mmHg. What is the volume of argon at STP?

11.49 In a laboratory, 75.0 mL of hydrogen gas is collected by water displacement at a temperature of 18°C. The barometric pressure is 751 mmHg. What would be the volume of hydrogen at STP?

11.50 Why is it necessary to compare volumes of gas only at the same temperature and pressure?

11.9 Avogadro's law

11.51 Do 3.03 g of hydrogen and 48.0 g of oxygen occupy the same volume at 50°C and 730. mmHg?

11.52 A 1.50-mole sample of CO_2 gas at 18°C and 1.00 atm occupies a volume of 36 L. What volume would a 1.50-mole sample of SO_2 gas take up at the same temperature and pressure?

11.10 Molar gas volume

11.53 Calculate the volume of 3.0 moles of nitrogen gas at STP.

11.54 Calculate the volume of a cylinder that contains 5.67 g of hydrogen gas at STP.

11.55 Calculate the density of nitrogen gas at STP.

11.56 What volume will be occupied by the following weights of each gas at STP:
   a. 15.4 g O_2 **c.** 395 mg CO
   b. 7.56 g HCl   **d.** 18.1 g NH_3

11.57 What is the molar mass of a gas with a density of 1.16 g/L at STP?

11.58 What is the mass of 10.0 L of H_2 at STP?

11.59   **a.** Does 1.00 mole of a gas always occupy 22.4 L?
   b. What are the restrictions?

11.11 Ideal Gas Law

11.60 Calculate the volume of 4.0 moles of nitrogen gas at 754 mmHg and 85°C.

11.61 How many moles of propane gas will be present in a 2.55-L cylinder if the temperature is 35°C and the pressure is 15.1 atm?

11.62 Calculate the molar weight of unknown gas if 987 mL at 18°C and 754 mmHg weighs 6.29 g.

11.63 How many grams of helium gas are present in a 22.4-liter cylinder at 18°C and 100. atm?

11.64 What pressure will be exerted by 12.5 g of oxygen confined to a volume of 540. mL at 25°C?

11.65 Calculate the mass of 20.0 L of methane gas (CH_4) at 25°C and 11.0 atm?

11.66 1.65 g of a gas occupies a volume of 3.00 liters at 75°C and 1.05 atm. What is the molar weight of the gas?

11.67 Calculate the density of methane (CH_4) gas at 45°C and 754 mmHg.

11.68   **a.** What volume will a mixture of 16.0 g of O_2 and 14.0 g of N_2 occupy at STP?
   b. What is the partial pressure of the O_2? Of the N_2?
   c. What is the relationship between the sum of the partial pressures and the total pressure?
   d. The mole fraction of a substance in a mixture is defined as the ratio of the number of moles of that substance divided by the total number of moles in the mixture. Can you find a relationship between the partial pressure of each gas and its mole fraction and the total pressure of the mixture?

11.12 Kinetic Theory of Gases

11.69 State the assumptions of the Kinetic Theory of Gases.

11.70 **a.** What is meant by an ideal gas?
b. Under what conditions of temperature and pressure does a real gas most approach ideal behavior?
c. Under what conditions of temperature and pressure does a real gas deviate most from ideal behavior?

11.71 Using the Kinetic Theory of Gases, explain why the pressure in an automobile tire increases after the car has been driven at high speed.

11.72 Using the Kinetic Theory of Gases, explain why an aerosol can should not be heated above the warning temperature listed on the can.

11.73 Using Charles' law, explain why a given gas decreases in density as its temperature rises.

11.74 Explain what happens to the particles making up a gas when the temperature of the gas is increased.

11.75 A sample of nitrogen gas is at 0°C. To what temperature must the gas be heated to double its kinetic energy?

11.13 Gas stoichiometry

11.76 How many liters of hydrogen chloride gas can be formed from the reaction of 3.5 L of chlorine with excess hydrogen at 20°C and 1.10 atm?

11.77 How many liters of oxygen are required to react with 3.0 L of carbon monoxide to form carbon dioxide at STP?

11.78 Zinc metal reacts with hydrochloric acid to produce zinc chloride and hydrogen gas.
a. How many liters of hydrogen gas can be produced from 2.0 moles of zinc at STP?
b. How many liters of hydrogen gas can be produced from 41.3 g of zinc at 18°C and 1.00 atm?
c. How many grams of zinc are needed to produce 10.0 L of H_2 at STP?
d. How many grams of zinc are needed to produce 10.0 L of H_2 at 25°C and 755 mmHg?

11.79 Glucose ($C_6H_{12}O_6$) is metabolized in living systems according to the overall reaction:

$$C_6H_{12}O_6(s) + 6O_2(g) \longrightarrow 6CO_2(g) + 6H_2O(l)$$

What is the volume of CO_2 gas produced from 28.0 g of glucose at 1.00 atm and 37°C?

11.80 Ammonia can be produced from nitrogen:

$$N_2(g) + 3H_2(g) \longrightarrow 2NH_3(g)$$

a. What volume of nitrogen is required to react with 1.5 L of hydrogen at 25°C and 1.0 atm?
b. What volume of NH_3 will be produced from 1.5 L of H_2 at 25°C and 1.0 atm.
c. If 2.0 L of nitrogen and 3.0 L of hydrogen are mixed together at 25°C and 1.0 atm, which reactant is limiting for the formation of NH_3 according to the above equation?

LIQUIDS AND SOLIDS
CHAPTER 12

12
Mg
Magnesium
24.31

INTRODUCTION

12.1 It may seem strange to you, but the structures of liquids and solids, which we can readily see and touch, defy understanding more often than the structure and behavior of gases, which often are invisible and difficult to handle. The reason for this is that whereas the attractive forces between molecules of gases are so small that we can consider each molecule to be moving independently of others (Section 11.12), the intermolecular forces of liquids and solids are much stronger and the behavior of individual molecules is greatly influenced by surrounding molecules. It is these intermolecular attractive forces that account for the existence of liquids and solids.

An understanding of the properties of liquids and solids demands a consideration of these forces. However, before doing so we must look at two concepts that play an important role in determining the strength of intermolecular forces, i.e., molecular shape and molecular polarity.

MOLECULAR SHAPE

12.2 Many molecular properties, including the nature and strength of intermolecular forces and the ability to exhibit physiological activity, depend on a molecule having a particular shape. Because visualization of molecular shape is most readily done by considering Lewis structures, you may wish to review Section 6.13 before proceeding.

Atoms are three-dimensional. We have envisioned them as spheres or clouds. Molecules also have three-dimensional shapes; the particular shape depends on how the atoms in the molecule are arranged. Figure 12.1 depicts the typical molecular shapes that we will discuss. Because Lewis structures are written on paper, they have only two dimensions. However, Lewis structures can tell us the three-dimensional arrangement in space of electron pairs (electron-pair geometry) and then the molecular shape.

Linear
The atoms in the
molecule are *lined* up

Bent
The atoms in the molecule
are not lined up, but rather
bent away from a straight line

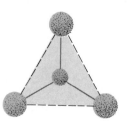

Triangular
The three outer atoms
form a *triangle*

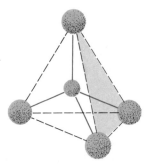

Tetrahedral
Drawing imaginary lines between
the four outer atoms produces the
four-sided figure called a *tetrahedron*

Pyramidal
This arrangement of
atoms is clearly a
pyramid

FIGURE 12.1

Molecular shapes. The shape
of the molecule is determined
by the arrangement of the
atoms.

The Valence Shell Electron-Pair Repulsion theory (VSEPR, pronounced "vesper" for short) allows us to predict the three-dimensional
arrangement of electrons around a central atom. (The central atom
concept is discussed in Section 6.13.) The basis for this theory is the
idea that electron pairs keep as far away from each other as possible.
Table 12.1 shows the geometries associated with the maximum possible separation in space for two, three, and four points around a
central point.

The geometry that will keep two points farthest apart is a linear

TABLE 12.1

GEOMETRIES OF MAXIMUM SEPARATION OF POINTS (•)
IN SPACE AROUND A CENTRAL POINT (●)

Number of Points	Arrangement		Angle of Separation	
2		Linear, i.e., along a line	180°	
3		Planar triangular; i.e., in one plane in space, and a triangle is formed by connecting the dots (•)	120°	
4		Tetrahedral; i.e., connecting the dots (•) forms the geometric figure called a tetrahedron	109°28′	

geometry, a separation of 180°. For three points, maximum separation is achieved by a planar triangular geometry, a separation of 120°. For four points, the greatest separation occurs in a tetrahedral geometry, a separation of 109°28′.

Given any Lewis structure, we can determine the number of sets of electron pairs around a central atom. For example,

Four sets   Four sets   Two sets

Two sets   Four sets   Three sets

Notice that we are not simply counting electron pairs, but rather we are considering how the pairs are grouped in sets. A set can consist of one pair (either a single bond or a nonbonding pair), two pairs (a double bond), or three pairs (a triple bond). Let us call this number of sets of electron pairs around a central atom the **set number**. *The set number can also be determined by counting the number of atoms bonded to the central atom plus the number of nonbonding pairs on the central atom.*

The electron distributions around the central atoms for the Lewis structures shown above are summarized as follows:

	Set Number	=	Number of Bonded Atoms	+	Number of Nonbonding Pairs
CH_4	4	=	4	+	0
NH_3	4	=	3	+	1
HCN	2	=	2	+	0
CO_2	2	=	2	+	0
H_2O	4	=	2	+	2
H∖C=O∕H	3	=	3	+	0

The **set number** tells you the **electron-pair geometry** directly:

Set Number	Geometry
2	Linear
3	Planar triangular
4	Tetrahedral

The molecular shape may or may not be the same as the electron-pair geometry. If there are *no nonbonding electron pairs* on the central atom, the **electron-pair geometry** and **molecular shape** *are identical*. For example,

CH_4

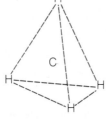

The four bonds (electron pairs) are arranged tetrahedrally; the four H atoms are also at the corners of a tetrahedron with C in the center; the molecule is tetrahedral

CH_2O

The three bonds are in a plane pointed to the corners of a triangle; the two H atoms and the O atom are at the corners of a triangle with C in the center; the molecule is planar triangular

| HCN | --H—C≡N-- | The single bond and triple bond are along a line. H, C, and N are all along the line; the molecule is linear |
| CO₂ | $-\ddot{O}=C=\ddot{O}-$ | The two double bonds are along a line; the C and both O's are along a line; the molecule is linear |

If there are nonbonding electron pairs around the central atom, then not every electron pair is associated with a bonded atom and the electron-pair geometry and molecular shape differ. *Only the arrangement of bonded atoms determines* **molecular shape**. For example,

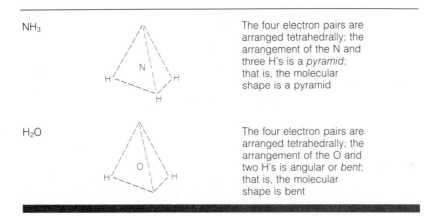

| NH₃ | | The four electron pairs are arranged tetrahedrally; the arrangement of the N and three H's is a *pyramid*; that is, the molecular shape is a pyramid |
| H₂O | | The four electron pairs are arranged tetrahedrally; the arrangement of the O and two H's is angular or *bent*; that is, the molecular shape is bent |

Electron pairs cannot be seen, but electron-pair geometry can be readily predicted theoretically by using VSEPR theory. However, molecular shape, the arrangement of atoms in a molecule, is real and can be determined experimentally. Similarly, when we build molecular models we see the shape defined by the atoms—we cannot see the electron pairs. Look back at Figure 12.1 This is why only the atoms are considered in determining molecular shape. Table 12.2 summarizes some relationships among set number, number of nonbonding electron pairs, electron-pair geometry, and molecular shape.

SAMPLE EXERCISE 12.1

What are the electron-pair geometry and molecular shape of the following compounds? **a**. CH₂Cl₂, **b**. H₂S, **c**. SO₃

SOLUTION

Write the Lewis structures according to the guidelines of Section 6.13 and

TABLE 12.2   ELECTRON-PAIR GEOMETRY AND MOLECULAR SHAPE*

Set Number	Number of Nonbonding Pairs	Electron-Pair Geometry	Molecular Shape	Example
2	0	Linear	Linear	CO_2
3	0	Planar triangular	Planar triangular	H_2CO
4	0	Tetrahedral	Tetrahedral	CH_4
4	1	Tetrahedral	Pyramidal	NH_3
4	2	Tetrahedral	Bent	H_2O

*Set number governs electron-pair geometry. Notice that when the number of nonbonding pairs is zero, molecular shape is identical to electron-pair geometry.

determine the set numbers and number of nonbonding electron pairs around the central atom. The geometries can be determined from these numbers and Table 12.2.

a.  There are four sets of electrons around the central carbon, so the set number is four and the electron-pair geometry is tetrahedral. Because there are no nonbonding electrons, the molecular shape is also tetrahedral.

b. 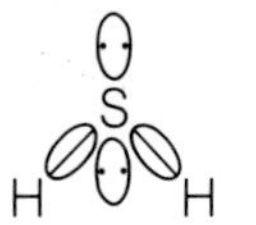 Electron-pair geometry is tetrahedral because once again there are four sets of electrons. However, there are only two bonded atoms, so the molecular shape is defined by the shape, which is bent.

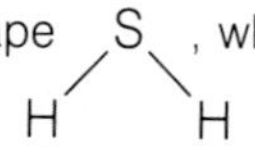

c. 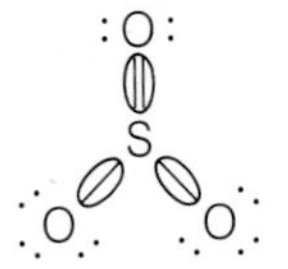 Electron-pair geometry is planar triangular because there are three sets of electrons. Since there are no nonbonding electrons on the central atom, the molecular shape is also planar triangular.

PROBLEM 12.1

What is the electron-pair geometry and molecular shape of the following compounds?

a. CO_2

b. SO_2

c. CH_3Br

MOLECULAR POLARITY

12.3   In Section 6.15 it was established that many covalent bonds are polar because of electronegativity differences between bonded atoms. When bond dipoles exist in a molecule, the molecule as a whole *may* have a dipole, that is, a + end and a − end; such a molecule is a **polar molecule**. You will see (Section 12.4) that the strongest intermolecular forces arise between polar molecules.

Whether or not a molecule is polar depends both on the existence of bond dipoles within the molecule and on the molecular shape. For example, in Section 6.15 we established that HCl has the bond dipole $^{\delta+}$H—Cl$^{\delta-}$, or $\overset{\longmapsto}{\text{H—Cl}}$. Because this bond dipole involves the entire molecule, the bond dipole and molecular dipole are the same thing. Therefore, this molecule is polar.

The bond dipoles in water (the molecular shape of which is bent, as we have seen) can be designated as follows:

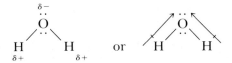

Recall that the more electronegative element draws the electrons toward itself and becomes slightly negative. In water, the fact that both bond dipoles point in the same general direction leads to a molecular dipole that can be marked as:

Water molecules are polar because their molecular shape is such that the effects of the two bond dipoles add together and create a permanent molecular dipole. That is, there exists a region of higher electron density (the negative end) and a region of lower electron density (the positive end).

In some molecules, because of the molecular shape, bond dipoles exert their effects in opposite directions and cancel each other out, leaving the molecule as a whole with no molecular dipole. The three cases in which bond dipoles cancel each other out follow:

Z—A—Z   **1.**   Molecular shape is linear, with two identical atoms Z bonded to a central atom A. For example, carbon dioxide

$$\overset{\longleftarrow\,\ \ \longrightarrow}{:O\!=\!C\!=\!O:}$$

2. Molecular shape is planar triangular, with three identical atoms Z arranged around a central atom A. For example, boron trifluoride

3. Molecular shape is tetrahedral, with four identical atoms Z tetrahedrally arranged around a central atom A. For example, carbon tetrachloride

An experimental test for molecular polarity can be carried out by placing a compound between two charged electrical plates as shown in Figure 12.2. Polar molecules will line up in the electrical field and a nonpolar molecule will not. When molecules line up, there is an observable change in the voltage between the two plates.

Problem 12.2

Which of the molecules in Problem 12.1 have a dipole moment?

INTERMOLECULAR FORCES

12.4 Intermolecular forces are the forces that act between ("inter" means "between") molecules; that is, they are the "glue" that holds molecules together in a liquid and solid. The covalent bond which you studied in Chapter 6 is an intramolecular force that holds the atoms together within a molecule ("intra" means "within"). Intermolecular forces are much weaker than covalent and ionic bonds. For example, the intermolecular forces in liquid water can be broken at the boiling point (100°C) and separate gas molecules (steam) form, but the covalent bonds are preserved in the water molecules making up the gaseous water. Temperatures much higher than 100°C are necessary to decompose water to hydrogen and oxygen.

The three types of intermolecular forces are

1. Dipole-dipole interactions

2. Dispersion (London) forces

3. Hydrogen bonds

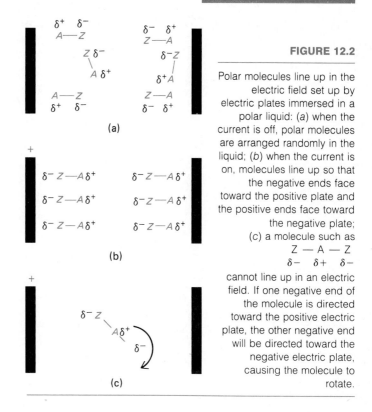

FIGURE 12.2

Polar molecules line up in the electric field set up by electric plates immersed in a polar liquid: (a) when the current is off, polar molecules are arranged randomly in the liquid; (b) when the current is on, molecules line up so that the negative ends face toward the positive plate and the positive ends face toward the negative plate; (c) a molecule such as

$$Z - A - Z$$
$$\delta- \quad \delta+ \quad \delta-$$

cannot line up in an electric field. If one negative end of the molecule is directed toward the positive electric plate, the other negative end will be directed toward the negative electric plate, causing the molecule to rotate.

Dipole-dipole interactions

These interactions occur between molecules that have a permanent dipole moment, as shown in Figure 12.3. Notice that the molecules arrange themselves so that the + poles of the molecules are attracted to the − poles of other molecules. All polar molecules are attracted to one another through this **dipole-dipole interaction**. The strength of the dipole-dipole attraction depends on the magnitude of the dipole moment. Nonpolar molecules do not exhibit this interaction because they do not have permanent dipoles.

Dispersion (London) forces

Figure 12.4 illustrates the development of these intermolecular forces. Electrons in molecules are in constant motion. At any instant the electron distribution within a molecule may be unsymmetrical, so that an "instantaneous dipole" develops (Figure 12.4b). This instantaneous dipole induces an unsymmetrical charge distribution in a nearby molecule, causing a net attractive force to develop between the two molecules (Figure 12.4c). An instant later the electron distributions shift, setting up instantaneous dipoles with other molecules (Figure 12.4d).

FIGURE 12.3

Dipole-dipole interaction is the mutual attraction between the + end of one polar molecule and the − end of another polar molecule. The dipole-dipole interactions are represented by dashed lines.

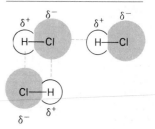

Molecules

X   Y

(a) Both molecules X and Y show uniformly distributed electron clouds.

(b) The electron cloud of X shows a concentration of negative charge to the right and hence a positive area to the left.

(c) The negative side of X causes a shift in the electron density of Y and an "induced dipole."

(d) This represents a shift of electron density in the opposite direction.

FIGURE 12.4

Dispersion forces arise by distortions of electron distributions in molecules which approach one another. Molecules in this figure are pictured as "blobs" of electron density which are initially uniformly distributed and which become distorted. Distortions create "instantaneous dipoles"; that is, + and − ends to the molecules which exist only for a fleeting moment. The events in (a) to (d) happen in a small fraction of a second.

These are called **dispersion forces** or **London forces** for Fritz London who first offered a mathematical understanding of these interactions. Dispersion forces exist between all kinds of molecules, whether polar or nonpolar.

Dispersion forces are the only intermolecular attractive force between nonpolar molecules and therefore are the only reason why gases such as H_2, He, and CH_4 can be liquefied.

The strength of dispersion forces depends on the number of electrons in a molecule and the ease with which their distribution can be disturbed. Molecules with greater numbers of atoms and hence greater molecular weights have larger numbers of electrons. Therefore, we see an increase in dispersion forces as molecular weight (and number of electrons) increases.

Electron distributions in molecules are more easily disturbed by neighboring molecules if the "surfaces" of the electron clouds can more nearly touch (Figure 12.5). For any given molecular weight, a molecule that has the shape of a spaghetti strand will have greater dispersion forces than a molecule that is ball-shaped.

Hydrogen bonds

The intermolecular force that is called the **hydrogen bond** occurs between a hydrogen covalently bonded to N, O, or F in one molecule and a nonbonding pair of electrons on N, O, or F in another molecule. Water, H_2O, is the most commonly encountered molecule that shows hydrogen bonding. The electron pair in the covalent bond between O and H is shifted toward the more electronegative oxygen, leaving the hydrogen partially positive in charge. Because of the small size of the hydrogen atom, its positive charge is concentrated, and this leads to a strong intermolecular attraction with a nonbonding pair of electrons on the oxygen of a nearby water molecule. This attraction is known as the hydrogen bond.

$$\underset{H}{\overset{\delta-}{O}}-\underset{}{\overset{\delta+}{H}}\cdots\underset{H}{:O:}\diagdown^{H}$$
Hydrogen bond

Each oxygen offers two nonbonding pairs of electrons for hydrogen bonding with hydrogens of two nearby water molecules. The hydrogen bond is stronger than other dipole-dipole attractions or London forces, but much weaker than the normal covalent bond. In Figure 12.6 the covalent bonds are shown by the usual dash (—) and the hydrogen bonds by dots (· · ·).

Many of the special physical properties of water (compared with molecules of similar molecular weight and polarity) arise from the

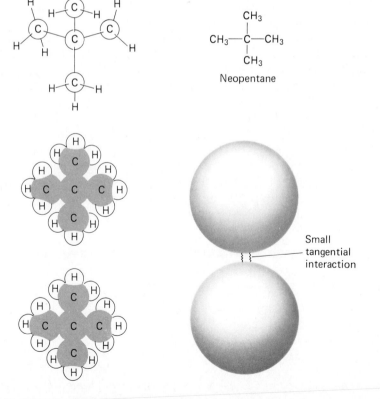

CH_3—CH_2—CH_2—CH_2—CH_3

n-Pentane

Attraction along elongated surface

CH_3—$\overset{\displaystyle CH_3}{\underset{\displaystyle CH_3}{\overset{|}{\underset{|}{C}}}}$—$CH_3$

Neopentane

Small tangential interaction

FIGURE 12.5

The effect of molecular shape on the extent of dispersion forces. We can envision molecules of *n*-pentane "touching" and interacting along their entire long surface. Neopentane is more like a sphere, and there is less "contact." The difference in intermolecular attraction is reflected in the boiling points of these two compounds. The boiling point of *n*-pentane is 36.2°C, whereas that of neopentane is lower, 9.5°C, although both have the same molecular weight.

FIGURE 12.6

Hydrogen bonding in water. Intermolecular hydrogen bonds (represented by •••) form between the slightly positive hydrogen of an O—H bond in one molecule and the electronegative oxygen of another molecule.

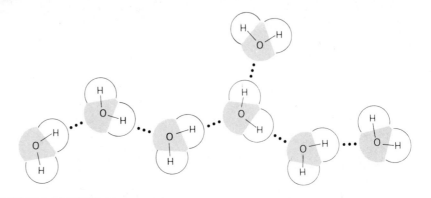

strong hydrogen bonding between water molecules. Water can climb upward in a narrow tube, a phenomenon known as capillary action. Glass tubes or the networks of pores in soil contain oxygen atoms as part of their molecular structure. Hydrogen bonds form between the water molecules and these oxygen atoms in the narrow area, and this helps to pull the water up the tube or pore. A blood sample is drawn up into a capillary tube by this same action.

Water has a much higher boiling point than compounds of similar, or even higher, molecular weight which lack hydrogen bonding. Compare H_2S, molecular weight 34, boiling point $-61°C$. Whereas most substances have a greater density in the solid state than in the liquid state, water is unique in that ice is less dense than liquid water from 0 to 4°C. This property, which results from a unique hydrogen bonding network in water, accounts for the fact that a body of water freezes from the top to the bottom allowing aquatic life to survive below the top ice layer.

Hydrogen bonding is very important in biological molecules, because hydrogen bonds affect the three-dimensional shape of proteins and act as the "hook-and-eyes" that hold together the two chains of the genetic material DNA (Figure 12.7).

SAMPLE EXERCISE

12.2

Which of the following molecules are capable of forming hydrogen bonds with one another?

a. NH_3 **b.** CH_4 **c.** HF   **d.** H—C$=$O
                                                   |
                                                   H

SOLUTION

Molecules with N—H, O—H, or F—H bonds show hydrogen bonding.

a. NH_3, because it has a polar N—H bond, will show hydrogen bonding.

b. CH_4 has only a nonpolar C—H bond and will *not* show hydrogen bonding.

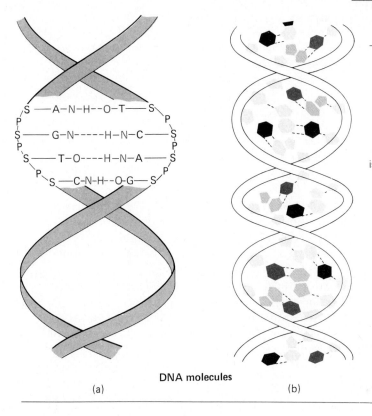

FIGURE 12.7

Genes are composed of very complex DNA (deoxyribonucleic acid) molecules which are held together in part by hydrogen bonds, shown as dashed lines here. The making and breaking of hydrogen bonds is an essential process in cell division and in the body's synthesis of proteins. (a) Deoxyribonucleic acid molecules have intertwined sugar (S)-phosphate (P) "backbones" held together by hydrogen bonding between portions (colored symbols) of so-called nitrogenous bases (designated A, T, C, G). (b) The molecular shapes of the sugar molecules and nitrogenous bases are portrayed to offer a better picture of the molecular complexity of DNA.

FIGURE 12.7

DNA molecules

(a)

(b)

c. HF has a polar H—F bond and will show hydrogen bonding.

d. In this case, although an oxygen with its nonbonding pair of electrons is present, the hydrogen is not bonded to N, O, or F, and so hydrogen bonding is not possible.

Problem 12.3

Which of the following molecules are capable of forming hydrogen bonds with one another?

a. H_2S

b.

c.

d.

CONDENSATION OF GASES

12.5 Intermolecular forces cause gas molecules to be pulled closer together into the liquid state as a gas is cooled or as the pressure on a gas is increased. The stronger the intermolecular forces that are present, the less cooling or compression is required to cause liquefaction.

Although an "ideal" gas would have no attractive forces between its molecules, all real gases have attractive intermolecular forces. When a gas is cooled, the motion of its molecules is slowed, and eventually after sufficient cooling the attractive forces are able to "pull" the molecules into a liquid. The temperature at which the gas begins to condense into a liquid could be called the **condensation point,** but it is more common to call it the **boiling point.** For any substance the change in state from gas to liquid (condensation) occurs at the same temperature as the change in state from liquid to gas (boiling) (Section 1.3). Boiling point depends on pressure. When the pressure on a gas is increased, the distance between molecules is decreased, and less cooling is necessary before the intermolecular attractions can force the molecules into a liquid.

Just as for the case of increased pressure pushing molecules together, molecules with strong intermolecular forces will need less cooling before they liquefy. The boiling points of such substances are higher than the boiling points of those with weaker intermolecular forces. Water has a boiling point of $+100°C$ and methane (CH_4) has one of $-162°C$ because, in addition to the always present dispersion forces, water molecules are held together by hydrogen bonds, the strongest of the intermolecular attractions, and dipole-dipole interactions, whereas methane molecules are attracted to one another only by dispersion forces.

SAMPLE EXERCISE

12.3

Predict the relative values of the boiling points of the following compounds, based on the intermolecular forces present in each:

a.

$$\begin{array}{c} H \\ \diagdown \\ C{=}O \\ \diagup \\ H \end{array}$$

Formaldehyde

b.

$$H{-}\underset{\underset{H}{|}}{\overset{\overset{H}{|}}{C}}{-}O{-}H$$

Methyl alcohol

c.   H—C≡C—H

Acetylene

SOLUTION

All compounds have dispersion forces. Acetylene has only dispersion forces, and we therefore expect it to have the lowest boiling point ($-82°C$); i.e., it must be cooled the most in order for the weak attractions to pull the molecules together. The dispersion forces in acetylene and formaldehyde are about equal because of the similar molecular weights. However, formaldehyde is polar because of the C=O bond dipole. Dipole-dipole interactions will cause it to liquefy at a higher temperature ($-21°C$) than acetylene.

Methyl alcohol has an O—H bond, which signals the existence of the strongest intermolecular force, the hydrogen bond; its boiling point is 65°C.

Problem 12.4

In each pair pick the compound with the higher boiling point:

a. NH_3 or NF_3

b. H_2O or H_2S

c. Cl_2 or Br_2

A MODEL OF LIQUIDS AND SOLIDS

12.6 Comparative molecular models of gases, liquids, and solids are shown in Figure 12.8. These models are based on the distinguishing properties of gases, liquids, and solids outlined in Table 1.1 and on the property of compressibility. Table 12.3 describes how these models of liquids and solids explain their characteristic properties.

FIGURE 12.8

Comparison of the behavior of particles in the (*a*) gaseous, (*b*) liquid, and (*c*) solid states of matter (see also Table 12.3). (*a*) Molecules are far apart; attractive forces between molecules are very small or nonexistent. Molecules are free to move in random motion. (*b*) Molecules are closer together; attractive forces are larger. Molecules can freely move but cannot separate from each other. (*c*) Molecules or ions are touching; attractive forces are very large and hold particles in a fixed orderly arrangement. Particles can vibrate only about fixed positions.

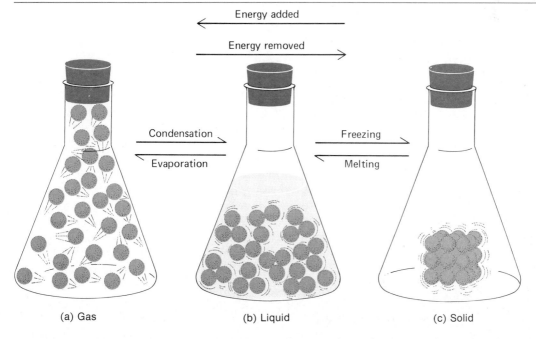

Energy added

Energy removed

Condensation

Evaporation

Freezing

Melting

(a) Gas   (b) Liquid   (c) Solid

TABLE 12.3	OBSERVABLE PROPERTIES OF LIQUIDS AND SOLIDS AND THEIR MOLECULAR EXPLANATION	
	Observable Property	Explanation from Model
	Liquids	
	Liquids have definite volume	Forces of attraction are strong enough to hold molecules together in a definite volume
	Liquids have an indefinite shape	Particles in a liquid are free to move (but not independently of each other), so that a liquid takes the shape of its container
	Liquids have low compressibility	Particles in a liquid lie close together with very little space between them
	Solids	
	Solids have a definite volume	Forces of attraction are very strong, holding particles in a definite volume
	Solids have a definite shape	Particles are held in fixed positions by strong attractive forces
	Solids have very low compressibility	Particles are touching; increasing pressure cannot squeeze particles closer together

PHYSICAL PROPERTIES OF LIQUIDS

12.7 Among the more important physical properties of liquids are vapor pressure, boiling point, heat of vaporization, and surface tension.

Vapor pressure

Liquids left in an open container slowly evaporate or vaporize; i.e., the liquid's molecules escape into the gas phase. The model of liquids can aid in the understanding of this phenomenon. Some molecules near the surface of the liquid have more than the average amount of kinetic energy, which they have gained through collisions with other molecules. These more energetic molecules can overcome the attractive forces of neighboring molecules and escape into the gas phase. Since it is the more energetic molecules that are leaving, the average kinetic energy, and therefore the temperature, decreases. It is for this reason that evaporation is said to be a cooling process. Perspiration evaporates from your skin, absorbing heat from the body. Sweating followed by evaporation is one of the body's cooling mechanisms. Liquids with low intermolecular attractions vaporize more readily at the same temperature than liquids with high intermolecular attractions. For example, ether, which does not show hydrogen bonding, evaporates much more readily than water, which has the strong hydrogen-bonding intermolecular attractions. Alcohol, a substance which evaporates more readily than water, is rubbed on the skin of a patient with a high fever to introduce an additional evaporative cooling mechanism.

When a liquid is placed into a closed container, the molecules that escape into the gas phase find themselves enclosed in a fixed space. As more and more molecules enter the gas phase, a pressure develops in the closed container (Figure 12.9). Since they cannot escape from the enclosed volume, the gas molecules eventually strike the liquid surface and a few are "recaptured" into the liquid phase. After a period of time the rate of condensation becomes equal to the rate of vaporization. Vaporization and condensation continue, but there is no *net* change in the number of molecules leaving the liquid for the gas phase. The pressure of the gas at this point is known as the **vapor pressure** of the liquid.

The value of the vapor pressure depends on the temperature and the nature of the liquid. Liquids with low intermolecular forces, such as ethers and gasoline, have high vapor pressures at room temperature. Liquids with high intermolecular forces, such as water and alcohols, have low vapor pressures at room temperature. The vapor pressure does *not* depend on the volume of liquid present. If the surface area of the liquid is increased, the rate of vaporization is initially faster, but the gas molecules have a larger liquid surface to strike and be "recaptured," and so the rate of condensation also increases. The point of dynamic equilibrium occurs at the same vapor pressure as with the original surface area.

The volume of the enclosed gas space also does not affect the vapor pressure. A larger enclosure allows more molecules to vaporize before dynamic equilibrium is established. However, a larger number of gas molecules spread over a larger volume yields the same pressure as a smaller number of gas molecules over a smaller volume.

Temperature, on the other hand, does affect the vapor pressure

FIGURE 12.9

The liquid is placed in the closed container and begins to evaporate (↑). As evaporation proceeds, more and more gas molecules (•) accumulate, and this leads to an increase in pressure. Gas molecules also begin to condense (↓). When the rates of evaporation and condensation are equal, the pressure remains constant because the net number of molecules in the gas phase remains constant. This constant pressure is called the **vapor pressure** of the liquid at the specified temperature (See also Figure 14.1).

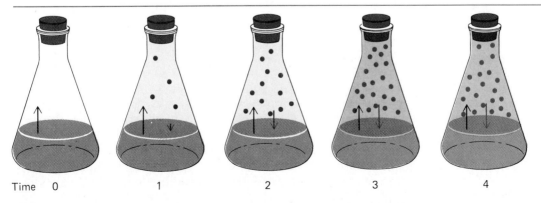

Time   0   1   2   3   4

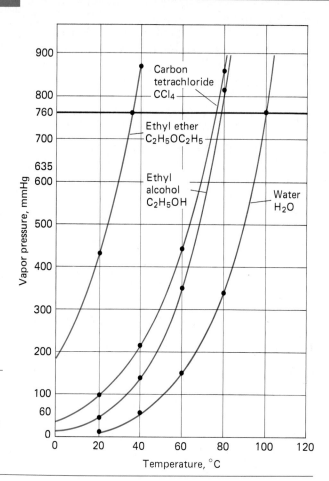

FIGURE 12.10

The vapor pressure of a liquid increases as temperature increases. The change in vapor pressure as a function of temperature is shown for several common liquids.

of a liquid. Vapor pressure increases with increasing temperature (see Figure 12.10). Kinetic energy and temperature are proportional. An increase in temperature gives rise to a corresponding increase in kinetic energy, and hence more molecules near the surface of the liquid are able to overcome the attractive forces of their neighbors and escape into the gas phase. Also at higher temperatures, molecules in the gas phase, having greater kinetic energy, are less likely to be recaptured into the liquid. The result is a greater number of gas molecules constrained within the same volume and hence a higher vapor pressure.

Boiling

A liquid *boils* when its vapor pressure is equal to the external pressure on the surface of the liquid. When this condition exists, we say we are at the **boiling point** of the liquid. During evaporation only molecules at the surface escape into the vapor phase, but at the boiling point some molecules *within* the liquid have sufficient energy to overcome the intermolecular attractive forces of their neighbors, so that bubbles of vapor form within the liquid. The bubbles rise in the liquid, and the vapor is released at the surface. It is the formation of vapor bubbles within the liquid itself that characterizes boiling and distinguishes it from evaporation.

Liquids with high intermolecular attractions, such as water, require a relatively high temperature before their vapor pressure equals the external pressure; hence these liquids are found to have a high boiling point. Liquids with low intermolecular attractions, such as ethers, have a lower boiling point.

The so-called normal boiling point is the temperature at which the vapor pressure is equal to an external pressure of one atmosphere. For example, water boils at 100°C when the external (atmospheric) pressure is 1 atm or 760 mmHg, because the vapor pressure of water is 760 mmHg at 100°C (see Figures 12.10 and 12.11).

The boiling point of a liquid can be reduced by lowering the external pressure (by water aspirator or vacuum pump), because then the vapor pressure of the liquid is equal to the external pressure at a lower temperature (Figure 12.11a). One application of this principle is in the food industry, where water is removed from such substances as coffee by boiling the liquid under a reduced pressure at a temperature lower than the normal boiling point where the product might decompose.

The boiling point of a liquid can be increased by raising the external pressure, because then the vapor pressure of the liquid is equal to the external pressure at a higher temperature (Figure 12.11b). A home pressure cooker works on this principle. By maintaining a pressure above one atmosphere inside the pressure cooker, the temperature of the liquid can rise above 100°C, thus allowing the food to cook in a shorter time.

Although heat must be continuously supplied, the temperature of a boiling liquid remains constant. If we add more heat (raise the flame) to an uncovered pot of boiling water, we find that the water boils faster but the temperature remains constant. If we remove the heat, the boiling process slows down and eventually stops. Recall that the molecules escaping from an evaporating or boiling liquid are those with the highest kinetic energy. Therefore, as these higher-energy molecules escape, the average kinetic energy of the remaining liquid molecules is lowered, and this causes the temperature to drop unless heat is added to the liquid.

Heat of vaporization

The quantity of heat needed to convert a fixed mass of liquid at a fixed temperature to the gaseous state is known as the **heat of vaporization.** Common units for heat of vaporization are calories per

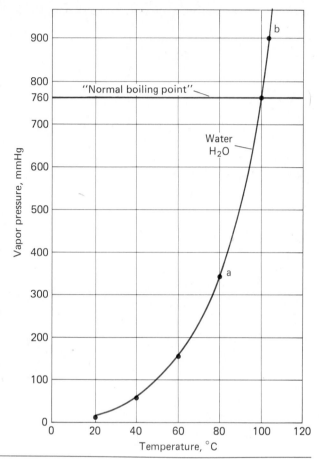

FIGURE 12.11

The "normal boiling point" of water is 100°C because at that temperature the vapor pressure of water is 760 mmHg, the same value as standard atmospheric pressure. When the vapor pressure of a liquid equals the external (atmospheric) pressure, a liquid boils. (*a*) If the external pressure is lowered, a liquid can boil at a lower temperature. If the external pressure is only 350 mmHg, water will boil at 80°C because at that temperature its vapor pressure equals 350 mmHg. (*b*) If the external pressure is raised, a liquid must be heated to a higher temperature before it boils. If the external pressure is 900 mmHg, the temperature of the water must be raised to 103°C in order for the vapor pressure of the water to equal 900 mmHg.

gram (cal/g) and kilocalories per mole (kcal/mol). When an amount of heat equal at least to the heat of vaporization is supplied to a boiling liquid, the liquid continues to boil at a constant temperature.

When the heat of vaporization is expressed in the units of kilocalorie per mole, it is known as the **molar heat of vaporization**. This quantity gives a measure of the strength of the intermolecular forces in the liquid. Table 12.4 lists the heats of vaporization for various liquids at their normal boiling points. The difference between the molar heat of vaporization for water (9.72 kcal/mol) and that for methane (2.21 kcal/mol) reflects the strong hydrogen-bonding attractions between water molecules in contrast to the weaker dispersion forces between methane molecules. A greater amount of heat is required to evaporate a gram of water than the same mass of any other common liquid. This property adds to the idealness of water as our body liquid

		Heat of vaporization	
Substance	Normal Boiling Point, °C	cal/g	kcal/mole
Methane	−161	138	2.21
Ethyl ether	34.6	89.8	6.64
Ethyl alcohol	78.3	204	9.38
Water	100	540.	9.72
Sodium chloride	1465	698	40.8

HEATS OF VAPORIZATION FOR SOME COMMON SUBSTANCES AT THEIR NORMAL BOILING POINTS

TABLE 12.4

because the evaporation of only a small amount of perspiration leads to significant cooling. The excess heat of an "overheated" person goes into the evaporation process and body temperature is maintained.

When a gas condenses to the liquid state, heat called the **heat of condensation** is given off in an amount exactly equal to the heat of vaporization. The heat of condensation must be removed for the gas to condense to a liquid at a constant temperature. A steam burn is severe because of the large amount of heat that is liberated when the steam vapor condenses to liquid water. Surgical and laboratory equipment is sterilized in an atmosphere of steam rather than boiling water because steam can be heated to temperatures above 100°C and because of the large amount of heat emitted when steam condenses.

SAMPLE EXERCISE

12.4

How much heat is required to vaporize 2.5 mol of ethyl alcohol at its normal boiling point?

SOLUTION

The heat needed to convert a liquid to a gas at its boiling point is the heat of vaporization. Since we are given moles of ethyl alcohol, we will use the molar heat of vaporization (9.38 kcal/mol) given in Table 12.4.

$$2.5 \text{ mol of ethyl alcohol} \times \frac{9.38 \text{ kcal}}{\text{mol of ethyl alcohol}} = 23 \text{ kcal}$$

Problem 12.5

How much heat is liberated when 22.5 g of steam is condensed to liquid water at 100°C?

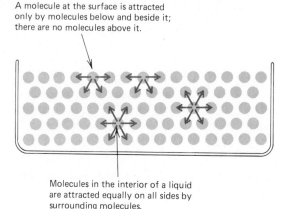

A molecule at the surface is attracted only by molecules below and beside it; there are no molecules above it.

Molecules in the interior of a liquid are attracted equally on all sides by surrounding molecules.

FIGURE 12.12

Molecules at the surface boundary of a liquid experience an unbalanced attractive force toward the interior of the liquid. This leads to a net pull inward and a smaller surface area. The result is the phenomenon of **surface tension.**

Surface tension

Intermolecular attractions lead to some interesting properties at the surface of a liquid. The molecules in the interior of a liquid, which are completely surrounded by other molecules, feel a balanced intermolecular attraction. However, molecules at the surface feel an unbalanced force because they are not attracted to other liquid molecules on one side (Figure 12.12). As a result the surface molecules experience a pull into the liquid known as surface tension. The surface tension is especially strong in water, where the hydrogen-bonding attractive force is present. The tension at the surface can balance a steel needle if it is carefully placed on the water surface. If the needle is pushed below the surface it will immediately sink because of the greater density of the steel needle compared with water.

An insect such as a water strider can walk on water balanced by the surface tension. Water tends to form spherical droplets because a sphere has the lowest surface area for a given volume, thereby maximizing the inner attractive forces.

Table 12.5 summarizes the relationship of intermolecular forces to the physical properties of liquids.

TABLE 12.5

RELATIONSHIP BETWEEN INTERMOLECULAR FORCES AND PHYSICAL PROPERTIES OF LIQUIDS

	Rate of Evaporation	Vapor Pressure	Boiling Point	Heat of Vaporization	Surface Tension
Strong intermolecular forces	Low	Low	High	High	High
Weak intermolecular forces	High	High	Low	Low	Low

CRYSTALLINE VERSUS AMORPHOUS SOLIDS

12.8   As you know, all solids have a definite volume and shape. However, we can classify solids into two distinct categories based on the arrangement of the particles of which the solid is composed. Some solids, known as **crystalline solids,** are made up of a highly regular repeating pattern of particles, whereas others, known as **amorphous solids,** do not have any regular repeating pattern in their structure.

If you have ever looked at salt crystals carefully, you probably have noticed that they appear to be in the shape of little cubes with very smooth surfaces. The visible geometry in a crystalline solid arises from the arrangement of the particles (atoms, ions, molecules) within the solid. The particular three-dimensional arrangement of particles in a crystalline solid is known as the **crystal lattice.** Figure 12.13 shows a representation of the crystal lattice in sodium chloride.

The portion of the crystal lattice which when repeated in three dimensions can generate the entire lattice is called the **unit cell.** From a two-dimensional point of view, the unit cell can be thought of as the repeating pattern in a sheet of wallpaper. Figure 12.13*b* shows the unit cell in sodium chloride. Every crystalline solid has a definite crystal lattice and a definite melting point.

Amorphous solids have their particles packed in an irregular manner; thus they lack any regular overall shape or form. Examples of amorphous solids are glass; plastics, such as polystyrene; asphalt; tar; and rubber. Amorphous solids can be distinguished from crystalline solids experimentally by the fact that they do not have a definite melting point. When you heat an amorphous solid, it simply softens over a wide range in temperature.

FIGURE 12.13

(*a*) The crystal lattice of sodium chloride. The colored cube outlines the unit cell. (*b*) The unit cell of NaCl. This pattern repeated in three dimensions gives the crystal lattice.

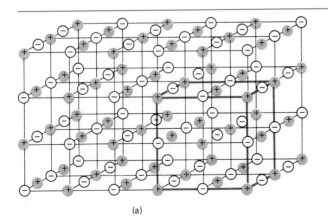

(a)

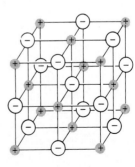

(b)

CLASSES OF CRYSTALLINE SOLIDS

12.9 Crystalline solids are divided into classes depending on the nature of the particles that make up the crystal lattice. The particles can be ions, atoms, or molecules.

Ionic crystals A crystalline solid in which the lattice is made up of oppositely charged ions is said to be an **ionic crystal** (Figure 12.13). All ionic compounds fit into this class (review Section 6.9). Ionic crystals melt at high temperatures because of the very strong electrostatic forces which exist between oppositely charged particles. Also, the very strong electrostatic forces make it difficult for the individual ions to slide past one another. Therefore, ionic solids are very hard.

Molecular crystals Crystalline solids in which molecules occupy the lattice points are known as **molecular crystals** (Figure 12.14). Examples of molecular crystals include ice, in which an H_2O molecule occupies each lattice point; "dry ice" (solid carbon dioxide), in which a CO_2 molecule is at each lattice point; and solid hydrogen, which has an H_2 molecule at each lattice point. The attractive forces in these solids are the intermolecular forces discussed in Section 12.4. Even the strongest intermolecular forces are considerably weaker than the electrostatic attraction which is the ionic bond. Because of the weaker attractive forces between molecules, molecular solids possess lower

FIGURE 12.14

Ice, a molecular crystal. A water molecule occupies each lattice point. Hydrogen bonds (———) hold the molecules in place.

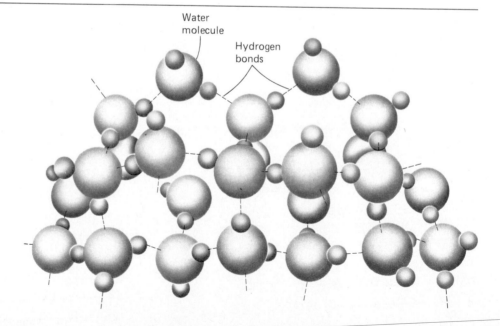

Water molecule

Hydrogen bonds

melting points than ionic solids. However, solids such as ice have higher melting points than hydrogen, because the hydrogen bonding forces present in ice are much stronger than the weak dispersion forces in hydrogen. Because of the weak attractions between molecules which enable one molecule to slide past another, molecular crystals tend to be soft.

Covalent crystals   Crystalline solids in which the lattice points are occupied by atoms covalently bonded to one another are called **co-valent crystals** (Figure 12.15). Each atom in a covalent crystal is covalently bonded to at least two other atoms. The entire crystal is actually one large molecule. Because of the strength of covalent bonds, covalent crystals possess very high melting points and are very hard. Examples of covalent crystals include diamond, quartz (silicon dioxide), and carborundum.

Metallic crystals   Crystalline solids in which the lattice points are occupied by metallic cations surrounded by moving electrons are called **metallic crystals** (Figure 12.16). This picture of a metallic crystal arises

FIGURE 12.15

Sand (SiO_2), a covalent crystal. Every silicon atom (small) is linked to four oxygen atoms arranged in a tetrahedral array. Atoms occupy the lattice points. Covalent bonds hold the atoms in place. (*Redrawn from Figure 9.8 in William L. Masterson and Emil J. Slowinski, Chemical Principles, 1977, W. B. Saunders Company, Philadelphia, by permission of both the authors and the publisher.*)

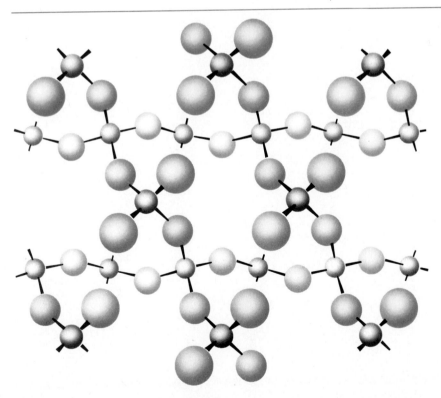

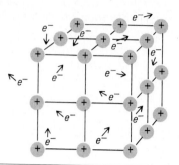

FIGURE 12.16

A metallic crystal. Metal cations occupy the lattice points. The valence electrons move freely in an "electron sea."

from the fact that metal atoms lose one or more of their valence electrons, and the electrons are then "free" to move through the entire lattice. The attractive forces in these crystals are the attractions between the positive metal ions and the mobile electrons. The strength of the attractive forces varies according to the size of the ion and the number of electrons ionized from each atom. Mercury's low melting point ($-38°C$) suggests that it has very weak forces, whereas tungsten (melting point 3400°C) clearly has very strong attractive forces. The

TABLE 12.6   A COMPARISON OF CRYSTALLINE SOLIDS

Type of Crystal	Example	Particles Occupying Lattice Sites	Attractive Forces	Melting Points	Water Solubility	Electrical Conductivity
Ionic	Sodium chloride, NaCl	Cations and anions	Ionic bond	High	Generally water-soluble	Conduct in molten state or in solution
Molecular	Ice, $H_2O(s)$	Molecules	Hydrogen bonds, dipole-dipole and dispersion forces	Low	Polar molecules are water soluble; nonpolar molecules dissolve in organic solvents	Nonconductors in pure state
Covalent	Diamond, C	Atoms	Covalent bonds	Very high	Not soluble	Usually nonconductors; a few are semiconductors
Metallic	Iron, Fe	Cations	Attraction between the cations and "free" valence electrons	Ranges from low to high	Not soluble	Good conductors in solid or molten state

excellent electrical conductivity of a metal can be accounted for by the freely moving electrons in the lattice of the metallic solid. Table 12.6 summarizes this section.

PROPERTIES OF SOLIDS

12.10 The temperature at which a crystalline solid is converted to a liquid is known as the **melting point.** The melting point of a crystalline solid is the same as the **freezing point** of the liquid.

$$\text{Solid} \underset{\text{freezing}}{\overset{\text{melting}}{\rightleftharpoons}} \text{liquid}$$

In contrast to boiling points, the melting points of most solids do not change greatly with a change in external pressure.

If we heat a solid that is initially at a temperature below its melting point, the temperature increases until it reaches the melting point. As the solid is heated, the particles in the lattice vibrate more rapidly about their fixed positions, though they remain in their same relative positions, as if they were "marching in place." At the melting point the vibrations become so rapid that the particles begin to move apart, and the lattice begins to break down. Though the temperature remains constant, heat must be added continuously to the system to feed the process whereby the lattice is broken down and the solid is converted to a liquid. Solids in which the forces between particles are stronger have higher melting points. More heat must be added to pull particles apart. In Table 12.6 you can see a reflection of the strength of bonds in the relative melting points of the classes of solids.

The amount of heat needed to convert 1 g of a solid to a liquid at the melting point is called the **heat of fusion.** The heat needed for the liquefaction of 1 mol of solid is known as the **molar heat of fusion** (Table 12.7). This is the amount of energy needed to break down the attractive forces in the solid at the melting point. Each gram of ice in an icepack will absorb 80 cal before melting to liquid water at 0°C.

During the reverse process of solidification, an amount of heat that is equal in magnitude to the heat of fusion must be removed or liberated. This quantity of heat liberated, which is exactly equivalent to the heat of fusion, may be called the **heat of solidification.** Vineyard owners protect their grapes by using water's high heat of solidification (80 cal/g). When a nighttime freeze is expected in the fall, a mist of water is sprayed over the grapevines. As the water freezes, 80 cal/g of heat is released and this prevents the temperature of the grapes from falling below 0°C.

Although the particles in a solid are more restricted in movement than those in a liquid, many solids have a measurable vapor pressure

TABLE 12.7 HEATS OF FUSION OF VARIOUS SUBSTANCES AT THEIR MELTING POINTS

Solid	Melting Point, °C	Heat of Fusion	
		cal/g	kcal/mole
Ice	0.0	80.	1.44
Ethyl alcohol	−117	24.9	1.15
Methane	−183	14	0.22
Benzene	6	30.1	2.65
Sodium chloride	804	124	7.19
Cu	1083	49	3.11
Ni	1453	74	4.34

and can evaporate directly from the solid to the vapor state without passing through the liquid state. This process is called **sublimation.**

$$\text{Solid} \xrightleftharpoons[\substack{\text{deposition} \\ \text{(condensation)}}]{\text{sublimation}} \text{vapor}$$

The odor of mothballs arises because of the vapor pressure of the solid white naphthalene. Naphthalene molecules are able to escape into the gaseous state and enter our nostrils. The purplish vapor in a closed bottle of solid iodine crystals presents direct evidence for the sublimation of iodine. The slow disappearance of snow in the winter, although the temperature remains below the melting point of ice, visibly demonstrates the sublimation of ice.

Crystalline solids that are molecular crystals provide the most likely candidates for sublimation because they generally contain the weakest attractive forces between their particles.

HEAT CHANGES AND PHASE CHANGES

12.11 When a solid is heated, its temperature rises until it reaches the melting point, and the temperature remains constant until the solid is totally converted to a liquid. If heating continues, the temperature of the liquid increases until it reaches the boiling point, and then the temperature remains constant until the liquid is totally converted to a gas. If the gas continues to be heated, its temperature rises. Figure 12.17 illustrates the changes in temperature of a pure substance as it is heated, beginning with a solid and continuing to the gaseous state as has just been described. If we are given the specific heat (review

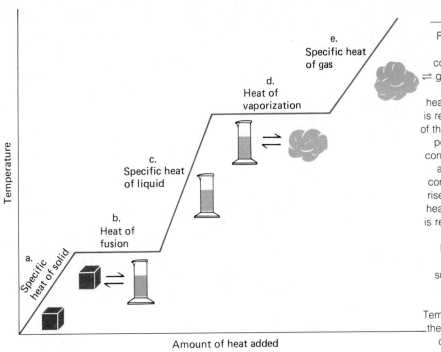

FIGURE 12.17

Plot of temperature versus added heat for the conversions: solid $\rightleftharpoons$ liquid $\rightleftharpoons$ gas. (*a*) Temperature rises steadily as the solid is heated. The slope of the line is related to the specific heat of the solid. (*b*) At the melting point, temperature remains constant until enough heat is absorbed to melt the solid completely. (*c*) Temperature rises steadily as the liquid is heated. The slope of the line is related to the specific heat of the liquid. (*d*) At the boiling point, temperature remains constant until a sufficient amount of heat is absorbed to vaporize the liquid completely. (*e*) Temperature rises steadily as the gas is heated. The slope of the line is related to the specific heat of the gas.

Section 3.6) and the heats of fusion and vaporization for a particular substance, we can calculate the amount of heat required for a fixed mass of that substance to undergo any given temperature change.

SAMPLE EXERCISE

12.5

Calculate the amount of heat required to convert 25.0 g of ice initially at $-15.0°C$ to steam at 125°C.
Given:

Specific heat ice = 0.500 cal/g°C

Specific heat water = 1.00 cal/g°C

Specific heat steam = 0.480 cal/g°C

Also see Tables 12.4 and 12.7.

Step 1 Calculate the amount of heat required to raise the temperature of ice from $-15.0°C$ to its melting point of 0°C (review Section 3.6).

   Heat = mass × temperature change × specific heat

   Heat = 25.0 g × (0°C − (−15.0°C)) × 0.500 $\dfrac{cal}{g × °C}$ = 188 cal

Step 2   Use the heat of fusion of ice to calculate the amount of heat necessary to melt 25 g of ice.

$$25.0 \ \cancel{g} \times 80.0 \ \frac{cal}{\cancel{g}} = 2000 \ cal$$

Step 3   In a manner similar to Step 1, calculate the heat required to raise the temperature of water from 0°C to its boiling point (100°C).

$$\text{Heat} = 25.0 \ \cancel{g} \times (100°\cancel{C} - 0°\cancel{C}) \times 1.00 \ \frac{cal}{\cancel{g} \times °\cancel{C}} = 2{,}500 \ cal$$

Step 4   Use the heat of vaporization of water to calculate the amount of heat necessary to vaporize 25.0 g of water.

$$25.0 \ \cancel{g} \times \frac{540 \ cal}{\cancel{g}} = 13{,}500 \ cal$$

Step 5   As in Steps 1 and 3, calculate the heat required for the indicated temperature change.

$$\text{Heat} = 25.0 \ \cancel{g} \times (125°\cancel{C} - 100°\cancel{C}) \times 0.480 \ \frac{cal}{\cancel{g} \times °\cancel{C}} = 300 \ cal$$

The total heat input for Steps 1 through 5 is:

Step 1	188 cal	Heating ice
Step 2	2,000 cal	Ice → water
Step 3	2,500 cal	Heating water
Step 4	13,500 cal	Water → steam
Step 5	300 cal	Heating steam
	18,488 cal	Total

Notice that the largest portion of the added heat is necessary to convert the liquid water to steam at the boiling point.

SUMMARY

Intermolecular forces play a very important role in determining the properties of liquids and solids (Section 12.1). One origin of attractive intermolecular forces is molecular polarity, which occurs when the molecular shape does not lead to a canceling out of bond dipoles (Section 12.3). The shape of a molecule can be predicted using the VSEPR theory, which is based on the idea that electron pairs prefer to keep as far apart as possible (Section 12.2).

    The three types of intermolecular forces are dispersion, or London, forces, which are the weakest; dipole-dipole interactions; and hydrogen bonds, which are the strongest. All molecules experience dispersion forces. Dipole-dipole interactions occur only for polar molecules. Hydrogen bonding requires an O—H, N—H, or H—F bond (Section 12.4).

Intermolecular attractions cause the condensation of gases and the solidification of liquids (Section 12.5). The shape and volume characteristics of liquids and solids can be explained on the molecular level (Section 12.6).

Among the more important physical properties of liquids are vapor pressure, boiling point, heat of vaporization, and surface tension. The magnitudes of these properties for various liquids are related to the strength of the intermolecular forces in the liquid (Section 12.7).

All solids have either a crystalline (regular) or amorphous structure (Section 12.8). Crystalline solids are divided into classes based on the component particles in their crystal lattices. The classes are ionic, molecular, covalent, and metallic (Section 12.9). The physical properties of solids such as melting point, heats of fusion, and vapor pressure depend on the forces holding together the particles in the crystal lattice (Section 12.10).

From a knowledge of specific heat, heat of vaporization, and heat of fusion, one can calculate the amount of heat energy required or released as the temperature of any pure substance is changed from one value to another (Section 12.11).

CHAPTER ACCOMPLISHMENTS

After completing this chapter you should be able to

12.1   Introduction

1.   Compare the strengths of the intermolecular forces in liquids and solids with the strength in gases.

12.2   Molecular shape

2.   State the essential idea of VSEPR theory.
3.   State the geometries of maximum separation for two, three, or four sets of electron pairs.
4.   Given the Lewis structure of a molecule, predict the electron-pair geometry.
5.   Given the Lewis structure of a molecule, predict the molecular shape.

12.3   Molecular polarity

6.   Given a Lewis structure of a molecule and a table of electronegativities, predict whether a molecule will be polar or nonpolar.
7.   Describe an experimental test of molecular polarity.

12.4   Intermolecular forces

8.   Describe and distinguish among the three intermolecular forces.
9.   State the relationship between the strength of the dipole-dipole force and the molecular polarity in a molecule.
10.   State the relationship between the strength of the dispersion forces and the molecular weight and the shape of a molecule.
11.   Given the Lewis structure of a molecule, predict the existence of hydrogen bonding, dipole-dipole, and dispersion intermolecular forces.

12.5   Condensation of gases

12.   State the relationship between the boiling point of a liquid and the strength of the intermolecular forces within the liquid.

12.6   A model of liquids and solids

13.   Explain the observable distinguishing properties of liquids and solids by using the molecular model of liquids and solids.

12.7   Physical properties of liquids

14.   Describe the process by which a constant vapor pressure for a liquid is attained in a closed container.
15.   State the relationship between the vapor pressure of a liquid and the strength of the intermolecular forces in the liquid.
16.   State the relationship between the temperature of a liquid and its vapor pressure.
17.   Define boiling point.
18.   Describe the relationship between the boiling point of a liquid and the external pressure.
19.   Distinguish between evaporation and boiling in a liquid.
20.   State the relationship between the boiling point of a liquid and the strength of the intermolecular forces in the liquid.
21.   Define heat of vaporization of a liquid, and state the units in which it is commonly expressed.
22.   State the relationship between the molar heat of vaporization and the strength of intermolecular forces within a liquid.
23.   Given the heat of vaporization (condensation), calculate the amount of heat needed (released) to vaporize (condense) a given amount of liquid (gas).
24.   Explain how the phenomenon of surface tension arises.

12.8   Crystalline versus amorphous solids

25.   Distinguish between crystalline and amorphous solids.

12.9   Classes of crystalline solids

26.   Name the four classes of crystalline solids.
27.   Distinguish among the four classes of crystalline solids with re-

spect to the nature of the particles occupying their lattice points and the nature of the attractive forces between these particles.

28. State the relationship between the melting point of a solid and the strength of the interparticle attractions within the solid.

12.10 Properties of solids

29. Define the heat of fusion of a solid and state the units in which it is commonly expressed.
30. Given the heat of fusion (solidification), calculate the amount of heat needed (released) to melt (solidify) a given amount of solid (liquid).
31. Define sublimation and give an example of a material that exhibits this phenomenon.

12.11 Heat changes and phase changes

32. Given appropriate specific heats and the heats of fusion and vaporization, calculate the amount of heat needed for a specified mass of a substance to undergo a given temperature change.

PROBLEMS

12.1 Introduction

12.6 **a.** In which physical state are the attractive forces the weakest?
b. Could an ideal gas be condensed to a liquid?

12.2 Molecular shape

12.7 What is the electron-pair geometry of the following compounds?
   a. CF_4 **c.** PH_3
   b. BCl_3 **d.** H_2Se

12.8 What is the molecular shape of each of the compounds in Problem 12.7?

12.9 **a.** Why does moving electron pairs further apart lower their energy?
b. In the case of a linear arrangement of electron pairs, what prevents their moving apart to an angle of 200°?

12.10 What is the electron-pair geometry and molecular shape of each of the following compounds?
   a. H_2CO **c.** CH_4O
   b. SO_3 **d.** OF_2

12.3 Molecular polarity

12.11 **a.** Write out the Lewis structures of CH_2Cl_2 and CCl_4.
b. Compare the molecular polarities of CH_2Cl_2 and CCl_4.

12.12 List HF, HCl, HBr, and HI in order of increasing molecular polarity.

12.13 **a.** Draw Lewis structures for $BeCl_2$ (assume covalent bonding), H_2S, and H_2O.
b. Which of the molecules in part (**a**) probably has the smallest molecular polarity? Explain.
c. Which of these molecules has the largest molecular polarity?

12.14 **a.** Draw a Lewis structure for NH_3.
b. Mark the bond dipoles in NH_3.
c. Mark the molecular dipole in NH_3.

12.15 Describe how the molecular polarity of a molecule can be measured experimentally.

12.16 On the basis of molecular shape only, which of the compounds in Problem 12.8 have a net zero dipole moment?

12.17 Does any compound in Problem 12.10 have a zero dipole moment? If so, which one?

12.18 Pick out the polar molecules from the following list:

a. $CHCl_3$

e. (a structure showing H and H attached to C=Ö:)

b. CH_4

c. O_2

d. ClF

f. SO_2

g. CO

h. PCl_3

12.19 Which compound do you predict to have the larger dipole moment, NH_3 or PH_3? Explain your answer.

12.4 Intermolecular forces

12.20 a. Explain the difference between the terms intermolecular and intramolecular force.
b. Give one example of each type of force.

12.21 a. State the three intermolecular attractive forces.
b. Describe the molecular conditions necessary for the existence of each of these forces in molecules.

12.22 State the intermolecular attractive forces present in samples of the following substances:

a. F_2

b. H_2O

c. CH_2F_2

d. HF

12.23 For each example in Problem 12.22 determine which type of attractive force will be *most* significant in determining the physical properties (such as boiling point) of the sample.

12.24 Determine the intermolecular attractive forces present in:

a. NH_3 b. HCN c. Ne

12.25 For each substance in Problem 12.24, determine which type of attractive force will be *most* significant in determining the physical properties (such as boiling point) of the substance.

12.26 a. In which types of molecules do we find dispersion forces?
b. What are the two factors that determine the strength of dispersion forces?

12.27 Explain why the boiling point of the inert gases decreases as we go up the periodic group.

12.28 Which of the following molecules are capable of hydrogen bonding between themselves?

a. H_3C-O-H

b. (structure: $H-C-Cl$ with Cl above and Cl below)

c. NF_3

d. H_2S

e. $H-N-Cl$ (with Cl below N)

f. HI

12.29 Which of these substances have dipole-dipole attractive forces between their molecules?

a. CF_4

b. HCl

c. $S=C=S$

d. $H-C\equiv N$

e. BF_3

f. PF_3

g. Cl_2

h. SO_2

12.30 Compounds A and B have the same molecular weight, but the boiling point of A (80°C) is considerably higher than B (-24°C). Explain why this is so.

(A)   (B)

12.31 Would you predict C_5H_{12} or $C_{10}H_{22}$ to have the stronger intermolecular forces? Explain.

12.5 Condensation of gases

12.32 Explain how increasing the pressure on a gas allows it to be liquefied at a higher temperature than it could be if the gas were at standard atmospheric pressure.

12.33 Explain why an increase in intermolecular attractions leads to an increase in the boiling point of a liquid.

12.34 Predict the order of increasing boiling point for the halogens, F_2, Cl_2, Br_2, and I_2.

12.35 Arrange the following molecules in order of increasing boiling point:

CH_3F, CH_4, He, CH_2F_2,

12.36 Which substance do you predict to be higher boiling, H_2O or NaCl? State your reasons.

12.6 A model of liquids and solids

12.37 Use the models of liquids and solids developed in Figure 12.8 to explain why liquids but not solids can flow.

12.38 Using the model of a liquid developed in Figure 12.8 and the model of a gas developed in Section 11.12, explain why diffusion is much more rapid in a gas than in a liquid.

12.39 Explain how the model of liquids and solids in Figure 12.8 accounts for the low compressibility of these physical states compared to gases.

12.7 Physical properties of liquids

12.40 Why do liquids with low intermolecular attractive forces have high vapor pressures?

12.41 **a.** Why does the kinetic energy of an isolated liquid decrease as liquid evaporates into the gas phase?
b. Will the remaining liquid evaporate as fast as the initial liquid?

12.42 Why is the process by which perspiration evaporates from your skin a cooling process?

12.43 Indicate which substance in each pair you would expect to have the higher vapor pressure at the same temperature:

   a. CH_4 *or* H_2O

   b.

$$H-\overset{\overset{\displaystyle H}{|}}{\underset{\underset{\displaystyle H}{|}}{C}}-OH \quad or \quad \overset{\displaystyle H}{\underset{\displaystyle H}{\diagdown}}C=O$$

   c.

$$H-\overset{\overset{\displaystyle H}{|}}{\underset{\underset{\displaystyle H}{|}}{C}}-\overset{\overset{\displaystyle H}{|}}{\underset{\underset{\displaystyle H}{|}}{C}}-\overset{\overset{\displaystyle H}{|}}{\underset{\underset{\displaystyle H}{|}}{C}}-H \quad or \quad H-\overset{\overset{\displaystyle H}{|}}{\underset{\underset{\displaystyle H}{|}}{C}}-\overset{\overset{\displaystyle H}{|}}{\underset{\underset{\displaystyle O}{||}}{C}}-H$$

12.44 **a.** Explain why increasing the surface area *does* increase the rate of evaporation of a liquid.
b. Explain why increasing the surface area *does not* increase the vapor pressure of a liquid.

12.45 Why does increasing the temperature of a liquid increase its vapor pressure?

12.46 At 20°C the vapor pressure of substance X is 700 mmHg and that of substance Y is 335 mmHg.
   a. Which substance has the higher vapor pressure?
   b. Which substance will have the lower normal boiling point?
   c. Which substance has the stronger intermolecular forces?

12.47 Describe at least two differences between the processes of evaporation and boiling.

12.48 The vapor pressure of a substance at 25°C is 835 mmHg. If the atmospheric pressure is 758 mmHg, will the substance be a liquid or gas at 25°C?

12.49 **a.** What is meant by the statement that water has a normal boiling point of 100°C?
   b. Can water be made to boil at temperatures other than 100°C? Explain.

12.50 A newly discovered liquid is very sensitive to heat and in fact decomposes at temperatures above 75°C before it can be converted to a gas. Describe how you would distill this liquid without decomposition.

12.51 **a.** What name is given to the quantity of heat required to convert a mole of liquid to a gas at its boiling point?
   b. A liquid and gas can coexist at the boiling point. Which state has the greater amount of energy? Where does the greater amount of energy come from?
   c. Which physical state has the stronger attractive forces at the boiling point?
   d. On a molecular level what is the function of the molar heat of vaporization?

12.52 Liquid A has a molar heat of vaporization of 17.3 kcal/mole. Liquid B has a molar heat of vaporization of 87 kcal/mole.
   a. Which liquid has the larger intermolecular forces?
   b. Which liquid would you predict to have the lower vapor pressure at 25°C?
   c. Which liquid would you predict to have the lower boiling point?

12.53 Using Table 12.4, calculate how much heat is required to vaporize 31.1 g of H_2O at 100°C.

12.54 Using Table 12.4, calculate how much heat is liberated when 3.4 moles of steam are condensed to liquid water at 100°C.

12.55 Predict what effect an increased temperature would have on the surface tension of a liquid.

12.56 Igniting a warmed dish containing alcohol gives a much more vigorous display of flames than igniting the cold dish. Explain.

12.57 Cheese boards typically have covering domes. Why should cheese not be left uncovered?

12.58 At 760 mmHg and 100°C large bubbles begin to form within a sample of water. What is contained in the bubbles?

12.59 Typical atmospheric pressure in Mexico City is 580 mmHg.
    a. Use Figure 12.11 to determine the typical boiling point of water in Mexico City.
    b. Will an egg take a longer or shorter period of time to reach the state of hard boiled in Mexico City as compared to at sea level?

12.8 Crystalline versus amorphous solids

12.60 Describe two physical properties which distinguish crystalline solids from amorphous solids.

12.61 Given an unknown white powdery solid, how might you determine whether it is crystalline or amorphous?

12.62 A purplish solid with shiny flat surfaces is given to you. You determine that its melting point is 33 to 34°C. Is this solid likely to be crystalline or amorphous?

12.9 Classes of crystalline solids

12.63 Indicate the species that occupy the lattice points for each of the following types of crystalline solids:
    a. Ionic
    b. Molecular
    c. Covalent
    d. Metallic

12.64   **a.** Give specific examples for each type of crystalline solid.
    b. Indicate the particles that would occupy the lattice points for each of your examples.

12.65   **a.** Indicate the attractive forces involved in each of the four types of crystalline solids.
    b. Which type of crystalline solid generally has the weakest attractive forces?

12.66 Ionic solids and metallic solids are made up of charged particles. However, metallic solids conduct electricity, whereas ionic solids do not conduct electricity in the solid state. Explain.

12.67 Explain why the melting point of sodium chloride is so much higher than that of pure hydrogen iodide (mp −51°C), a compound of higher molecular weight than NaCl.

12.68 Diamond and graphite are both forms of carbon. Diamond has a covalent crystal structure whereas graphite has a molecular crystal arrangement. Use this fact to explain why diamond is very hard whereas graphite is soft.

12.10 Properties of solids

12.69   **a.** What name is given to the quantity of heat required to convert a mole of solid to a liquid at its melting point?
    b. A solid and liquid can coexist at the melting point. Which state has the greater amount of energy? Where does the greater amount of energy come from?
    c. Which physical state has the larger attractive forces at the melting point?
    d. On a molecular basis what is the function of the molar heat of fusion?

12.70 Using Table 12.7, calculate the amount of heat liberated by the solidification of 41.1 g of water at 0.0°C?

12.71 The heat of fusion for a substance is much less than the heat of vaporization. From a knowledge of the strengths of intermolecular forces in each state, explain this statement.

12.72 The heat required to convert a solid directly to a gas is called the heat of sublimation. Would you predict the heat of sublimation to be larger or smaller than the heat of vaporization?

12.73 When we detect the odor of a solid, what must we actually be smelling?

12.11 Heat changes and phase changes

12.74 How many calories are required to convert 46.0 g of solid ethyl alcohol at $-153°C$ to vapor at 79°C? The melting point of the alcohol is $-117°C$ and the boiling point is 79°C. The specific heat of the liquid alcohol is 0.535 cal/(g·°C) and for the solid alcohol 0.232 cal/(g·°C). See Tables 12.4 and 12.7 for heats of vaporization and fusion.

12.75 Use the data in this chapter to calculate the number of kilocalories of heat required to convert 1.50 moles of ice initially at $-6.0°C$ to vapor at 110.°C.

12.76 A sponge bath using isopropyl alcohol (rubbing alcohol) (mol. wt. 60.1) can be used to lower the temperature of a person suffering from a high fever. Evaporation of the alcohol consumes heat which comes from the patient. The molar heat of vaporization for isopropyl alcohol is 10.1 kcal/mole (assume that it is the same at room temperature as at the boiling point). How much heat is removed in the evaporation of 150. g of the alcohol?

SOLUTIONS
CHAPTER

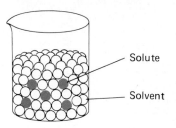

13

Al

Aluminum

26.98

SOLUTIONS DEFINED

13.1 Up until now we have been discussing *pure substances,* one of the major categories of matter. We will now discuss solutions, probably the most important *mixtures,* the other major category of matter. **Solutions** are *homogeneous* mixtures of two or more substances. In order for the mixture to be truly homogeneous, the particles of the intermixed substances must be ionic or molecular in size (Section 13.3).

In describing solutions we identify and give different names to the substance dissolved and the substance doing the dissolving. The substance dissolved is called the **solute.** The substance doing the dissolving is the **solvent.** In general, the solute is the component that is present in lesser amount in the solution. We can envision the solute particles as surrounded by the solvent particles.

$$\text{Solute} + \text{solvent} = \text{solution} \qquad (13.1)$$

Because there are three physical states of matter and each state in principle can be either the solute or solvent in a solution, there are conceivably nine types of solutions. Table 13.1 shows the nine types and gives a common example of each type that actually exists. There are no common examples of solids or liquids truly dissolved in a gas. The most common solutions are those in which the solvent is a liquid, and those are the solutions which will concern us in this chapter. In discussing mixtures of gases in Chapter 11, we dealt with the other commonly encountered solutions, gases dissolved in gases.

Now let us consider properties of solutions in which the solvent is a liquid.

1. As in all mixtures.
    a. The composition is variable (Section 13.5).
    b. The solute and solvent may be separated by physical means.

Solute

Solvent

TABLE 13.1　　　　　THE NINE SOLUTION TYPES BASED ON THE PHYSICAL STATE OF SOLUTE AND SOLVENT, WITH EXAMPLES

Solute	Solvent		
	Gas	Liquid*	Solid
Gas	$O_2(g)$ in $N_2(g)$ **Air**	$CO_2(g)$ in $H_2O(l)$ **Soda**	$H_2(g)$ in $Pd(s)$ **Hydrogenation catalyst**†
Liquid	No examples exist	Alcohol(l) in $H_2O(l)$ **Martini**	$Hg(l)$ in $Ag(s)$ **Dental fillings** (amalgam)
Solid	No examples exist	$NaCl(s)$ in $H_2O(l)$ **Salt water** (saline solution)	$Zn(s)$ in $Cu(s)$ **Brass**‡

* This section is placed in a box because the most common solutions are those in which the solvent is a liquid.
† This catalyst is used in hydrogenation reactions in organic chemistry (Chapter 18).
‡ All metal alloys like brass are solid-in-solid solutions.

2. Because solutions are homogeneous mixtures.
    a. The solute particles must be uniformly distributed among the solvent particles throughout the solution.
    b. The solute particles will not settle out.
    c. The solution has the same chemical and physical properties in every part.

3. A true solution is transparent, though it may be colored. Cola, white wine, and apple juice are examples of colored solutions (see Figure 13.1).

SOLUTION TERMINOLOGY

13.2　In addition to the terms "solute" and "solvent," there are several other words that are commonly used to describe solutions.

Solubility　**Solubility** is a statement of how much solute can dissolve in some given amount of solvent at a given temperature. A quantitative statement of solubility is a ratio

$$\frac{\text{Maximum amount solute}}{\text{Amount solvent}}$$

and usually has the units maximum grams solute per 100 g solvent (see Table 13.2). Qualitatively we speak of solutes as **soluble** (some unspecified amount dissolves) or **insoluble** (the solute does *not* dissolve). The modifiers, such as *slightly* or *very* soluble, loosely describe

FIGURE 13.1

Some commonly encountered solutions displaying the properties described in Section 13.1. (*Photography by Bryan Lees.*)

smaller or larger amounts of solute dissolving in a given amount of solvent.

Saturated Solutions   A **saturated solution** is one in which no more solute can dissolve in a given amount of solvent at a given temperature.

SOLUBILITY LIMITS OF SOME SATURATED SOLUTIONS   **TABLE 13.2**

Solute	Solubility, Maximum Grams Solute 100 g H_2O	
	20°	60°
NaCl	36.0	37.3
KBr	65.2	85.5
KNO_3	31.6	110.0
$AgC_2H_3O_2$	1.04	1.89
$K_2Cr_2O_7$	13.1	50.5
$KMnO_4$	6.4	22.2
$AgNO_3$	222	525
$BaSO_4$	0.00023	0.00036

Table 13.2 lists the solubility limits of some saturated solutions. To prepare a saturated solution of NaCl, one dissolves at least 36.0 g of NaCl in 100 g of water at 20°C. At 20°C, 36.0 g is the maximum amount of NaCl that will dissolve; any excess NaCl will settle to the bottom of the container (see Figure 13.2).

Although a saturated solution containing excess solid appears to the eye to be static (exhibiting lack of activity), the solute particles are actually involved in a dynamic (active) process. The solid solute is dissolving (going into) solution, and at the same time some solute particles are coming out of solution to join the pure solid. For a solution in which water is the solvent, we can represent this process as

$$\text{Undissolved solute } (s) \rightleftharpoons \text{dissolved solute } (aq)$$

where (s) indicates a solid and (aq) means an *aqueous* or water solution. The two half-headed arrows ($\rightleftharpoons$) indicate that the process is reversible and can proceed in either direction. Because the two opposing processes are occurring at equal rates, we say that there is a *dynamic equilibrium* (Section 14.1) between undissolved and dissolved solute. Usually one can actually see any undissolved solute in a saturated solution. Exactly at the solubility limit (for NaCl, 36.0 g NaCl/100 g H₂O at 20°C), the amount of undissolved solute is too small to be seen (Figure 13.2b).

Unsaturated solutions   If the amount of solute dissolved is *less than* the solubility limit, then the solution is **unsaturated.**

FIGURE 13.2

Unsaturated and saturated solutions of sodium chloride at 20°C: (a) 20.0 g is less than the maximum amount of NaCl that can dissolve in 100 g of H₂O at 20°C; (b) 36.0 g is exactly the solubility limit— a small speck of undissolved solid in equilibrium with the dissolved solute may or may not be visible; (c) in this solution 36.0 g of NaCl is in solution— the extra 9.0 g appear as undissolved solid on the bottom of the flask.

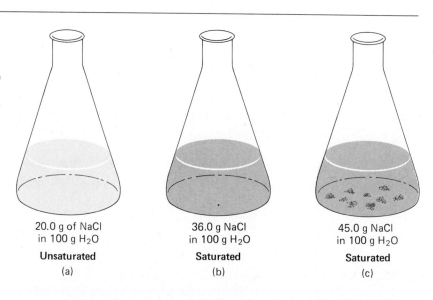

20.0 g of NaCl
in 100 g H₂O
Unsaturated
(a)

36.0 g NaCl
in 100 g H₂O
Saturated
(b)

45.0 g NaCl
in 100 g H₂O
Saturated
(c)

Problem 13.1

Consult Table 13.2 and decide whether the following solutions are saturated or unsaturated.

a. 15 g KBr in 100 g H_2O at 20°C

b. 15 g $KMnO_4$ in 100 g H_2O at 60°C

c. 115 g $AgNO_3$ in 100 g H_2O at 20°C

Use Table 13.2 to decide whether the following solutions are saturated or unsaturated:

a. 19 g of NaCl in 50 g H_2O at 20°C

b. 135 g of KNO_3 in 150 H_2O at 60°C

c. 1.5 g of $KMnO_4$ in 20 g H_2O at 20°C

SOLUTION

Because Table 13.2 gives solubility limits in grams solute per 100 g H_2O, we must calculate the number of grams solute per 100 g H_2O in the solutions described. This can be done by setting up the *given* ratio of given grams solute to given grams H_2O equal to the desired ratio of x g solute to 100 g H_2O.

a. Given   Desired

$$\frac{19 \text{ g NaCl}}{50 \text{ g } H_2O} = \frac{x \text{ g NaCl}}{100 \text{ g } H_2O}$$

$$50x = 1900$$

$$x = 38 \text{ g}$$

At 20°C, 38 g NaCl per 100 g H_2O would be a saturated solution because 38 is greater than 36.

b. Given   Desired

$$\frac{135 \text{ g } KNO_3}{150 \text{ g } H_2O} = \frac{x \text{ g } KNO_3}{100 \text{ g } H_2O}$$

$$150x = 13,500$$

$$x = 90 \text{ g}$$

At 60°C, 90 g KNO_3 per 100 g H_2O is an unsaturated solution.

c. Given   Desired

$$\frac{1.5 \text{ g KMnO}_4}{20 \text{ g H}_2\text{O}} = \frac{x \text{ g KMnO}_4}{100 \text{ g H}_2\text{O}}$$

$$20x = 150$$

$$x = 7.5 \text{ g}$$

At 20°C, 7.5 g KMnO$_4$ per 100 g H$_2$O is a saturated solution.

Dilute versus concentrated solutions   In comparing unsaturated solutions that have identical components, the terms dilute and concentrated can be used to relate the amounts of solute in a given amount of solvent qualitatively. A **dilute solution** has a comparatively small amount of solute; a **concentrated solution,** a relatively large amount. For example, 1 or 2 g of NaCl in 100 g of water would be a dilute solution, whereas 30 g of NaCl in 100 g of water is concentrated because 30 g is approaching the solubility limit.

Because different solutes have different solubility limits, it is not very useful to use the terms dilute and concentrated in comparing solutions which contain different solutes or solvents.

Problem 13.2

In comparison with a solution in which 3.0 g KBr are dissolved in 5.0 g of H$_2$O, is the solution in part a of Problem 13.1 dilute or concentrated?

Supersaturated solutions   It is possible to prepare a solution that contains a dissolved amount of solute that exceeds the normal solubility limit. Such a solution is described as **supersaturated.** In order to prepare and maintain a supersaturated solution, one must work carefully and must protect the solution from disturbances. For example, the normal solubility limit of sodium acetate is 46.5 g per 100 g H$_2$O at 20°C. If we mix 75 g of sodium acetate and 100 g H$_2$O and heat the mixture to 50°C, the solid will dissolve (solubility at 50°C is 83 g sodium acetate per 100 g H$_2$O). If we then slowly and *carefully* cool the solution to 20°C, the 75 g of solute will remain in solution and the solution is supersaturated. If a small "seed crystal" of sodium acetate is added to the supersaturated solution, or if the container is shaken, the solute in excess of the normal solubility limit will fall out of solution. The result will be a saturated solution (see Figure 13.3).

Miscible and immiscible   The terms "miscible" and "immiscible" are used in describing mixtures of liquids with liquids. **Miscible** liquids dissolve in each other in all proportions. For example, grain alcohol and water are totally miscible. They dissolve in one another not only on a 50 : 50 basis but even when one exceeds the other greatly. This

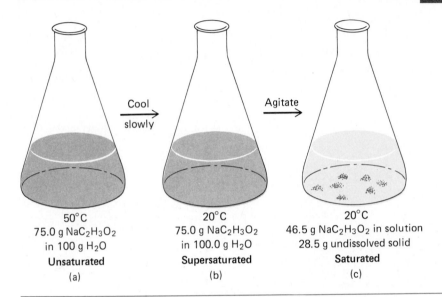

| 50°C | 20°C | 20°C |
| 75.0 g NaC₂H₃O₂ | 75.0 g NaC₂H₃O₂ | 46.5 g NaC₂H₃O₂ in solution |

$50°C$
$75.0 \text{ g } NaC_2H_3O_2$
in $100 \text{ g } H_2O$
Unsaturated
(a)

$20°C$
$75.0 \text{ g } NaC_2H_3O_2$
in $100.0 \text{ g } H_2O$
Supersaturated
(b)

$20°C$
$46.5 \text{ g } NaC_2H_3O_2$ in solution
28.5 g undissolved solid
Saturated
(c)

FIGURE 13.3

Preparation of a supersaturated solution. (*a*) At 50°C, 75 g of sodium acetate dissolves to form an unsaturated solution. (Solubility $=$ 83 g $NaC_2H_3O_2/$ 100 g H_2O at 50°C.) (*b*) If this solution is carefully cooled to 20°C, the solution is supersaturated because an amount of solute exceeding the solubility limit at 20°C is in solution. (Solubility $=$ 46.5 g $NaC_2H_3O_2/100$ g H_2O at 20°C.) (*c*) If solution (*b*) is disturbed, the excess solid over and above the solubility limit falls out and the result is a saturated solution.

is what is meant by "in all proportions." In contrast, oil and water are totally **immiscible;** i.e., they are completely insoluble in one another regardless of the proportions involved.

SOLUTION FORMATION

13.3   You can visualize how a typical ionic compound like NaCl dissolves in water by remembering several ideas that have already been encountered.

1.   Ionic compounds are constructed of ions that are held together in orderly arrays in the solid state.

2.   Water molecules are freely moving in the liquid state.

3.   Water molecules are dipoles with the oxygen end negative and the hydrogen end positive.

4.   Cation-anion (positive-negative) attraction is a low-energy (stable) condition.

Figure 13.4*a* represents solid NaCl just at the moment it has been added to a beaker of water. The positive cations and negative anions are held together by electrostatic attraction.

At the surface of the solid the moving water molecules can collide with the ions; these collisions can gradually "chip" away the ions from the crystal. The collisions are particularly effective when the negatively

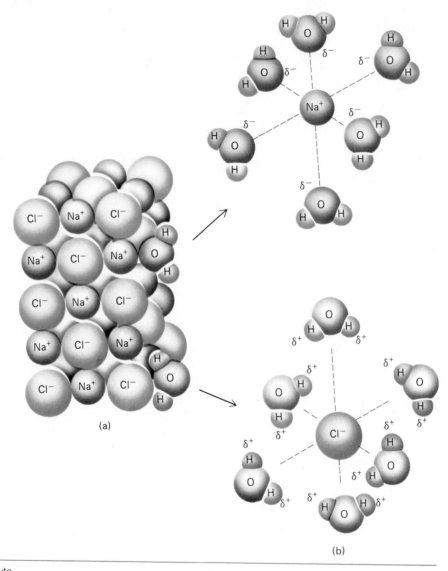

FIGURE 13.4

Dissolving of an ionic solute such as NaCl in water. (*a*) Solid NaCl is a collection of ions tightly bound together. Water molecules collide with the solid when it is put into water. (*b*) Water molecules have succeeded in "chipping away" ions from the solid and have surrounded the ions.

charged end of the water dipole collides with a positively charged ion

$$\delta^+ \text{(H)} \quad \delta^- $$
$$\text{O} \text{----} \text{(Na}^+\text{)}$$
$$\delta^+ \text{(H)}$$

or the positive end collides with a negative ion.

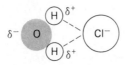

In these cases the water molecules exert an ion-dipole attraction which helps to pull apart the array of ions.

When the ions are separated from the crystal, the water molecules surround them (see Figure 13.4*b*). Notice that the positively charged ends of the water dipoles point toward Cl^- and the negatively charged ends of the water dipoles point toward the Na^+. This process is known as **hydration** when the solvent is water. Hydration energy is the energy that is released through ion-dipole attractions. **Solvation** is the general name of the process by which solvent particles surround solute particles for any solvent.

The dissolving and hydration process can be represented by a chemical equation of the form

$$NaCl(s) \rightleftharpoons Na^+(aq) + Cl^-(aq)$$

The double-headed arrow is used because the process is reversible (Figure 13.5). As the solution becomes more concentrated, the attraction between the ions overcomes the ion-dipole attraction because there are not enough water molecules for efficient hydration, so some solid comes out of solution.

When an ionic compound dissolves in water, it does so by the process just described. However, different ionic compounds dissolve to different extents. Table 13.2 shows quite large differences in solubility, ranging from the very soluble silver nitrate to the only slightly soluble silver acetate to the practically insoluble barium sulfate. When

FIGURE 13.5

The dissolving process is reversible: NaCl (*s*) $\rightleftharpoons$ $Na^+(aq)$ + $Cl^-(aq)$. (*a*) When solid NaCl is first placed in water, only separation of the ions occurs. (*b*) When some ions begin to accumulate in solution, they begin to clump together again into the solid. (*c*) When the rates of the two processes are equal, there is no net change in the number of ions in solution, and the solution is saturated.

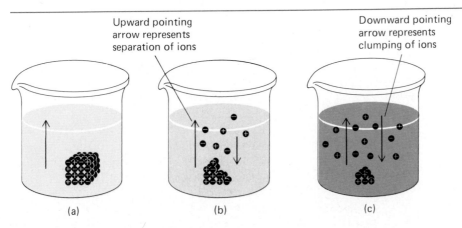

Upward pointing arrow represents separation of ions

Downward pointing arrow represents clumping of ions

(a)   (b)   (c)

ionic compounds dissolve only to the very small extent that $BaSO_4$ does (about 2.3×10^{-4} g per 100 g of H_2O), the compound for all practical purposes is classified as insoluble. That is, in the representation of the reversible dissolving process, $BaSO_4(s) \rightleftharpoons Ba^{2+}(aq) + SO_4^{2-}(aq)$, there is almost no tendency for the ions to separate; the process from left to right ($\rightarrow$) essentially does not occur. Similarly, since the solubility of ZnS in water at 25°C is only about 3×10^{-11} g per 100 g of H_2O, we find that ZnS is classified as insoluble according to the solubility rules given later in Table 13.3; $BaSO_4$ is also classified as insoluble in that table.

One factor that determines just how soluble an ionic compound will be is the magnitude of its hydration energy compared with the energy associated with the attraction between its ions in the crystalline solid (see Figure 13.6). In many ionic solids, the ions are held together too tightly in the crystal for water molecules to "entice" them away through ion-dipole attractions.

It is useful to be able to classify given ionic compounds as soluble or insoluble. Table 13.3 in Section 13.10 summarizes the solubility behavior of many commonly encountered compounds.

FACTORS INFLUENCING SOLUBILITY

13.4   There are several factors which influence the extent to which a given solute dissolves in a given solvent.

Nature of solute and solvent   We have examined the solution process for an ionic compound in water. This process depends on the ionic nature of the solute and the dipole nature of the solvent, which together make interaction and intermingling of particles possible.

Solvation in general involves the solvent molecules "sneaking in between" the solute particles. For this reason, substances which have similar intermolecular forces tend to dissolve in one another. Substances with dissimilar intermolecular forces do not tend to form solutions. For example, water and grain alcohol are completely miscible because of the similarity of their intermolecular forces. Both are polar and both form hydrogen bonds. Figure 13.7 shows how they are able to mix intimately.

Gasoline, which can be represented as C_8H_{18}, will not dissolve in either water or alcohol because gasoline is nonpolar. It will, however, dissolve in pentane, C_5H_{12}, or hexane, C_6H_{14}, both of which, like gasoline, display weak dispersion forces (Section 12.4) as their principal intermolecular forces.

This generalization about the way in which similarity of inter-

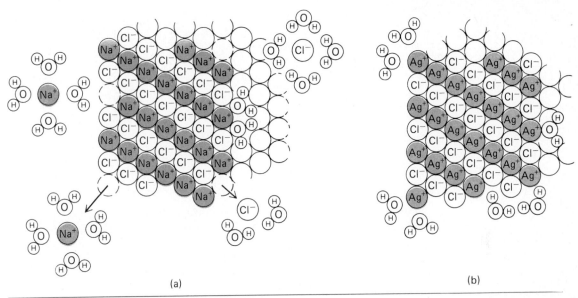

(a)   (b)

FIGURE 13.6

The solubility of ionic compounds in water depends on the relative magnitudes of the attractive forces between ions and hydration energy. (a) In the case of NaCl the water molecules are able to overcome the attractions between ions and "lure" Na^+ and Cl^- into solution. (b) The water molecules find it very difficult to pull Ag^+ and Cl^- away from the crystal because the attractions between the ions are stronger than in NaCl.

molecular forces in solute and solvent leads to solubility is often referred to as the principle of "like dissolves like." Though the principle provides a somewhat useful guideline, the solution process does not depend on relative polarities alone. It is fortunate that it is easy to test for solubility in the laboratory.

Temperature   Temperature affects both the speed and the extent of the solution process. Molecules move more quickly with higher kinetic energies at higher temperatures; hence the intermingling process of solution formation is speeded up. Sugar dissolves more quickly in hot tea than in iced tea.

In Chapter 14 we will see that temperature has a great effect on equilibrium processes (Section 14.6). In the solution process, temperature exerts its effect on the equilibrium between undissolved and dissolved solute. For solid solutes in liquid solvents, the effect is usually such that the solubility increases as temperature increases (Figure 13.8). Not only does sugar dissolve faster in hot tea, but also more sugar dissolves in a given amount of hot tea than in the same amount of iced tea. For gaseous solutes in liquid solvents, increasing temperature decreases solubility. Cold soda and cold beer maintain their fizz (CO_2 gaseous solute) more readily than do warm soda and warm beer.

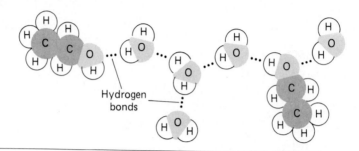

FIGURE 13.7

Hydrogen bonding between
     O   and
H   H
   Water
            O   H
     H   C   H
        H   C   molecules.
           H   H
          Ethyl
          alcohol
Compare this figure to Figure
12.6, which shows hydrogen
bonding in pure water. The
carbon chain in ethyl alcohol
is easily accommodated
within the hydrogen-bonding
network.

Pressure Pressure has almost no effect on the solubility of solids in liquids. The solubility of a gas in a liquid, on the other hand, is directly proportional to the partial pressure of the gas above the solution. "Carbonated" beverages are bottled in such a way that the partial pressure of CO_2 above the solution is greater than 1 atm. This increases the solubility of CO_2 in H_2O (see Figure 13.9). When the bottle is opened, CO_2 gas escapes, thus lowering the partial pressure of CO_2 above the solution. Hence the solubility of CO_2 in the soda is reduced.

The external pressure on the body affects the solubility of gases

FIGURE 13.8

Solubility of various
compounds in water. Notice
that the solubility of all solids
except for Li_2SO_4 increases
as temperature increases.
The solubility of the gas HCl
decreases as temperature
increases.

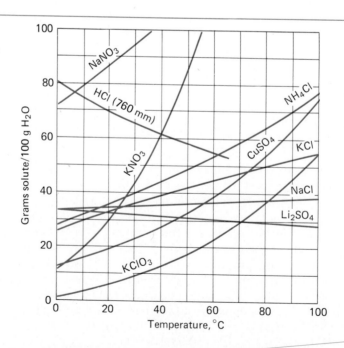

in the blood. Deep sea divers are subjected to high pressures at low depths, and this increases the amount of nitrogen (from the air they breathe) dissolved in their blood. If they come back to the lower surface pressure too rapidly, the nitrogen gas leaves solution in much the same way CO_2 gas bubbles leave an opened soda. Gas bubbles in the bloodstream can cause pain and even death. The bubbles act as embolisms (clots) and block blood circulation. This condition is know as the "bends."

The effect of pressure on gas solubility is put to good use in **hyperbaric therapy.** Patients are placed in a chamber in which the pressure can be maintained at more than twice atmospheric pressure. This chamber can be used to reduce the pressure on divers slowly, thus avoiding the "bends."

Often the hyperbaric chamber is filled with pure oxygen. The increased partial pressure of oxygen leads to greater solubility of oxygen in blood. This is helpful in some cases of hypoxia or oxygen deficiency, such as in carbon monoxide poisoning. Hyperbaric chambers have also been used to treat infections, such as gangrene, caused by anaerobic bacteria. Oxygen is toxic to these bacteria. X-ray radiation of cancer cells is sometimes carried out in a hyperbaric chamber because under high oxygen pressure the difference in sensitivity of cancer and normal cells to the radiation is increased.

Particle size of solid solutes   Because the solution process occurs at the surface of a solid solute, increasing the surface area speeds up the process. Finely pulverized solids offer more surface area per unit mass than big chunks of solid. Particle size affects only the speed of solution

FIGURE 13.9

Pressure affects the solubility of a gas in a liquid. (a) The cap maintains a high partial pressure of CO_2 above the soda which holds CO_2 in solution; (b) Removal of the cap allows CO_2 gas to escape, thus diminishing the partial pressure of gas above the solution, and CO_2 bubbles escape from the solution.

formation. The total amount of mass that dissolves is set by other factors and does not depend on whether a fine powder quickly dissolves or a large chunk dissolves slowly.

Stirring   Stirring also affects only the speed of the solution process. It does so by mechanically bringing the solute and solvent into more intimate contact.

CONCENTRATION EXPRESSIONS

13.5   Because they are mixtures, solutions have a variable composition. When we state what the composition of a solution is, we call that statement the concentration of the solution. The **concentration** of a solution is the amount of solute that is present in some given amount of solvent or given amount of solution.

Concentration is a ratio of the form:

$$\frac{\text{Amount of solute}}{\text{Amount of solvent}} \quad \text{or more often} \quad \frac{\text{amount of solute}}{\text{amount of solution}}$$

Previously (Section 13.2) we have seen solubility given as a ratio. Notice the difference between the solubility, which is the "maximum" amount of solute per amount solvent at a given temperature, and the concentration, in which the amount of solute can take on *any* value up to the solubility limit. Also, the most often used concentration expressions are those in which the amount of solute is given per amount of *solution* rather than per amount of *solvent*.

The various concentration expressions differ in the *units* used for the amounts of solute and solution. The molarity expression used most often by chemists employs the units of moles for the solute and liters for solution. In percentage expressions the units for the solute and solution may be either grams (weight) or milliliters (volume). This gives rise to three possible percentage expressions: percentage weight per weight (% w/w), percentage volume per volume (% v/v), and percentage weight per volume (% w/v).

Percentage weight/weight (% w/w)

If you were given 100 g of a 10% (by weight) sugar solution, you could easily figure out how much sugar and how much water were mixed together by recognizing that the percentage refers to the **percentage of solute** in the total weight of the solution. Hence 10 percent of 100 g of solution = 0.1 × 100 g = 10 g sugar. The other 90 percent of the solution is water, hence 90 g water. 10 g solute +

90 g solvent = 100 g solution. Remember, Equation (13.1) says that grams solution equals grams of solvent plus grams of solute. Percentage is always part of the whole divided by the whole times 100. (Review Section 2.6, which concerns percentage calculations.) For calculations involving solutions, you will find the following equations useful:

$$\% \text{ w/w of solute} = \frac{\text{grams solute}}{\text{grams solution}} \times 100\% \qquad (13.2)$$

$$\% \text{ w/w of solute} = \frac{\text{grams solute}}{\text{grams solute} + \text{grams solvent}} \times 100\% \qquad (13.3)$$

SAMPLE EXERCISE

13.2

In a laboratory, 233 g of a NaCl solution was evaporated to dryness; i.e., the water was boiled off. What was left was 28.4 g of solid NaCl.

a. What was the percentage by weight of NaCl in the solution?

b. How much water was boiled away?

SOLUTION

a. Use Equation (13.2):

$$\% \text{ w/w of solute} = \frac{28.4 \text{ g NaCl}}{233 \text{ g solution}} \times 100\% = 12.2 \text{ percent}$$

b. Grams of solution equals grams of solute plus grams of solvent [Equation (13.1)]. Thus,

233 g = 28.4 g + ? g water

233 − 28.4 = 204.6 g water (Sig figs require that the answer be 205 g.)

SAMPLE EXERCISE

13.3

What is the percentage by weight of potassium hydroxide in a solution prepared by dissolving 5.0 g KOH in 55 g of water?

SOLUTION

Use Equation (13.3):

$$\% \text{ w/w of solute} = \frac{5.0 \text{ g KOH}}{5.0 \text{ g KOH} + 55 \text{ g H}_2\text{O}} \times 100\% = 8.3 \text{ percent}$$

In order to do problems involving the preparation of some given amount of solution of a given percentage, you must use your basic knowledge of what percentage means and your knowledge of Equation (13.1).

In many problems it is convenient to recognize the ratio of grams of solute per 100 g of solution as a conversion factor. That is, every percent weight/weight gives you a conversion factor relating grams of solute to grams of solution, for example

$$\frac{5 \text{ g solute}}{100 \text{ g solution}} \text{ for 5\% w/w}$$

SAMPLE EXERCISE

13.4

How many grams of sugar and water would you use to prepare 700. g of a 7.0 percent (w/w) solution?

SOLUTION

The grams of sugar needed can be obtained using the conversion factor method.

Step 1 The *given* is 700. g of solution.

Step 2 The *new* unit is grams of sugar.

Step 3 The percent weight/weight provides the conversion factor 7.0 g of sugar/100 g of solution.

Step 4 Set up the usual format:

Given × conversion factor = new quantity and unit

$$700. \text{ g solution} \times \frac{7.0 \text{ g sugar}}{100. \text{ g solution}} = 49 \text{ g sugar}$$

The grams of water solvent can be obtained by using Equation (13.1):

$$700. \text{ g solution} = 49 \text{ g sugar} + ? \text{ g water}$$

$$700. - 49 = 651 \text{ g water}$$

Problem 13.3

In a laboratory, 12.0 g of $AgNO_3$ is dissolved in 26.0 g of H_2O. What is the percentage by weight of $AgNO_3$ in the solution?

Problem 13.4

How would you prepare 500.0 g of a 32.0 percent solution of KBr in water?

Percentage weight/volume (% w/v)

Weight/volume percentage is the one used most often in medicine because the most common solutions are those in which the solute is a solid, most easily measured in grams, and the solution is a liquid, most easily measured in volume units, e.g., milliliters.

The numerical value expressed in a weight per volume percent is equal to the number of grams of solute present in 100 mL of solution. For example, a 10% w/v dextrose solution has 10 g of dextrose per 100 mL of solution. A 0.9% w/v salt solution (physiological saline) contains 0.9 g of salt per 100 mL of solution. This value can be readily calculated through the equation

$$\% \text{ w/v of solute} = \frac{\text{grams of solute}}{\text{milliliters of solution}} \times 100\%$$

SAMPLE EXERCISE

13.5

What is the percent weight/volume of KOH in 60. mL of a solution which contains 5.0 g of KOH?

SOLUTION

Use Equation (13.4):

$$\% \text{ w/v of KOH} = \frac{5.0 \text{ g of KOH}}{60. \text{ mL of solution}} \times 100\%$$

$$= 8.3\% \text{ w/v}$$

There is 8.3 g of KOH per 100 mL of solution.

As with weight/weight problems the given percent weight/volume can be used as a conversion factor.

SAMPLE EXERCISE

13.6

Physiological saline solution is a 0.900% w/v NaCl solution used for storing red blood cells as well as for other medical purposes. How much salt (in grams) is needed to prepare 250. mL of physiological saline solution?

SOLUTION

Use the Unit Conversion Method.

Step 1   The given is 250. mL of solution.

Step 2   The new unit is grams of NaCl.

Step 3   The percent weight/volume provides the conversion factor 0.900 g of NaCl/100 mL of solution.

Step 4   Set up the usual format:

Given × conversion factor = new quantity and unit

$$250. \text{ mL of soln} \times \frac{0.900 \text{ g of NaCl}}{100 \text{ mL of soln}} = 2.25 \text{ g of NaCl}$$

Problem 13.5

How many milliliters of 7.0% w/v glucose solution would contain 49 g of glucose?

Percentage volume/volume (% v/v)

This expression is encountered only for solutions which are mixtures of liquids. For example, a water solution of alcohol might be labeled 45% v/v. This indicates 45 mL of alcohol per 100 mL of solution. In general, the numerical value of percent volume/volume equals the number of milliliters of solute present in 100 mL of solution.

Parts per million (ppm)

To understand the expression "parts per million," it is useful to realize that a statement of percentage is actually a statement of parts per hundred. A 5 percent solution indicates 5 parts of solute per 100 parts of solution. A solution which has the concentration 5 ppm contains 5 parts of solute per 1 million parts of solution.

The parts per million expression is most useful for very dilute solutions wherein percentage expressions would involve very small numbers. For example, drinking water typically contains only 1 mg of fluoride ion (F^-) per liter of solution. In percent weight/volume this would be

$$\frac{0.001 \text{ g } F^-}{1000 \text{ mL solution}} \times 100\% = 0.0001 \text{ \% w/v}$$

This concentration in parts per million is 1 ppm:

$$\frac{0.001 \text{ g } F^-}{1000 \text{ mL soln}} = \frac{1 \text{ g } F^-}{1{,}000{,}000 \text{ mL soln}} = \frac{1 \text{ part}}{1 \text{ million parts}} = 1 \text{ ppm}$$

(Multiply the left-hand term by 1000/1000 = 1 to get the right-hand term.)

For even more dilute solutions the expression "parts per billion" (ppb) is employed.

Molarity

"Molarity" is the concentration expression used most frequently by chemists because moles are the units of matter most conveniently manipulated in the laboratory in doing chemical reactions (see Chapter 10). Of course, moles of material are always readily related to grams of material through the molar weight (see Section 8.6). **Molarity** (abbreviated capital M) is defined as moles of solute per liter of solution:

$$\text{Molarity } (M) = \frac{\text{moles solute}}{\text{liters of solution}} \qquad (13.5)$$

Whenever one is told the number of moles of solute or the amount of solute in grams that is dissolved in a specified volume of solution, *the molarity can be calculated by dividing the number of moles of solute by the numbers of liters of solution.*

Determine the molarity of the following solutions:

a.   4 moles of NaCl are dissolved in enough water to make 2 L of solution.

b.   58.8 g of KOH is dissolved in enough water to make 700.0 mL of solution.

SOLUTION

a.   Because the units given are moles of solute and liters of solution, the problem is completely straightforward and we can immediately divide:

$$\frac{4 \text{ moles NaCl}}{2 \text{ L of solution}} = \frac{2 \text{ moles}}{\text{L}} = \text{Molarity}$$

b.   In this case grams of solute must be converted to moles of solute and milliliters of solution to liters of solution before we divide.

$$58.8 \text{ g KOH} \times \frac{1 \text{ mole KOH}}{56.0 \text{ g KOH}} = 1.05 \text{ mole KOH}$$

$$700.0 \text{ mL} \times \frac{1 \text{ liter}}{1000 \text{ mL}} = 0.7000 \text{ L}$$

$$\frac{1.05 \text{ moles KOH}}{0.7000 \text{ L}} = \frac{1.50 \text{ moles}}{\text{L}} = \text{Molarity}$$

Note carefully the distinction between "moles" and "molarity." The number of moles of material is an absolute amount and can be directly converted to grams. Molarity is a *ratio* which tells the number of moles distributed through a volume of solution. It is convenient to recognize that the total number of moles in a given solution can be obtained by multiplying $M \times V$ (in liters). This relationship comes about by rearranging the equation by which *M*olarity is calculated:

$$M = \frac{\text{moles solute}}{\text{Volume (in liters)}} \qquad (13.5)$$

Multiplying each side of the equation by V yields

$$M \times V \textbf{(in liters)} = \textbf{moles of solute} \qquad (13.6)$$

We speak of 1 M, 2 M, 6 M solutions (read "1 molar," "2 molar," etc.). The statement of molarity provides a conversion factor between moles of solute and liters of solution. For example

Molarity:	1 M	2 M	6 M
Conversion factor:	$\dfrac{1 \text{ mole}}{1 \text{ L}}$	$\dfrac{2 \text{ moles}}{1 \text{ L}}$	$\dfrac{6 \text{ moles}}{1 \text{ L}}$

Of course, the reciprocals are also conversion factors. Recognizing the molarity as a conversion factor enables us to handle many different types of concentration problems with the Unit Conversion Method.

SAMPLE EXERCISE
13.8

How many moles of KBr are there in 35.8 mL of a 0.172 M solution?

SOLUTION

Use the Unit Conversion Method, with the molarity as a conversion factor.

Step 1 The given is 35.8 mL of solution (0.172 M is also given).

Step 2 The new is moles KBr.

Step 3 The necessary conversion factors are $\dfrac{1 \text{ L}}{1000 \text{ mL}}$ (to use M, the volume of the solution must be in liters) and $\dfrac{0.172 \text{ moles KBr}}{1 \text{ L solution}}$

Step 4 Set up the usual format:

Given × conversion factors =

$$35.8 \text{ mL} \times \frac{1 \text{ L}}{1000 \text{ mL}} \times \frac{0.172 \text{ mole KBr}}{1 \text{ L}} = 0.00616 \text{ mole KBr}$$

This problem can also be solved by remembering that

$$M \times V(\text{in liters}) = \text{moles solute}$$

The unit conversion method is stressed because it has more diverse applications than the $M \times V$ formula.

Of course, conversion factors of the type

$$\frac{\text{molar weight}}{1 \text{ mole}}$$

continue to be important and are often used in conjunction with the molarity conversion factor.

SAMPLE EXERCISE
13.9

How many grams of NaOH are there in 2.50 L of 0.343 M solution?

SOLUTION

Step 1 The given is 2.50 liters of solution (0.343 M is also given).

Step 2   The new is grams NaOH.

Step 3   The conversion factors are

$$\frac{0.343 \text{ mole NaOH}}{1 \text{ liter}} \quad \text{and} \quad \frac{40.0 \text{ g NaOH}}{1 \text{ mole NaOH}}$$

Step 4

$$2.50 \, \cancel{L} \times \frac{0.343 \, \cancel{\text{mole NaOH}}}{1 \, \cancel{L}} \times \frac{40.0 \text{ g NaOH}}{1 \, \cancel{\text{mole NaOH}}} = 34.3 \text{ g NaOH}$$

In order to prepare a solution of some given molarity in the laboratory, one weighs out the appropriate amount of solute on a balance and transfers it to a volumetric flask in which the volume of the solution can be measured (see Figure 13.10). Solvent is then added until the desired volume of solution is achieved. Remember, in molar solutions we are concerned with the volume of the solution, *not* with the volume of solvent.

SAMPLE EXERCISE

13.10

How would you prepare 0.500 L of a 0.140 *M* solution of CuSO₄ (molar weight = 160. g)? You have available solid CuSO₄, a balance, and a 500-mL volumetric flask.

SOLUTION

Using the Unit Conversion Method, determine the number of grams of CuSO₄ required to prepare this solution.

Step 1   The given is 0.500 L of solution (0.140 *M* is also given).

FIGURE 13.10

Preparation of 500. mL of 0.140 *M* CuSO₄.

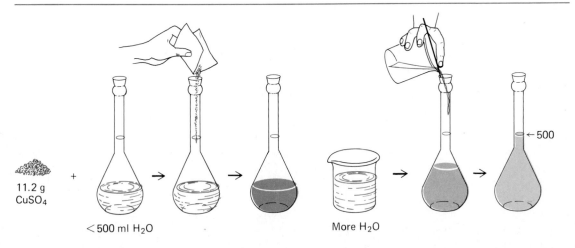

11.2 g
CuSO₄

+

<500 ml H₂O

More H₂O

←500

Step 2   The new is grams of $CuSO_4$.

Step 3   The conversion factors are

$$\frac{0.140 \text{ mole } CuSO_4}{1 \text{ liter}} \quad \text{and} \quad \frac{160. \text{ g } CuSO_4}{1 \text{ mole } CuSO_4}$$

Step 4

$$0.500 \, \cancel{L} \times \frac{0.140 \, \cancel{\text{mole } CuSO_4}}{1 \, \cancel{L}} \times \frac{160. \text{ g of } CuSO_4}{1 \, \cancel{\text{mole } CuSO_4}} = 11.2 \text{ g } CuSO_4$$

Weigh out the 11.2 g of $CuSO_4$ and transfer it to the 500-mL volumetric flask (500 mL = 0.500 L). Add some water, and swirl it to dissolve the solute. When the solute is dissolved, add water exactly to the etched mark which gives 0.500 L of solution (see Figure 13.10).

The calculation of the number of grams of solute necessary to make a given volume of a solution of given molarity always involves the multiplication

$$\mathbf{M} \times \mathbf{V} \text{ (in liters)} \times \textbf{molar weight} = \textbf{grams solute} \quad \textbf{(13.7)}$$

See Step 4 in Sample Exercise 13.10. Compare Equation 13.7 with Equation 13.6.

Problem 13.6

How many grams of NaOH are needed to make 125 mL of 2.44 M solution?

All problems involving molarity can be solved by using one or more of the following conversion factors or their reciprocals:

$$\overset{M}{\underset{1 \text{ L solution}}{\frac{\text{given moles}}{}}} \qquad \frac{1 \text{ L}}{1000 \text{ mL}} \qquad \frac{\text{molar weight solute}}{1 \text{ mole solute}}$$

SAMPLE EXERCISE

13.11

Find the number of milliliters of 1.32 M solution that contain 6.72 g of NaF (molar weight = 42.0 g)?

SOLUTION

This problem requires all the previously listed factors for its solution.

Step 1   The given is 6.72 g NaF (1.32 M is also given).

Step 2   The new is milliliters of solution.

Step 3   Set up the conversion factors in the usual form: $\dfrac{\textbf{New} \text{ units}}{\textbf{Given} \text{ units}}$

$$\frac{\text{1 mole NaF}}{\text{42.0 g NaF}} \qquad \textbf{given} \text{ units} \qquad \text{This factor gives moles NaF.}$$

$$\frac{\text{1 L solution}}{\text{1.32 moles NaF}} \qquad \begin{array}{l}\textbf{new} \text{ units} \\ \text{This factor gives liters of solution.}\end{array}$$

from which milliliters can easily be determined.

Step 4   The format is

$$6.72 \text{ g NaF} \times \frac{\text{1 mole NaF}}{\text{42.0 g NaF}} \times \frac{\text{1 L solution}}{\text{1.32 moles NaF}} \times \frac{\text{1000 mL}}{\text{1 L}} = 121 \text{ mL solution}$$

Problem 13.7

How many liters of 0.643 *M* solution contain 114 g of NaF?

DILUTION

13.6   Another method of preparing solutions of some desired molarity is through the process of dilution. **Dilution** involves adding solvent (usually water) to a concentrated solution. The addition of water produces a more dilute solution because the *same amount of solute* is distributed through a larger amount of solvent or solution. For example, if 1 L of water is added to 1 L of a 6 *M* solution, the solution becomes a 3 *M* solution:[1]

$$\begin{array}{ccc} & \text{Original} & & & \text{Final} \\ 6\,M = & \dfrac{6 \text{ moles}}{1 \text{ L solution}} & + \text{ 1 L water} \longrightarrow & \dfrac{6 \text{ moles}}{2 \text{ L solution}} = 3\,M \end{array}$$

See Figure 13.11.

Calculations involving dilution problems center on the fact that the *same amount of solute* is present in both the original solution and the final diluted solution.

$$\textbf{moles solute}_{\text{original solution}} = \textbf{moles solute}_{\text{diluted solution}} \qquad \textbf{(13.8)}$$

Furthermore, recall from Equation (13.6) that the number of moles of solute present in any solution can always be determined by multiplying $M \times V$ (in liters):

$$M \times V = \text{moles solute}$$

$$\frac{\text{moles solute}}{\text{1 liter solution}} \times 1 \text{ liter solution} = \text{moles solute}$$

---

[1] Combining 1 L of solution with 1 L of water does not produce exactly 2 L of solution. However, this assumption is sufficiently accurate to 2 sig figs.

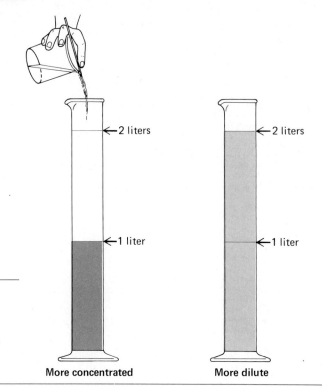

FIGURE 13.11

The dilution process. When water is added to a concentrated solution, the solution becomes more dilute. The same amount of solute is distributed over a larger amount of solution.

Therefore, we can rewrite Equation (13.8) as

$$M_o \times V_o = M_d \times V_d \qquad (13.9)$$

where *o* stands for the original solution and *d* stands for the diluted solution.

SAMPLE EXERCISE

13.12

What is the molarity of the solution prepared by adding 2 L of water to 1.5 L of a 0.5 *M* KOH solution?

SOLUTION

Whenever water (solvent) is added to a solution, the problem should be recognized as a dilution problem. Then use Equation (13.9).

Original Solution		Diluted Solution
$M_o = 0.5\ M$		$M_d = ?$
$V_o = 1.5\ L$		$V_d = 1.5\ L + 2.0\ L = 3.5\ L$
$M_o \times V_o$	=	$M_d \times V_d$
$0.5\ M \times 1.5\ L$	=	$M_d \times 3.5\ L$

Divide each side of the equation by 3.5 L to isolate M_d.

$$M_d = \frac{0.5\ M \times 1.5\ \cancel{L}}{3.5\ \cancel{L}}$$

$$M_d = 0.2\ M$$

[Because volume appears on both sides of Equation (13.9), the units of volume must be *consistent*, but they do not necessarily have to be liters. In this case, you could have used 1500 mL and 3500 mL and obtained the same result. However, one *cannot* use liters on one side and milliliters on the other.]

SAMPLE EXERCISE

13.13

What volume of a 0.34 M $MgCl_2$ solution is required to make 450. mL of a 0.10 M solution by dilution?

SOLUTION

The word *dilution* signals the use of Equation (13.9).

Original Solution   Diluted Solution

Original Solution		Diluted Solution
$M_o = 0.34\ M$		$M_d = 0.10\ M$
$V_o = ?$		$V_d = 450.\ mL$
$M_o \times V_o$	$=$	$M_d \times V_d$
$0.34\ M \times V_o$	$=$	$0.10\ M \times 450.\ mL$

Divide each side by 0.34 M to isolate V_o:

$$V_o = \frac{0.10\ \cancel{M} \times 450.\ mL}{0.34\ \cancel{M}}$$

$$V_o = 132\ mL$$

Significant figures demand that the answer be reported as 130 or 1.3 × 10^2 mL.

Problem 13.8

In a laboratory, 500. mL of water is added to 250. mL of 0.75 M NaOH solution. What is the molarity of the resulting solution?

ELECTROLYTES

13.7 In Section 13.3 and Figure 13.4 we developed a picture of what a solution of an ionic compound is really like. For example, a solution of NaCl is actually Na$^+$ ions and Cl$^-$ ions intermingled between (and hydrated by) H_2O molecules. Experimentally, it is found that a solution of an ionic compound conducts electricity. In fact, it is the presence of ions in a solution that allows that solution to conduct

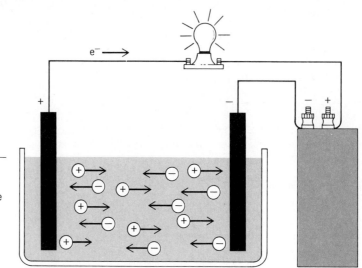

FIGURE 13.12

Ions in solution carry the current (electrons) from one electrode to the other. If there are no ions in solution, the circuit is open; that is, there is no way for current to span the space between the electrodes.

electricity in the first place. Solutions that conduct electricity are classified as **electrolyte solutions.**

Electric current is a movement of electric charge. In metal wires electrons carry the charge (see Figure 13.12). If we attach electrodes (metal rods) to the poles of a battery and immerse the electrodes in a solution containing ions, the ions carry the electric charge from one pole to the other to complete the circuit and make the light glow. If there are no ions, no current flows and the light does not glow. *A solution that conducts electricity definitely contains ions.*

Soluble ionic compounds are **strong electrolytes** because they completely dissociate in water. **Dissociation** is the name given to the process described in Section 13.3 whereby ions that are closely "associated" in the solid crystal disassociate, or break apart, and become independent in solution. The formula for the ionic compound tells how many ions are present per formula unit or how many moles of ions there are per mole of compound. (You might wish to review Sections 8.4 and 8.7 at this point.) For example,

$$KOH(s) \rightleftharpoons K^+(aq) + OH^-(aq)$$

$$Na_2SO_4(s) \rightleftharpoons 2Na^+(aq) + SO_4^{2-}(aq)$$

$$Al(NO_3)_3(s) \rightleftharpoons Al^{3+}(aq) + 3NO_3^-(aq)$$

The breaking apart or dissociation of ions is complete ($\rightarrow$). The tendency to recombine ($\leftarrow$) can be ignored unless the solubility limit is exceeded.

Most molecular compounds are *non*electrolytes; their solutions do *not* conduct electricity. This is so because molecular compounds are not composed of ions, and thus there can be no dissociation in water. If there are no ions, then the solution is a nonelectrolyte. However, some molecular compounds react with water in such a way as to produce ions. This process is called **ionization.** If by reaction with water a molecular compound produces many ions per mole of compound, it is a *strong* electrolyte. If only a few ions are produced per mole of compound, the compound is a *weak* electrolyte (Figure 13.13).

One particularly important class of molecular compounds that undergoes ionization in water is **acids.** Acids are molecular compounds in which hydrogen is bound to an electronegative element. Their water solutions exhibit certain special properties which will be discussed fully in Chapter 15. The water solutions of acids are electrolytes because acids react in water to form ions, i.e., *acids ionize in water.* For example, in water HCl molecules form H^+ ions and Cl^- ions:

$$HCl(g) \overset{H_2O}{\rightleftharpoons} H^+(aq) + Cl^-(aq)$$

We can ignore the tendency to recombine because it is so small in this case. Thus we say that HCl (hydrochloric acid) ionizes 100 percent. Every molecule in solution ionizes, and HCl therefore is a strong electrolyte. Hydrochloric acid is also termed a **strong acid** because of its 100 percent ionization. There are six common strong acids (strong electrolytes): HCl, HBr, HI, HNO_3, H_2SO_4, and $HClO_4$. All other common acids are weak.

FIGURE 13.13

Strong and weak electrolytes and nonelectrolytes. (*a*) Sodium chloride is a strong electrolyte, as shown by the brightly glowing light bulb. (*b*) Vinegar has only a very small number of ions in solution and hence is a weak electrolyte, as witnessed by the dimly glowing bulb. (*c*) Pure water is a nonelectrolyte, and therefore the light bulb does not glow.

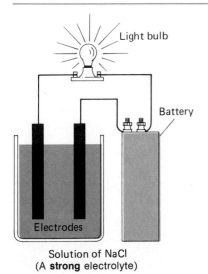

Solution of NaCl
(A **strong** electrolyte)
(a)

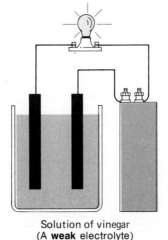

Solution of vinegar
(A **weak** electrolyte)
(b)

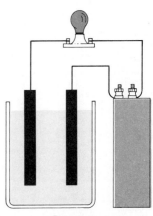

Pure water
(A **nonelectrolyte**)
(c)

Vinegar is a solution of acetic acid. $HC_2H_3O_2$. Acetic acid ionizes only to a very limited extent:

$$HC_2H_3O_2 \rightleftharpoons H^+(aq) + C_2H_3O_2{}^-(aq)$$

This representation ($\rightleftharpoons$) indicates that a few molecules of $HC_2H_3O_2$ will form ions in water, but most molecules of $HC_2H_3O_2$ do not react. Because only a small number of ions are produced, $HC_2H_3O_2$ is a weak electrolyte (see Figure 13.13b). Most acids are weak electrolytes and thus weak acids. Learn the preceding list of the six strong acids. All others, then, are weak. Aside from acids, almost all other molecular compounds are nonelectrolytes. One notable exception is ammonia, NH_3, which is a weak electrolyte.

PARTICLES IN SOLUTION

13.8 Many chemical reactions are conducted in solution. The reactants are solutes. Products may be solutes, or they may leave the solution if they are either insoluble solids or gases. To paint an accurate picture of a reaction in solution, it is necessary to know the nature of the solute particles as well as the solubilities of the reactants and products.

From the preceding discussion about electrolyte solutions we can obtain a summary of the nature of solute particles.

Summary of the nature of solute particles

TYPE 1   The solute particles of *ionic compounds* are *ions*. Therefore, an ionic compound in solution should be represented as the sum of its hydrated ions. For example, NaCl in solution is represented as $Na^+(aq) + Cl^-(aq)$ because the ions are separate and independent and surrounded by water molecules (*aq*).

TYPE 2   The solute particles of *molecular compounds that are non-electrolytes* are *molecules*. Sugar is a nonelectrolyte. The solid is composed of molecules $C_{12}H_{22}O_{11}$. A sugar solution is a mixture of water molecules and sugar molecules. The process of dissolving the nonelectrolyte sugar in water can be represented as

$$C_{12}H_{22}O_{11}(s) \longrightarrow C_{12}H_{22}O_{11}(aq)$$

TYPE 3   The solute particles of *molecular compounds that are strong electrolytes* are *ions*. Therefore, such a compound should be represented as the sum of the ions which it forms. HCl for example, should be expressed as $H^+(aq) + Cl^-(aq)$ if it is in solution.

TYPE 4   The solute particles of *molecular compounds that are weak electrolytes* are mixtures of *molecules* and *ions*. Because the number of molecules is much greater than the number of ions, we usually represent weak electrolytes by their molecular formula. Acetic acid, for example, is represented as $HC_2H_3O_2(aq)$.

IONIC EQUATIONS

13.9 Equipped with the knowledge of the nature of solute particles, we can now represent chemical reactions in solution by ionic equations rather than by what we shall call the "traditional" equations that you learned to use in Chapter 9.

For example, in Chapter 9 (Section 9.8), we represented the single replacement reaction between magnesium metal and hydrochloric acid to yield magnesium chloride and hydrogen gas as

$$Mg + 2HCl \longrightarrow MgCl_2 + H_2 \uparrow$$

Traditional equation

Figure 13.14 shows a student dropping Mg metal into HCl solution. The result is bubbles of H_2 gas and a clear solution of $MgCl_2$. The physical picture of this reaction is much better described by an ionic equation in which we are careful to indicate appropriate solute particles for materials in solution. Thus the ionic equation for the reaction would be

Ionic equation

$$Mg(s) \qquad + \quad \underbrace{2H^+(aq) + 2Cl^-(aq)} \longrightarrow \underbrace{Mg^{2+}(aq) + 2Cl^-(aq)} + \qquad H_2 \uparrow$$

| Solid does not dissolve; Mg "disappears" because it reacts | See the Summary, type 3 in Section 13.8 | See Summary, type 1 | The gas escapes; it is not in solution |

Notice that this ionic equation also tells us that magnesium and hydrogen were involved in chemical changes, but that chloride ion was unchanged. Mg metal becomes Mg^{2+} ions; H^+ ions become H_2 gas; Cl^- ions remain Cl^- ions. Ions that are unchanged in chemical reactions are called **spectator ions.** Spectators simply "watch" the other ions react.

We can totally represent the chemical changes that occur in the preceding reaction by a so-called net ionic equation, which shows only the reacting species, not spectators.

$$Mg(s) + 2H^+(aq) \longrightarrow Mg^{2+}(aq) + H_2 \uparrow$$ **Net ionic equation**

We have previously represented the double-replacement reaction between solutions of NaCl and $AgNO_3$ in the following way:

$$NaCl + AgNO_3 \longrightarrow AgCl \downarrow + NaNO_3$$ **Traditional equation**

The corresponding ionic equation shows the nature of the solute particles in solution:

FIGURE 13.14

A strip of magnesium metal reacts with a solution of HCl to produce H_2 gas (bubbles) and a solution of $MgCl_2$. (*Photograph by Bryan Lees.*)

Ionic equation

$$\underbrace{Na^+(aq) + Cl^-(aq) + Ag^+(aq) + NO_3^-(aq)}_{\text{See Summary, type 1.}} \longrightarrow \underset{\substack{\text{This solid}\\\text{is } not \text{ in}\\\text{solution}}}{AgCl\downarrow} + \underbrace{Na^+(aq) + NO_3^-(aq)}_{\text{Type 1}}$$

The *net* ionic equation is written by identifying the spectator ions and realizing that because they appear on both sides of the equation they can be canceled out. The spectators are Na^+ ions and NO_3^- ions. Thus

Net ionic equation

$$Cl^-(aq) + Ag^+(aq) \longrightarrow AgCl\downarrow$$

USING THE SOLUBILITY RULES

13.10 In the preceding discussion of the Mg + HCl reaction, the information that the product $MgCl_2$ is soluble, coupled with the recognition of $MgCl_2$ as an ionic compound, suggested the use of Summary, type 1, to represent it properly. In the other preceding example it was noted that the AgCl was insoluble. Usually you will not be told whether products are soluble. However, you can use the solubility rules in Table 13.3 to decide whether or not a given ionic compound is soluble.

TABLE 13.3   SOLUBILITY RULES*

Ion	Rules
Group IA, NH_4^+	1. All ionic compounds in which the cation is a Group IA element or NH_4^+ are *soluble*
NO_3^-	2. All nitrates are *soluble*
Cl^-, Br^-, I^-	3. All chlorides, bromides, and iodides are *soluble,* except for AgX and PbX_2† (where X stands for Cl, Br, or I)
SO_4^{2-}	4. All sulfates are *soluble* except for $CaSO_4$,† $SrSO_4$, $BaSO_4$, $PbSO_4$, and Ag_2SO_4†
S^{2-}	5. All sulfides are *insoluble* except those of the Group IA or IIA elements or NH_4^+
	6. All other compounds are *insoluble*

* Mercury compounds are not covered by these rules as written because of the complexity of the mercury(I) ion, which is beyond the scope of this book.
† Actually slightly soluble.

SAMPLE EXERCISE

13.14

Classify the following compounds as soluble or insoluble:

a. $MgSO_4$

b. CuS

c. $CaCO_3$

d. $Fe(NO_3)_3$

e. K_2CO_3

f. $PbSO_4$

SOLUTION

Step 1 Consult the solubility rules in Table 13.3.

Step 2 Classify each compound with respect to its cation and anion components.

Step 3 Apply the appropriate rule.

Cation	Anion	Rule
a. Mg^{2+}	SO_4^{2-}	**4.** All sulfates are *soluble;* this is not an exception
b. Cu^{2+}	S^{2-}	**5.** All sulfides are *insoluble;* this is not an exception
c. Ca^{2+}	CO_3^{2-}	**6.** This compound does not fit any of the first five categories; such compounds are *insoluble*
d. Fe^{3+}	NO_3^-	**2.** All nitrates are *soluble*
e. K^+	CO_3^{2-}	**1.** K^+ is in Group IA
f. Pb^{2+}	SO_4^{2-}	**4.** $PbSO_4$ is *insoluble*

SAMPLE EXERCISE

13.15

How should the compounds in Sample Exercise 13.14 be represented in an ionic equation if they are products of a reaction?

SOLUTION

Soluble ionic compounds form ions in solution. Insoluble compounds are shown with the precipitate arrow ($\downarrow$). Thus

a. $Mg^{2+}(aq) + SO_4^{2-}(aq)$

b. $CuS\downarrow$

c. $CaCO_3\downarrow$

d. $Fe^{3+}(aq) + 3NO_3^-(aq)$

e. $2K^+(aq) + CO_3^{2-}(aq)$

f. $PbSO_4\downarrow$

Net ionic equations for double-displacement reactions between ionic compounds can be written by the stepwise procedure shown for AgCl in Section 13.9. That is, we can write the **traditional,** then the **ionic,** and finally eliminate spectators to reach the **net.** However, we can go directly from the **traditional** to the **net** by making use of the fact that the **net ionic equation** for these reactions will always be the combination of ions that form the precipitate.

SAMPLE EXERCISE

13.16

Write a net ionic equation for the reaction between solutions of Na_3PO_4 and $Fe_2(SO_4)_3$.

SOLUTION

Step 1   Write the traditional equation by "switching the cation-anion partners."

$$Na_3PO_4 + Fe_2(SO_4)_3 \longrightarrow Na_2SO_4 + FePO_4$$

Balance

$$2Na_3PO_4 + Fe_2(SO_4)_3 \longrightarrow 3Na_2SO_4 + 2FePO_4$$

Step 2   Evaluate the solubility of the products.
$$Na_2SO_4 \text{ is soluble (rule 1).}$$
$$FePO_4 \text{ is insoluble (rule 6).}$$

Step 3   The net ionic equation is the combination of ions producing the precipitate:

$$Fe^{3+}(aq) + PO_4{}^{3-}(aq) \longrightarrow FePO_4 \downarrow$$

Problem 13.9

What is the net ionic equation for the combination of solutions of K_2SO_4 and $Ba(NO_3)_2$?

Problem 13.10

Is there a net ionic equation for the combination of solutions of $CuCl_2$ and KNO_3? Explain.

COLLIGATIVE PROPERTIES OF SOLUTIONS

13.11   In Section 12.7 we discussed the physical properties of liquids. The physical properties of liquids are altered in predictable ways by the presence of dissolved solutes. Thus, for example, boiling points of solutions are typically higher than boiling points of the pure solvent, and solution freezing points are typically lower. Both of these effects stem from the fact that the vapor pressure of a solution is lower than that of a pure solvent. Furthermore, solutions demonstrate a property, osmotic pressure, which is not present in a pure liquid. The numerical values of the vapor pressure lowering, the boiling point elevation, the freezing point depression, and the osmotic pressure depend *only* on the concentration of solute particles in solution, and not on the identity of the solute particles.

Such properties, which do not depend on the nature of the dissolved species, but only on the *number of dissolved particles* are called **colligative properties.** Glance back at Section 13.8 to remind yourself that ionic compounds and some ionizable molecular compounds produce more particles in solution than do molecular nonelectrolytes.

Vapor pressure of a solution

Recall from Section 12.7 and Figure 12.11 that vapor pressure increases with temperature. Figure 13.15 shows the typical variation of the vapor pressure of a pure solvent and a solution. Notice that the vapor pressure of the solution is always *lower* than that of the pure solvent.

The lowering of vapor pressure by the presence of solute particles is readily explained by remembering that the development of vapor pressure depends on the ability of molecules to escape from the surface of a liquid. As Figure 13.16 shows, in a solution solute particles literally get in the way of solvent molecules and interfere with their escape. The greater the number of solute particles, the greater the interference effect, and therefore, the greater is the lowering of vapor pressure.

Boiling-point elevation of a solution

Lowering of vapor pressure leads directly to the elevation of the nor-

The black line shows the change in vapor pressure of pure water as temperature varies. The colored line plots vapor pressure of a water solution versus temperature. Vapor pressure of the solution is lower than that for the pure solvent at every temperature. Because the normal boiling temperature is that at which vapor pressure equals 760 mmHg, solutions boil at higher temperatures; the intersection of 760 mmHg and the solution curve occurs at a higher temperature (B in the figure). Freezing points are lowered because the solution curve and vapor pressure curve for ice intersect at a lower temperature (F in the figure).

FIGURE 13.15

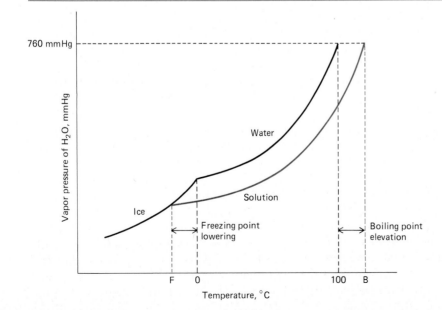

FIGURE 13.16

(a) In the pure solvent every space on the surface is occupied by a solvent molecule. (b) In a solution some spaces are occupied by solute particles. Thus there is less surface area from which solvent molecules can escape, and vapor pressure is lowered.

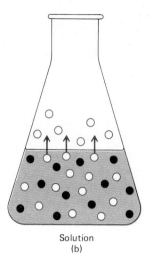

Pure solvent
(a)

Solution
(b)

○ Solvent molecule
● Solute molecule

mal boiling point of a solution compared with that of the pure solvent. Recall that the normal boiling point is defined as the temperature at which the vapor pressure is equal to an external pressure of 760 mmHg. Because the vapor pressure of a solution is lower at all temperatures than that of the pure solvent, a higher temperature is required to reach a vapor pressure of 760 mmHg. This idea is shown graphically in Figure 13.15.

Adding salt to a pot of boiling water leads to a slightly increased boiling temperature because the salt lowers the vapor pressure of the solution and a higher temperature is necessary in order to reestablish the vapor pressure of the solution equal to the atmospheric pressure. Adding more salt to the solution will increase the boiling temperature still further. Mole for mole, salt has twice the effect on boiling point as sugar because sugar is a molecular solid and produces 1 mole of sugar molecules in solution, whereas salt produces 2 moles of ions (1 mole of Na^+ and 1 mole of Cl^-) per mole of compound. Variations in colligative properties depend on numbers of particles.

Freezing-point depression of a solution

A liquid freezes at the temperature at which the vapor pressure of the liquid equals the vapor pressure of the solid state of the substance. For water the vapor pressure of liquid water and the vapor pressure of ice are equal at 0°C. Figure 13.15 shows that the lowering of the vapor pressure curve for a liquid solution results in the intersection of that curve with the solid state curve (for pure solvent) at a lower temperature. This means that solutions freeze at lower temperatures than pure solvents.

This colligative property of freezing point depression is put to good practical uses in cold winter weather. Pure water in an automobile radiator would freeze in winter and damage the engine. "Antifreeze" is simply a solution. The solute (usually ethylene glycol, $HOCH_2CH_2OH$) depresses the freezing point of the solution to temperatures far below 0°C.

The same principle is employed to melt ice on roadways. Remember melting and freezing occur at the same temperature. Whereas pure water freezes or ice melts at 0°C, a solution undergoes these transitions at lower temperatures. Thus "salt" sprinkled on icy roads forms a solution with the ice that melts at a lower temperature. The principle that the magnitude of effects on colligative properties depends on numbers of particles is employed by using $CaCl_2$ rather than NaCl. $CaCl_2$ produces 3 moles of particles (1 Ca^{2+} and $2Cl^-$) per mole of compound, while NaCl forms 2 moles. Thus $CaCl_2$ is $1\frac{1}{2}$ times more effective at melting ice on streets and roadways than NaCl.

OSMOTIC PRESSURE OF SOLUTIONS

13.12 Osmotic pressure is a property that you may have not encountered previously. Because of this probable unfamiliarity and because the development of osmotic pressure has many physiological consequences, we will devote a whole section to this colligative property. Osmotic pressure depends on the phenomenon of *osmosis* and the concept of particle flow through an osmotic membrane.

An osmotic membrane can be thought of as a thin sheet with small holes in it which allow the passage of water and other small molecules, the dimensions of which are not significantly larger than water. Particles larger than a water molecule such as a sugar molecule or a Na^+ or Cl^- ion cannot pass through an osmotic membrane (see Figure 13.17).

Osmosis

Osmosis is defined as the flow of water across an osmotic membrane from a more dilute solution (or pure water) into a more concentrated solution. This represents flow from a region where there are relatively more water molecules to where there are fewer. It is important to realize that water flows across the osmotic membrane in both directions, as shown in Figure 13.17. However, there is a *net* flow of water from a solution with a larger relative number of water molecules (dilute solution or pure solvent) to one with relatively fewer water molecules (concentrated solution).

FIGURE 13.17

An osmotic membrane can separate particles on the basis of size. Its "holes" allow the passage of H_2O molecules in both directions, but the membrane prevents the passage of glucose molecules, Na^+ ions, and Cl^- ions, which are larger than water molecules.

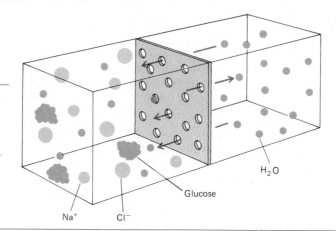

H₂O

Glucose

Na⁺   Cl⁻

Once again we can use the idea of solute particles "getting in the way" to explain this net flow of water. Figure 13.18*a* shows solute particles blocking some water molecules from passing through an osmotic membrane. This blockage is less effective in a dilute solution than in a concentrated solution because there are fewer solute particles in a dilute solution. The net flow of water from the dilute to the concentrated chamber increases the volume of the more concentrated solution, as shown in Figure 13.18*b*. The net movement of water cannot continue indefinitely because the increased height of the more concentrated solution causes a downward pressure from the added weight of liquid, which eventually balances the opposing tendency of water to flow into the more concentrated solution. When this balance of forces occurs, there is no further net movement of water. At this point the concentrated solution has become more dilute and the dilute solution more concentrated. Another way of saying this is that the concentration gradient has been diminished. A **concentration gradient** is a *difference* in concentration between solutions. The greater the difference or gradient, the greater the potential for a net flow of water in the direction dilute (more water) → concentrated (less water).

Osmotic pressure

When a solution is separated from pure water by an osmotic membrane, the pressure that develops from the excess height of the solution side as the water flows in is called the **osmotic pressure** of the solution. The osmotic pressure of a solution can also be defined as the external pressure which must be applied to prevent a net flow of water across a membrane from pure water to the solution. Figure 13.19 shows this application of pressure. The osmotic pressure measures the tendency for water to flow from pure water across an osmotic

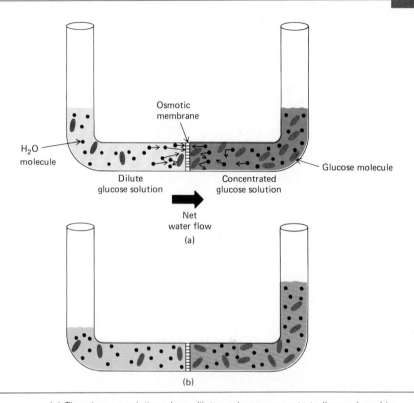

FIGURE 13.18

(a) The glucose solutions (one dilute and one concentrated) are placed in compartments separated by an osmotic membrane. Only water molecules can fit through the "holes" of the osmotic membrane. Larger glucose molecules cannot pass through. Also, glucose molecules block the passage of water molecules by deflecting them. The more glucose, the more the interference. Thus the net flow is from the dilute solution where there is more water and less blocking glucose into the more concentrated solution. (b) This shows the compartments after time has elapsed. The direction of net flow is obvious from the altered water levels.

membrane into a solution. This tendency depends only on the concentration of solute particles in the solution; that is, it is a colligative property. The greater the concentration of solute particles, the larger the solution's osmotic pressure and the greater the tendency for water to flow into it.

It is the difference in osmotic pressure of two solutions with different solute concentrations that causes the net flow of water that we've been discussing. In comparing solutions, a more dilute solution with a lower osmotic pressure is said to be **hypotonic** compared with a more concentrated solution with a higher osmotic pressure. In turn, the more concentrated solution is called **hypertonic** compared with

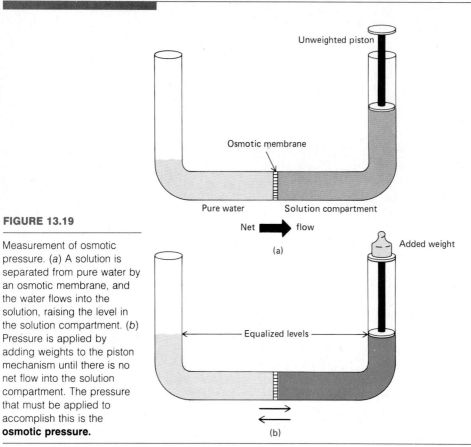

FIGURE 13.19

Measurement of osmotic pressure. (a) A solution is separated from pure water by an osmotic membrane, and the water flows into the solution, raising the level in the solution compartment. (b) Pressure is applied by adding weights to the piston mechanism until there is no net flow into the solution compartment. The pressure that must be applied to accomplish this is the **osmotic pressure.**

the dilute solution. Solutions of exactly the same particle concentration have identical osmotic pressures and are said to be **isotonic**. We can summarize these ideas schematically:

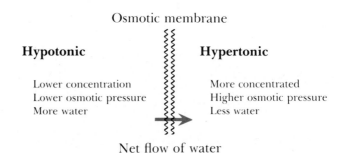

Osmosis in foods and agriculture

We witness the effects of osmosis in many foods. Prunes swell up if placed in water (a hypotonic medium to the prune), cucumbers shrink and become pickles in brine (a hypertonic medium to the cucumber), eggplants give off their water when sprinkled with salt, and hams are dehydrated and therefore preserved by "cooking" them in a brine solution. The movement of water in each case is through cell membranes. In agriculture osmosis helps one understand how water flows from the ground root system (hypotonic) to the concentrated sap cells (hypertonic) in the upper regions of a tree.

SOLUTION STOICHIOMETRY

13.13 Because chemical reactions often occur in solution, it is necessary and convenient to be able to do stoichiometric conversions involving volumes of solutions of known molarity. It would be a good idea to review Section 10.7 and Figure 10.1 before proceeding.

The key to solving solution stoichiometry problems is to remember that for any solution:

$$M \times V \text{ (in liters)} = \text{moles of solute} \qquad (13.6)$$

that is, M is a conversion factor relating moles of solute to liters of solution. Thus if you are given the molarity and volume of some reactant or product, you can easily determine the number of moles of that reactant or product and then proceed to carry out the stoichiometry problem, as was done in Chapter 10.

SAMPLE EXERCISE

13.17

How many milliliters of 0.52 M AgNO$_3$ solution are needed to react completely with 250. mL of a 0.36 M MgBr$_2$ solution according to the equation

$$2AgNO_3 + MgBr_2 \longrightarrow 2AgBr + Mg(NO_3)_2$$

SOLUTION

Step 1   The given is 250. mL of 0.36 M MgBr$_2$.

Step 2   The new is milliliters of 0.52 M AgNO$_3$.

Step 3   The central focus in all stoichiometry problems is the mole ratio obtained from the balanced equation which relates **new** material to **given** material. In this case, the central focus is

$$\boxed{\frac{2 \text{ moles AgNO}_3}{1 \text{ mole MgBr}}}$$

Step 4   To use this factor, we recall that $M \times V$ (in liters) will give us moles:

$$250 \text{ mL} \times \frac{1 \text{ L}}{1000 \text{ mL}} \times \frac{0.36 \text{ mole MgBr}_2}{1 \text{ L}} = 0.090 \text{ mole MgBr}_2$$

Step 5 The complete setup is now

$$0.090 \text{ mole MgBr}_2 \times \boxed{\frac{2 \text{ moles AgNO}_3}{1 \text{ mole MgBr}_2}} \times \frac{1 \text{ L}}{0.52 \text{ mole AgNO}_3} \times \frac{1000 \text{ mL}}{1 \text{ L}}$$

$$= 3.5 \times 10^2 \text{ mL of } 0.52 \ M \text{ AgNO}_3$$

Notice again the use of molarity (*M*) as the conversion factor between moles and volume (in liters).

SAMPLE EXERCISE

13.18

How many grams of AgCl precipitate from 34 mL of 0.24 *M* AgNO$_3$ solution when excess NaCl solution is added?

SOLUTION

The reaction is represented by the equation

$$\text{AgNO}_3 + \text{NaCl} \longrightarrow \text{AgCl} \downarrow + \text{NaNO}_3$$

Step 1 The given is 34 mL of 0.24 *M* AgNO$_3$.

Step 2 The new is grams AgCl.

Step 3 The central focus will be

$$\boxed{\frac{1 \text{ mole AgCl}}{1 \text{ mole AgNO}_3}}$$

Step 4 Moles of AgNO$_3$ can be calculated from the *V* and *M* of AgNO$_3$.

Step 5 The formula weight for AgCl (143.3) relates grams and moles.

Step 6 Integrating all the aforementioned information and beginning as always with the given, you can express the setup as follows:

$$34 \text{ mL} \times \frac{1 \text{ L}}{1000 \text{ mL}} \times \frac{0.24 \text{ mole AgNO}_3}{\text{L}} \times \boxed{\frac{1 \text{ mole AgCl}}{1 \text{ mole AgNO}_3}} \times \frac{143.3 \text{ g AgCl}}{1 \text{ mole AgCl}} =$$

$$1.2 \text{ g AgCl}$$

SUMMARY

Solutions are homogeneous mixtures of two or more substances. The most common solutions are those in which the solvent, or dissolving medium, is a liquid. Solutes, which are the substances dissolved, may be either gases, liquids, or solids (Section 13.1). Given the solubility of a particular solute in a particular solvent at some specified temperature, one can determine whether a given solution is saturated, unsaturated, dilute, or concentrated (Section 13.2).

Ionic compounds dissolve in water through the action of water molecules surrounding cations and anions in the hydration process

(Section 13.3). The solution process in general involves an intermingling of solute and solvent particles, and hence solubility depends greatly on the similarity of the intermolecular forces of the components of the solution. Temperature, pressure, particle size, and stirring also affect the solution process (Section 13.4).

The composition of a solution is called its concentration. Concentration is a ratio of the amount of the solute to the amount of solution. Percentage by weight/weight, weight/volume and ppm, and molarity are common expressions of concentration (Section 13.5). Solutions of specified concentration can be made by directly dissolving an appropriate amount of solute in solvent, or more dilute solutions can be made from more concentrated solutions by the dilution process (Section 13.6).

Solutions that contain ions conduct electricity and are known as electrolytes. Solutions are classified as strong electrolytes, weak electrolytes, or nonelectrolytes depending on the concentration of ions present. Soluble ionic compounds form strong electrolyte solutions because they completely dissociate in water. Most molecular compounds are nonelectrolytes, but they may form strong or weak electrolyte solutions through ionization reactions. Acids are examples of molecular compounds that produce electrolyte solutions (Section 13.7). Solute particles are either ions (electrolyte solutions) or molecules (nonelectrolytes) (Section 13.8).

The physical picture of a reaction in solution is better described by an ionic equation than by a "traditional" equation (Section 13.9). To write ionic equations you must know how to use the solubility rules (Section 13.10).

There are four so-called colligative properties of solutions; vapor pressure lowering, boiling point elevation, freezing point depression, and osmotic pressure. These properties depend only on the concentration of solute particles (Section 13.11). Solutions separated by osmotic membranes exhibit osmosis and osmotic pressure develops (Section 13.12).

Stoichiometric calculations involving solutions can be carried out in the usual manner (Chapter 10) by remembering that moles of solute can always be calculated by multiplying molarity M times volume V in liters (Section 13.13).

CHAPTER ACCOMPLISHMENTS

After completing this chapter you should be able to

13.1 Solutions defined

1. Define the term "solution."

2. Name and distinguish between the two components of a solution.
3. Give a specific example of each of the seven known types of solutions shown in Table 13.1.

13.2   Solution terminology

4. Define and distinguish among the terms "solubility," "soluble," and "insoluble."
5. Define and distinguish among saturated, unsaturated, and supersaturated solutions.
6. Define and, by example, distinguish between dilute and concentrated solutions.
7. Indicate the type of mixtures to which the terms "miscible" and "immiscible" are applied.
8. Given solubility data, state whether a solution is saturated or unsaturated.

13.3   Solution formation

9. Describe the formation of a water solution of an ionic solid.

13.4   Factors influencing solubility

10. State the factors which affect the extent of solubility of a given solute in a given solvent.
11. State the factors which affect only the speed by which a given solute dissolves in a given solvent.
12. Given the Lewis structures of two molecular compounds, predict whether one will dissolve in the other.
13. Describe the relationship between the partial pressure of a gas and its solubility in a liquid.

13.5   Concentration expressions

14. Define the term "concentration," and distinguish between it and the term "solubility."
15. Given the grams of solute and the grams of solvent or solution, calculate the percentage weight/weight of the solution.
16. Calculate the number of grams of solute and solvent needed to prepare a given mass of solution of some given percentage weight/weight.
17. Given the grams of solute and the volume of a solution, calculate the percentage weight/volume of the solution.
18. Calculate the number of grams of solute needed to prepare a given volume of solution of some given percentage weight/volume.
19. Define the term "molarity."
20. Given the volume and molarity of a solution, calculate the number of moles of solute in that solution.

21.   Given the volume and molarity of a solution, calculate the number of grams of specified solute in that solution.

22.   Given a specific solute, describe how you would prepare a given volume of a solution having some specified molarity.

23.   Calculate the volume of a solution with some specified molarity that would contain a given mass of a specified solute.

13.6   Dilution

24.   Given the volume and molarity of an original solution, calculate the molarity of a new solution prepared by adding a given amount of solvent to the original solution.

25.   Calculate the volume of a more concentrated solution of given molarity which can be diluted to prepare a specified volume of a new solution of lower molarity.

13.7   Electrolytes

26.   Define electrolyte solution.

27.   Define and give specific examples of strong and weak electrolytes and nonelectrolytes.

28.   Distinguish between the processes of dissociation and ionization.

29.   Give examples of strong and weak acids.

13.8   Particles in solution

30.   Describe the basic particles found in solutions of ionic compounds and in solutions of molecular electrolytes and nonelectrolytes.

13.9   Ionic equations

31.   Given a traditional equation, write a net ionic equation for a reaction in which a gas or precipitate is formed.

13.10   Using the solubility rules

32.   Use the solubility rules in Table 13.3 to predict whether a given compound is soluble.

13.11   Colligative properties of solutions

33.   Name four colligative properties.

34.   Describe the effect of added solute on the vapor pressure, boiling point, and freezing point of a liquid solvent.

13.12   Osmotic pressure of solutions

35.   Given the concentrations of two solutions separated by an osmotic membrane, tell the direction in which the net flow of water will occur.

36.   Distinguish among isotonic, hypotonic, and hypertonic solutions.

13.13 Solution stoichiometry

37.   Given a balanced chemical equation and the volume and molarity of one reactant, calculate the volume (or molarity) of a second reactant given the molarity (or volume) of the second reactant.

38.   Given a balanced chemical equation and the volume and molarity of one reactant combining with an excess of the other reactants, calculate the mass of a product that can be produced.

PROBLEMS

13.1 Solutions defined

13.11   Define in your own words what is meant by a solution.

13.12   A compound, like a solution, is homogeneous. Describe how a solution differs from a compound.

13.13   Give specific examples other than those in Table 13.1, for each of the three types of solution in which the solvent is a liquid.

13.14   a.   Are all solutions mixtures?
  b.   Are all mixtures solutions? Explain.

13.15   Which of the following examples would you classify as solutions?
  a.   Helium gas dispersed in air
  b.   Scotch whiskey
  c.   Urine
  d.   Fog
  e.   Soda water
  f.   Bronze (tin homogeneously dispersed in copper)
  g.   Blood
  h.   Milk

13.16   Indicate a possible solute and a possible solvent for each of the examples you classified as a solution in Problem 13.15.

13.17   Why is dusty air not considered a solution?

13.2 Solution terminology

13.18   a.   Describe how you would experimentally determine the solubility of a solute in a given solvent.
  b.   Describe how you would experimentally determine whether a given solution is saturated or unsaturated.

13.19   a.   Does a saturated solution necessarily have to be a concentrated solution? Explain.
  b.   Give a specific example from Table 13.2 of a solution that is saturated, yet dilute.

13.20   Use Table 13.2 to decide whether the following solutions are saturated or unsaturated:
  a.   18 g of NaCl in 100 g H_2O at 20°C
  b.   12 g of $KMnO_4$ in 100 g H_2O at 60°C
  c.   1.7 g of $AgC_2H_3O_2$ in 100 g H_2O at 60°C
  d.   71 g of KBr in 100 g of H_2O at 20°C

13.21   Consult Table 13.2 and decide whether the following solutions are saturated or unsaturated:
  a.   17 g KBr in 25 g H_2O at 60°C
  b.   7.5 g $KMnO_4$ in 50 g H_2O at 60°C
  c.   50. g NaCl in 150 g H_2O at 60°C
  d.   41 g KNO_3 in 125 g H_2O at 20°C

13.22   Other than by adding more solute, describe how a dilute solution of NaCl can be made more concentrated.

13.23   Describe how you would prepare a supersaturated solution of potassium nitrate in water.

13.24   You are given three solutions: one is unsaturated, one is saturated, and one is supersaturated. Describe how you would experimentally distinguish among these three solutions.

13.25   a.   To which type of mixture do the terms "miscible" and "immiscible" apply?
  b.   Give an example of two miscible substances other than alcohol and water.
  c.   Give an example of two immiscible substances other than oil and water.

13.3 Solution formation

13.26   a.   What attractive forces must be broken

down in the dissolving of an ionic solid?

b. What new attractive forces are formed in the dissolving of an ionic solid?

c. Explain why NaCl does not dissolve in a nonpolar solvent such as gasoline.

13.4   Factors influencing solubility

13.27   a. How does the nature of a solute affect whether it will dissolve in a given solvent?

b. What type of solutes will dissolve best in water solution?

c. What type of solutes will dissolve best in gasoline?

13.28 Explain why a solute such as sugar dissolves faster in hot water than in cold water.

13.29   a. Explain why a bottle of soda goes "flat" if left open to the air.

b. Would keeping the bottle warm help the soda to stay fizzier longer? Explain.

13.30   a. What factor enables ground-up solutes to dissolve faster than the same solute in the form of chunks?

b. Give two practical examples where solutes are finely ground before dissolving them in a solvent.

13.31   a. Describe how stirring enables a solute to dissolve more quickly.

b. Does stirring affect the solubility of a solute in a given solvent?

13.32   a. Write out Lewis structures for methyl alcohol, CH_3OH, and hydrogen chloride, HCl.

b. Predict whether HCl will dissolve in CH_3OH. Explain your answer.

13.5   Concentration expressions

13.33 Describe and illustrate by example the difference between the terms "solubility" and "concentration."

13.34 A solution was prepared by dissolving 18.3 g of KNO_3 in 75.0 g of water at 20°C. What is the percentage weight/weight of this solution?

13.35 In a laboratory, 179 g of a silver nitrate solution is evaporated to dryness. After all the water is evaporated, 39.6 g of silver nitrate remains.

a. What is the percentage weight/weight of $AgNO_3$ in the original solution?

b. How much water was boiled off?

13.36 An antifreeze solution was prepared by dissolving 250. g of methyl alcohol in 650. g of water. What is the concentration of this solution in percentage weight/weight?

13.37 Describe how you would prepare 250. g of a 5.00 percent $BaCl_2$ solution weight/weight.

13.38 How many grams of sugar and water would you use to prepare 20.0 g of a 0.90 percent weight/weight solution?

13.39 Using Table 13.2, calculate the weight/weight percentage of a saturated aqueous potassium nitrate solution at 20°C.

13.40 Using Table 13.2, calculate the weight/weight percentage of a saturated aqueous $KMnO_4$ solution at 60°C.

13.41 What mass of sugar and water would be needed to prepare 90.0 g of a 15.0 percent weight/weight sugar solution?

13.42 75.0 mL of solution was prepared by dissolving 13.8 g of KNO_3 in water at 20°C. What is the percentage weight/volume of this solution?

13.43 In a laboratory, 179 mL of a silver nitrate solution is evaporated to dryness. After all the water is evaporated, 69.3 g of silver nitrate remains. What is the percentage weight/volume of $AgNO_3$ in the original solution?

13.44 How many grams of sugar would you use to prepare 20.0 mL of a 0.50% w/v solution?

13.45 How many milliliters of 4.0% w/v $Ba(NO_3)_2$ solution would contain 1.25 g of barium nitrate?

13.46 An antifreeze solution was prepared by dissolving 250. mL of methyl alcohol in 700. mL of water. What is the concentration of this solution in percent volume/volume?

13.47 Which solution is more concentrated: a 0.01 percent weight/volume glucose solution or a 50-ppm solution?

13.48 How much NaCl would be needed to prepare 800. mL of a 0.900 percent physiological saline solution?

13.49 Describe the differences in the expression

of the concentration of a solution by percentage weight/volume or by molarity.

13.50 Calculate the molarity of a solution which contains 2.50 moles of KNO_3 dissolved in 5.00 L.

13.51 How many moles of NaCl are present in 100. mL of 0.125 M solution?

13.52 How many grams of glucose, $C_6H_{12}O_6$, are present in 1.50 L of a 0.400 M solution?

13.53 Given the equipment available in your laboratory, describe how you would prepare 0.500 L of a salt solution, 0.900 M NaCl.

13.54 Find the number of grams of NaOH present in 375 mL of a 0.300 M solution.

13.55 How many grams of solute are needed to prepare 100.0 mL of a 0.250 M silver nitrate solution?

13.56 Calculate the molarity of an HCl solution with density 1.18 g/mL. The solution is 36.0 percent HCl (weight/weight).

13.57 A solution of $CaCl_2$, of density 1.05 g/mL contains 6.00 percent $CaCl_2$ (weight/weight). Calculate the molarity of the solution.

13.58 In an experiment, 0.175 mole of H_2SO_4 is needed for a reaction. The H_2SO_4 solution available is 18.0 M. How many milliliters of this solution must be measured out?

13.59 A student needs to measure out 15.2 g of $NaHCO_3$. The only $NaHCO_3$ present in the laboratory is in a 0.100 M solution. How many milliliters of the solution must the student measure out in order to have the required number of grams in solution?

13.6   Dilution

13.60 What is the molarity of 50.0 mL of a 0.50 M NaOH solution after it has been diluted to 300. mL?

13.61 If 300.0 mL of water is added to 200.0 mL of a 0.500 M Na_2SO_4 solution, what is the molarity of the resulting solution?

13.62 What volume of a 1.25 M KNO_3 solution is required to make 1.00 L of a 0.100 M solution by dilution?

13.63 To what volume must you dilute 80.0 mL of 3.0 M $CuSO_4$ to have a 0.50 M solution?

13.64 A student needs to prepare 200.0 mL of a 4.0 M HCl solution. The student has available a 12.0 M HCl solution. What volume of the concentrated solution must be measured out to prepare the 4.0 M solution by dilution?

13.7   Electrolytes

13.65 State the evidence for the existence of ions in a solution of NaCl.

13.66 Describe how you would experimentally show that HCl gas ionizes when dissolved in water.

13.67 Hydrogen bromide is a covalent substance which acts as a strong electrolyte in aqueous solution; $HC_2H_3O_2$ is a covalent substance which acts as a weak electrolyte in aqueous solution. Explain how one covalent substance can be a strong electrolyte and the other weak.

13.68 Using NaCl and HCl as specific examples, describe the differences between the processes of dissociation and ionization.

13.69 An aqueous solution of HCl conducts electricity. However, a solution of HCl in benzene does not conduct electricity. Explain what is happening in the two solutions.

13.70 Solid sodium chloride, although made up of ions, does not conduct electricity. Molten sodium chloride is, however, a strong electrolyte. What condition other than the presence of ions is necessary for the conduction of electricity?

13.71 Give the names and formulas of three strong and three weak acids.

13.8   Particles in solution

13.72 What solute particles will be found in aqueous solutions of the following:

a.	KBr	e.	H_2SO_3
b.	$Ca(OH)_2$	f.	Glucose, $C_6H_{12}O_6$
c.	HNO_3	g.	Methyl alcohol, CH_3OH
d.	$HC_2H_3O_2$	h.	$(NH_4)_3PO_4$

13.9   Ionic equations

13.73 Write balanced net ionic equations for the following reactions:

a. $K_2SO_4(aq) + Ba(NO_3)_2(aq) \longrightarrow$
$$KNO_3(aq) + BaSO_4 \downarrow$$

b. $Zn(s) + H_2SO_4(aq) \longrightarrow$
$$ZnSO_4(aq) + H_2 \uparrow$$

c. $NaCl(aq) + AgNO_3(aq) \longrightarrow$
$$AgCl \downarrow + NaNO_3(aq)$$

d. $(NH_4)_2S(aq) + Pb(NO_3)_2(aq) \longrightarrow$
$$NH_4NO_3(aq) + PbS \downarrow$$

e. $FeCl_3(aq) + LiOH(aq) \longrightarrow$
$$LiCl(aq) + Fe(OH)_3 \downarrow$$

f. $HCl(aq) + Na_2CO_3(aq) \longrightarrow$
$$NaCl(aq) + H_2O + CO_2 \uparrow$$

13.10 Using the solubility rules

13.74 Using Table 13.3, indicate which of the following compounds are soluble in water:

a.	K_2SO_4	**e.**	$Al(OH)_3$
b.	$CaCl_2$	**f.**	$CsOH$
c.	Na_2S	**g.**	$NH_4C_2H_3O_2$
d.	$Mg_3(PO_4)_2$		

13.75 What substances would be formed in solution from the following solutes:

a.	Na_2SO_4	**d.**	$Ca_3(PO_4)_2$
b.	FeS	**e.**	$Mg(NO_3)_2$
c.	KOH	**f.**	$(NH_4)_2CO_3$

13.11 Colligative properties of solutions

13.76 Name four colligative properties.

13.77 Comparing pure water and a 1 M aqueous glucose solution, which has the higher (**a**) vapor pressure, (**b**) boiling point, (**c**) freezing point, (**d**) osmotic pressure?

13.78 With respect to the same properties, how would a 1 M NaCl solution compare with the two solutions mentioned in Problem 13.77?

13.12 Osmotic pressure of solutions

13.79 Compartments A and B are separated by an osmotic membrane. Given the following concentrations of solutions in the compartments, indicate whether there will be a net flow of water, and if so, in which direction.

a. A contains pure water; B contains 1 percent glucose.

b. A contains 1 M glucose; B contains 1 M NaCl.

c. A contains 1 M NaCl; B contains 0.5 M KBr.

d. A contains 0.5 M KBr; B contains 1 M glucose.

13.80 Red blood cells are safely stored in 0.9 percent NaCl solution because this solution is isotonic with the interior of the cells. What does this mean?

13.81 Describe each of the following solutions as hypertonic, hypotonic, or isotonic with respect to 0.9 percent NaCl (0.15 M NaCl):

a. 2 percent NaCl
b. 0.15 M glucose
c. 0.35 M glucose
d. 0.12 M $CaCl_2$
e. 0.30 M glucose

13.13 Solution stoichiometry

13.82 How many milliliters of a 0.100 M H_2SO_4 solution are needed to react completely with 150. mL of 0.250 M NaOH solution according to the equation

$H_2SO_4(aq) + 2\ NaOH(aq)$
$$\longrightarrow Na_2SO_4(aq) + 2H_2O(l)$$

13.83 Find the molarity of a nitric acid solution if 17.5 mL is required to react with 13.9 mL of a 0.755 M NaOH solution:

$NaOH(aq) + HNO_3(aq)$
$$\longrightarrow NaNO_3(aq) + H_2O(l)$$

13.84 What volume of a 1.00 M K_2CrO_4 solution is needed to react with 50.0 mL of a 0.600 M $BaCl_2$ solution according to the equation

$BaCl_2(aq) + K_2CrO_4(aq)$
$$\longrightarrow BaCrO_4 \downarrow + 2KCl(aq)$$

13.85 According to the equation in Problem 13.84, what mass of $BaCrO_4$ can be obtained from the reaction of 50.0 mL of 0.600 M $BaCl_2$?

13.86 Find the molarity of a silver nitrate solution if 65.4 mL is required to react completely with 24.9 mL of a 0.500 M NaCl solution according to the equation

$AgNO_3(aq) + NaCl(aq)$
$$\longrightarrow AgCl \downarrow + NaNO_3(aq)$$

13.87 What mass of silver chloride will precipitate from the reaction in Problem 13.86?

CHEMICAL EQUILIBRIUM
CHAPTER

14

Si

Silicon

28.09

REVERSIBLE REACTIONS

14.1 Throughout this book we have written and discussed chemical reactions as if they began with an initial set of materials (reactants) and went completely to a final set of materials (products). Actually, in many cases the products can react to re-form the original reactants. Reactions in which the products can themselves react to re-form reactants are called **reversible reactions.**

You have previously seen examples of *reversible* processes. For example, in Section 12.7 you encountered the two opposing processes, vaporization and condensation, which occur whenever a liquid is in a closed container. You saw that, in a closed container, eventually the rate of condensation equals the rate of vaporization, and the vapor pressure remains constant thereafter (Figure 14.1). It is important to realize that in such a case we are not dealing with a *static* (inactive) mixture of liquid and vapor. Rather, active vaporization and condensation are still taking place, although the processes are difficult to see because they are occurring at the same rate. We call this a state of **dynamic (active) equilibrium.** This term is used whenever two opposing processes coexist at the same rate.

A dynamic equilibrium also exists between a dissolved and undissolved substance in a **saturated** solution.

$$NaCl(s) \rightleftharpoons Na^+(aq) + Cl^-(aq)$$

$$Sugar(s) \rightleftharpoons sugar(aq)$$

The two half-headed arrows ($\rightleftharpoons$) indicate that the process is reversible and can proceed in either direction. The rate at which the solid NaCl dissociates into ions is equal to the rate at which ions emerge from solution, and the rate at which sugar dissolves is equal to the rate at which it comes out of solution.

With this concept in mind of two opposing processes, one that is

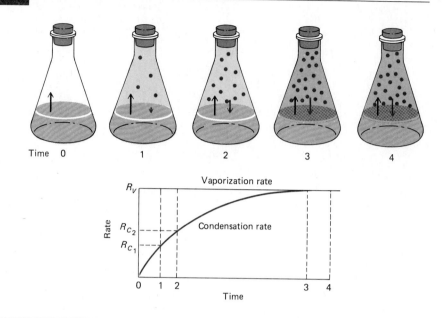

FIGURE 14.1

Establishment of liquid-vapor equilibrium:

Liquid $\underset{\text{condensation}}{\overset{\text{vaporization}}{\rightleftharpoons}}$ vapor

At time zero, only vaporization (↑) is occurring. Vaporization continues at a steady rate, as shown by the straight horizontal line on the graph. As vapor forms, condensation slowly starts (↓). At time 1, the condensation rate is R_{c1} and the rate of vaporization is R_v. At time 2, the condensation rate is R_{c2} and the vaporization rate remains R_v. Eventually the rate of condensation equals the rate of vaporization and there is a dynamic equilibrium. At equilibrium the liquid and vapor concentration remain fixed.

exactly the reverse of the other, let us consider an example of a reversible chemical reaction.

If we place 0.1 mole of N_2O_4, a colorless gas, into a sealed 1-L tube and heat it to 100°C, we obtain visible evidence of the formation of a reddish brown gas, which on analysis proves to be NO_2. Thus we have evidence that the reaction $N_2O_4 \rightarrow 2NO_2$ is occurring. The intensity of the reddish brown color increases rapidly in the beginning, but after a period of time the color reaches a maximum. The contents of the container at this point are found to contain both NO_2 and N_2O_4.

If we reverse the experiment and start with 0.1 mole of NO_2, the reddish brown color begins to diminish. This is evidence that the reaction $2NO_2 \rightarrow N_2O_4$ is occurring. After a period of time the reddish brown color reaches a minimum and does not diminish further. We conclude from these experiments that the reaction is reversible because:

1. If we start with N_2O_4, NO_2 forms, and if we start with NO_2, N_2O_4 forms.

2. The reactant is not totally converted to product in either experiment. Evidence for this lies in the color of the mixture.

    a. If we begin with colorless N_2O_4, a reddish brown coloration as dark as pure NO_2 will never develop.

    b. If we begin with NO_2 a totally colorless gas indicative of pure N_2O_4 will not be produced.

The reversible reaction between N_2O_4 (dinitrogen tetroxide) and NO_2 (nitrogen dioxide) can be written in the following way:

$$N_2O_4(g) \rightleftharpoons 2NO_2(g) \qquad (14.1)$$

This is an example of dynamic equilibrium for a chemical system, i.e., a chemical equilibrium. In order to understand chemical equilibrium in more detail we need to discuss what we mean by the rate of a chemical reaction.

RATES OF REACTION

14.2 The **rate,** or **speed,** of a chemical reaction is a measure of the change in concentration of either a reactant or product in a given amount of time. Reaction rate is change in concentration per change in time, just as travel rate is change in distance per change in time. The rate of decomposition of dinitrogen tetroxide can be determined by measuring the increase in reddish-brown color per unit time. The increase in color is related to the increase in NO_2 concentration.

For a particular reaction at a given temperature, the answer to the question of how fast or slow the rate of reaction is depends on the concentration of the reacting species. Before a reaction can occur, the reacting particles must collide with each other. The *greater the concentrations of reactants, the greater the number of collisions* per unit time. An analogy can be made to a game of billiards; the greater the number of billiard balls on the table, the greater the number of possible and probable collisions with a moving ball. Mathematically, we can express this as

$$\text{Rate} \propto \text{concentration of reacting species}$$

or

$$\text{Rate} = k \, [\text{concentration}]$$

A proportionality expression can be converted to an equation through the use of a constant (see Appendix 1F). In this case k is called a **rate constant.**

Now let us consider the reaction represented by Equation (14.1) in terms of the effect of concentration on reaction rate. The rate of the forward reaction depends on the concentration of N_2O_4, and the rate of the reverse reaction on the NO_2 concentration. If we begin with pure N_2O_4, then initially the concentration of N_2O_4 is large, there are many collisions of N_2O_4 molecules, and so the rate of decomposition to NO_2 is fast. As time passes, the concentration of N_2O_4 decreases, the chances for collisions decrease, and therefore the rate of decomposition decreases. However, the concentration of NO_2 is increasing, and thereby the rate of the reverse reaction increases. At some point the rate of the reverse reaction becomes equal to the rate of the forward reaction and the concentrations of reactants and products no longer change. This process is illustrated diagrammatically and graphically in Figure 14.2. When the rate of the forward reaction equals the rate of the reverse reaction, we say that the reaction is at **chemical equilibrium,** and the concentrations of reactants and products at equilibrium are called **equilibrium concentrations.** It is important to remember that the equilibrium condition is a dynamic (active) condition.

At equilibrium

$\text{Rate}_{\text{forward}} = \text{Rate}_{\text{reverse}}$

EQUILIBRIUM CONSTANT

14.3   Now let us apply the concept of the dependence of rate on concentration to a reversible reaction such as:

$$A + B \rightleftharpoons C + D$$

For this reversible reaction in which A and B combine in a single step to give C and D, and C and D can likewise combine to form A and B,

FIGURE 14.2

The $N_2O_4 \rightleftharpoons 2NO_2$ equilibrium. At time zero, beginning with pure, colorless N_2O_4 only the forward reaction occurs at a rate designated by the arrow (↑). (1) As red-brown NO_2 forms, the reverse reaction can occur, but at first the rate is slow because of the small concentration of NO_2. (2) The rate of the forward reaction decreases because N_2O_4 decreases. The rate of the reverse reaction increases because NO_2 increases. (3 and 4) Equilibrium is obtained: rate$_{\text{forward}}$ = rate$_{\text{reverse}}$ and the concentration of N_2O_4 and NO_2 remain fixed. Compare this figure to Figure 14.1.

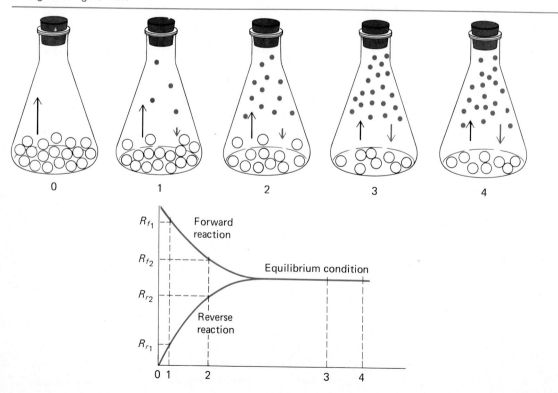

we can write individual expressions for the rates of the *forward* and *reverse* reactions at any instant of time:

$$\text{Rate}_{\text{forward}} \propto \text{concentration } A \times \text{concentration } B$$

$$\text{Rate}_{\text{reverse}} \propto \text{concentration } C \times \text{concentration } D$$

In the usual manner, an equal sign and proportionality constant can replace the symbol $\propto$. Here the proportionality constant is the *rate constant* and it is fixed in value for a *particular reaction at a given temperature.*

$$\text{Rate}_{\text{forward}} = k_{\text{forward}}[A][B] \qquad (14.2)$$

$$\text{Rate}_{\text{reverse}} = k_{\text{reverse}}[C][D] \qquad (14.3)$$

The bracketing symbol [ ] means that the concentration of a substance is expressed in moles per liter. For example, $[A]$ is read, "concentration of A in moles per liter."

At the condition of equilibrium remember that

$$\textbf{Rate}_{\textbf{forward}} = \textbf{Rate}_{\textbf{reverse}}$$

Therefore, because the left-hand terms of Equations 14.2 and 14.3 are equal, the right-hand terms must also be equal.

$$k_{\text{forward}}[A][B] = k_{\text{reverse}}[C][D]$$

Rewriting this to isolate the constants to one side, we have

$$\frac{k_{\text{forward}}}{k_{\text{reverse}}} = \frac{[C][D]}{[A][B]} \qquad \textit{only at equilibrium}$$

Because k_{forward} and k_{reverse} are both constants, their ratio is also a constant which is called the **equilibrium constant,** and it is often symbolized by K_{eq}.

$$\underset{\substack{\text{Equilibrium} \\ \text{constant}}}{\frac{k_{\text{forward}}}{k_{\text{reverse}}}} = K_{\text{eq}} = \underset{\substack{\text{Equilibrium} \\ \text{expression}}}{\frac{[C][D]}{[A][B]}}$$

Each reversible reaction has a unique K_{eq}, which has a fixed value for a particular temperature. The concentrations in the equilibrium expression are the equilibrium concentrations. As you shall see (Section 14.5), the magnitude of K_{eq} offers valuable information about the amounts of products and reactants present at equilibrium in a reversible reaction.

The coefficients of an equation for a reversible reaction must be reflected in the equilibrium expression. The numerator of the equilibrium expression contains each product concentration raised to a power equal to the coefficient of that product in the balanced reaction

(1 is understood in the preceding example). The denominator contains the reactant concentrations raised to their respective powers (again 1 for the preceding example). For the general balanced equation

$$aA + bB \rightleftharpoons cC + dD$$

the equilibrium expression is written

$$K_{eq} = \frac{[C]^c[D]^d}{[A]^a[B]^b}$$

SAMPLE EXERCISE

14.1

Write an equilibrium-constant expression for the equilibrium:

$$N_2O_4 \rightleftharpoons 2NO_2$$

SOLUTION

Step 1   In the numerator write the concentration of each product raised to a power equal to its coefficient in the balanced equation

$$K_{eq} = \frac{[NO_2]^2}{?}$$

Step 2   In the denominator write the reactant concentrations each raised to a power equal to its coefficient in the balanced equation.

$$K_{eq} = \frac{[NO_2]^2}{[N_2O_4]}$$

SAMPLE EXERCISE

14.2

When the color in the reaction vessel remained the same, indicating that equilibrium had been established for the gas mixture reaction, a chemist found that the concentration of N_2O_4 in the vessel was 0.040 mole/L and that the concentration of NO_2 was 0.12 mole/L. Calculate the equilibrium constant for the $N_2O_4(g) \rightleftharpoons NO_2(g)$ reaction.

SOLUTION

Use the equilibrium-constant expression written in Sample Exercise 14.1:

$$K_{eq} = \frac{[NO_2]^2}{[N_2O_4]}$$

Fill in the appropriate numerical concentrations and do the operations indicated by the exponents and fraction line:

$$K_{eq} = \frac{(0.12)^2}{0.04} = \frac{0.014}{0.040} = 0.36$$

In Section 14.5 you will see how knowing the value of K_{eq} provides information about the product and reactant concentrations at equilibrium.

Problem 14.1

Write an equilibrium-constant expression for each of the following equilibria:

a. $H_2(g) + I_2(g) \rightleftharpoons 2HI(g)$

b. $N_2(g) + 3H_2(g) \rightleftharpoons 2NH_3(g)$

RULES FOR WRITING K_{eq}

14.4 Certain conventions or rules have been established for writing correct equilibrium expressions.

Rule 1 An equilibrium-constant expression can only be written for a balanced chemical equation.

Rule 2 At a particular temperature each reaction has a unique equilibrium-constant value. The value of the equilibrium constant changes with the temperature.

Rule 3 The equilibrium expression contains the concentration terms for *gases and substances in solution only.*

Rule 4 The concentration term ([solid] or [liquid]) for a pure liquid or solid does not appear in the normal equilibrium expression. The concentration of a pure liquid or solid is a constant, and so it is automatically incorporated into the equilibrium constant. For example, if we consider pure water, for which we know the density to be 1.0 g/mL, we can calculate the concentration in moles per liters to be 55 *M*.

$$\frac{1.0 \text{ g}}{\text{mL}} \times \frac{1000}{1000} = \frac{1000 \text{ g}}{1000 \text{ mL}} = \frac{1000 \text{ g}}{1 \text{ L}} \times \frac{1 \text{ mole}}{18 \text{ g}} = 55 \frac{\text{moles}}{\text{L}}$$

A sample of pure water always has this concentration. *All pure solids or liquids have constant concentrations.*

As a demonstration of Rule 4, consider the equilibrium

$$CO_2(g) + H_2(g) \rightleftharpoons CO(g) + H_2O(l)$$

Using the general method of writing K_{eq}, we write

$$K = \frac{[CO][H_2O]}{[CO_2][H_2]}$$

But $[H_2O]$ is a constant, so if we divide each side of this equation by $[H_2O]$, the left side will still be a constant:

$$\frac{K}{[H_2O]} = K_{eq} = \frac{[CO]}{[CO_2][H_2]}$$

As a matter of practicality, simply omitting the concentration of a solid or liquid when one writes the equilibrium expression has the desired effect of incorporating these concentrations in K_{eq}.

SAMPLE EXERCISE

14.3

Write the equilibrium-constant expression for

$$CH_3Cl(g) + 3Cl_2(g) \rightleftharpoons CCl_4(l) + 3HCl(g)$$

SOLUTION

Step 1 Put the product concentrations raised to the power of their coefficients in the numerator. Do not include CCl_4 because it is a pure liquid.

$$K_{eq} = \frac{[HCl]^3}{?}$$

Step 2 The denominator contains the reactant concentrations raised to the power of the coefficients in the balanced equation:

$$K_{eq} = \frac{[HCl]^3}{[CH_3Cl][Cl_2]^3}$$

SAMPLE EXERCISE

14.4

An equilibrium exists in a saturated aqueous solution of silver chloride between the solid silver chloride and the dissolved ions:

$$AgCl(s) \rightleftharpoons Ag^+(aq) + Cl^-(aq)$$

Write the equilibrium expression.

SOLUTION

Solid silver chloride is omitted from the equilibrium expression, leaving terms only in the numerator.

$$K_{eq} = [Ag^+][Cl^-]$$

The equilibrium constant for the equilibrium between a solid and its dissolved ions is known as the **solubility product constant,** abbreviated K_{sp}. In this case,

$$K_{sp} = [Ag^+][Cl^-]$$

SAMPLE EXERCISE

14.5

Write the equilibrium expression for

$$4NH_3(g) + 5O_2(g) \rightleftharpoons 4NO(g) + 6H_2O(g)$$

SOLUTION

Step 1

$$K_{eq} = \frac{[NO]^4[H_2O]^6}{?}$$

The concentration of water is included in K_{eq} because water is present as a gas in this reaction.

Step 2

$$K_{eq} = \frac{[NO]^4[H_2O]^6}{[NH_3]^4[O_2]^5}$$

Problem 14.2

Write the equilibrium-constant expression for each of the following equilibria:

a. $CO(g) + 2H_2(g) \rightleftharpoons CH_3OH(l)$

b. $NH_4Cl(s) \rightleftharpoons NH_3(g) + HCl(g)$

c. $3Fe(s) + 4H_2O(g) \rightleftharpoons Fe_3O_4(s) + 4H_2(g)$

Problem 14.3

Write the solubility-product expression K_{sp} for the equilibria in the following saturated solutions:

a. $CaCO_3(s) \rightleftharpoons Ca^{2+}(aq) + CO_3^{2-}(aq)$

b. $PbCl_2(s) \rightleftharpoons Pb^{2+}(aq) + 2Cl^-(aq)$

c. $Ba_3(PO_4)_2(s) \rightleftharpoons 3Ba^{2+}(aq) + 2PO_4^{3-}(aq)$

INTERPRETING THE VALUE OF K_{eq}

14.5 What is the purpose or usefulness of knowing the value of K_{eq} for a reaction? Consider the simple example

$$R \rightleftharpoons P$$

for which the equilibrium expression is

$$K_{eq} = \frac{[P]}{[R]}$$

Because the equilibrium constant gives us a measure of the concentration of product $[P]$ compared with reactant $[R]$ concentrations at equilibrium, the value of K_{eq} tells us how far a reaction has proceeded toward product at the equilibrium point.

TABLE 14.1   COMPARISON OF K_{eq} VALUE AND REACTANT AND PRODUCT CONCENTRATIONS AT EQUILIBRIUM

K_{eq}	Description of Equilibrium Condition
1×10^{30} 1×10^{20} 1×10^{10}	Essentially all products at equilibrium
More products than reactants at equilibrium	
1×10^{2} 1 1×10^{-2}	Significant amounts of products and reactants at equilibrium
More reactants than products at equilibrium	
1×10^{-10} 1×10^{-20} 1×10^{-30}	Essentially all reactants at equilibrium; reaction has not occurred to any significant extent

If the equilibrium constant is large ($>1 \times 10^{2}$), then the equilibrium system is made up mostly of product. We say that the reaction has proceeded far to the right. If $[P]/[R] = 100$, then clearly the numerator, $[P]$, is considerably greater than the denominator, $[R]$.

If the equilibrium constant is small ($<1 \times 10^{-2}$), the equilibrium condition consists mainly of reactants. The equilibrium is said to be far to the left. If $[P]/[R] = 0.01$, then clearly the denominator, $[R]$, is considerably greater than the numerator, $[P]$.

For reactions in which the numerical value of K_{eq} is between 1×10^{-2} and 1×10^{2}, the equilibrium state is a mixture containing significant quantities of products and reactants. Table 14.1 compares K_{eq} values and product and reactant concentrations at equilibrium, and Sample Exercise 14.6 offers some illustrations.

SAMPLE EXERCISE

14.6

Given K_{eq} describe qualitatively the equilibrium composition for each of the following reactions:

a. $N_2O_4(g) \rightleftharpoons 2NO_2(g)$ K_{eq} (at 100°C) $= 3.6 \times 10^{-1}$
b. $N_2(g) + 3H_2(g) \rightleftharpoons 2NH_3(g)$ K_{eq} (at 25°C) $= 5 \times 10^{8}$
c. $H_2(g) + Cl_2(g) \rightleftharpoons 2HCl(g)$ K_{eq} (at 25°C) $= 2.4 \times 10^{33}$
d. $Cl_2(g) \rightleftharpoons 2Cl(g)$ K_{eq} (at 1000°C) $= 1.2 \times 10^{-6}$
e. $N_2(g) + O_2(g) \rightleftharpoons 2NO(g)$ K_{eq} (at 25°C) $= 1 \times 10^{-30}$

SOLUTION

a. By the guidelines given in Table 14.1, the equilibrium composition will consist of significant quantities of products and reactants because K_{eq} is between 1×10^{-2} and 1×10^{2}.
b. The equilibrium system will consist mostly of products.
c. The equilibrium state will contain essentially only product. The reaction is said to have gone to completion.

d. There will be a much greater amount of reactant than product in the equilibrium state.

e. Essentially only reactants will remain at equilibrium. The reaction does not proceed as written; the reverse reaction goes to completion.

Problem 14.4

Given K_{eq} and the guidelines in Table 14.1, describe qualitatively the equilibrium composition for each of the following reactions:

a. $I_2(g) + Cl_2(g) \rightleftharpoons 2ICl(g)$ $K_{eq} = 2.0 \times 10^5$

b. $SO_2(g) + NO_2(g) \rightleftharpoons SO_3(g) + NO(g)$ $K_{eq} = 2.5 \times 10^{-1}$

c. $ZnO(s) + H_2(g) \rightleftharpoons Zn(s) + H_2O(g)$ $K_{eq} = 2 \times 10^{-16}$

LE CHÂTELIER'S PRINCIPLE

14.6   It is possible to change the point at which equilibrium is achieved, which is called the **position of equilibrium,** and therefore the equilibrium concentrations of products and reactants by altering the conditions of concentration, pressure, or temperature that a system in equilibrium experiences. This means that we can sometimes "shift" a reaction in such a way as to produce more of a desired product.

Shifts in equilibrium are governed by **Le Châtelier's principle,** which states that *if a stress is applied to a system in equilibrium, the equilibrium will shift so as to relieve that stress.* Let us examine how this principle works and affects the yield of ammonia (NH_3) in the reaction

$$N_2(g) + 3H_2(g) \rightleftharpoons 2NH_3(g)$$

The starting point in the following discussion is always the equilibrium condition; i.e., we start with equilibrium concentrations of N_2, H_2, and NH_3.

Effect of a change in concentration   Increasing the concentration of either N_2 or H_2 will cause the equilibrium to shift to the product side away from the stress of added reactant concentration. The equilibrium will reestablish itself with a higher concentration of ammonia and a lower concentration of the reactant that was not added. A higher yield of ammonia is obtained.

Equilibrium: $N_2 + 3H_2 \rightleftharpoons 2NH_3$

Stress by adding N_2:

           Shift $\longrightarrow$ to consume N_2 and relieve stress

Such shifting also consumes H_2 and produces more NH_3.

If the equilibrium is disturbed by the removal of ammonia, the system will react to this stress of lowered product concentration by shifting to the product side. The result will be an increase in the yield of ammonia. More N_2 and H_2 will be consumed.

Equilibrium $$N_2 + 3H_2 \rightleftharpoons 2NH_3$$

Stress by removing NH_3:

Shift $\longrightarrow$ to replace NH_3 and relieve stress

Such shifting also consumes N_2 and H_2.

If the equilibrium is disturbed by adding ammonia, the system will react to this stress of added product concentration by shifting to the reactant side. The result will be an increase in the equilibrium concentrations of N_2 and H_2.

Equilibrium: $$N_2 + 3H_2 \rightleftharpoons 2NH_3$$

Stress by adding NH_3:

Shift $\longleftarrow$ to consume NH_3 and relieve stress

Such shifting also increases N_2 and H_2.

Effect of temperature   You have previously seen (Section 10.12) that heat can be treated as if it were a product in an exothermic reaction or a reactant in an endothermic reaction. Increasing temperature is then thought of as providing more heat for the reaction and decreasing the temperature as providing less heat for the reaction.

The reaction of hydrogen and nitrogen to produce ammonia is exothermic:

$$N_2 + 3H_2 \rightleftharpoons 2NH_3 + heat$$

Increasing the temperature of this reaction causes the equilibrium to shift to the reactant side, thereby decreasing the yield of ammonia. On the other hand, the reaction responds to the stress of a decreased temperature by shifting the equilibrium to the product side, thus increasing the yield of ammonia.

Equilibrium: $$N_2 + 3H_2 \rightleftharpoons 2NH_3 + heat$$

Stress by increased T means more heat:

Shift $\longleftarrow$ to consume heat

Such shifting also consumes NH_3.

Stress by decreased T means less heat:

$$\text{\textbf{Shift}} \longrightarrow \text{to produce heat}$$

Such shifting also produces NH_3.

Effect of pressure   A change in pressure affects the equilibrium of only those reactions which undergo a change in volume in proceeding from reactant to product. In practice, this means that pressure changes affect those reactions which show a change in the number of moles of gaseous materials. The stress of an increased pressure (decreased V) is relieved by a shift in the equilibrium to the side having fewer gas molecules. A decrease in pressure (increased V) shifts the equilibrium to the side having a greater number of gas molecules. Consider the ammonia example,

$$N_2(g) + 3H_2(g) \rightleftharpoons 2NH_3(g)$$

$$\text{1 mole } N_2 + \text{3 moles } H_2 \qquad \text{2 moles } NH_3$$

Here 4 moles of gaseous reactants are converted into 2 moles of gaseous products. An increase in pressure P will shift the equilibrium to increase the yield of ammonia. Decreased P has the opposite effect.

Equilibrium: $N_2(g) + 3H_2(g) \rightleftharpoons 2NH_3(g)$

Stress by increasing P and decreasing V:

$$\text{\textbf{Shift}} \longrightarrow \text{to produce fewer moles of gas and relieve } P$$

Such shifting also produces NH_3.
Stress by decreasing P and increasing V:

$$\text{\textbf{Shift}} \longleftarrow \text{to produce more moles of gas and restore } P$$

Such shifting also consumes NH_3.
    Table 14.2 provides a summary of the ammonia example. Le Châtelier's principle indicates that the best equilibrium yield of ammonia should be obtained from using an excess concentration of one of the reactants at a low temperature and a high pressure. Fritz Haber applied those ideas to the commercial synthesis of ammonia, a very important chemical in the manufacture of farm fertilizers.

SAMPLE EXERCISE

14.7

State the effect, if any, of the following changes on the position of the equilibrium for the reaction

$$CH_4(g) + Cl_2(g) \rightleftharpoons CH_3Cl(g) + HCl(g) + 26.4 \text{ kcal}$$

a.   Increasing the concentration of CH_4

TABLE 14.2

EFFECTS OF CONCENTRATION, T, AND P CHANGES ON THE POSITION OF THE EQUILIBRIUM $N_2(g) + 3H_2(g) \rightleftharpoons 2NH_3(g) + \text{heat}$

Change	Shift	Result Is an Increase in	
		Reactants	Products
Increase $[N_2]$	$\longrightarrow$		X
Increase $[H_2]$	$\longrightarrow$		X
Increase $[NH_3]$	$\longleftarrow$	X	
Decrease $[N_2]$	$\longleftarrow$	X	
Decrease $[H_2]$	$\longleftarrow$	X	
Decrease $[NH_3]$	$\longrightarrow$		X
Increase temperature	$\longleftarrow$	X	
Decrease temperature	$\longrightarrow$		X
Increase pressure	$\longrightarrow$		X
Decrease pressure	$\longleftarrow$	X	

b. Removing CH_3Cl as it is formed
c. Increasing the reaction temperature
d. Decreasing the total pressure

SOLUTION

a. The stress of added CH_4 is relieved by a shift in the equilibrium ($\rightarrow$) to products.
b. The effect of removing CH_3Cl is to shift the equilibrium ($\rightarrow$) to products to produce more CH_3Cl.
c. The effect of increasing the reaction temperature is to shift the reaction ($\leftarrow$) toward absorbing heat, i.e., back to reactants.
d. Because there is no change in the number of moles of gas (2 moles gaseous reactants, 2 moles gaseous products), a pressure change will have no effect on the equilibrium position.

SAMPLE EXERCISE

14.8

What is the effect of adding a chloride ion to the equilibrium

$$AgCl(s) \rightleftharpoons Ag^+(aq) + Cl^-(aq)$$

SOLUTION

Increasing the concentration of chloride ions will shift the equilibrium to $AgCl(s)$ and decrease the concentration of Ag^+ in solution.

Problem 14.5

State the effect, if any, of the following changes on the position of the equilibrium for the reaction

$$C_2H_2(g) + H_2(g) \rightleftharpoons C_2H_4(g) + 41.2 \text{ kcal}$$

a. Increasing the concentration of H_2

b. Adding C_2H_4

c. Increasing the total pressure

d. Decreasing the reaction temperature

Problem 14.6

For the equilibrium

$$BaSO_4(s) \rightleftharpoons Ba^{2+}(aq) + SO_4^{2-}(aq)$$

state the effect of the following on the equilibrium condition:

a. Addition of Ba^{2+} ion

b. Removal of SO_4^{2-} ion

c. Addition of $BaSO_4(s)$

PREDICTING THE OCCURRENCE OF REACTIONS

14.7   Let us return to a consideration of ionic equations (Section 13.9) and the occurrence of reactions in solutions in order to see another application of the principles of equilibrium. In general, we say that a chemical reaction between ions in solution occurs only if one of the products is:

1.   A gas

2.   A precipitate (insoluble solid)

3. H_2O (or some other molecular nonelectrolyte)

Otherwise, the combination of reactants is merely a physical mixing together.

Single- and double-displacement reactions (review Section 9.8) occur only in solution. You have already learned to predict the occurrence of a single-replacement reaction based on the Activity Series. Notice that these reactions always produce either a gas or an insoluble solid. For example,

$$Mg(s) + 2H^+(aq) + 2\cancel{Cl}^-(aq) \longrightarrow Mg^{2+}(aq) + 2\cancel{Cl}^-(aq) + \boxed{H_2 \uparrow}$$

$$Mg(s) + Cu^{2+}(aq) + \cancel{SO_4^{2-}}(aq) \longrightarrow Mg^{2+}(aq) + \cancel{SO_4^{2-}}(aq) + \boxed{Cu \downarrow}$$

(A cancel indicates a spectator ion.)

The mixing together of solutions of ionic compounds may or may not result in double-displacement reactions, depending on the solubilities of the predicted products. Previous examples (Section 9.8) have always been of the kind where a precipitate formed, and hence there was a reaction. For example, in Sample Exercise 9.9 we completed the equation for the reaction between $MgCl_2$ and K_2CO_3:

Traditional equation

$$MgCl_2 + K_2CO_3 \longrightarrow MgCO_3 + 2KCl$$

Using Table 13.3 we see that of the above ionic compounds only $MgCO_3$ is insoluble, so we can write this equation:

Ionic equation

$$Mg^{2+}(aq) + 2Cl^-(aq) + 2K^+(aq) + CO_3^{2-}(aq) \longrightarrow$$
$$MgCO_3 \downarrow + 2K^+(aq) + 2Cl^-(aq)$$

K^+ and Cl^- ions are spectators, so

Net ionic equation

$$Mg^{2+}(aq) + CO_3^{2-}(aq) \longrightarrow MgCO_3 \downarrow$$

To predict the occurrence of a double-displacement reaction between soluble ionic compounds, *evaluate the solubilities* of the predicted products. A visible reaction occurs if one product is insoluble.

SAMPLE EXERCISE

14.9

Predict whether or not reactions occur upon combination of solutions of the following:

a. $CuCl_2 + AgNO_3 \longrightarrow$
b. $CuCl_2 + KNO_3 \longrightarrow$

SOLUTION

Step 1   Predict the products by the method of switching cation-anion partners, as described in Section 9.8.

Step 2   Evaluate the solubilities of the products using Table 13.3.

a. $CuCl_2 + AgNO_3 \longrightarrow Cu(NO_3)_2 + 2AgCl \downarrow$
                                                          Soluble,   Insoluble,
                                                          rule 2   rule 3

Therefore, a reaction has occurred and we can write equations in the traditional or ionic forms.

b. $CuCl_2 + KNO_3 \longrightarrow Cu(NO_3)_2 + 2KCl$
                                                          Soluble,   Soluble,
                                                          rule 2   rule 1

Therefore, there is *no* reaction and we would represent this fact by writing,
$CuCl_2 + KNO_3 \longrightarrow NR$

For the reaction of a *strong* acid with a *strong* base, the net ionic equation is always

$$H^+(aq) + OH^-(aq) \rightleftharpoons H_2O$$

The K_{eq} for the formation of H_2O from H^+ and OH^- ions is 1×10^{14}. Thus we see that the reaction goes to completion. **Whenever a molecular nonelectrolyte is formed from its ions, K_{eq} is large and the equilibrium lies far to the right.**

SAMPLE EXERCISE

14.10

Explain in terms of the principles of equilibria why each of the following reactions goes to completion:

a. $3KOH + Al(NO_3)_3 \longrightarrow 3KNO_3 + Al(OH)_3 \downarrow$
b. $CaC_2 + 2HCl \longrightarrow CaCl_2 + C_2H_2 \uparrow$
c. $2KOH + H_2SO_4 \longrightarrow K_2SO_4 + 2H_2O$

SOLUTION

a. A *precipitate* forms, which removes one product from solution; thus the reaction shifts to the right to replace it. Large K_{eq} are associated with precipitate formation. In this case, K_{eq} for the formation of $Al(OH)_3$ is 2×10^{23}.
b. Acetylene (C_2H_2) *gas* escapes, and the equilibrium shifts to replenish it.
c. This is the reaction between a strong base (KOH) and a strong acid (H_2SO_4) to form *water*. K_{eq} for this reaction is 1×10^{14}.

SUMMARY

Many processes in nature are reversible. This includes chemical reactions. Reactions in which products have the ability to react to reform reactants are called **reversible reactions.** Reversible processes tend to achieve a state of dynamic equilibrium in which the two processes coexist at the same rate (Section 14.1). The rate of a single-step chemical reaction is directly proportional to the concentration of the reactants (Section 14.2).

From the relationship between reaction rate and concentration and the fact that at equilibrium the rates of forward and reverse reactions are equal, the equilibrium-constant expression can be developed (Section 14.3). Correct equilibrium-constant expressions can easily be written by following an established set of rules (Section 14.4).

The magnitude of the equilibrium constant K_{eq} indicates the relative amounts of reactants and products present in an equilibrium mixture (Section 14.5).

Le Châtelier's principle states that if we apply a stress to a system in equilibrium, the equilibrium will shift so as to relieve the stress.

Common stress factors are changes in concentration, temperature, or pressure (Section 14.6).

The principles of equilibrium explain why reactions typically go to completion when a gas, precipitate, or molecular nonelectrolyte such as water is formed (Sections 14.7 and 14.8).

CHAPTER ACCOMPLISHMENTS

After completing this chapter you should be able to

14.1   Reversible reactions

1.   Relate the terms "reversible reaction" and "dynamic equilibrium."
2.   Qualitatively describe the changes in initial concentrations of reactants and products in a reversible reaction.

14.2   Rates of reaction

3.   Define rate of reaction.
4.   State a relationship between reactant concentration and reaction rate.
5.   State the relationship between the reaction rates of forward and reverse reactions and chemical equilibrium.

14.3   Equilibrium constant

6.   Explain how the equilibrium-constant expression is derived from the forward and reverse rate expressions for one-step reactions.
7.   Write the equilibrium-constant expression for a reaction given a balanced chemical equation.

14.4   Rules for writing K_{eq}

8.   Determine the value of the equilibrium constant expression given the equilibrium concentrations.
9.   State the convention for the treatment of pure solids and pure liquids in equilibrium-constant expressions.
10.   Given a balanced chemical equation, write the equilibrium-constant expression, using the common convention regarding pure liquids and solids.
11.   Write the solubility-product expression for the equilibrium in a saturated solution of some specified ionic solute.

14.5   Interpreting the value of K_{eq}

12.   Given a value of K_{eq}, qualitatively describe the equilibrium composition of a given reaction.

14.6   Le Châtelier's principle

13.   Given a chemical equation, describe the effect on the equilibrium concentrations of a change in concentration of one species.

14. Given an exothermic or endothermic chemical reaction, describe the effect of a change in temperature on the equilibrium concentrations.

15. Given a chemical equation involving one or more gases, describe the effect on the equilibrium concentrations of a change in total pressure on the reaction.

16. Given a saturated aqueous solution of an ionic compound, describe the effect of an increase in concentration of one ion on the solubility of the other ion.

14.7 Predicting the occurrence of reactions

17. State the conditions necessary for the occurrence of reactions in solution.

18. Given two reactants in solution, predict whether a double-displacement reaction will occur.

19. Write an ionic equation and a net ionic equation for the reaction of an acid with a base.

14.8 "Going to completion"

20. State the three conditions under which a reaction can be said to go to completion.

PROBLEMS

14.7 Describe two nonchemical examples of a dynamic equilibrium.

14.8 A dynamic equilibrium exists according to the following equation

$$A(g) + B(g) \rightleftharpoons C(g) + D(g)$$

A is a yellowish colored gas, *D* is a bluish colored gas, and *B* and *C* are colorless gases.

a. Predict the observable changes that occur when 0.1 mole of *A* and 0.1 mole of *B* are placed in an empty container.

b. Predict the observable changes that occur if the reaction is begun with 0.1 mole of *C* and 0.1 mole of *D*.

c. Describe the appearance of the system if the reaction is left at equilibrium.

14.9 a. Describe how the evaporation of water in a closed container leads to a dynamic-equilibrium condition.

b. Would the evaporation of water in an open container lead to a dynamic equilibrium? Explain.

14.10 Explain how the concentrations of reactants and products stay constant at the state of dynamic equilibrium.

14.11 What is the difference between *equal* product and reactant concentrations and *constant* product and reactant concentrations?

14.2 Rates of reaction

14.12 Given the single-step reaction $A + B \rightarrow C$.

a. Explain why the rate of the reaction increases if the concentration of *A* increases.

b. What would happen to the rate of the reaction if the concentration of *A* were decreased?

14.13 What statement can be made concerning the rates of the forward and reverse reactions at the equilibrium condition.

14.3 Equilibrium constant

14.14 Write an equilibrium-constant expression for each of the following equilibria:

a. $H_2(g) + Cl_2(g) \rightleftharpoons 2HCl(g)$

b. $PCl_5(g) \rightleftharpoons PCl_3(g) + Cl_2(g)$

c. $CO(g) + H_2O(g) \rightleftharpoons CO_2(g) + H_2(g)$
d. $4NH_3(g) + 5O_2(g) \rightleftharpoons 4NO(g) + 6H_2O(g)$
e. $3O_2(g) \rightleftharpoons 2O_3(g)$
f. $CH_4(g) + 2O_2(g) \rightleftharpoons CO_2(g) + 2H_2O(g)$
g. $4HCl(g) + O_2(g) \rightleftharpoons 2Cl_2(g) + 2H_2O(g)$

14.15 Write an equilibrium-constant expression for a reaction in which the rate of the forward reaction is given by $\text{rate}_{\text{forward}} = k[A]^2$ and the rate of the reverse reaction is given by $\text{rate}_{\text{reverse}} = k'[B]$.

14.16 What is the equilibrium-constant expression for each of the following:

a. $H_2(g) + Br_2(g) \rightleftharpoons 2HBr(g)$
b. $CH_4(g) + Cl_2(g) \rightleftharpoons CH_3Cl(g) + HCl(g)$
c. $2NO_2(g) \rightleftharpoons N_2(g) + 2O_2(g)$
d. $4NH_3(g) + 3O_2(g) \rightleftharpoons 2N_2(g) + 6H_2O(g)$

14.17 **a.** Write the equilibrium constant expression for the following equilibrium:

$$2HI(g) \rightleftharpoons H_2(g) + I_2(g)$$

b. Both HI and H_2 are colorless gases, whereas $I_2(g)$ is purple. How might you know when equilibrium in the above reaction has been reached?
c. A chemist analyzed an equilibrium mixture of HI, H_2 and I_2. She found the concentrations to be

$$[HI] = 0.27M$$
$$[H_2] = 0.86M$$
$$[I_2] = 0.86M$$

Determine the numerical value of K_{eq}.

14.18 Write an equilibrium-constant expression for a reaction in which the rate of the forward reaction is given by $\text{rate}_{\text{forward}} = k[A]^2[B]$ and the rate of the reverse reaction is given by the $\text{rate}_{\text{reverse}} = k'[C][D]$.

14.4 Rules for writing K_{eq}

14.19 Write an equilibrium-constant expression for each of the following equilibria:
a. $C(s) + H_2O(g) \rightleftharpoons CO(g) + H_2(g)$
b. $4H_2O(g) + 3Fe(s) \rightleftharpoons Fe_3O_4(s) + 4H_2(g)$
c. $2H_2O(l) \rightleftharpoons 2H_2(g) + O_2(g)$
d. $MnO_2(s) + 4HCl(aq) \rightleftharpoons MnCl_2(aq) + Cl_2(g) + 2H_2O(l)$
e. $H_2CO_3(aq) \rightleftharpoons CO_2(g) + H_2O(l)$

14.20 What is the equilibrium-constant expression for each of the following:
a. $PCl_5(s) \rightleftharpoons PCl_3(l) + Cl_2(g)$
b. $4Al(s) + 3O_2(g) \rightleftharpoons 2Al_2O_3(s)$
c. $H_2O(l) \rightleftharpoons H_2O(g)$
d. $4HNO_3(l) \rightleftharpoons 4NO_2(g) + 2H_2O(g) + O_2(g)$
e. $H_2SO_3(l) \rightleftharpoons SO_2(g) + H_2O(l)$

14.21 Write the equilibrium constant expression for each of the following physiological equilibria:
a. α-D-Glucose(aq) $\rightleftharpoons$ β-D-glucose(g)
b. $CH_3OH(aq) + CH_3COOH(aq) \rightleftharpoons$ $CH_3COOCH_3(aq) + H_2O(l)$
c. Hemoglobin(aq) + oxygen(g) $\rightleftharpoons$ oxyhemoglobin(aq)
d. Double-strand DNA $\rightleftharpoons$ 2 single-strand DNA

14.22 Write the solubility-product expressions K_{sp} for the equilibria in the following saturated solutions:
a. $AgF(s) \rightleftharpoons Ag^+(aq) + F^-(aq)$
b. $PbBr_2(s) \rightleftharpoons Pb^{2+}(aq) + 2Br^-(aq)$
c. $Ca_3(PO_4)_2(s) \rightleftharpoons 3Ca^{2+}(aq) + 2PO_4^{3-}(aq)$
d. $BaSO_4(s) \rightleftharpoons Ba^{2+}(aq) + SO_4^{2-}(aq)$
e. $As_2S_3(s) \rightleftharpoons 2As^{3+}(aq) + 3S^{2-}(aq)$

14.23 What is the solubility-product expression K_{sp} for the equilibrium in each of the following:
a. $AgBr(s) \rightleftharpoons Ag^+(aq) + Br^-(aq)$
b. $Cu(OH)_2(s) \rightleftharpoons Cu^{2+}(aq) + 2OH^-(aq)$
c. $Al(OH)_3(s) \rightleftharpoons Al^{3+}(aq) + 3OH^-(aq)$
d. $Mg_3(PO_4)_2(s) \rightleftharpoons 3Mg^{2+}(aq) + 2PO_4^{3-}(aq)$
e. $Bi_2S_3(s) \rightleftharpoons 2Bi^{3+}(aq) + 3S^{2-}(aq)$

14.24 Calculate the value of K_{eq} at 127°C for the reaction

$$Ni(CO)_4(g) \rightleftharpoons Ni(s) + 4CO(g)$$

given the following set of equilibrium concentrations at this temperature:

$$[Ni(CO)_4] = 1.80 \text{ moles/L}$$
$$[CO] = 0.800 \text{ mole/L}$$

14.5 Interpreting the value of K_{eq}

14.25 Use Table 14.1 and the given K_{eq} to describe the equilibrium composition of each of the following reactions qualitatively:
a. $2NO_2(g) \rightleftharpoons N_2(g) + 2O_2(g)$
$$K_{eq}(\text{at } 25°C) = 6.7 \times 10^{16}$$

b. $CH_3OH(g) \rightleftharpoons CO(g) + 2H_2(g)$
$$K_{eq}(\text{at } 100°C) = 7.37 \times 10^{-8}$$
c. $N_2(g) + O_2(g) \rightleftharpoons 2NO(g)$
$$K_{eq}(\text{at } 2400°C) = 3.4 \times 10^{-3}$$
d. $2SiO(g) \rightleftharpoons 2Si(l) + O_2(g)$
$$K_{eq}(\text{at } 1727°C) = 9.62 \times 10^{-13}$$

14.26 In each of the following cases use Table 14.1 and the given K_{eq} to describe the equilibrium composition qualitatively:
a. $2NO(g) \rightleftharpoons N_2(g) + O_2(g)$
$$K_{eq}(\text{at } 25°C) = 2.2 \times 10^{30}$$
b. $PCl_5(g) \rightleftharpoons PCl_3(g) + Cl_2(g)$
$$K_{eq}(\text{at } 127°C) = 1.19 \times 10^{-2}$$
c. $2HCl(g) \rightleftharpoons H_2(g) + Cl_2(g)$
$$K_{eq}(\text{at } 2727°C) = 1.27 \times 10^{-4}$$
d. $2Na_2O(s) \rightleftharpoons 4Na(l) + O_2(g)$
$$K_{eq}(\text{at } 427°C) = 1.84 \times 10^{-25}$$

14.27 From the following solubility-product expressions pick:
a. The reaction that is most complete from left to right as written.
b. The reaction that is least complete from left to right as written.
    1. $AgCl(s) \rightleftharpoons Ag^+(aq) + Cl^-(aq)$
$$K_{sp}(\text{at } 25°C) = 1.7 \times 10^{-10}$$
    2. $PbSO_4(s) \rightleftharpoons Pb^{2+}(aq) + SO_4^{2-}(aq)$
$$K_{sp}(\text{at } 25°C) = 1.3 \times 10^{-8}$$
    3. $Fe(OH)_3(s) \rightleftharpoons Fe^{3+}(aq) + 3OH^-(aq)$
$$K_{sp}(\text{at } 25°C) = 6 \times 10^{-38}$$

14.28 In each of the following physiological reactions, use Table 14.1 and the given K_{eq} to describe the equilibrium composition qualitatively:
a. $ATP + H_2O \rightleftharpoons ADP + HPO_4^{2-}$
$$K_{eq}(\text{at } 25°) = 1.3 \times 10^5$$
b. Arginine $+ HPO_4^{2-} \rightleftharpoons$
arginine-phosphate $+ H_2O$
$$K_{eq} \text{ (at } 25°) = 7.9 \times 10^{-6}$$
c. Glucose-6-phosphate $+$ arginine $\rightleftharpoons$
glucose $+$ arginine-phosphate
$$K_{eq} \text{ (at } 37°) = 1.5 \times 10^{-3}$$
d. $ATP + CH_3COOH + $ coenzyme A $\rightleftharpoons$
AMP $+$ pyrophosphate $+$ acetylcoenzyme A
$$K_{eq} \text{ (at } 37°) = 1.0$$

14.6 Le Châtelier's principle

14.29 State the effect, if any, of the following changes on the position of equilibrium for the reaction

$4NH_3(g) + 3O_2(g) \rightleftharpoons 2N_2(g)$
$\qquad\qquad\qquad + 6H_2O(g) + 366$ kcal

a. Increasing the concentration of oxygen
b. Adding N_2 to the equilibrium mixture
c. Removing H_2O as it is formed
d. Increasing the reaction temperature
e. Decreasing the total pressure

14.30 State the effect, if any, of the following changes on the position of equilibrium for the reaction

$$2SO_3(g) + 47 \text{ kcal} \rightleftharpoons 2SO_2(g) + O_2(g)$$

a. Increasing the concentration of SO_3
b. Adding O_2 to the equilibrium mixture
c. Removing O_2 as it is formed
d. Increasing the reaction temperature
e. Decreasing the total pressure

14.31 Consider the equilibrium

$$C(s) + H_2O(g) \rightleftharpoons CO(g) + H_2(g) + \text{heat}$$

a. What should be done to the total pressure of the system to maximize the equilibrium concentration of H_2?
b. What should be done to the pressure of the system to maximize the equilibrium concentration of H_2O?

14.32 Consider the equilibrium

$$11.0 \text{ kcal} + 2F_2(g) + O_2(g) \rightleftharpoons 2OF_2(g)$$

a. Should the temperature be increased or decreased to maximize the equilibrium yield of product?
b. Should the pressure be increased or decreased to maximize the equilibrium yield of the product?

14.33 For the equilibrium

$$Ca_3(PO_4)_2(s) \rightleftharpoons 3Ca^{2+}(aq) + 2PO_4^{3-}(aq)$$

state the effect on the equilibrium condition of the following:
a. Addition of a Ca^{2+} ion
b. Addition of a PO_4^{3-} ion
c. Removal of a PO_4^{3-} ion
d. Addition of $Ca_3(PO_4)_2(s)$

14.34 Describe the effect on the dynamic equilibrium

$$AgCl(s) \rightleftharpoons Ag^+(aq) + Cl^-(aq)$$

of adding NaCl to a saturated solution of AgCl.

14.35 Will $BaSO_4$ ($K_{sp} = 1.5 \times 10^{-19}$) be more soluble in sulfuric acid solution or in pure water? Explain.

14.36 The equilibrium between carbonic acid (H_2CO_3) and bicarbonate ion is very important in the regulation of the acid-base concentration in blood. State the effect if any of the following changes in the position of the equilibrium:

$$H_2CO_3(aq) \rightleftharpoons H^+(aq) + HCO_3^-(aq)$$

a. Adding H_2CO_3 to the equilibrium mixture
b. Removing H^+ as it is formed
c. Adding H^+ to the equilibrium mixture
d. Adding HCO_3^- to the equilibrium mixture
e. Removing H_2CO_3 as it is formed

14.37 Reconsider Problem 14.36, part (c). What happens to the bicarbonate (HCO_3^-) concentration upon addition of H^+ and the consequent shift?

14.38 An equilibrium that exists in solutions of carbon dioxide such as soda and beer is

$$H_2CO_3(aq) \rightleftharpoons CO_2(g) + H_2O(l)$$

Use Le Châtelier's principle to explain why an open bottle of beer goes flat.

14.7 Predicting the occurrence of reactions

14.39 Predict whether or not the following double-displacement reactions will occur. Write a balanced net ionic equation for those which do occur.

a. $NaBr(aq) + AgNO_3(aq) \longrightarrow$
b. $HNO_3(aq) + KOH(aq) \longrightarrow$
c. $NH_4Cl(aq) + MgBr_2(aq) \longrightarrow$
d. $Ba(OH)_2(aq) + H_2SO_4(aq) \longrightarrow$
e. $Ca(NO_3)_2(aq) + K_3PO_4(aq) \longrightarrow$

14.8 "Going to completion"

14.40 List the three types of products that drive an equilibrium all the way to the product side.

14.41 How does the formation of a precipitate drive a reaction to completion?

ACIDS AND BASES
CHAPTER

INTRODUCTION

15.1 We have daily contacts with acids and bases. Acids are necessary for digestion of proteins in the stomach (hydrochloric acid); they are found on salads in vinegar (acetic acid), in fruits such as lemons (citric acid), in battery acid (sulfuric acid); and the drug, aspirin, is acetylsalicylic acid. Bases occur in lye (sodium hydroxide), window cleaner (ammonia solution), soaps and detergents (mixtures of various bases), and antacids (aluminum hydroxide, magnesium hydroxide, and sodium bicarbonate). Figure 15.1 depicts some familiar acid and base products.

FIGURE 15.1

Some commonly encountered acids and bases. Acids: vinegar is acetic acid; aspirin is acetylsalicylic acid; vitamin C is ascorbic acid. Bases: Drano is lye, which is essentially NaOH; milk of magnesia is $Mg(OH)_2$; the bicarbonate anion (HCO_3^-) of baking soda acts as a Brønsted-Lowry base. (*Photograph by Bryan Lees.*)

TABLE 15.1 COMMON LABORATORY ACIDS AND BASES

Name		Formula	Strength*	Use or Occurrence
ACIDS	Acetic acid	$HC_2H_3O_2$	W	Vinegar
	Boric acid	H_3BO_3	W	Eyewash
	Carbonic acid	H_2CO_3	W	Carbonated water
	Hydrobromic acid	HBr	S	Manufacture of chemicals
	Hydrochloric acid	HCl	S	Gastric juice
	Hydrocyanic acid	HCN	W	Exterminating rodents
	Hydrofluoric acid	HF	W	Etching glass
	Hydroiodic acid	HI	S	Manufacture of iodine compounds
	Nitric acid	HNO_3	S	Test for proteins
	Perchloric acid	$HClO_4$	S	Manufacture of explosives
	Phosphoric acid	H_3PO_4	M	Soft drink additive
	Sulfuric acid	H_2SO_4	S	Battery acid
BASES	Aluminum hydroxide	$Al(OH)_3$	S†	Antiperspirant
	Ammonia	NH_3	W	Respiratory stimulant
	Calcium hydroxide	$Ca(OH)_2$	S†	In cement
	Lithium hydroxide	LiOH	S	Photographic developer
	Magnesium hydroxide	$Mg(OH)_2$	S†	Milk of magnesia
	Potassium hydroxide	KOH	S	Manufacture of liquid soap
	Sodium hydroxide	NaOH	S	Lye
	Zinc hydroxide	$Zn(OH)_2$	S†	

* S = strong; M = medium; and W = weak. See Table 15.3 for a more quantitative classification.
† These bases are strong in that to whatever extent they dissolve, they dissociate. However, they are at best only slightly soluble.

Table 15.1 lists the names and formulas of some acids and bases and cites their occurrence or use in food, medicine, agriculture, or industry. The name of the acid actually applies to the aqueous solution. Hydrochloric acid, for example, is formed by dissolving hydrogen chloride, a gas, in water. Notice that with the exception of ammonia, all the bases in the table are ionic compounds, which is to say, combinations of metal ions and hydroxide ions. To function effectively in a modern society requires some awareness of the physical and chemical properties of acidic and basic solutions. An understanding of buffers, sweet and sour soils, acid indigestion and acid rain requires a knowledge of acidity and basicity.

Table 15.2 lists some general properties of acids and bases. Although we will discuss some of these properties in more detail as this chapter unfolds, you can begin to recognize various foods, drugs, agricultural products, and other commonly encountered substances as acidic or basic by using this table.

We have delayed a discussion of acids and bases to this point because it required a knowledge of solutions and ionic equations (Sec-

PROPERTIES OF ACIDS AND BASES   **TABLE 15.2**

Acids	Turn litmus (a dye) red
	Taste sour (not a recommended general test)
	Oxyacids produced by the reaction of water and a nonmetallic oxide
	React with a base in a neutralization reaction
Bases	Turn litmus blue (remember *b* for blue, *b* for base)
	Taste bitter
	Feel slippery (soapy)
	Metal hydroxide bases produced by the reaction of water and a metallic oxide
	React with an acid in a neutralization reaction

tions 13.8 and 13.9) and equilibrium concepts (Chapter 14). As you proceed in this chapter and whenever you encounter acids and bases, you will find it helpful to form the following associations:

Associate **acid with H$^+$** (hydrogen ion or proton, that is, an H atom without its one electron).

Associate **base with OH$^-$** (hydroxide ion).

Let us see how these associations hold within common definitions of acids and bases.

THE ARRHENIUS DEFINITION

15.2   As scientific study of acids and bases proceeded through the nineteenth and twentieth centuries, various definitions of these substances were formulated. These different definitions do not contradict one another. The newer definitions, such as that of Brønsted and Lowry (Section 15.3), are broader and more general definitions that include the older ones, such as that of Arrhenius. We will examine both definitions, because sometimes one offers a clearer explanation of some observed phenomenon, and sometimes the other definition provides more clarity.

In 1884 Svante Arrhenius defined acids and bases in terms of the ions which they release in aqueous solution:

Acids release H$^+$ into water.

Bases release OH$^-$ into water.

Considering the metal hydroxide (MOH) bases, the origin of OH$^-$ ions in solutions of these compounds is obvious. These are ionic compounds made up of metal cations and hydroxide anions in the solid state. When ionic compounds dissolve, the ions separate and are solvated by water (see Section 13.3). We can represent this **dissociation** by using a general equation for the solubility equilibrium in a saturated base solution:

$$MOH(s) \rightleftharpoons M^+(aq) + OH^-(aq)$$

Problem 15.1

Use equations to show how $KOH(s)$ and $Ca(OH)_2(s)$ release OH^- ions into solution.

Acids are molecular compounds which undergo ionization in aqueous solution; indeed, as you saw in Section 13.8, they are the principal examples of molecular electrolytes. **Ionization** is the reaction of a molecular compound with water to produce ions. In the case of the ionization of molecular compounds of the general formula HA, one of the ions produced is H^+; therefore, we classify HA as an **Arrhenius acid.**

$$HA(g \text{ or } l) \rightleftharpoons H^+(aq) + A^-(aq)$$

This reaction is more correctly represented as

$$HA(g \text{ or } l) + H_2O \rightleftharpoons H_3O^+(aq) + A^-(aq)$$
Hydronium ion

which shows that HA ionizes by reaction with H_2O. For simplicity, we often just write $H^+(aq)$, as before. However, you should remember that an H^+ in aqueous solution will always form a coordinate covalent bond with a lone pair of electrons on water's oxygen. This arrangement is the hydronium ion.

BRØNSTED-LOWRY DEFINITION

15.3 If you examine Table 15.1 again, you will find that all the acids and bases there easily fit the Arrhenius definitions except for ammonia. That is, the bases are of the form $M(OH)_n$ and the acids are of the form HA. Because of NH_3 and some other materials with acidic or basic properties that do not obviously fit the Arrhenius definitions, J. N. Brønsted and T. M. Lowry offered new definitions in 1923:

Acids are H^+ (proton) donors.

Bases are H^+ (proton) acceptors.

The Brønsted-Lowry definition of acids is essentially identical to the Arrhenius definition. An acid donates H^+ or releases H^+ into the solution.

In the Brønsted-Lowry theory, for compounds of the type MOH, M^+ is regarded strictly as a spectator ion. The true basic species is the hydroxide anion (OH^-), because it is OH^- that will quite readily accept H^+ (a proton) to form water.

$$OH^- + H^+ \longrightarrow H_2O$$
$$\text{Hydroxide} \quad \text{Proton} \qquad \text{Water}$$

Hydroxide ion (OH^-) is clearly a base because it accepts H^+. MOH is a base because in water it dissociates into M^+ and OH^- ions. It is convenient to ignore M^+ when writing Brønsted-Lowry acid-base expressions that involve metal hydroxide bases.

The Brønsted-Lowry definition of bases allows us to understand ammonia's ($\ddot{N}H_3$) basicity by recognizing that the lone pair on nitrogen is available for bonding with H^+. $\ddot{N}H_3$ can accept H^+,

$$\ddot{N}H_3 + H^+ \rightleftharpoons \overset{H}{\underset{+}{\ddot{N}H_3}} \equiv NH_4{}^+$$

For an aqueous solution of NH_3, water is the source of H^+,

$$H_3N: + H:\ddot{O}-H \rightleftharpoons H_3\overset{+}{N}:H + {}^-:\ddot{O}-H$$

This equation shows us that NH_3 is a base in the Arrhenius sense also because the reaction of NH_3 with H_2O releases OH^- ions in solution. Ammonia is a weak base because the equilibrium lies to the left.

Notice in the ammonia equilibrium

$$NH_3(g) + H_2O(l) \rightleftharpoons NH_4{}^+(aq) + OH^-(aq)$$

that water acts as a Brønsted-Lowry acid because it donates H^+ to NH_3. When we examine the preceding equilibrium from right to left ($\leftarrow$), we observe that the ammonium ion ($NH_4{}^+$) acts as an acid (i.e., it gives up H^+) and hydroxide (OH^-) acts as a base (i.e., it accepts H^+). The $NH_4{}^+$ ion is said to be the **conjugate acid** of the base NH_3. The hydroxide ion (OH^-) is the **conjugate base** of the acid H_2O. Every acid has a conjugate base, the species remaining after the acid has donated a hydrogen ion. Every base has a conjugate acid, the species formed after the base has accepted a hydrogen ion.

$$HA + B \rightleftharpoons HB^+ + A^-$$
$$\text{Acid} \quad \text{Base} \quad \underset{\text{acid}}{\text{Conjugate}} \quad \underset{\text{base}}{\text{Conjugate}}$$

Water, which we saw act as a Brønsted-Lowry acid in the preceding example, can also act as a Brønsted-Lowry base. Water acts as a base when it accepts H^+ to form hydronium ions. For example, when hydrogen chloride gas is dissolved in water, the following reaction occurs:

$$HCl(g) + H_2O \rightleftharpoons H_3O^+(aq) + Cl^-(aq)$$

Here HCl is an acid because it donates a hydrogen ion. H_2O is a base in this reaction because it accepts a hydrogen ion. H_3O^+ is the conjugate acid of the base H_2O, and Cl^- is the conjugate base of the acid HCl. A substance such as H_2O which can act as both an acid and a base is called **amphoteric.**

SAMPLE EXERCISE

15.1

Identify the acid-conjugate base and base-conjugate acid pairs for the equilibrium

$$HCl + NH_3 \rightleftharpoons NH_4^+ + Cl^-$$

SOLUTION

We recognize HCl as an acid from Table 15.1 or by noticing that it has given up H^+ as the reaction proceeds from left to right. Cl^- is the conjugate base, the species left when H^+ is released from the acid HCl.

We recognize NH_3 as a base from Table 15.1 or by noticing that it accepts H^+ as the reaction proceeds from left to right. NH_4^+ is the conjugate acid, the species formed when the base NH_3 accepts H^+.

Problem 15.2

Label the acid, base, conjugate acid, and conjugate base for the equilibrium

$$OH^- + HNO_3 \rightleftharpoons NO_3^- + H_2O$$

Clearly our experience is that water does not display the characteristics of either an acid or a base as they are described in Table 15.2. This brings us to the important point that any substance that donates H^+ can be classified *theoretically* as an acid. However, unless the concentration of H^+ produced reaches some lower limit ($10^{-6} M$), we do not witness the properties in Table 15.2. It is easy to recognize the acids and bases listed in Table 15.1 by their properties because they produce sufficiently large concentrations of H^+ or OH^- in solution. Let us now examine the idea of acid and base strength and see how it is related to the amount of H^+ or OH^- in solution.

ACID AND BASE STRENGTH

15.4 In Section 13.7 you saw that a strong electrolyte is one that has many ions in solution. Similarly, the criterion for a strong acid is that it ionizes to a great extent and produces many ions, specifically H^+ ions (or, more correctly, H_3O^+). The six strong acids you encountered in Section 13.7 essentially ionize 100 percent. For example, because the equilibrium

$$HCl \rightleftharpoons H^+ + Cl^-$$

lies so far to the right, we may write

$$HCl \longrightarrow H^+ + Cl^-$$

Weak acids, on the other hand, ionize to only a limited extent. For example, acetic acid ionizes less than 5 percent:

$$HC_2H_3O_2 \rightleftharpoons H^+ + C_2H_3O_2^-$$

Because the equilibrium lies to the left and only a small number of H^+ ions (H_3O^+) are produced, acetic acid is a weak acid.

The equilibrium constant corresponding to the ionization of an acid is called the **ionization constant** and is designated K_a. For $HA \rightleftharpoons H^+ + A^-$,

$$K_a = \frac{[H^+][A^-]}{[HA]}$$

Table 15.3, which lists the relative strengths of selected acids and bases, also gives some selected K_a values. Notice that it is arranged in order of decreasing acid strength reading from top to bottom. As with other equilibrium constants, a large value for K_a indicates that the equilibrium lies far to the right, and a small value indicates that the equilibrium position is to the left. Thus strong acids have high K_a values and weak acids, low ones. In particular, compare the K_a values of HCl (strong) and acetic acid (weak).

Table 15.3 also shows that if acids have more than one ionizable H^+, we consider the equilibria for removing one H^+ at a time. For example, H_2SO_4 is **diprotic,** which means it has two ionizable hydrogens. We represent the equilibria for their ionization

$$H_2SO_4 \rightleftharpoons H^+ + HSO_4^-$$

$$HSO_4^- \rightleftharpoons H^+ + SO_4^{2-}$$

There is a greater tendency for the first H^+ to ionize than the second. Therefore, we see that H_2SO_4 is a stronger acid than HSO_4^-. In general, there is always a greater tendency for the first H^+ to ionize in any acid with more than one ionizable H^+.

Base strength follows the same principle as electrolyte strength

TABLE 15.3   RELATIVE STRENGTHS OF ACIDS AND BASES

K_a	Acid Name	Acid Formula	Formula	Base Name	K_b
1×10^{10}	Perchloric	$HClO_4$	$\rightleftharpoons H^+ + ClO_4^-$	Perchlorate ion	1×10^{-24}
	Hydroiodic	HI	$\rightleftharpoons H^+ + I^-$	Iodide ion	
	Hydrobromic	HBr	$\rightleftharpoons H^+ + Br^-$	Bromide ion	
1×10^7	Hydrochloric	HCl	$\rightleftharpoons H^+ + Cl^-$	Chloride ion	1.4×10^{-21}
	Nitric	HNO_3	$\rightleftharpoons H^+ + NO_3^-$	Nitrate ion	
	Sulfuric	H_2SO_4	$\rightleftharpoons H^+ + HSO_4^-$	Hydrogen sulfate ion	
	Hydronium ion	H_3O^+	$\rightleftharpoons H^+ + H_2O$	Water	
1.2×10^{-2}	Hydrogen sulfate ion	HSO_4^-	$\rightleftharpoons H^+ + SO_4^{2-}$	Sulfate ion	
	Phosphoric	H_3PO_4	$\rightleftharpoons H^+ + H_2PO_4^-$	Dihydrogen phosphate ion	
1.8×10^{-5}	Hydrofluoric	HF	$\rightleftharpoons H^+ + F^-$	Fluoride ion	1.4×10^{-11}
	Acetic	$HC_2H_3O_2$	$\rightleftharpoons H^+ + C_2H_3O_2^-$	Acetate ion	
	Carbonic	H_2CO_3	$\rightleftharpoons H^+ + HCO_3^-$	Bicarbonate ion	2.4×10^{-8}
	Hydrosulfuric	H_2S	$\rightleftharpoons H^+ + HS^-$	Hydrogen sulfide ion	
	Dihydrogen phosphate ion	$H_2PO_4^-$	$\rightleftharpoons H^+ + HPO_4^{2-}$	Hydrogen phosphate ion	
	Boric	H_3BO_3	$\rightleftharpoons H^+ + H_2BO_3^-$	Dihydrogen borate ion	
	Ammonium ion	NH_4^+	$\rightleftharpoons H^+ + NH_3$	Ammonia	1.8×10^{-5}
4.0×10^{-10}	Hydrocyanic	HCN	$\rightleftharpoons H^+ + CN^-$	Cyanide ion	
	Bicarbonate	HCO_3^-	$\rightleftharpoons H^+ + CO_3^{2-}$	Carbonate ion	
	Hydrogen phosphate ion	HPO_4^{2-}	$\rightleftharpoons H^+ + PO_4^{3-}$	Phosphate ion	5.9×10^{-3}
1.0×10^{-15}	Hydrogen sulfide ion	HS^-	$\rightleftharpoons H^+ + S^{2-}$	Sulfide ion	
	Water	H_2O	$\rightleftharpoons H^+ + OH^-$	Hydroxide ion	5.5×10^1

Left side: K_a — Increasing (top), Acid strength, Decreasing (bottom). Right side: K_b — Decreasing (top), Base strength, Increasing (bottom).

or acid strength; i.e., extensive dissociation or ionization producing many ions means that the base is strong, and little dissociation or ionization yielding few ions means it is weak. In the case of bases, it is the hydroxide ion (OH^-) concentration that is of interest. Therefore, soluble metal hydroxides, such as the Group IA hydroxides, are strong bases because these ionic compounds dissociate 100 percent as they dissolve.

Magnesium hydroxide is classified as a strong base because all the formula units that do enter solution do so by completely dissociating by the process

$$Mg(OH)_2 \rightleftharpoons Mg^{2+} + 2OH^-$$

However, the Group IIA hydroxides, such as $Mg(OH)_2$, have limited solubilities in aqueous solution. So, in contrast to soluble strong bases, few formula units dissolve and a saturated solution of magnesium hydroxide contains only a few hydroxide ions. Consequently, unlike a metal hydroxide base such as NaOH (lye), $Mg(OH)_2$ solutions can be ingested and are used to treat acidic stomach conditions and peptic ulcers. Bases neutralize acids, as was pointed out in Table 15.2 and will be discussed in Section 15.8.

The equilibrium

$$NH_3(aq) + H_2O \rightleftharpoons NH_4^+(aq) + OH^-(aq)$$

lies to the left. Therefore, because there is little ionization and only a small number of OH^- ions are produced, ammonia is a weak base. The equilibrium constant corresponding to this reaction is the so-called K_b (basicity constant) of NH_3 and equals 1.8×10^{-5}. Notice that this equilibrium constant gives a measure not only of the extent to which OH^- ions are produced (the Arrhenius criterion for basicity), but also of how well a base *accepts* H^+ (The Brønsted-Lowry criterion).

Table 15.3 also gives us a measure of base strength. In this case, the measure of base strength is in terms of the Brønsted-Lowry test of how well the base accepts H^+. Recall that there is a conjugate base corresponding to every acid.

$$\underset{\text{Acid}}{HA} \rightleftharpoons \underset{\text{Conjugate base}}{H^+ + A^-}$$

Therefore, because Table 15.3 ranks the tendency of acids to give up H^+ (tendency for the reaction to proceed →), it also tells us the tendency for the conjugate base to accept H^+ (tendency for the reaction to proceed ←).

$$HA \underset{\text{base reaction}}{\overset{\text{acid reaction}}{\rightleftharpoons}} H^+ + A^-$$

Strong acids have weak conjugate bases. For example, because HCl completely ionizes to $H^+ + Cl^-$, Cl^- has essentially no tendency to accept H^+ and therefore is a very weak base. Conversely, very weak acids have strong conjugate bases. Water has only a very slight tendency to ionize:

$$HOH \rightleftharpoons H^+ + OH^-$$

The conjugate base of water is the strong base OH^-.

Problem 15.3

Arrange the following list of acids in order of decreasing acid strength (use Table 15.3):

HNO_3

H_3BO_3

HF

H_2CO_3

H_2O

Problem 15.4

Arrange the conjugate bases of the acids in Problem 15.3 in order of increasing base strength.

IONIZATION OF WATER

15.5 Although water is a molecular compound, it contains a very small concentration of ions. In viewing water as an acid (HA), we say that it undergoes the ionization reaction

$$\text{HOH}(l) \Longleftrightarrow \text{H}^+(aq) + \text{OH}^-(aq)$$
$$\quad\text{(HA)} \qquad\qquad\qquad\qquad \text{(A}^-\text{)}$$

As for any acid, this equilibrium is more correctly shown as

$$\underset{\text{Acid}}{\text{HOH}(l)} + \underset{\text{Base}}{\text{H}_2\text{O}(l)} \Longleftrightarrow \underset{\text{Acid}}{\text{H}_3\text{O}^+(aq)} + \underset{\text{Base}}{\text{OH}^-(aq)}$$

This representation shows us that the equilibrium is possible because of the amphoteric nature of water. Reference to Table 15.3 tells us that the equilibrium lies to the left because H_3O^+ and OH^- are the stronger acid-base pair. Therefore, in pure water the concentrations of hydronium and hydroxide ions are very small.

As we proceed in our discussion, we will use the simpler expression for the water-ionization equilibrium, that is,

$$\text{HOH}(l) \Longleftrightarrow \text{H}^+(aq) + \text{OH}^-(aq)$$

The equilibrium-constant expression for this reaction is

$$K_{eq} = [\text{H}^+][\text{OH}^-]$$

(Remember, pure liquids and solids do not appear in K_{eq} expressions.) Because this equilibrium expression is so important in acid-base chemistry, it is given the special symbol K_w. Also, the product of the concentrations, $[\text{H}^+][\text{OH}^-]$, is sometimes called the **ion product** of water.

$$K_w = [\text{H}^+][\text{OH}^-]$$

In pure water at 25°C, $[\text{H}^+] = [\text{OH}^-]$, and each is found to have the concentration 1.0×10^{-7} M. Thus the value of K_w can be found by substitution:

$$K_w = (1.0 \times 10^{-7})(1.0 \times 10^{-7})$$

$$K_w = 1.0 \times 10^{-14}$$

The addition of acids to water will raise the concentration of H^+ above $1.0 \times 10^{-7}\ M$. The addition of bases will raise the concentration of OH^- above $1.0 \times 10^{-7}\ M$. But the *product of the concentrations* of H^+ and OH^- ions in water must remain constant and equal to K_w. This is so because the water equilibrium $HOH(l) \rightleftharpoons H^+(aq) + OH^-(aq)$ is present in all aqueous solutions and the equilibrium-constant expression ($K_w = [H^+][OH^-] = 1.0 \times 10^{-14}$) for this equilibrium must always be satisfied. In accord with Le Châtelier's principle, when $[H^+]$ increases, $[OH^-]$ decreases, and when $[OH^-]$ increases, $[H^+]$ decreases, and the ion product, $[H^+][OH^-]$, always remains 1.0×10^{-14}.

Neutral solution:

$[H^+] = [OH^-] = 1.0 \times 10^{-7}\ M$

Acidic solution:

$[H^+] > 1.0 \times 10^{-7}\ M$

$[OH^-] < 1.0 \times 10^{-7}\ M$

Basic solution:

$[OH^-] > 1.0 \times 10^{-7}\ M$

$[H^+] < 1.0 \times 10^{-7}\ M$

SAMPLE EXERCISE

15.2

An aqueous solution is found to have $[H^+] = 1.0 \times 10^{-4}\ M$. Is this solution acidic, basic, or neutral?

SOLUTION

Since $1 \times 10^{-4}\ M$ is greater than $1 \times 10^{-7}\ M$, the solution is acidic.

SAMPLE EXERCISE

15.3

What is $[OH^-]$ in Sample Exercise 15.2?

SOLUTION

If we are given $[H^+]$ or $[OH^-]$ and asked to determine $[OH^-]$ or $[H^+]$, we make use of the fact that $K_w = [H^+][OH^-] = 1.0 \times 10^{-14}$. In this example, $[H^+]$ is given as $1.0 \times 10^{-4}\ M$ and the $[OH^-]$ is unknown. Substituting in the K_w expression,

$$K_w = [H^+][OH^-] = 1.0 \times 10^{-14}$$

$$[OH^-] = \frac{1.0 \times 10^{-14}}{[H^+]}$$

$$[OH^-] = \frac{1.0 \times 10^{-14}}{1.0 \times 10^{-4}}$$

$$[OH^-] = 1.0 \times 10^{-10}\ M$$

Consistent with this being an acidic solution, $[OH^-]$ turns out to be less than $1.0 \times 10^{-7}\ M$.

Problem 15.5

What is $[OH^-]$ in a solution in which $[H^+] = 3.2 \times 10^{-9}\ M$? Is this solution acidic, basic, or neutral?

pH

15.6 The concentrations of H^+ and OH^- mentioned or calculated in the previous section were somewhat unwieldy exponential numbers. In order to be able to express the acidity or basicity of solutions without the use of exponential numbers, chemists devised the pH scale of acidities. The pH of a solution is defined as

$$pH = -\log[H^+]$$

which is said, "pH equals the negative logarithm of the hydrogen ion concentration in moles per liter." The pH is very easy to calculate for solutions in which $[H^+]$ is an exact power of 10, that is, $1 \times$ the power of 10. In this case,

$$[H^+] = 1.0 \times 10^{-x}$$

$$pH = x$$

The logarithm of 1.0×10^{-x} is $-x$. The negative logarithm is $-(-x) = x$.

SAMPLE EXERCISE

15.4

Determine the pH of the following solutions, in which

a. $[H^+] = 1.0 \times 10^{-7}$
b. $[H^+] = 1.0 \times 10^{-11}$
c. $[H^+] = 0.001$

SOLUTION

a. pH $= 7$. The logarithm of 1.0×10^{-7} is -7; the negative log is $-(-7) = 7$
b. pH $= 11$
c. First write 0.001 in scientific notation, $0.001 = 1 \times 10^{-3}$. Then as before, pH $= 3$.

Problem 15.6

Determine the pH of the following solutions, in which

a. $[H^+] = 1.0 \times 10^{-5}$

b. $[H^+] = 0.000010$

Let us summarize the pH characteristics of neutral, acidic, and basic solutions:

Neutral solution:

$$[H^+] = 1.0 \times 10^{-7} M$$

$$pH = 7.00$$

Acidic solution:

$$[H^+] > 1.0 \times 10^{-7} M$$

$$pH < 7.00$$

Basic solution:

$$[H^+] < 1.0 \times 10^{-7}$$

$$pH > 7.00$$

Thus the pH scale extends from 0 to 14, with values less than 7, indicating an acidic solution, and values greater than 7, a basic solution.

For most solutions, the hydrogen ion concentration is *not* an exact power of 10. That is, in most solutions $[H^+]$ is not equal to 1.0×10^{-x}, but rather $[H^+] = N \times 10^{-x}$, where N is some number other than 1. In order to determine the pH in this case, you must be able to manipulate logarithms. There are two ways: with a calculator and using tables.

Those of you who have a logarithm button on your calculator can perform these manipulations quite simply. Most calculators take "logs" by two steps:

Enter number.

Press log button.

The logarithm is then displayed on the screen. Change the sign to determine the pH. For example,
Given:

$$[H^+] = 4.0 \times 10^{-3}$$

Enter 4.0×10^{-3} (or 0.0040).

Press log.

Display is -2.3979.

pH $= 2.40$.

If your calculator is not equipped to do logarithms, then you will need a log table to calculate pH. Table 15.4 is a two-place log table adequate for our needs. Every number written in scientific notation has two parts, the decimal number (between 1 and 10) and the power of 10 (Section 2.4). For example,

$$3.9 \times 10^{-6}$$

Decimal   Power of 10
number

The logarithm of this number is the sum of the log of the decimal number, which we get from Table 15.4, and the log of the power of 10. *The log of a power of 10 is the exponent.* Thus, if

$$[H^+] = N \times 10^{-x}$$

$$\log [H^+] = \log N + \log 10^{-x}$$

$$\log [H^+] = \log N + (-x)$$
From log table

$$pH = -\log [H^+]$$

For the specific example $[H^+] = 3.9 \times 10^{-6}$ we need to look up

TABLE 15.4   TWO-PLACE LOGARITHMS

Tenths
Row
Ones
Column

	0.0	0.1	0.2	0.3	0.4	0.5	0.6	0.7	0.8	0.9
1	0.00	0.04	0.08	0.11	0.15	0.18	0.20	0.23	0.26	0.28
2	0.30	0.32	0.34	0.36	0.38	0.40	0.41	0.43	0.45	0.46
3	0.48	0.49	0.51	0.52	0.53	0.54	0.56	0.57	0.58	0.59
4	0.60	0.61	0.62	0.63	0.64	0.65	0.66	0.67	0.68	0.69
5	0.70	0.71	0.72	0.72	0.73	0.74	0.75	0.76	0.76	0.77
6	0.78	0.79	0.79	0.80	0.81	0.81	0.82	0.83	0.83	0.84
7	0.85	0.85	0.86	0.86	0.87	0.88	0.88	0.89	0.89	0.90
8	0.90	0.91	0.91	0.92	0.92	0.93	0.93	0.94	0.94	0.95
9	0.95	0.96	0.96	0.97	0.97	0.98	0.98	0.99	0.99	1.00

the log of 3.9 and add it to -6. To use the table to determine the log of 3.9, we locate 3 in the vertical (ones) column and 0.9 in the horizontal (tenths) row: 0.59 appears in the box at the intersection of the 3 in the ones column and 0.9 in the tenths row. Thus the log of 3.9 is 0.59; $\log 3.9 \times 10^{-6} = 0.59 + -6 = -5.41$. So for $[H^+] = 3.9 \times 10^{-6} M$, the $\log [H^+] = -5.41$, and the pH equals 5.41.

Check that the table gives you the same answer as a calculator by using the table to determine the pH of a solution where $[H^+] = 4.0 \times 10^{-3} M$, the example treated previously by the calculator method. In the table 0.60 appears at the intersection of the 4 in the ones column and 0.0 in the tenths row. $0.60 + -3 = -2.40$. Thus we obtain pH $= 2.40$, as before.

SAMPLE EXERCISE

15.5

Determine the pH of a solution in which $[H^+] = 2.3 \times 10^{-5} M$.

SOLUTION

Step 1 Recognize that this $[H^+]$ concentration is of the form

$$[H^+] = N \times 10^{-x}$$

Therefore, $\log [H^+] = \log N + (-x)$

Step 2 Look up the log of 2.3 in Table 15.4.

$$\log 2.3 = 0.36$$

Step 3

$$\log [H^+] = 0.36 + (-5) = -4.64$$

Step 4

$$pH = -\log [H^+] = -(-4.64) = 4.64$$

SAMPLE EXERCISE

15.6

Determine the pH of a solution in which $[H^+] = 0.032$.

SOLUTION

Step 1 Write the concentration in scientific notation to conform to

$$[H^+] = N \times 10^{-x}$$
$$[H^+] = 3.2 \times 10^{-2}$$

Step 2 Look up log 3.2 in Table 15.4.

$$\log 3.2 = 0.51$$

Step 3

$$\log [H^+] = 0.51 + (-2) = -1.49$$

Step 4

$$pH = 1.49$$

Given the hydroxide ion concentration $[OH^-]$, we can also determine the pH of a solution. Using the given $[OH^-]$ and K_w, we calculate $[H^+]$ and then pH.

SAMPLE EXERCISE

15.7

Determine the pH of a solution in which $[OH^-] = 5.0 \times 10^{-5}$ M.

SOLUTION

Calculate $[H^+]$ using K_w (see Sample Exercise 15.3). Calculate pH as in previous exercises.

$$K_w = [H^+][OH^-] = 1.0 \times 10^{-14}$$

$$[H^+][5.0 \times 10^{-5}] = 1.0 \times 10^{-14}$$

$$[H^+] = \frac{1.0 \times 10^{-14}}{5.0 \times 10^{-5}} = 2.0 \times 10^{-10}$$

$$\log [H^+] = \log 2.0 + (-10)$$

$$= 0.30 - 10 = -9.70$$

$$pH = -(-9.70)$$

$$= 9.70$$

Problem 15.7

Determine the pH of the following solutions with the given concentrations:

a. $[H^+] = 6.8 \times 10^{-6}$ M

b. $[OH^-] = 7.1 \times 10^{-1}$ M

MEASUREMENT OF pH

15.7 Most bodily fluids such as blood and urine have a normal pH range (Table 15.5). Continued deviation from a particular range usually indicates some pathological condition and therefore offers a means of diagnosis. Medical laboratories require a simple and quick method of measuring pH.

Plants grow best in soil with a specific pH range and do not show maximum growth or fruit production outside this range (Table 15.6). Farmers must be able to measure the pH of the soil (actually the suspension formed from soil and water) in which each crop is planted, so they can adjust it if necessary.

NORMAL pH RANGE FOR SOME BODILY FLUIDS   **TABLE 15.5**

Fluid	Normal pH Range
Liver bile	7.4–8.0
Blood	7.35–7.45
Gastric juice	1–2
Pancreatic juice	7–8
Saliva	6.4–7.0
Sweat	4.5–7.5
Tears	7.0–7.4
Urine	4.5–7.5

Many chemical reactions yield different products depending on the pH condition, thus the laboratory worker must have a way of quickly measuring the pH.

The pH of a colorless or lightly colored solution or suspension is often determined by the use of an **indicator,** a dye that changes color as pH changes. Frequently the dye is impregnated on paper to which a solution to be tested can easily be applied. Indicators are weak acids which have one color in the acid form (symbolized HIn) and another color in the conjugate base form (In^-).

$$HIn \rightleftharpoons H^+ + In^-$$

   Color at   Color at
   lower pH   higher pH

When H^+ is removed from the equilibrium (by reaction with some base), there is a shift in the position of equilibrium to the In^- side

pH RANGE FOR OPTIMUM PLANT GROWTH   **TABLE 15.6**

Plant	pH Required
Apples	5.0–6.5
Cherries	6.0–7.5
Strawberries	5.0–6.5
Peas	6.0–7.5
Snap beans	6.0–7.5
Tomatoes	5.5–7.5
Tulips	6.0–7.0
Roses	6.0–8.0
Azaleas	4.5–6.0
Hydrangeas	5.0–6.0

RED	pH 2.0
VERY STRONGLY ACID	
ORANGE	pH 4.0
STRONGLY ACID	
YELLOW	pH 6.0
WEAKLY ACID	
GREEN	pH 8.0
WEAKLY ALKALINE	
BLUE	pH 10.0
STRONGLY ALKALINE	

FIGURE 15.2

Color chart for universal
indicator paper. Solutions of
different pH turn the paper
different colors.

(Le Châtelier's principle). When the concentration of In⁻ exceeds that
of HIn, the color is that of the high pH range. For the dye named
litmus, a commonly used indicator, this is blue. If the solution is now
made acidic (that is, H^+ is added), the equilibrium shifts to the HIn
side, giving the color of the low pH range which, for litmus, is red.

Litmus changes color at pH 5 to 8, but other indicators change
at other pH values. Each changes over a unique pH range depending
on its acid ionization constant.

The use of litmus paper, which is red at any pH less than 5 and
blue at any pH greater than 8, enables us to determine only whether
a solution is acidic or basic. A universal pH paper can be prepared
by impregnating paper with several dyes that change color at narrow
and different pH ranges (Figure 15.2). The pH of an unknown solution is then determined by comparing the color developed on pH
paper from a drop of the solution with that on a chart which relates
color to pH.

A pH meter (Figure 15.3) is used to determine pH values more
accurately and to determine those of highly colored solutions. The
pH meters work by accurately measuring H^+ ion concentration. Table
15.7 lists the pH values of some commonly encountered materials.

REACTIONS OF ACIDS AND BASES

15.8   Probably the most common and important reaction of acids
and bases is their reaction with one another, the **neutralization reaction,** which we will discuss first.

FIGURE 15.3

A pH meter. The electrodes
are immersed in blood, so
that the pH displayed is that
of blood (7.41). (*Photograph
by Bryan Lees.*)

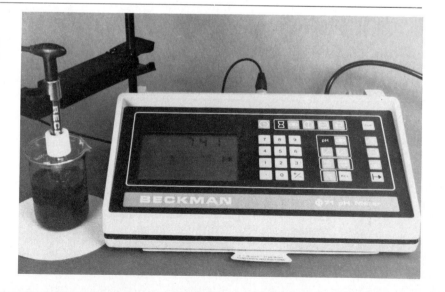

APPROXIMATE pH VALUES OF SOME COMMON MATERIALS   **TABLE 15.7**

pH	Solution
0	Battery acid
1	Stomach acid
2	Lemon juice, lime juice
3	Vinegar, wine, soft drinks, beer, orange juice, pickles
4	Tomatoes, grapes
5	Black coffee, rainwater
6	Urine, milk, saliva
7	Pure water, blood
8	Seawater
9	Clorox, phosphate detergent
10	Soap, milk of magnesia
11	Household ammonia
12	Hair remover
13	Oven cleaner

Neutralization

When acids and bases react with one another, ionic compounds, called **salts,** form. Table salt, NaCl, is only one example of the class of compounds known as salts. **Salts** are ionic compounds formed from the reaction of an acid with a base: the cation of the salt comes from the base, and the anion comes from the acid.

If the base reacting in the neutralization reaction is a hydroxide base, water is also formed. Consider the neutralization reaction between HCl and NaOH:

$$\underset{\text{Acid}}{HCl(aq)} + \underset{\text{Base}}{NaOH(aq)} \longrightarrow \underset{\text{Salt}}{NaCl(aq)} + \underset{\text{Water}}{H_2O}$$

The salt gets its cation (Na^+) from the base and its anion (Cl^-) from the acid. Water forms from the hydroxide of the base and H^+ from the acid.

Strong acid-strong base neutralization reactions

For a strong acid-strong base reactant pair, which HCl and NaOH are, the reaction can be viewed as a double displacement (see Section 9.8), in which cations and anions switch partners.

$$H\!-\!Cl + Na^+OH^- \longrightarrow Na^+Cl^- + H\!-\!OH$$

Other examples of strong acid–strong base neutralization reactions are

$$HNO_3(aq) + LiOH(aq) \longrightarrow LiNO_3(aq) + H_2O$$

$$2HCl(aq) + Ca(OH)_2(aq) \longrightarrow CaCl_2 + 2H_2O$$

$$\underset{\text{Acid}}{H_2SO_4(aq)} + \underset{\text{Base}}{2NaOH(aq)} \longrightarrow \underset{\text{Salt}}{Na_2SO_4} + \underset{\text{Water}}{2H_2O}$$

Because the strong acid is completely ionized, and the strong base completely dissociates in aqueous solution, the following ionic equations (review Section 13.9) apply:

Traditional equation

$$\underset{\text{Strong acid}}{HA} + \underset{\text{Strong base}}{MOH} \longrightarrow \underset{\text{Salt (base cation + acid anion)}}{MA} + H_2O$$

Full ionic $H^+(aq) + A^-(aq) + M^+(aq) + OH^-(aq) \longrightarrow$

$$M^+(aq) + A^-(aq) + H_2O(l)$$

Net ionic $H^+ + OH^- \longrightarrow H_2O$

The net ionic equation for neutralization for any strong acid–strong base combination is $H^+ + OH^- \longrightarrow H_2O$. The pH at neutralization of a strong acid–strong base pair is 7.0.

In terms of Brønsted-Lowry theory this strong acid–strong base reaction is a proton transfer reaction in which the equilibrium lies far to the right.

$$\underset{\text{Strong acid}}{\text{(H) Cl}} + \underset{\text{Strong base}}{\text{OH}^-} \rightleftharpoons \underset{\text{Weak base}}{\text{Cl}^-} + \underset{\text{Weak acid}}{\text{HOH}}$$

A single-headed arrow is typically used for the strong acid–strong base reaction because the equilibrium is so far to the right.

$$HCl + OH^- \longrightarrow Cl^- + HOH$$

Notice again that in this Brønsted-Lowry equation NaOH is represented simply as OH^- because Na^+ is merely a spectator.

Neutralization reactions of weak acids or weak bases

The neutralization reactions of weak acids or weak bases are best considered in terms of Brønsted-Lowry theory because the weak species is not fully ionized in solution. For example, consider the reaction between the weak acid, acetic acid, and a strong base:

$$\underset{\text{Weak acid}}{\text{(H) C}_2\text{H}_3\text{O}_2} + \underset{\text{Strong base}}{(Na^+)^-OH} \rightleftharpoons \underset{\text{Salt}}{(Na^+)C_2H_3O_2^-} + H_2O$$

The acid donates H^+ and the base accepts it; water forms because the base is hydroxide. The salt, sodium acetate, forms in the usual manner, cation from base, anion from acid. The sodium ion is shown here in parentheses because ordinarily the metal ion of MOH is not represented in Brønsted-Lowry reactions.

Now consider the neutralization reaction of a weak acid and weak base:

$$\text{(H)} \; C_2H_3O_2 \; + \; \ddot{N}H_3 \rightleftharpoons \; C_2H_3O_2{}^- NH_4{}^+$$

Weak acid   Weak base   Salt

Again a salt forms wherein the cation comes from the base and the anion from the acid. No water forms because the base is not a hydroxide base. The reaction between a strong acid, such as HCl, and NH_3 gives a similar result:

$$\text{(H)} \; Cl \; + \; \ddot{N}H_3 \rightleftharpoons \; NH_4{}^+ Cl^-$$

Strong acid   Weak base   Salt

Neutralization reactions are the relief of indiscreet eaters and stomach ulcer sufferers. Food overindulgence or stressful situations can lead to an oversecretion of hydrochloric acid in the stomach, for which an antacid is needed for relief. $Al(OH)_3$ and $Mg(OH)_2$ are the two most common hydroxide-base antacids. They function by neutralizing some of the excess H^+ ions in the stomach.

$$2HCl + Mg(OH)_2 \longrightarrow MgCl_2 + 2H_2O$$

$$3HCl + Al(OH)_3 \longrightarrow AlCl_3 + 3H_2O$$

A mixture of the aluminum and magnesium compounds is often desirable because aluminum compounds tend to cause constipation, whereas magnesium compounds have a laxative effect. Figure 15.4 depicts a variety of common antacid products.

Reactions of acids with carbonates and bicarbonates

Other antacid medications employ the reactions of acids with basic bicarbonate and carbonate anions. Acids stronger than H_2CO_3 (see Table 15.3) react with carbonates and bicarbonates to form carbonic acid, H_2CO_3, most of which (because of the instability of H_2CO_3) decomposes to CO_2 and H_2O. We can write an equation for the reaction between hydrochloric acid and sodium bicarbonate which also shows the carbonic acid equilibrium leading to water and carbon dioxide.

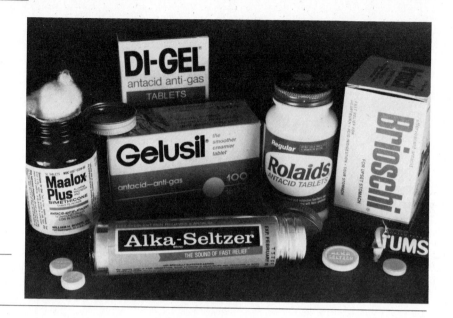

$$HCl(aq) + NaHCO_3(aq) \longrightarrow NaCl(aq) + H_2CO_3(aq)$$

$$\Updownarrow$$

$$H_2O + CO_2(g)$$

Keeping in mind that $Na^+(aq)$ and $Cl^-(aq)$ are spectator ions and that the position of the carbonic acid equilibrium is far to the side of $H_2O + CO_2$, we can write the net ionic equation for the reaction of any acid with a bicarbonate as

$$H^+(aq) + HCO_3{}^-(aq) \longrightarrow H_2O + CO_2(g)$$

Similarly, for a carbonate the reaction would be

$$2H^+(aq) + CO_3{}^{2-}(aq) \longrightarrow H_2O + CO_2(g)$$

Many popular antacids remove excess H^+ through a bicarbonate (Alka Seltzer, Bromo Seltzer) or carbonate (Tums, Pepto-Bismol) ingredient (Figure 15.4). They make you burp because CO_2 gas is released in the stomach.

Acid-base reactions are also important in baking. In baking powder, sodium bicarbonate (baking soda) is combined with acidic ingredients. When wet, this mixture releases carbon dioxide, causing the rising of such baked goods as biscuits and certain breads.

Reactions of acids with active metals

Another example of an acid-base reaction which impacts on food handling is the reaction of acid with some metals. Acids react with Group IA, IIA, and IIIA metals to form hydrogen gas and a salt of the metallic ion. For example,

$$2Al(s) + 6HCl(aq) \longrightarrow 2AlCl_3(aq) + 3H_2(g)$$

| Metal | Acid | Salt of the metal | Hydrogen |

Some transition metals such as iron and zinc react in a similar way, whereas others such as gold, silver, copper, and mercury do not react with acids to form hydrogen gas (see Section 9.8 and Table 9.2).

Acidic foods such as wine or tomatoes should not be brought into contact with cast iron or carbon steel utensils because the acid in these foods will react with the iron. In stainless steel, iron is alloyed in such a way that it no longer reacts with acid. There is also increasing concern that the weak acid H_2CO_3 in soda is reacting with aluminum metal soda cans, albeit slightly, to produce undesirable concentrations of aluminum ions in soda.

DETERMINATION OF ACID-BASE CONCENTRATIONS—TITRATION

15.9 The neutralization reaction discussed in Section 15.8 can be used to determine the unknown concentration of an acid or base given a known concentration of base or acid. The procedure is called **titration.** Consider the neutralization reaction between HCl and NaOH again:

$$HCl + NaOH \longrightarrow NaCl + H_2O$$

or in ionic equation form:

$$H^+(aq) + Cl^-(aq) + Na^+(aq) + OH^-(aq) \longrightarrow$$
$$Na^+(aq) + Cl^-(aq) + H_2O(l)$$

At the point of neutralization, called the **equivalence point,** the number of moles of H^+ from the acid exactly equals the number of moles of OH^- from the base. This is particularly apparent from the net ionic equations for the reactions of any strong acid and strong base:

$$H^+ + OH^- \longrightarrow H_2O$$

In doing titration calculations you must always remember that at the equivalence point

$$\text{moles of } H^+ = \text{moles of } OH^-$$

In doing a titration experiment, a certain volume of the unknown acid or base is placed in the flask with an indicator (and/or it is attached to a pH meter). The indicator must be one that undergoes a color change at the equivalence point. In practice, the indicator will rarely change color exactly at the equivalence point but rather at a close pH called the **endpoint.** A burette is then used for the addition of base to unknown acids or for the addition of acid to unknown bases until the end point is reached. The burette enables one to deliver very accurately measured volumes of solutions.

FIGURE 15.5

Acid-base titration procedure. A base (or acid) of known concentration is added dropwise to an acid (or base) of unknown concentration until the indicator just changes color. This is the endpoint. From the magnitude of the added volume of known concentration, the unknown concentration can be calculated. (a) These are initial conditions. (b) During titration a color change may be noted as a drop of base hits the solution. It is temporary and disappears with stirring. (c) This is the point at which the color change becomes permanent.

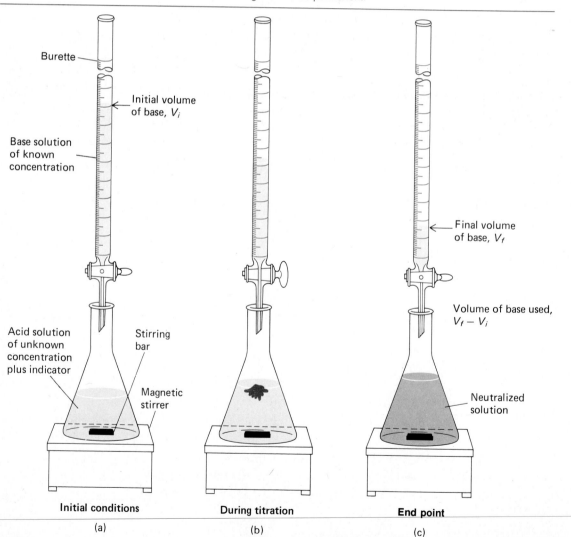

Burette

Initial volume of base, V_i

Base solution of known concentration

Acid solution of unknown concentration plus indicator

Stirring bar

Magnetic stirrer

Final volume of base, V_f

Volume of base used, $V_f - V_i$

Neutralized solution

Initial conditions

During titration

End point

(a)

(b)

(c)

Figure 15.5 shows the measured addition of a volume of base of known concentration until the known volume of unknown acid is neutralized, as witnessed by the change in indicator color at the end-point. Ideally, if an indicator is judiciously selected and careful technique employed, the equivalence point and the endpoint will be nearly identical. We know the volume and concentration of added base; then we can calculate the moles of OH^-, and this is equal to the number of moles of H^+ neutralized. Titration calculations always involve a calculation of the number of moles of H^+ or OH^- neutralized and then a calculation of the molarity of H^+ or OH^-.

Calculation of the number of moles of added OH^- from the known concentration M and volume is done by remembering from Section 13.5 that $M \times V$ (in liters) = moles of OH^-, and at the equivalence point moles of OH^- = moles of H^+. Then the molarity of H^+ is calculated by dividing the number of moles by the volume (in liters) of the unknown acid (see Sample Exercise 15.8).

SAMPLE EXERCISE

15.8

Using a pH meter a chemist finds that 42.50 mL of a 0.100 M NaOH solution is required to titrate (neutralize) 31.00 mL of a hydrochloric acid solution of unknown concentration. What is the molarity of the hydrochloric acid?

SOLUTION

Step 1   Calculate the moles of added OH^- from M of $OH^- \times V$ (in liters) = moles of OH^-.

$$\frac{0.100 \text{ mole}}{1 \text{ L}} \times 0.0425 \text{ L} = 0.00425 \text{ mole of } OH^-$$

(0.100 M NaOH is 0.100 M in OH^-; 42.5 mL $\times$ 1 L/1000 mL = 0.0425 L).

Step 2   At the equivalence point, moles of OH^- = moles of H^+. Therefore, 0.00425 mole of H^+ was neutralized.

Step 3   Calculate the acid molarity by dividing moles by liters of solution. There is 0.00425 mole of H^+ in 0.00425 mole of HCl:

$$31.00 \text{ mL of HCl} \times \frac{1 \text{ L}}{1000 \text{ mL}} = 0.03100 \text{ L of HCl solution}$$

$$M \text{ of HCl} = \frac{\text{moles of HCl}}{\text{liter of solution}}$$

$$M \text{ of HCl} = \frac{0.00425 \text{ mole of HCl}}{0.03100 \text{ L HCl}} = 0.137 \ M$$

In doing titration calculations you must sometimes take into account the fact that the molarity of the acid is not always equal to the molarity of the H^+ ion and the molarity of the base is not always equal to the molarity of OH^- ion. This lack of equality occurs for diprotic and triprotic acids, for example, H_2SO_4 and H_3PO_4, and for metal hydroxides which have more than one hydroxide anion, for example, $Mg(OH)_2$. The subscript of the hydrogen ion or hydroxide ion relates the molarity of the ion to the molarity of the acid or base. Thus 1 M H_2SO_4 is 2 M in H^+ because there are 2 moles of H^+ per mole of H_2SO_4, and 1 M H_3PO_4 is 3 M in H^+. Multiply the acid molarity by the subscript of hydrogen to determine the molarity of H^+. Similarly, 0.1 M $Mg(OH)_2$ is 0.2 M in OH^- because there are 2 moles of OH^- per mole of $Mg(OH)_2$.

Problem 15.8

What is the molarity of H^+ in each of the following acid solutions?

a.   0.5 M H_3PO_4

b.   0.2 M H_2SO_4

c.   0.45 M HBr

SAMPLE EXERCISE

15.9

A student neutralizes 29.10 mL of a potassium hydroxide solution with 15.30 mL of a 0.500 M H_2SO_4 solution. What is the molarity of the KOH solution?

SOLUTION

Note that in this case an acid of known concentration is used to titrate a base of unknown concentration. The principles are just the same as those of the reverse type of titration.

Step 1   As before, we want to calculate moles of added known, in this case, H^+ from M of $H^+ \times V$ (in liters) = moles of H^+. 0.500 M H_2SO_4 is 1 M in H^+ because there are 2 moles of H^+ per 1 mole of H_2SO_4:

$$\frac{0.500 \; \text{mole of } H_2SO_4}{L} \times \frac{2 \text{ moles of } H^+}{1 \text{ mole of } H_2SO_4} = \frac{1 \text{ mole of } H^+}{L}$$

$$15.30 \; \text{mL solution} \times \frac{1 \text{ L}}{1000 \; \text{mL}} = 0.0153 \text{ L of solution}$$

$$M \times V = \frac{1 \text{ mole of } H^+}{L} \times 0.0153 \, L = 0.0153 \text{ mole of } H^+$$

Step 2   At the equivalence point, moles of H^+ = moles of OH^-. Therefore, 0.0153 mole of OH^- was neutralized.

Step 3   Calculate the base molarity by dividing moles by liters of solution. There is 0.0153 mole of OH^- in 0.0153 mole of KOH:

$$29.10 \; \text{mL of KOH} \times \frac{1 \text{ L}}{1000 \; \text{mL}} = 0.02910 \text{ L of KOH}$$

$$M = \frac{0.0153 \text{ mole KOH}}{0.02910 \text{ L KOH}} = 0.526 \text{ } M \text{ KOH}$$

A student neutralized 14.40 mL of a H_2SO_4 solution with 35.20 mL of an 0.200 M NaOH solution. What is the molarity of the H_2SO_4 solution?

SOLUTION

Step 1 Calculate the moles of added known:

$$M \text{ of } OH^- \times V \text{ (in liters)} = \text{moles of } OH^-$$

$$\frac{0.200 \text{ mole of } OH^-}{\cancel{L}} \times 0.0352 \cancel{L} = 0.00704 \text{ mole of } OH^-$$

Step 2 At the equivalence point, moles of OH^- = moles of H^+. Therefore, 0.00704 mole of H^+ was neutralized.

Step 3 Calculate the acid molarity by dividing moles of acid by liters of solution. First determine moles of acid from moles of H^+:

$$0.00704 \cancel{\text{mole of } H^+} \times \frac{1 \text{ mole of } H_2SO_4}{2 \cancel{\text{moles of } H^+}} = 0.00352 \text{ mole of } H_2SO_4$$

$$14.40 \cancel{\text{mL}} \text{ of solution} \times \frac{1 \text{ L}}{1000 \cancel{\text{mL}}} = 0.0144 \text{ L of solution}$$

$$M = \frac{0.00352 \text{ mole of } H_2SO_4}{0.0144 \text{ L of solution}} = 0.244 \text{ } M \text{ } H_2SO_4$$

WHAT IS A BUFFER?

15.10 Although, as Table 15.5 shows, the pH values of various bodily fluids can be very different, for example, pH 1 to 2 in gastric juice versus pH 7 to 8 in pancreatic juice, the normal range within any particular fluid is quite narrow. This is especially true in blood plasma, where the pH of a healthy individual must remain between 7.35 and 7.45. Should the blood pH fall to pH = 7.2, oxygenated hemoglobin (HbO_2), the carrier of O_2 from the lungs to all cells in the body, releases its oxygen, which leads to cell starvation and eventually bodily death.

The body uses a system of buffers to maintain the proper pH of bodily fluids within necessary narrow ranges. A **buffer** is a weak acid–weak base pair that by reacting with added amounts of a base or acid can resist large changes in the solution's pH. For example, if 1 mL of 1.0 M HCl is added to 100 mL of pure water, the pH of the water plummets 5 whole units from 7 to 2. If the same 1 mL of 1.0 M HCl is added to 100 mL of bicarbonate buffer (we will discuss the composition of this buffer in the next section), the pH of the buffer solution decreases from 7.4 to 6.9 a difference of only 0.5 pH units.

In general, buffers are composed of weak conjugate acid–base pairs: either (1) a weak acid (e.g., acetic acid) and its conjugate base (the acetate ion) or (2) a weak base (e.g., ammonia) and its conjugate acid (the ammonium ion). Many bodily fluid buffers are of the weak acid–conjugate base type and so we will concentrate our attention on these buffers, using the acetic acid–acetate ion system as a general example.

HOW A BUFFER FUNCTIONS

15.11 Before we look at how buffers work, let us see first how we prepare a buffer of the weak acid–conjugate base type. A solution of acetic acid contains both the molecular (un-ionized) acid ($HC_2H_3O_2$) and the conjugate base, the anion $C_2H_3O_2^-$.

$$HC_2H_3O_2(aq) \rightleftharpoons C_2H_3O_2^-(aq) + H^+(aq) \qquad (15.1)$$

The anion (base) concentration is very small because the equilibrium lies to the left. We can increase the concentration of anion by adding to the solution an acetate salt which completely dissociates. For example, sodium acetate:

$$C_2H_3O_2^-Na^+ \longrightarrow C_2H_3O_2^-(aq) + Na^+(aq)$$

A buffer is made by combining a solution of a weak acid of known concentration with a solution of known concentration of the salt (anion) of that acid. The H^+ concentration, and hence the pH, of the solution depends on the ratio of the acid to the anion. This can be seen easily by an examination of the equilibrium-constant expression for the acetic acid ionization shown in Equation (15.1).

$$K_a = \frac{[H^+][C_2H_3O_2^-]}{[HC_2H_3O_2]} \qquad (15.2)$$

We isolate (H^+) by multiplying both sides by $[HC_2H_3O_2]$ and dividing by $[C_2H_3O_2^-]$:

$$\frac{K_a[HC_2H_3O_2]}{[C_2H_3O_2^-]} = [H^+] \qquad (15.3)$$

Because K_a is a constant, the $[H^+]$ (and pH) changes only as the ratio of the concentrations change.

Let us now suppose that we have made a buffer solution such that the acetic acid and acetate ion concentrations are equal and quite high (~0.1 M). We can use Le Châtelier's principle to see the effect of added acid or base on the position of equilibrium and then qualitatively predict the effect on the acid/anion ratio and hence $[H^+]$.

Equilibrium: $HC_2H_3O_2(aq) \rightleftharpoons C_2H_3O_2^-(aq) + H^+(aq)$
Stress by adding H^+

The equilibrium shifts to the left to relieve the stress by consuming the added $H^+(aq)$. Acetate ion is also consumed and acetic acid is formed. However, since the concentrations of acid and acetate ion were initially large, the ratio after equilibrium is reestablished is not much affected and the hydrogen-ion concentration after equilibrium is reestablished is essentially the same as it was before acid was added.

Now let us consider the effect of added base.

Equilibrium: $HC_2H_3O_2(aq) \rightleftharpoons C_2H_3O_2^-(aq) + H^+(aq)$

Stress by adding OH^-

The added OH^- reacts with H^+ to form water,

$$OH^-(aq) + H^+(aq) \longrightarrow H_2O$$

thus decreasing the $H^+(aq)$ concentration. Adding base provides the stress of removing H^+ from the equilibrium. The original acetic acid equilibrium shifts to the right to relieve the stress by forming $H^+(aq)$. The concentration of acetic acid is also decreased and that of acetate ion is increased. However, the new ratio of acid to anion is relatively unaffected, and the hydrogen ion concentration after equilibrium is reestablished is essentially the same as it was before hydroxide was added.

The rearranged equilibrium-constant expression, in which $[H^+]$ is isolated, readily shows us the hydrogen ion concentration of a buffer solution as a function of acid and anion concentrations. The application of some simple math to this expression yields an equation that readily shows us the pH of a buffer solution. The new equation is simply the negative logarithmic form of

$$[H^+] = K_a \frac{[HC_2H_3O_2]}{[C_2H_3O_2^-]} \tag{15.3}$$

which reads:

$$-\log[H^+] = -\log K_a - \log \frac{[HC_2H_3O_2]}{[C_2H_3O_2^-]}$$

or

$$pH = pK_a + \log \frac{[C_2H_3O_2]}{[HC_2H_3O_2]} \tag{15.4}$$

where $pK_a = -\log K_a$. Equation (15.4) can be written for any weak acid in the form

$$pH = pK_a + \log \frac{\text{anion of the weak acid}}{\text{weak acid}} \tag{15.5}$$

This equation is known as the Henderson-Hasselbach equation for the biochemists who developed it.

If we know the identity of the acid, and thereby the pK_a, and the ratio of concentrations of anion to weak acid, the pH of the solution can be calculated. The pK_a or K_a must either be given or be calculated from the K_a in Table 15.3. You will see these data given and this equation used in Section 15.12.

BLOOD BUFFERS

15.12 Now we can discuss the body's mechanisms for protecting blood from dramatic pH changes. Three interconnected systems maintain the pH in blood: (1) the blood buffers, which actually serve to neutralize the added hydrogen and hydroxide ions which form from the body's metabolic reactions; (2) the lungs, which are involved in the excretion and inhalation of carbon dioxide, and thereby maintain the concentration of carbonic acid in blood; and (3) the kidneys, which excrete hydrogen ions and bicarbonate ions from the blood. In this section we will concentrate on the role of the blood buffers and witness the supporting role of the other two systems.

There are three major body buffers: (1) the H_2CO_3/HCO_3^- buffer, (2) the $H_2PO_4^-/HPO_4^{2-}$ buffer, and (3) the protein buffers. The carbonic acid–bicarbonate mixture is the major buffer in blood. As we have seen before, carbonic acid is an unstable weak acid which in aqueous solution, in this case blood, is always in equilibrium with $CO_2(aq)$.

$$H_2CO_3(aq) \rightleftharpoons CO_2(aq) + H_2O \qquad (15.6)$$

The position of this equilibrium is to the right. Dissolved $CO_2(aq)$ is also in equilibrium with $CO_2(g)$ in the lungs.

$$\underset{\text{Blood}}{CO_2(aq)} \rightleftharpoons \underset{\text{Lungs}}{CO_2(g)} \qquad (15.7)$$

A moment's thought (and perhaps some help from Le Châtelier) should convince you that the concentration of $H_2CO_3(aq)$ can be directly affected by that of $CO_2(g)$.

Carbonic acid is also in equilibrium with the other half of the buffer mixture, i.e., the bicarbonate anion, through the acid dissociation:

$$H_2CO_3(aq) \rightleftharpoons HCO_3^-(aq) + H^+(aq) \qquad (15.8)$$

The operation of the buffering action here is completely analogous to the acetic acid–acetate ion system discussed in Section 15.11. Acidic by-products (H^+) of the metabolic cycles can be neutralized by $HCO_3^-(aq)$, forming $H_2CO_3(aq)$ [equilibrium of Equation (15.8) shifts left]. When in excess, H_2CO_3 is removed from the body as $CO_2(g)$ [equilibria of Equations (15.6) and (15.7) shift right]. Excess base is neutralized by $H_2CO_3(aq)$ forming $HCO_3^-(aq)$ [equilibrium of Equa-

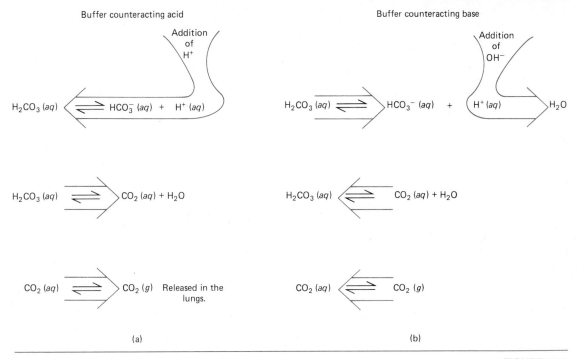

Buffer counteracting acid

Buffer counteracting base

(a)

(b)

Action of the carbonic acid-bicarbonate buffer upon addition of acid (a) or base (b). (a) Addition of H^+ shifts ($\Leftarrow$) the equilibrium to the left, producing H_2CO_3. This new H_2CO_3 constitutes a stress on the second equilibrium, shifting it to the right; $CO_2(aq)$ is thus produced, and this shifts the third equilibrium to the right. (b) Addition of OH^- consumes H^+ and shifts the first equilibrium to the right, thus diminishing H_2CO_3. To compensate, the second equilibrium shifts left to replace H_2CO_3. The third equilibrium shifts left to replace $CO_2(aq)$.

tion (15.8) shifts right]. Figure 15.6 summarizes the equilibrium shifts in the buffer system caused by additions of acid or base.

The concentrations of $HCO_3^-(aq)$ and $H_2CO_3(aq)$ in the blood of a healthy individual are 2.5×10^{-2} M and 1.25×10^{-3} M, respectively. The pK_a of carbonic acid is 6.1. Using these data and the Henderson-Hasselbach relationship [Equation (15.5)], we obtain

$$pH = pK_a + \log \frac{[HCO_3^-]}{[H_2CO_3]}$$

$$pH = 6.1 + \log \frac{2.5 \times 10^{-2}}{1.25 \times 10^{-3}}$$

$$pH = 6.1 + \log 20$$

$$pH = 6.1 + 1.3 = 7.4$$

The pH we should expect to find in normal blood is 7.4.

The capacity of a particular buffer to resist changes in pH, called the **buffering capacity,** depends on the ratio of anion to acid remaining fairly constant. For the carbonic acid–bicarbonate buffer this means a relative constancy in the ratio $[HCO_3^-]/[H_2CO_3]$ of about 20:1.

A key factor in the maintenance of this 20:1 ratio is the coupling of the H_2CO_3 concentration to the partial pressure of CO_2 in the lungs, as was shown in Equations (15.6) and (15.7). The CO_2 in our lungs provides a ready and essentially limitless source of more H_2CO_3 so that a shortage is unlikely. Furthermore, excess H_2CO_3 can be disposed of by shifting the equilibrium of Equations (15.6) and (15.7) to the right in the direction of $CO_2(g)$. Increased $CO_2(g)$ exhalation accomplishes this.

The concentration of HCO_3^- is regulated by the kidneys. If the HCO_3^- concentration drops, the kidneys remove H^+ from the blood and the H^+ concentration in urine increases. Removal of H^+ shifts the equilibrium, $H_2CO_3 \rightleftharpoons HCO_3^- + H^+$, to the right, thus replenishing $[HCO_3^-]$. As we discussed, the H_2CO_3 lost by this shift can be replaced by CO_2. The kidneys promote the excretion of excess concentrations of HCO_3^-.

The carbonic acid–bicarbonate system in blood has a high buffering capacity because of the coupled equilibria between H_2CO_3 and the unlimited supply of CO_2 in the lungs and the regulation of $[HCO_3^-]$ by the kidneys. If both $[HCO_3^-]$ and $[H_2CO_3]$ are maintained at nearly constant values, then the ratio $[HCO_3^-]/[H_2CO_3]$ remains constant and so does the pH of the solution.

Another buffer system is the $H_2PO_4^-/HPO_4^{2-}$ weak acid–conjugate base system. This is the major buffer in intracellular fluids. This buffer mixture functions through the equilibrium

$$H_2PO_4^-(aq) \rightleftharpoons HPO_4^{2-}(aq) + H^+(aq)$$

Weak acid	Weak base
neutralizes added	neutralizes added
base	acid

Problem 15.9

The pK_a of $H_2PO_4^-$ is 7.21; for normal cell concentrations $[HPO_4^{2-}] = 2.4 \times 10^{-3}$ M and $[H_2PO_4^-] = 1.5 \times 10^{-3}$ M. Use the Henderson-Hasselbach equation,

$$pH = pK_a + \log \frac{[HPO_4^{2-}]}{[H_2PO_4^-]}$$

to show that a pH of 7.4 is maintained within cells by this phosphate buffer system.

SUMMARY

Acids and bases are important classes of materials which we encounter daily. The H^+ ion is associated with acidity and the OH^- ion with basicity (Section 15.1). The Arrhenius definition of acids and bases directly reflects this correspondence between the type of ion in solution and the acidic or basic nature of the solution (Section 15.2). Brønsted and Lowry define acids as H^+ donors and bases as H^+ acceptors. Every acid has a conjugate base and every base a conjugate acid (Section 15.3).

Just as electrolytes are classified as strong or weak depending on the numbers of ions they release in solution, acids are classified as strong or weak depending on the numbers of H^+ ions produced. Bases are strong or weak depending on the numbers of OH^- ions produced per mole of base. Table 15.3 lists acid ionization constants (K_a) and basicity constants (K_b). These are equilibrium constants, and from their magnitude we can predict relative acid and base strengths. Strong acids have weak conjugate bases; weak acids have strong conjugate bases (Section 15.4).

Because of the slight tendency of water to ionize, every water solution contains H^+ ions and OH^- ions. The product of the concentration of these ions is a constant, $K_w = 1 \times 10^{-14}$ (Section 15.5). The pH scale was developed as a convenient way to express the acidity of a solution. pH is minus the logarithm of the hydrogen ion concentration. On this scale, pH = 7 corresponds to a neutral solution, a pH < 7 indicates an acid solution, and a pH > 7 means a basic solution (Section 15.6). pH can be measured by dyes called indicators or by a pH meter (Section 15.7).

Neutralization, probably the most important acid-base reaction, is the reaction between an acid and base to form a salt. If the base is a metal hydroxide, water is also a product of the neutralization reaction. Acids react with bicarbonate and carbonate salts to form carbon dioxide and water. Either the neutralization reaction or the bicarbonate-carbonate reaction is employed in antacid tablets. Acids also react with active metals (Section 15.8). The unknown concentration of an acid or base can be determined by doing a titration experiment in which the unknown is carefully neutralized by the addition of a base or acid of known concentration (Section 15.9).

A buffer is a solution which can resist changes in pH. Most often, we encounter buffers composed of a weak acid and its conjugate base (Section 15.10). The pH of such a solution is determined by the ratio of the conjugate acid–base components. Because this ratio in a buffer remains nearly constant upon addition of small amounts of outside acid or base, the pH is maintained within a narrow range. The Henderson-Hasselbach equation can be used to evaluate the pH of a buffer solution (Section 15.11).

The principal buffer system in blood is the H_2CO_3/HCO_3^- system. This buffer system is supported by the lungs which help to adjust H_2CO_3 concentrations by inhalation or exhalation of CO_2 and the kidneys which help regulate HCO_3^- concentrations (Section 15.12).

CHAPTER ACCOMPLISHMENTS

After completing this chapter you should be able to

15.1   Introduction

1.   List some common acids and bases and their application in every-day life.
2.   State the physical and chemical properties which distinguish acids from bases.

15.2   The Arrhenius definition

3.   State the Arrhenius definitions of acids and bases.
4.   Recognize acids and bases according to the Arrhenius definition.
5.   State the ions present in an aqueous solution of a given metallic hydroxide or a given acid.

15.3   Brønsted-Lowry definition

6.   State the Brønsted-Lowry definitions of acids and bases.
7.   Recognize acid-base pairs and their conjugate acid-base pairs in a given equation.
8.   State the definition and give an example of an amphoteric substance.

15.4   Acid and base strength

9.   Describe the relationship between the classification of acids and bases as strong or weak and the extent of their ionization.
10.   Describe what is meant by the acid and base ionization constants.
11.   State the relationship between the magnitude of the acid ionization constant and the strength of an acid and its conjugate base, and between the basicity constant and the strength of a base and its conjugate acid.
12.   Given a table of acid and base ionization constants, arrange a list of acids and bases in order of increasing or decreasing acid and base strength.

15.5   Ionization of water

13.   Write the equilibrium-constant expression corresponding to K_w.
14.   State the value of $[H^+]$ and $[OH^-]$ in pure water.
15.   State the value of K_w.
16.   Given a value of $[H^+]$ or $[OH^-]$, state whether the solution is acidic or basic.
17.   Given a value for $[H^+]$, calculate $[OH^-]$, or given a value of $[OH^-]$, calculate $[H^+]$.

15.6   pH

18.   Define pH.

19. Given the pH of a solution, state whether the solution is acidic, basic, or neutral.
20. Calculate the pH of a solution given the value of $[H^+]$ or $[OH^-]$.

15.7 Measurement of pH
21. State two methods by which the pH of a solution can be determined experimentally.

15.8 Reactions of acids and bases
22. Write an equation for the neutralization reaction between a given acid and base.
23. Write an equation for the reaction between a given acid and a bicarbonate or carbonate salt.

15.9 Titration
24. Describe how a titration is performed.
25. Given the volume of one solution (acidic or basic) and the volume and molarity of a second solution (basic or acidic) needed to titrate the first, calculate the molarity of the first solution.

15.10 What is a buffer?
26. State the general composition of buffers.

15.11 How a buffer functions
27. Describe how a buffer of the weak acid–conjugate base type is prepared.
28. State what the $[H^+]$ concentration of a given buffer depends on.
29. Describe the changes in equilibrium that occur when a small amount of acid is added to a given weak acid–conjugate base buffer.
30. Given concentrations of weak acid and conjugate base and an appropriate pK_a, use the Henderson-Hasselbach equation to calculate the pH of a given weak acid–conjugate base buffer.

15.12 Blood buffers
31. State the three systems that serve to regulate the pH in blood.
32. State the three equilibria that influence the concentration of H_2CO_3 in blood plasma.
33. Describe how excess acid and excess base are neutralized by the carbonic acid–bicarbonate buffer.
34. Describe why the carbonic acid–bicarbonate system in blood has a high buffering capacity.

PROBLEMS

15.1 Introduction

15.10 Describe a procedure for distinguishing between an acidic and a basic substance.

15.11 Give the names of two acids and two bases found in commercial products in your household.

15.2 The Arrhenius definition

15.12 **a.** Give two examples of an Arrhenius acid.
b. Give two examples of an Arrhenius base.

15.13 Show by an equation how the ionization of HNO_3 in aqueous solution leads to the formation of the hydronium ion.

15.14 Write equations to show the dissociation or ionization of the following compounds in water:
a. LiOH
b. H_2SO_4
c. $Ca(OH)_2$

15.3 Brønsted-Lowry definition

15.15 **a.** Distinguish between a Brønsted-Lowry acid and a Brønsted-Lowry base.
b. Distinguish between a Brønsted-Lowry acid and an Arrhenius acid.
c. Distinguish between a Brønsted-Lowry base and an Arrhenius base.

15.16 Show by equations how the ion HSO_4^- can act as an amphoteric substance.

15.17 Show by an equation how the ion NH_4^+ can act as a Brønsted-Lowry acid.

15.18 Identify the acid–conjugate base and the base–conjugate acid pairs in the following reactions:
a. $HBr + H_2O \rightleftharpoons H_3O^+ + Br^-$
b. $NH_3 + H_3O^+ \rightleftharpoons NH_4^+ + H_2O$
c. $HSO_4^- + OH^- \rightleftharpoons SO_4^{2-} + H_2O$
d. $HCO_3^- + H_3O^+ \rightleftharpoons H_2CO_3 + H_2O$
e. $NH_3 + H_3PO_4 \rightleftharpoons NH_4^+ + H_2PO_4^-$
f. $HS^- + H_2O \rightleftharpoons H_2S + OH^-$

15.19 Give the conjugate bases of the following substances:
a. H_2O **d.** NH_4^+
b. HF   **e.** H_2SO_4
c. $H_2PO_4^-$ **f.** NH_3

15.20 Give the conjugate acids of the following substances:
a. H_2O **d.** NH_3
b. F^- **e.** HSO_4^-
c. $H_2PO_4^-$ **f.** OH^-

15.21 Which of the following can act as an amphoteric substance? H_2CO_3, HCO_3^- or CO_3^{2-}? Write equations to demonstrate your answer.

15.22 Give the formula of the species formed when
a. HCl acts as a Brønsted acid.
b. H_2O acts as a Brønsted base.
c. F^- acts as a Brønsted base.
d. NH_4^+ acts as a Brønsted acid.

15.23 Identify the Brønsted-Lowry acid and the Brønsted-Lowry base in each case and complete the following equations by transferring a proton from the acid to the base:
a. $NH_4^+ + NH_3 \rightleftharpoons$
b. $H_2SO_4 + H_2O \rightleftharpoons$
c. $HCl + CN^- \rightleftharpoons$
d. $H_2CO_3 + NO_3^- \rightleftharpoons$
e. $HSO_4^- + OH^- \rightleftharpoons$
f. $HSO_4^- + NO_3^- \rightleftharpoons$

15.4 Acid and base strength

15.24 Arrange the following substances in order of increasing acid strength: H_3BO_3, HSO_4^-, $HClO_4$, H_2O, and HCN.

15.25 Arrange the conjugate bases of the acids in Problem 15.24 in order of increasing base strength.

15.26 **a.** Write an equilibrium-constant expression for the acid ionization of $HC_2H_3O_2$.
b. What quantities would have to be determined experimentally to evaluate this equilibrium constant?

15.27 **a.** Write an equilibrium-constant expression for the reaction of HSO_4^- as a Brønsted-Lowry base in aqueous solution.
b. What quantities would have to be determined experimentally to evaluate this equilibrium constant?

15.28 What is the difference between the terms "dilute" and "concentrated" and "weak" and "strong" with respect to basicity and acidity?

15.29 **a.** Write all steps showing *complete* ionization of H_2CO_3.
b. Write all steps showing *complete* ionization of H_3PO_4.

15.5 Ionization of water

15.30 **a.** Write an equilibrium-constant expression for the reaction of H_2O as a Brønsted-Lowry acid in pure water.
b. Write an equilibrium constant expression for the reaction of H_2O as a Brønsted-Lowry base in pure water.
c. What is the relationship between the equilibrium constants in parts (*a*) and (*b*).
d. What are the numerical values of these equilibrium constants?

15.31 Indicate whether the following solutions are acidic, basic, or neutral:
a. $[H^+] = 1.1 \times 10^{-3} M$
b. $[H^+] = 6.0 \times 10^{-11} M$
c. $[OH^-] = 2.3 \times 10^{-11} M$
d. $[OH^-] = 5.0 \times 10^{-7} M$
e. $[H^+] = 0.00013 M$
f. $[H^+] = 0.0000001 M$

15.32 Calculate the hydrogen-ion concentration in an aqueous solution in which the hydroxide-ion concentration is:
a. $2.5 \times 10^{-9} M$
b. $3.0 \times 10^{-6} M$
c. $0.0015 M$

15.33 Classify each of the solutions in Problem 15.32 as acidic, basic, or neutral.

15.34 Calculate the hydroxide-ion concentration in an aqueous solution in which the hydrogen-ion concentration is:
a. $5.0 \times 10^{-7} M$
b. $2.4 \times 10^{-3} M$
c. $0.000095 M$

15.35 Classify each of the solutions in Problem 15.34 as acidic, basic, or neutral.

15.36 Determine the pH of the following solutions, and indicate whether the solution is acidic, basic, or neutral:
a. $[H^+] = 1.0 \times 10^{-7} M$
b. $[H^+] = 1.0 \times 10^{-2} M$
c. $[H^+] = 1.0 \times 10^{-10} M$
d. $[H^+] = 0.0001 M$

15.37 Determine the pH of the following solutions, and indicate whether the solution is acidic, basic, or neutral:
a. $[H^+] = 5.0 \times 10^{-3} M$
b. $[H^+] = 9.7 \times 10^{-8} M$
c. $[H^+] = 0.047 M$
d. $[H^+] = 0.91 M$
e. $[OH^-] = 3.4 \times 10^{-6} M$
f. $[OH^-] = 1.2 \times 10^{-9} M$
g. $[OH^-] = 0.031 M$
h. $[OH^-] = 0.000020 M$

15.38 Indicate whether the following solutions are acidic, basic, or neutral:
a. Seawater: pH = 8.1
b. Vinegar: $[H^+] = 1.6 \times 10^{-3} M$
c. Coffee: pH = 5.42
d. Orange juice: $[H^+] = 2.0 \times 10^{-4} M$
e. Blood: pH = 7.4
f. Soda water: pH = 2.8
g. Stomach fluid: pH = 1.8
h. Ammonia water: $[H^+] = 1.9 \times 10^{-11} M$

15.39 A solution is prepared by dissolving 25.0 g of NaOH in water and then adding water sufficient to bring the volume to 1.00 L in a volumetric flask. What are $[OH^-]$, $[H^+]$, and the pH?

15.40 A solution is prepared by diluting 10.0 mL of a 6.00 M HCl solution to 0.500 L. Calculate $[H^+]$, $[OH^-]$, and the pH.

15.7 Measurement of pH

15.41 Describe how a pH meter could be used to indicate the equivalence point of a titration reaction between a strong acid and a strong base.

15.42 Describe the relationship between an indicator equilibrium, Le Châtelier's principle, and the color change observed in acidic and basic solution.

15.8 Reactions of acids and bases

15.43 Complete and balance the following neutralization reactions:
a. $KOH + HCl \longrightarrow$
b. $Ba(OH)_2 + HNO_3 \longrightarrow$
c. $NaOH + H_3PO_4 \longrightarrow$
d. $Ca(OH)_2 + H_2SO_4 \longrightarrow$
e. $Ca(OH)_2 + H_3PO_4 \longrightarrow$

15.44 Indicate the acid and base needed to prepare the following salts:

a. $NaNO_3$
b. K_2SO_4
c. $CaCl_2$
d. $AlPO_4$
e. $Ca(C_2H_3O_2)_2$
f. $(NH_4)_2SO_4$

15.45 Complete and balance the following equations:

a. $HNO_3 + KHCO_3 \longrightarrow$
b. $HCl + Li_2CO_3 \longrightarrow$
c. $Zn + HCl \longrightarrow$
d. $Na_2CO_3 + HC_2H_3O_2 \longrightarrow$
e. $Mg + H_2SO_4 \longrightarrow$

15.46 Write a balanced net ionic equation for each of the following reactions:

a. $HCl + Ca(OH)_2 \longrightarrow CaCl_2 + H_2O$
b. $H_2SO_4 + KOH \longrightarrow K_2SO_4 + H_2O$
c. $H_3PO_4 + NaOH \longrightarrow Na_3PO_4 + H_2O$
d. $H_3PO_4 + Ca(OH)_2 \longrightarrow Ca_3(PO_4)_2 + H_2O$

15.9 Determination of acid-base concentrations—titration

15.47 A student found that 68.30 mL of an HCl solution was required to neutralize 31.75 mL of 0.150 M NaOH completely. What is the molarity of the acid solution?

15.48 A chemist finds that 39.52 mL of a calcium hydroxide solution is required to titrate 18.90 mL of a 0.200 M H_3PO_4 solution. What is the molarity of the basic solution?

$$3\,Ca(OH)_2 + 2H_3PO_4 \longrightarrow Ca_3(PO_4)_2 + 6H_2O$$

15.49 How many milliliters of 0.250 M H_2SO_4 are required to neutralize 25.10 mL of 0.100 M NaOH?

$$2NaOH + H_2SO_4 \longrightarrow Na_2SO_4 + 2H_2O$$

15.50 How many milliliters of 0.150 M $HClO_4$ are needed to titrate 35.0 mL of a 0.215 M LiOH?

15.51 Explain why a 0.30 M H_2SO_4 solution is 0.60 M in H^+.

15.52 In a laboratory, 57.3 g of NaOH is dissolved in water, and the solution is diluted to 2.50 L. Then 30.0 mL of this NaOH solution is used to titrate 42.5 mL of a HCl solution. What is the molarity of the HCl solution?

15.53 In an experiment, 35.0 mL of 1.00 M HCl is

required to titrate a Drano (active basic ingredient NaOH) solution. How many moles of NaOH are present in the Drano solution?

15.54 In another experiment, 10.0 g of vinegar (active acid ingredient is acetic acid, $HC_2H_3O_2$) is titrated with 65.40 mL of 0.150 M NaOH.

$$HC_2H_3O_2 + NaOH \longrightarrow NaC_2H_3O_2 + H_2O$$

a. How many moles of $HC_2H_3O_2$ are present in 10.0 g of the vinegar?
b. How many grams of $HC_2H_3O_2$ are present in 10.0 g of vinegar?
c. What is the percentage by weight of acetic acid in the vinegar solution?

15.55 The active sour ingredient in vinegar is acetic acid, $HC_2H_3O_2$. In order to test the acid molarity of a vinegar solution, a laboratory technician titrated 24.50 mL of vinegar with 0.4789 M NaOH. She found that it took 46.09 mL of NaOH to neutralize the vinegar completely. What is the acid molarity of the vinegar? Assume that the only acid present is $HC_2H_3O_2$.

15.56 Gastric juice, found in the stomach, contains hydrochloric acid. A 30.10-mL sample of gastric juice is titrated with 0.1050 M NaOH. The chemist finds that 17.20 mL of NaOH is needed to neutralize the gastric juice acid. What is the molarity of acid in gastric juice? Assume that HCl is the only substance in gastric juice that reacts with NaOH.

15.10 What is a buffer?

15.57 Define a buffer solution.

15.58 What substance would you need to add to hydrosulfuric acid, H_2S, to prepare a buffer solution?

15.11 How a buffer functions

15.59 Using the appropriate equilibrium equations, describe how a hydrosulfuric acid—sodium hydrogen sulfide mixture can act as a buffer.

15.60 Describe how the $[H^+]$ concentration of a given buffer varies when the acid/anion concentration is:

a. Doubled
b. Decreased to one-third of its original value

15.61 Qualitatively describe the changes in the

concentrations of $[H^+]$, $[H_2CO_3]$, and $[HCO_3^-]$ when a small amount of hydrochloric acid is added to a carbonic acid–bicarbonate buffer.

15.62 Calculate the pH of the acetic acid–acetate ion buffer that contains $[HC_2H_3O_2] = 0.15\ M$ and $[C_2H_3O_2^-] = 0.10\ M$ (pK_a for acetic acid is 4.7).

15.12 Blood buffers

15.63 Calculate the pH of a blood sample in which the bicarbonate concentration is twice the carbonic acid concentration (pK_a for H_2CO_3 in blood plasma $= 6.1$).

15.64 Explain how the $[H_2CO_3]$ concentration in blood plasma would vary if:

  a. The concentration of $CO_2(aq)$ increased.

  b. The partial pressure of $CO_2(g)$ in the lungs increased.

  c. The concentration of bicarbonate decreased.

15.65 What would be the pH of blood if $[H_2CO_3]$ and $[HCO_3^-]$ were present in equal concentrations?

15.66 **a.** Describe the role of the lungs in regulating blood pH.

  b. Describe the role of the kidneys in regulating blood pH.

15.67 Although the concentration of $[H_2CO_3]$ in blood plasma is only one-twentieth of the $[HCO_3^-]$ concentration, blood has a high buffering capacity toward base. Explain.

OXIDATION REDUCTION CHAPTER

16

S

Sulfur

32.06

INTRODUCTION

16.1 Chemical reactions involve rearrangements of electron structure. For example, when a compound forms, the valence electrons of the combining atoms interact and a new arrangement of electrons results.

$$C + 2H_2 \longrightarrow CH_4$$

$$\cdot \overset{\cdot}{\underset{\cdot}{C}} \cdot + 2\ H \overset{x}{\underset{x}{}} H \longrightarrow H \overset{x}{\underset{x}{}} \overset{H}{\underset{H}{\overset{\cdot x}{\underset{\cdot x}{C}}}} \overset{x}{\underset{x}{}} H$$

$$2Na + Cl_2 \longrightarrow 2NaCl$$

$$2Na^x + \ :\overset{..}{\underset{..}{Cl}}:\overset{..}{\underset{..}{Cl}}: \longrightarrow 2\ Na^+ \ :\overset{..}{\underset{..}{Cl}}\overset{x}{}$$

Previously (Chapter 9) in dealing with chemical reactions and equations, we have not always kept track of the electrons. In this chapter you will see the advantages of doing so.

ELECTRON-TRANSFER REACTIONS

16.2 Very often a chemical reaction involves a transfer of electrons. This is particularly obvious in the case of the formation of an ionic compound such as NaCl. Sodium atoms transfer their electrons to chlorine, and the result is sodium ions and chloride ions. We could represent the transfer of electrons in terms of the loss of electrons by sodium and the gain of electrons by chlorine:

$$Na \longrightarrow Na^+ + e^- \text{ Sodium loses an electron} \quad \textbf{oxidation}$$

$$Cl_2 + 2e^- \longrightarrow 2Cl^- \text{ Chlorine gains electrons} \quad \textbf{reduction}$$

Oxidation is the name given to the process in which there is a *loss of electrons*. **Reduction** is the name given to the process in which there is a *gain of electrons*. The two processes always occur simultaneously and in such a way that

Number of **electrons lost** = number of **electrons gained**

Therefore, two sodium atoms must be losing electrons for every one chlorine molecule (with its two chlorine atoms) which gains electrons.

$$2Na \longrightarrow 2Na^+ + 2e^- \qquad \text{oxidation half-reaction}$$
$$Cl_2 + 2e^- \longrightarrow 2Cl^- \qquad \text{reduction half-reaction}$$
$$2Na + Cl_2 + 2\cancel{e^-} \longrightarrow 2Na^+ + 2Cl^- + 2\cancel{e^-}$$

Every electron-transfer reaction is the sum of an oxidation half-reaction in which electrons are lost and a reduction half-reaction in which electrons are gained. The sum of the two half-reactions is the total reaction with which we are familiar, written either as an ionic or a traditional equation (Section 13.9). Notice that the electrons cancel out because there are equal numbers gained (left side) and lost (right side). Electron-transfer reactions are called **redox** reactions because they are the sum of *red*uction and *ox*idation half-reactions.

All reactions in which atoms or molecules form ions and/or ions form atoms or molecules are redox or electron-transfer reactions. In other words, if a species undergoes a change in charge, there must have been an electron transfer or redox reaction.

WRITING HALF-REACTIONS

16.3 The net ionic equation (Section 13.9) for a redox reaction shows the species participating in the two half-reactions most readily. For example, consider again (Section 9.8) the reaction between an active metal and an acid to produce an ionic compound and hydrogen gas. For the reaction of calcium metal and hydrobromic acid,

Traditional equation $$Ca(s) + 2HBr(aq) \longrightarrow CaBr_2(aq) + H_2(g)$$

Ionic equation $$Ca(s) + 2H^+(aq) + 2Br^-(aq) \longrightarrow Ca^{2+}(aq) + 2Br^-(aq) + H_2(g)$$

Net ionic equation $$Ca(s) + 2H^+(aq) \longrightarrow Ca^{2+}(aq) + H_2(g)$$

The two half-reactions here are the oxidation of calcium atoms to Ca^{2+} ions and the reduction of H^+ ions to hydrogen molecules:

$$Ca(s) \longrightarrow Ca^{2+}(aq) + 2e^- \qquad \textbf{oxidation}$$
$$2H^+(aq) + 2e^- \longrightarrow H_2(g) \qquad \textbf{reduction}$$

The calcium half-reaction is recognized as an oxidation because metals lose electrons and become cations. Notice also that two electrons are required on the right of the calcium half-reaction to balance the equation with respect to charge. Once the oxidation half-reaction has been pinpointed, the remaining half-reaction must be reduction. Hydrogen ions must gain electrons (i.e., be reduced) to form H_2 molecules.

SAMPLE EXERCISE 16.1

Identify the following half-reactions as oxidations or reductions:

a. $Li^+ + e^- \longrightarrow Li$
b. $2I^- \longrightarrow I_2 + 2e^-$
c. $H_2 \longrightarrow 2H^+ + 2e^-$
d. $Fe^{3+} + e^- \longrightarrow Fe^{2+}$

SOLUTION

a. *Reduction.* Lithium ion gains one e^-, which appears on the left.
b. *Oxidation.* Iodide loses two e^-, which appear on the right.
c. *Oxidation.* Hydrogen loses two e^-, which appear on the right.
d. *Reduction.* Iron ion gains one e^-, which appears on the left.

Problem 16.1

Identify the following half-reactions as oxidations or reductions:

a. $Cr \longrightarrow Cr^{3+} + 3e^-$

b. $Cl_2 + 2e^- \longrightarrow 2Cl^-$

c. $S^{2-} \longrightarrow S + 2e^-$

d. $Hg^{2+} + 2e^- \longrightarrow Hg$

The material that is oxidized is also referred to as the **reducing agent** because it provides the electrons that are gained by the material that is reduced. In the example of calcium in acid, calcium is being oxidized (losing electrons) and is thereby acting as a reducing agent because it provides the electrons which reduce H^+. Similarly, the material that is reduced is referred to as the **oxidizing agent** because it accepts the electrons lost in oxidation. Therefore, H^+ is an oxidizing agent for Ca, from which it gains electrons and is thereby reduced. Remember that

Material **oxidized** = **reducing agent**
(Electrons lost)

Material **reduced** = **oxidizing agent**
(Electrons gained)

SAMPLE EXERCISE 16.2

Given the following ionic equation, write the oxidation and reduction half-reactions, and identify the oxidizing and reducing agents.

$$Zn(s) + Cu^{2+}(aq) + SO_4^{2-}(aq) \longrightarrow Zn^{2+}(aq) + SO_4^{2-}(aq) + Cu(s)$$

SOLUTION

Step 1 Write the *net* ionic equation; that is, cross out the spectator ions:

$$Zn(s) + Cu^{2+}(aq) \longrightarrow Zn^{2+}(aq) + Cu(s)$$

Now only the materials that have been changed appear in the equation.

Step 2 $Zn(s)$ must lose electrons (be oxidized) to become Zn^{2+}. Therefore, the oxidation half-reaction is

$$Zn(s) \longrightarrow Zn^{2+}(aq) + 2e^-$$

Electrons on the right side show the loss or oxidation.

Step 3 $Cu^{2+}(aq)$ must gain electrons (be reduced) to become $Cu(s)$. Therefore, the reduction half-reaction is

$$Cu^{2+}(aq) + 2e^- \longrightarrow Cu(s)$$

Electrons on the left side show the gain or reduction.

Step 4 The material oxidized = the reducing agent = $Zn(s)$
The material reduced = the oxidizing agent = $Cu^{2+}(aq)$

Problem 16.2

Given the following ionic equation, write the oxidation and reduction half-reactions, and identify the oxidizing and reducing agents.

$$Ni(s) + 2H^+(aq) + 2I^-(aq) \longrightarrow Ni^{2+}(aq) + 2I^-(aq) + H_2\uparrow$$

As in the preceding examples, it is often easy to pick out the substances oxidized and reduced in a net ionic equation. The investigator looks for those ions or atoms that undergo a change in charge during the course of the reaction, keeping in mind that a neutral atom has a charge of zero. However, this readily observable change of charge does not always occur during oxidation or reduction reactions. For example, the sulfite ion (SO_3^{2-}) may be oxidized to the sulfate ion (SO_4^{2-}). In this case, there is no change in charge to tell us that the process is oxidation. The fact that one oxygen is gained by sulfur indicates oxidation of sulfur (gain of oxygen is an alternate definition of oxidation), but this test is not always available, and also, we cannot tell how many electrons have been lost. A general method for recognizing oxidized and reduced species and determining numbers of electrons gained or lost involves what are called **oxidation numbers.**

OXIDATION NUMBERS

16.4 Every atom, whether in the elemental state or combined in a molecule or ion, can be assigned an oxidation number based on a set of rules determined by convention. The oxidation number has no physical meaning. Rather it is a bookkeeping device for conveniently locating electron transfers during chemical reactions.

The oxidation number of a monatomic ion is identical to the charge on the ion. By convention we have indicated the charge on an ion by its magnitude followed by the sign, for example, Mg^{2+}. The oxidation number will be distinguished from this by writing first the sign, then the magnitude; for example, for the magnesium ion, the oxidation number is $+2$. Free atoms are electrically neutral and are therefore assigned an oxidation number of zero. The assignment of oxidation numbers to atoms covalently bonded is based on the assumption that the bonding electrons belong completely to the more electronegative of the two bonded atoms. Read through the following rules and see how they are applied in subsequent examples.

1. The oxidation number of an atom in its elemental state is zero. For example, the oxidation number of K is zero, and the oxidation number of H in H_2 is zero.

Rules for assigning oxidation numbers

2. The oxidation number of any monatomic ion is equal to the charge on the ion. For example, for Al^{3+}, the charge is $3+$, and the oxidation number is $+3$. For O^{2-}, the charge is $2-$, and the oxidation number is -2.

3. *The oxygen (-2) rule.* In almost all its common compounds, oxygen has the oxidation number -2. (Hydrogen peroxide, H_2O_2, and its derivatives are the only notable exceptions. In these cases, the oxidation number is -1.)

4. *The hydrogen ($+1$) rule.* In its compounds with other nonmetals, hydrogen has the oxidation number $+1$. When hydrogen is combined with a metallic cation to form a metal hydride, the H atom is assigned the oxidation number -1, but these compounds are encountered infrequently.

5. The sum of the oxidation numbers of atoms in a neutral compound must be zero.

6. The sum of the oxidation numbers of atoms in a polyatomic ion must equal the charge on the ion.

Usually the last two rules are used in conjunction with one or more of the earlier ones. For example, to assign oxidation numbers to the atoms in sodium dihydrogenphosphate, NaH_2PO_4, we will use rules 2, 3, 4, and 6. This is an ionic compound composed of Na^+ ions and $H_2PO_4^-$ ions.

The oxidation number of Na is $+1$ rule 2

The oxidation number of H is $+1$ rule 4

The oxidation number of O is -2 rule 3

Rule 6 will enable us to ascertain the oxidation number of phosphorus by summing up oxidation numbers. The sum of the oxidation numbers of two hydrogens and four oxygens plus the oxidation number of phosphorus must equal $1-$, the charge on the polyatomic ion, $H_2PO_4^-$. Thus:

$$2(+1) + P + 4(-2) = -1$$
$$2 \quad + P + \quad -8 \; = -1$$
$$P \qquad\qquad = -1 + 8 - 2$$
$$P \qquad\qquad = +5$$

The oxidation number of P is $+5$ according to rule 6. Check by rule 5:

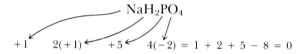

$$+1 \quad 2(+1) \quad +5 \quad 4(-2) = 1 + 2 + 5 - 8 = 0$$

SAMPLE EXERCISE

16.3

Determine the oxidation number of

a. Carbon in CO_2
b. Sulfur in SO_4^{2-}
c. Manganese in $KMnO_4$
d. Sulfur in SO_3^{2-}

SOLUTION

a. Rule 3 tells us that the sum of the oxidation numbers of oxygen in CO_2 would be -4, that is, $2(-2)$. By rule 5, the oxidation number of carbon must be such that the sum for the molecule is zero. Hence the oxidation number of carbon must be $+4$, that is, $+4 - 4 = 0$.

b. In SO_4^{2-} each oxygen has the oxidation number -2 (rule 3). The sum of the oxidation numbers of sulfur and oxygen must equal the charge on the sulfate ion, $2-$ according to rule 6, therefore

$$S + 4\,O \quad = -2$$
$$S + 4(-2) = -2$$
$$S + \quad (-8) = -2$$
$$S \qquad\quad = -2 + 8$$
$$S \qquad\quad = +6$$

The oxidation number of sulfur in SO_4^{2-} is $+6$.

c. The compound $KMnO_4$ is made up of the K^+ ion and the MnO_4^- ion. The K^+ ion has an oxidation number of $+1$ (rule 2), and each oxygen, -2 (rule

3). The sum of the oxidation numbers of potassium, manganese, and oxygen must equal zero for the neutral compound $KMnO_4$, therefore

$$K + Mn + 4\,O = 0$$
$$+1 + Mn + 4(-2) = 0$$
$$+1 + Mn + (-8) = 0$$
$$Mn = +8 - 1$$
$$Mn = +7$$

The oxidation number of Mn in $KMnO_4$ is $+7$.

d. In $SO_3{}^{2-}$ each oxygen has the oxidation number -2 (rule 3). The sum of the oxidation numbers of sulfur and oxygen must equal the charge on the sulfite ion, $2-$ according to rule 6, therefore

$$S + 3\,O = -2$$
$$S + 3(-2) = -2$$
$$S + (-6) = -2$$
$$S = -2 + 6$$
$$S = +4$$

The oxidation number of sulfur in $SO_3{}^{2-}$ is $+4$.

Problem 16.3

Determine the oxidation number of

a. Nitrogen in N_2O_5

b. Nitrogen in $NO_3{}^-$

c. Nitrogen in NO_2

DEFINITIONS REVISITED

16.5 The concept of oxidation number gives us easily applied working definitions of oxidation and reduction:

Oxidation = increase in oxidation number

Reduction = decrease in oxidation number

Look back at all previous examples of oxidation and reduction reactions to see that this is the case.

Now we have a method of recognizing oxidation or reduction even when there is no *obvious* change in charge or transfer of electrons. Notice that we can recognize $SO_3{}^{2-} \rightarrow SO_4{}^{2-}$ as oxidation because the oxidation number of sulfur increases from $+4$ to $+6$ [see Sample Exercise 16.3, parts (*b*) and (*d*)].

Also we will see that the magnitude of the change in oxidation number tells us the numbers of electrons lost or gained during oxidation or reduction.

SAMPLE EXERCISE
16.4

In the following unbalanced equation

$$Cu(s) + HNO_3(aq) \longrightarrow Cu(NO_3)_2(aq) + NO_2(g) + H_2O(l)$$

a. Determine the substances oxidized and reduced.
b. Indicate the oxidizing agent and reducing agent.

SOLUTION

Step 1 Assign an oxidation number to each atom in every species in the reaction, and write the number below the atom:

$$\begin{array}{ccccccc}
Cu & + & HNO_3 & \longrightarrow & Cu(NO_3)_2 & + & NO_2 & + & H_2O \\
0 & & +1\ +5\ -2 & & +2\ +5\ -2 & & +4\ -2 & & +1\ -2 \\
\text{Rule 1} & & \text{Rules 4, 6, 3} & & \text{Rules 2, 6, 3} & & \text{Rules 6, 3} & & \text{Rules 4, 3}
\end{array}$$

Step 2 Look for an increase in oxidation number for some atom. This will be the oxidation half-reaction. Copper has gone from 0 to +2:

$$Cu \longrightarrow Cu^{2+}$$

Because Cu is oxidized, it is the reducing agent.

Step 3 A decrease in oxidation number occurs for nitrogen (+5 to +4). Because N remains bound to oxygen throughout, we include oxygen in the half-reaction:

$$NO_3^- \longrightarrow NO_2$$

Because nitrate is reduced, it is the oxidizing agent.

Problem 16.4

In the following unbalanced equation,

a. Determine the substances oxidized and reduced.

b. Indicate the oxidizing and reducing agents.

$$FeCl_3(aq) + H_2S(aq) \longrightarrow FeCl_2(aq) + S(s) + HCl(aq)$$

BALANCING REDOX REACTIONS

16.6 In Chapter 9 we discussed balancing chemical equations by inspection. There are many equations that would be tedious or very difficult to balance by an essentially trial-and-error inspection process. An alternate balancing procedure involves using to advantage the fact that in redox reactions the total number of electrons gained in the reduction half-reaction must be equal to the number of electrons lost in the oxidation half-reaction. The **Ion-Electron Method** of balancing allows one to arrive at a balanced net ionic equation by adding together two half-reactions, each of which has been balanced, so that the number of electrons lost equals the number of electrons gained.

For example, in Section 16.2 we used the idea of adding together balanced half-reactions so that electrons lost equal electrons gained for the combination of Na and Cl_2. This was a simple example, but the method works even in complex situations.

Before establishing guidelines and exploring more complex examples, let us do one more simple example. Let us balance the equation $Al(s) + Cl_2(g) \rightarrow Al^{3+}(aq) + Cl^-(aq)$ by the Ion-Electron Method. Remember, both atoms and charges must be balanced. For this reaction the oxidation half-reaction is

$$Al \longrightarrow Al^{3+} + 3e^-$$

and the reduction half-reaction is

$$Cl_2 + 2e^- \longrightarrow 2Cl^-$$

(Review Sections 16.2 and 16.3 if this is not clear to you.) As the half-reactions are written, each Al is losing three e^-, while each Cl_2 is gaining two e^-. In order to achieve the condition that electrons lost equal electrons gained, we must multiply the oxidation half-reaction by 2 and the reduction half-reaction by 3. Thus

$$2(Al \longrightarrow Al^{3+} + 3e^-) = 2Al \longrightarrow 2Al^{3+} + 6e^-$$

$$3(Cl_2 + 2e^- \longrightarrow 2Cl^-) = 3Cl_2 + 6e^- \longrightarrow 6Cl^-$$

Now the addition of the two half-reactions will produce a balanced equation:

$$\begin{array}{rcl} 2Al & \longrightarrow & 2Al^{3+} + 6e^- \\ 3Cl_2 + 6e^- & \longrightarrow & 6Cl^- \\ \hline 2Al + 3Cl_2 + \cancel{6e^-} & \longrightarrow & 2Al^{3+} + 6Cl^- + \cancel{6e^-} \end{array}$$

The equation is balanced with respect to both mass and charge. Notice that the electrons cancel out because equal numbers of electrons are lost and gained. In this simple example, it was necessary to use only guidelines 8 and 9 from the following list because of the simplicity of the half-reactions and the fact that we were starting with an ionic rather than a traditional equation.

1. From the given traditional equation, write an ionic equation and cross out spectator ions to establish the unbalanced net ionic equation (review Section 13.9).

2. Assign oxidation numbers to each atom or ion involved in the net ionic equation (Section 16.4).

Guidelines for balancing redox equations by the ion-electron method

3. Determine the substances undergoing oxidation and reduction by noticing changes in oxidation number. Write down the separate oxidation and reduction half-reactions (Section 16.5).

4. Balance each half-reaction with respect to all elements except hydrogen and oxygen.

5. Balance with respect to oxygen by using water (H_2O) as a source of oxygen on whichever side of the equation is deficient in oxygen.

6. Balance with respect to hydrogen by using H^+ ions on whichever side of the equation is deficient in hydrogen. (This procedure applies to neutral and acidic solutions. Basic solutions require an alternate procedure; see Section 16.7.)

7. Add up the total charges on each side of each half-reaction. Balance the charges by adding electrons to the side with the more positive total charge, so that the charges are equal on each side of each half-reaction.

8. Multiply each half-reaction by the smallest whole number that allows the total number of electrons lost in oxidation to equal the total number gained in reduction. Remember that the multiplying coefficients multiply all elements in the half-reactions as well as multiplying the electrons.

9. Add the multiplied oxidation and reduction half-reactions together. Cancel ions and molecules that appear on both sides of the reaction. The electrons cancel out because the number of electrons lost in oxidation has been made equal to the total gained in reduction.

10. Check to be sure atoms of each element are balanced and that total electric charge is the same on both sides of the equation.

See how these guidelines are applied in Sample Exercise 16.5.

SAMPLE EXERCISE

16.5

Write a balanced ionic equation for the following oxidation-reduction reaction:

$$HNO_3(aq) + HI(aq) \longrightarrow NO(g) + I_2(aq) + H_2O(l)$$

SOLUTION

Step 1 Write the net ionic equation (review Section 13.9 if necessary):

Full ionic $H^+(aq) + NO_3^-(aq) + H^+(aq) + I^-(aq) \longrightarrow$
$$NO(g) + I_2(aq) + H_2O(l)$$

Net ionic $H^+(aq) + NO_3^-(aq) + I^-(aq) \longrightarrow NO(g) + I_2(aq) + H_2O(l)$

(There are no spectator ions, but H^+ need only appear once.)

Step 2 Assign oxidation numbers:

$$H^+ + NO_3^- + I^- \longrightarrow NO + I_2 + H_2O$$
$$\quad +1 \quad\ +5 \ -2 \quad -1 \qquad\quad +2\ -2 \quad 0 \quad +1\ -2$$

Step 3   Write down the separate oxidation and reduction half-reactions:

$$I^- \longrightarrow I_2 \qquad \textbf{oxidation} \text{ (increase in oxidation number)}$$
$$-1 \qquad 0$$

$$NO_3^- \longrightarrow NO \qquad \textbf{reduction} \text{ (decrease in oxidation number)}$$
$$+5 \qquad\quad +2$$

Step 4   Balance all elements except H and O:

$$2\,I^- \longrightarrow I_2 \quad \text{This balances I.}$$

$$NO_3^- \longrightarrow NO \quad \text{N is balanced.}$$

Step 5   Balance O by using H_2O as a source of O:

$$NO_3^- \longrightarrow NO + 2H_2O$$

Two oxygens (hence two H_2O) are required on the right to balance oxygen on the left.

Step 6   Balance H by using H^+:

$$4\,H^+ + NO_3^- \longrightarrow NO + 2H_2O$$

Step 7   Add up the charges on each side of each half-reaction:

$$2I^- \longrightarrow I_2$$
$$2- \qquad\ 0$$

$$\underbrace{4H^+ + NO_3^-}_{3+} \longrightarrow \underbrace{NO + 2H_2O}_{0}$$

Balance the charges by adding an appropriate number of electrons (negative charges) to the side with the larger positive charge:

$$2I^- \longrightarrow I_2 + 2e^-$$

$$4H^+ + NO_3^- + 3e^- \longrightarrow NO + 2H_2O$$

Step 8   Find the multiplying factors that will ensure that electrons lost equal electrons gained. In this case, the lowest common denominator of two e^- and three e^- is six e^-, so the multiplying factors are 3 for the oxidation half-reaction and 2 for the reduction:

$$3(2I^- \longrightarrow I_2 + 2e^-)$$

$$2(4H^+ + NO_3^- + 3e^- \longrightarrow NO + 2H_2O)$$

Step 9   Add the half-reactions together:

$$6I^- \longrightarrow 3I_2 + 6e^-$$
$$\underline{8H^+ + 2NO_3^- + 6e^- \longrightarrow 2NO + 4H_2O}$$
$$8H^+ + 2NO_3^- + 6I^- + \cancel{6e^-} \longrightarrow 3I_2 + 2NO + 4H_2O + \cancel{6e^-}$$
$$8H^+(aq) + 2NO_3^-(aq) + 6I^-(aq) \longrightarrow 3I_2(aq) + 2NO(g) + 4H_2O(l)$$

Step 10   Check for balance of atoms:

Reactants	Products
8 H, 2 N, 6 O, 6 I	8 H, 2 N, 6 O, 6 I

Check for balance of charge:

Reactants	Products
$8(1+) + 2(1-) + 6(1-) = 0$	$0 + 0 + 0 = 0$

The traditional balanced equation can be written by realizing that the H^+ ions must have come from HI and HNO_3. Associate six of the H^+ ions with six I^- ions to form six HI and two H^+ ions with two NO_3^- ions to form two HNO_3:

$$6HI(aq) + 2HNO_3(aq) \longrightarrow 3I_2(aq) + 2NO(g) + 4H_2O(l)$$

SAMPLE EXERCISE

16.6

Balance the following equation by the Ion-Electron Method:

$$KMnO_4(aq) + HCl(aq) \longrightarrow MnCl_2(aq) + Cl_2(g) + KCl(aq) + H_2O(l)$$

SOLUTION

Step 1   Write the net ionic equation:

Full ionic $K^+(aq) + MnO_4^-(aq) + H^+(aq) + Cl^-(aq) \longrightarrow$
$Mn^{2+}(aq) + 2Cl^-(aq) + Cl_2(g) + K^+(aq)$

Net ionic $MnO_4^-(aq) + H^+(aq) + Cl^-(aq) \longrightarrow$
$Mn^{2+}(aq) + Cl_2(g) + H_2O(l)$

(The K^+ ion undergoes no change. Cl^- may be omitted from the right side of the equation. Although some Cl^- is changed to Cl_2, some Cl^- ions are present only as spectator ions to balance the charge of K^+ ions.)

Step 2   Assign oxidation numbers:

$$MnO_4^- + H^+ + Cl^- \longrightarrow Mn^{2+} + Cl_2 + H_2O$$
$$\quad\; +7 \;\; -2 \qquad +1 \qquad -1 \qquad\;\; +2 \qquad 0 \qquad +1 \;\; -2$$

Step 3   Write down the separate oxidation and reduction half-reactions:

$Cl^- \longrightarrow Cl_2$ **oxidation** (increase in oxidation number)
$\;\, -1 \qquad\;\; 0$

$MnO_4^- \longrightarrow Mn^{2+}$ **reduction** (decrease in oxidation number)
$\;\, +7 \qquad\quad\; +2$

Step 4 Balance all elements except H and O:

$$2Cl^- \longrightarrow Cl_2 \qquad \text{This balances Cl.}$$

$$MnO_4 \longrightarrow Mn^{2+} \qquad \text{Mn is balanced.}$$

Step 5 Balance O by using H_2O as a source of O:

$$MnO_4^- \longrightarrow Mn^{2+} + 4H_2O$$

Step 6 Balance H by using H^+:

$$MnO_4^- + 8H^+ \longrightarrow Mn^{2+} + 4H_2O$$

Step 7 Add up the charges on each side of each half-reaction:

$$2Cl^- \longrightarrow Cl_2$$
$$2- \qquad\qquad 0$$

$$\underbrace{MnO_4^- + 8H^+}_{7+} \longrightarrow \underbrace{Mn^{2+} + 4H_2O}_{2+}$$

Balance the charges by adding an appropriate number of electrons to the side with the larger positive charge:

$$2Cl^- \longrightarrow Cl_2 + 2e^-$$

$$\underset{1-}{MnO_4^-} + \underset{8+}{8H^+} + \underset{5-}{5e^-} \longrightarrow \underset{=\quad 2+}{Mn^{2+} + 4H_2O}$$

Step 8 Find the multiplying factors that will ensure that **electrons lost = electrons gained.** In this case, the lowest common denominator of two e^- and five e^- is ten e^-, so the factors are 5 for the oxidation half-reaction and 2 for the reduction half-reaction:

$$5(2Cl^- \longrightarrow Cl_2 + 2e^-)$$

$$2(MnO_4^- + 8H^+ + 5e^- \longrightarrow Mn^{2+} + 4H_2O)$$

Step 9 Add the half-reactions together:

$$10Cl^- \longrightarrow 5Cl_2 + 10e^-$$
$$\underline{2MnO_4^- + 16H^+ + 10e^- \longrightarrow 2Mn^{2+} + 8H_2O}$$
$$2MnO_4^- + 16H^+ + 10Cl^- + \cancel{10e^-} \longrightarrow 5Cl_2 + 2Mn^2 + 8H_2O + \cancel{10e^-}$$
$$2MnO_4^-(aq) + 16H^+(aq) + 10Cl^-(aq) \longrightarrow$$
$$\qquad\qquad\qquad 5Cl_2(aq) + 2Mn^{2+}(aq) + 8H_2O(l)$$

Step 10 Check for balance of atoms:

Reactants	Products
2 Mn, 8 O, 16 H, 10 Cl	2 Mn, 8 O, 16 H, 10 Cl

Check for balance of charges:

Reactants	Products
$2(1-) + 16(1+) + 10(1-) = +4$	$2(2+) = +4$

The traditional equation can be constructed by using the coefficients we have determined for the ions and then completing the equation by inspection. Thus

$$2KMnO_4(aq) + 16HCl(aq) \longrightarrow$$

Coefficient of MnO_4^-

The ionic equation shows the apparent contradiction:* 16 H^+ and 10 Cl^-; in such a case, choose the larger coefficient.

$$2MnCl_2(aq) + 5Cl_2(g) + \_KCl(aq) + 8H_2O(l)$$

Coefficient of Mn^{2+}

Coefficient of Cl_2

Coefficient of H_2O

Balancing K by inspection will complete the equation:

$$2KMnO_4(aq) + 16HCl(aq) \longrightarrow 2MnCl_2(aq) + 5Cl_2(g) + 2KCl(aq) + 8H_2O(l)$$

Balance check: 2 K   2 Mn   8 O   16 H   16 Cl   2 Mn   4 + 10 + 2 = 16 Cl   2 K   16 H   8 O

* This comes about because Cl^- (10) is oxidized to Cl_2, while some remains unchanged.

Problem 16.5

Balance the following equation:

$$SnBr_2(aq) + KIO_3(aq) + HBr(aq) \longrightarrow SnBr_4(aq) + KI(aq) + H_2O(l)$$

    In the preceding examples, the solutions have been acidic, so the use of H^+ for balancing was quite legitimate and understandable. Under neutral conditions, we may still introduce H^+ for balancing purposes in guideline 6. Any artificially added H^+ will be canceled out when the half-reactions are added together. Sample Exercise 16.7 provides an example of neutral conditions.

SAMPLE EXERCISE

16.7

Balance the following equation by the Ion-Electron Method:

$$C_2H_6(g) + O_2(g) \longrightarrow CO_2(g) + H_2O(g)$$

SOLUTION

Step 1   Because there are no ions, the "net ionic equation" in this example is the same as the full equation:

$$C_2H_6(g) + O_2(g) \longrightarrow CO_2(g) + H_2O(g)$$

Step 2   Assign oxidation numbers:

$$C_2H_6 + O_2 \longrightarrow CO_2 + H_2O$$

       -3 $+1$ 0 $+4$ -2 $+1$ -2

C and O undergo changes in oxidation number.

Step 3

$$C_2H_6 \longrightarrow CO_2 \qquad \textbf{oxidation}$$
$$\underset{-3}{} \qquad \underset{+4}{}$$

$$O_2 \longrightarrow H_2O \qquad \textbf{reduction}$$
$$\underset{0}{} \qquad \underset{-2}{}$$

Step 4   Balance C:

$$C_2H_6 \longrightarrow 2CO_2$$

Step 5   Balance O by using H_2O as a source:

$$C_2H_6 + 4H_2O \longrightarrow 2CO_2$$

$$O_2 \longrightarrow 2H_2O$$

Step 6   Balance H by using H^+:

$$C_2H_6 + 4H_2O \longrightarrow 2CO_2 + 14H^+$$

$$O_2 + 4H^+ \longrightarrow 2H_2O$$

Step 7   Add up the charges on each side of each half-reaction:

$$\underbrace{C_2H_6 + 4H_2O}_{0} \longrightarrow \underbrace{2CO_2 + 14H^+}_{14+}$$

$$\underbrace{O_2 + 4H^+}_{4+} \longrightarrow \underbrace{2H_2O}_{0}$$

Balance the charges by using electrons:

$$C_2H_6 + 4H_2O \longrightarrow 2CO_2 + 14H^+ + 14e^-$$

$$O_2 + 4H^+ + 4e^- \longrightarrow 2H_2O$$

Step 8   Find the multiplying factor that will ensure **electrons lost = electrons gained.** The lowest common denominator of 14 and 4 is 28:

$$2(C_2H_6 + 4H_2O \qquad \longrightarrow 2CO_2 + 14H^+ + 14e^-)$$

$$7(O_2 + 4H^+ + 4e^- \longrightarrow 2H_2O)$$

Step 9   Add the half-reactions together and cancel common species:

$$2C_2H_6 + 8H_2O \longrightarrow 4CO_2 + 28H^+ + 28e^-$$
$$7O_2 + 28H^+ + 28e^- \longrightarrow 14H_2O$$
$$\overline{2C_2H_6 + 7O_2 + 8H_2O + \cancel{28H^+} + \cancel{28e^-} \longrightarrow 4CO_2 + 14H_2O + \cancel{28H^+} + \cancel{28e^-}}$$
$$\quad -8H_2O \qquad\qquad\qquad\qquad -8H_2O$$
$$2C_2H_6(g) + 7O_2(g) \longrightarrow 4CO_2(g) + 6H_2O(g)$$

Step 10   Check for balance of atoms:

Reactants	Products
4 C, 12 H, 14 O	4 C, 12 H, 14 O

Check for balance of charge—in this example, zero charge on both sides.

BASIC REDOX REACTIONS

16.7 In order to balance equations for redox reactions occurring in basic solution, it is necessary to modify guidelines 5 and 6, which deal with the balancing of oxygen and hydrogen, respectively. The modification is necessary because in a basic solution the sources of oxygen and hydrogen will be OH^- and H_2O rather than H^+ and H_2O as in acidic and neutral solutions. All other guidelines remain the same. The new guidelines for basic solutions are designated 5*b* and 6*b* and appear below:

5b. For *each* oxygen atom needed in a half-reaction, add *two* hydroxide ions (OH^-) to the side needing oxygen and one water molecule to the other side of the reaction.

6b. For half-reactions unbalanced in hydrogen atoms, add one water molecule for each hydrogen needed and one hydroxide ion to the other side of the equation.

Sample Exercise 16.8 applies these new guidelines. The appearance of the hydroxide ion (OH^-) on either side of the equation signals a basic solution.

SAMPLE EXERCISE

16.8

Balance the following equation:

$$KI(aq) + KMnO_4(aq) + H_2O(l) \longrightarrow MnO_2(s) + I_2(aq) + KOH(aq)$$

SOLUTION

Step 1 Write the net ionic equation:

Full ionic $\cancel{K^+(aq)} + I^-(aq) + MnO_4^-(aq) + H_2O(l) \longrightarrow$
$$MnO_2(s) + I_2(aq) + \cancel{K^+(aq)} + OH^-(aq)$$

Net ionic $I^-(aq) + MnO_4^-(aq) + H_2O(l) \longrightarrow$
$$MnO_2(s) + I_2(aq) + OH^-(aq)$$

Step 2 Assign oxidation numbers:

$$\underset{-1}{I^-} + \underset{+7\ -2}{MnO_4^-} + \underset{+1\ -2}{H_2O} \longrightarrow \underset{+4\ -2}{MnO_2} + \underset{0}{I_2} + \underset{-2\ +1}{OH^-}$$

Step 3 Write down the separate oxidation and reduction half-reactions:

$$\underset{-1}{I^-} \longrightarrow \underset{0}{I_2} \qquad \textbf{oxidation}$$

$$\underset{+7}{MnO_4^-} \longrightarrow \underset{+4}{MnO_2} \qquad \textbf{reduction}$$

Step 4   Balance the elements other than H and O:

$$2I^- \longrightarrow I_2$$

$$MnO_4^- \longrightarrow MnO_2$$

Step 5b   Balance O by using OH^- and H_2O as noted in guideline 5b:

$$MnO_4^- \longrightarrow MnO_2$$

Two oxygen atoms are needed on the right side, so we add four hydroxide ions (two for each oxygen needed). Four hydroxide ions (OH^-) on the right demands two water molecules on the left:

$$MnO_4^- + 2H_2O \longrightarrow MnO_2 + 4OH^-$$

Step 6b   Balance the H atoms. In this case, they are already balanced (four on each side).

Step 7   Add up the charges on each side of each half-reaction:

$$2I^- \longrightarrow I_2$$
$$\;\;2-\qquad\;\; 0$$

$$MnO_4^- + 2H_2O \longrightarrow MnO_2 + 4OH^-$$
$$\;\;1-\qquad\qquad\qquad\qquad 4-$$

Balance the charges by using electrons:

$$2I^- \longrightarrow I_2 + 2e^-$$

$$MnO_4^- + 2H_2O + 3e^- \longrightarrow MnO_2 + 4OH^-$$

Step 8   Find the multiplying factor to ensure electrons lost equal electrons gained:

$$3(2I^- \longrightarrow I_2 + 2e^-)$$

$$2(MnO_4^- + 2H_2O + 3e^- \longrightarrow MnO_2 + 4OH^-)$$

Step 9   Add the half-reactions together:

$$6I^- \longrightarrow 3I_2 + 6e^-$$
$$2MnO_4 + 4H_2O + 6e^- \longrightarrow 2MnO_2 + 8OH^-$$

$$\rule{8cm}{0.4pt}$$

$$2MnO_4^- + 4H_2O + 6I^- + \cancel{6e^-} \longrightarrow 3I_2 + 2MnO_2 + 8OH^- + \cancel{6e^-}$$
$$2MnO_4^-(aq) + 4H_2O(l) + 6I^-(aq) \longrightarrow 3I_2(aq) + 2MnO_2(s) + 8OH^-(aq)$$

Step 10   Check for balance of atoms:

Reactants	Products
2 Mn, 12 O, 8 H, 6 I	2 Mn, 12 O, 8 H, 6 I
(8 + 4)	(4 + 8)

Check for balance of charges:

Reactants	Products
$2(1-) + 6(1-) = -8$	-8

The original equation can be balanced by using the coefficients of the ions; thus

$$6KI(aq) + 2KMnO_4(aq) + 4H_2O(aq) \longrightarrow 2MnO_2(s) + 3I_2(aq) + 8KOH(aq)$$

8 K   6 I   2 Mn   12 O   8 H   2 Mn   12 O 6 I   8 K   8 H

Problem 16.6

Balance the following equation:

$$Cl_2(g) + KOH(aq) \longrightarrow KClO_3(aq) + KCl(aq) + H_2O(l)$$

AN ALTERNATE METHOD

16.8 The Ion-Electron Method is superior to other balancing methods because it offers the maximum amount of information about balancing all elements involved in the reaction. In the foregoing examples, all or nearly all the coefficients of the final balanced equations were obtained directly by following the guidelines.

An alternate balancing method is the Oxidation-Number Method. This method is applied directly to the traditional equation. It has two major disadvantages:

1. Electron transfer is obscured.

2. Coefficients are obtained for only the species actually oxidized or reduced.

Nonetheless, the method is included here as an alternative to the Ion-Electron Method. There are three steps in this method:

1. Assign oxidation numbers and determine which elements have been oxidized and which reduced.

2. Find multiplying factors that will ensure that the increase in oxidation number equals the decrease in oxidation number. Use these factors as coefficients of the oxidized and reduced elements.

3. Balance the rest of the equation by inspection.

SAMPLE EXERCISE

16.9

Balance the following equation by the Oxidation Number Method:

$$KMnO_4 + HCl \longrightarrow MnCl_2 + Cl_2 + KCl + H_2O$$

SOLUTION

Step 1 Assign oxidation numbers and identify the oxidized and reduced elements:

$$KMnO_4 \; + \; HCl \; \longrightarrow \; MnCl_2 \; + \; Cl_2 + \; KCl \; + \; H_2O$$

K: +1, Mn: +7, O: −2, H: +1, Cl: −1, Mn: +2, Cl: −1, Cl: 0, K: +1, Cl: −1, H: +1, O: −2

Mn $\longrightarrow$ Mn   **reduction** (decrease of 5 in oxidation number)
+7   +2

Cl $\longrightarrow$ Cl   **oxidation** (increase of 1 in oxidation number)
−1   0

Step 2 Find the multiplying factor that will ensure that the decrease in oxidation number equals the increase in oxidation number. In this case, the element oxidized must be multiplied by 5. Thus we can use the coefficients:

$$KMnO_4 + 5HCl \; \longrightarrow \; MnCl_2 + \tfrac{5}{2}Cl_2 + KCl + H_2O$$
                    (5 Cl)   No coefficient; this Cl has not changed

or we can multiply through by 2 to clear the fractional coefficient:

$$2KMnO_4 + 10HCl \; \longrightarrow \; 2MnCl_2 + 5Cl_2 + 2KCl + 2H_2O$$

Step 3 Balance by inspection (Section 9.4):

Reactants	Products
2 K, 2 Mn, 8 O, 10 H, 10 Cl	2 Mn, 4 Cl, 10 Cl, 2 K, 2 Cl, 4 H, 2 O

Both K and Mn are balanced, but O, H, and Cl are not. To balance Cl, increase the coefficient of HCl to 16:

$$2KMnO_4 + 16HCl \; \longrightarrow \; 2MnCl_2 + 5Cl_2 + 2KCl + 2H_2O$$

Now K, Mn, and Cl are balanced. Both H and O can be balanced simultaneously by changing the coefficient of H_2O to 8:

$$2KMnO_4 + 16HCl \; \longrightarrow \; 2MnCl_2 + 5Cl_2 + 2KCl + 8H_2O$$

Refer to Sample Exercise 16.6, in which this equation was balanced by the Ion-Electron Method.

SAMPLE EXERCISE

16.10

Balance the following equation using the Oxidation-Number Method:

$$Fe + H_2SO_4 \; \longrightarrow \; Fe_2(SO_4)_3 + SO_2 + H_2O$$

SOLUTION

Step 1

$$Fe + H_2SO_4 \longrightarrow Fe_2(SO_4)_3 + SO_2 + H_2O$$

$$\begin{array}{cccccccc} & Fe & H_2SO_4 & & Fe_2(SO_4)_3 & SO_2 & H_2O \\ & 0 & +1 \ +6 \ -2 & & +3 \ +6 \ -2 & +4 \ -2 & +1 \ -2 \end{array}$$

$$\begin{array}{ll} Fe \longrightarrow Fe & \textbf{oxidation} \text{ (increase of 3 in oxidation number)} \\ 0 \qquad +3 \end{array}$$

$$\begin{array}{ll} S \longrightarrow S & \textbf{reduction} \text{ (decrease of 2 in oxidation number)} \\ +6 \quad +4 \end{array}$$

Step 2   The lowest common denominator of 3 and 2 is 6; thus we use the coefficient 2 for Fe ($2 \times 3 = 6$) and 3 for S ($3 \times 2 = 6$):

$$2Fe + 3H_2SO_4 \longrightarrow Fe_2(SO_4)_3 + 3SO_2 + H_2O$$

Step 3   Only the Fe is balanced at this point. Balance the S by increasing the coefficient of H_2SO_4 to 6.

$$2Fe + 6H_2SO_4 \longrightarrow Fe_2(SO_4)_3 + 3SO_2 + H_2O$$

2 Fe 12 H 6 S 24 O   2 Fe 3 S 12 O 3 S 6 O 2 H 1 O

Balanced

Now Fe and S are balanced, but the count on H and O is

$$\begin{array}{cc} 12\ H & 2\ H \\ 24\ O & 19\ O \end{array}$$

Balance these by making the coefficient of water 6:

$$2Fe + 6H_2SO_4 \longrightarrow Fe_2(SO_4)_3 + 3SO_2 + 6H_2O$$

Problem 16.7

Balance the following equation using the Oxidation-Number Method:

$$HNO_3 + I_2 \longrightarrow NO_2 + HIO_3 + H_2O$$

ACTIVITY SERIES REVISITED

16.9   In Section 16.10 we will see that such common devices as batteries and such common industrial processes as chromplating are based on electron-transfer (redox) reactions. Before considering these applications, it will be useful to revisit and expand the concept of the activity series (Table 9.2) which was previously presented in Section 9.8.

Table 16.1 is an enlarged, more informative version of Table 9.2. We see that what is meant by a "more active metal," is a metal that more readily loses or donates electrons or is more easily oxidized. The

ACTIVITY SERIES OF METALS (OXIDATION HALF-REACTION)   **TABLE 16.1**

$$Li \rightarrow Li^+ \quad + e^-$$
$$K \rightarrow K^+ \quad + e^-$$
$$Ba \rightarrow Ba^{2+} \quad + 2e^-$$
$$Ca \rightarrow Ca^{2+} \quad + 2e^-$$
$$Na \rightarrow Na^+ \quad + e^-$$
$$Mg \rightarrow Mg^{2+} \quad + 2e^-$$
$$Al \rightarrow Al^{3+} \quad + 3e^-$$
$$Zn \rightarrow Zn^{2+} \quad + 2e^-$$
$$Fe \rightarrow Fe^{2+} \quad + 2e^-$$
$$Cd \rightarrow Cd^{2+} \quad + 2e^-$$
$$Ni \rightarrow Ni^{2+} \quad + 2e^-$$
$$Sn \rightarrow Sn^{2+} \quad + 2e^-$$
$$Pb \rightarrow Pb^{2+} \quad + 2e^-$$
$$H_2 \rightarrow 2H^+ \quad + 2e^-$$
$$Cu \rightarrow Cu^{2+} \quad + 2e^-$$
$$Ag \rightarrow Ag^+ \quad + e^-$$
$$Au \rightarrow Au^{3+} \quad + 3e^-$$

(Left margin: Increasingly stronger reducing agents → Ease of oxidation increases)

(Right margin: Increasingly stronger oxidizing agents)

more easily oxidized metals are the stronger reducing agents because they so readily donate the electrons necessary to reduce other substances. Likewise, metallic cations that more readily accept electrons and are thereby more easily reduced, are stronger oxidizing agents.

Whereas formerly (Section 9.8) you were only able to say that zinc metal would replace copper cations from their compounds because Zn appears above Cu in the activity series, you can now see that the reaction occurs because Zn is more readily oxidized (stronger reducing agent) than Cu, and Cu^{2+} is more readily reduced (stronger oxidizing agent), than Zn^{2+} (see Figure 9.3).

$$Zn(s) + Cu^{2+}(aq) + SO_4^{2-}(aq) \longrightarrow Zn^{2+}(aq) + SO_4^{2-}(aq) + Cu(s)$$

$$Zn \longrightarrow Zn^{2+} + 2e^- \qquad \textbf{oxidation}$$

$$Cu^{2+} + 2e^- \longrightarrow Cu \qquad \textbf{reduction}$$

In general, a spontaneous reaction will occur when the oxidation half-reaction of a relatively stronger reducing agent (Zn in the previous example) is coupled with the reduction half-reaction of a relatively stronger oxidizing agent (Cu^{2+} in the example). Notice that the half-reactions shown in Table 16.1 are oxidation half-reactions. Therefore, the reverse reactions must be reductions. For example,

$$Al \longrightarrow Al^{3+} + 3e^- \qquad \textbf{oxidation}$$

$$Al^{3+} + 3e^- \longrightarrow Al \qquad \textbf{reduction}$$

SAMPLE EXERCISE

16.11

Which reaction will occur spontaneously?

a. $Pb(s) + Mg(NO_3)_2(aq) \longrightarrow Pb(NO_3)_2(aq) + Mg(s)$
b. $Mg(s) + Pb(NO_3)_2(aq) \longrightarrow Mg(NO_3)_2(aq) + Pb(s)$?

SOLUTION

Consider the half-reactions in each case:

a.
$$Pb \longrightarrow Pb^{2+} + 2e^- \qquad \textbf{oxidation}$$
$$Mg^{2+} + 2e^- \longrightarrow Mg \qquad \textbf{reduction}$$

b.
$$Mg \longrightarrow Mg^{2+} + 2e^- \qquad \textbf{oxidation}$$
$$Pb^{2+} + 2e^- \longrightarrow Pb \qquad \textbf{reduction}$$

A spontaneous reaction will occur when the stronger reducing agent is oxidized and the stronger oxidizing agent is reduced. Consultation with Table 16.1 reveals that Mg is the stronger reducing agent and Pb^{2+} the stronger oxidizing agent. Hence reaction (*b*) is spontaneous.

USES OF REDOX

16.10 **Voltaic cells** Spontaneous redox (electron transfer) reactions of the type just described can be used to produce a flow of electrons (electricity) through the construction of an apparatus known as a **voltaic cell.** Such a voltaic cell is commonly called a battery. Figure 16.1 shows two possible ways that a voltaic cell employing the spontaneous reaction between Zn and Cu^{2+} can be made.

A voltaic cell cannot be made by simply putting zinc metal into a copper sulfate solution, because in that case electrons would be directly transferred from Zn to Cu^{2+} rather than induced to flow through a wire. Rather the reducing agent (Zn, the electron donor) must be separated from the oxidizing agent (Cu^{2+}, the electron acceptor). The design shown in Figure 16.1*a* achieves an electron flow in the following way. The metal atoms of the zinc electrode (anode) are oxidized to Zn^{2+} ions which enter the solution.[1] The electrons that are lost flow through the wire to the copper electrode (cathode), where they reduce Cu^{2+} ions to copper metal. To complete the circuit ions must be able to flow between the two solutions. The porous walls of the inner container allow this. Anions (in this case SO_4^{2-}) flow toward the anode and cations flow toward the cathode. In Figure 16.1*b* the necessary flow is established through a **salt bridge.** This figure

---

[1] All solutions referred to in this section are assumed to be 1 *M* unless otherwise specified.

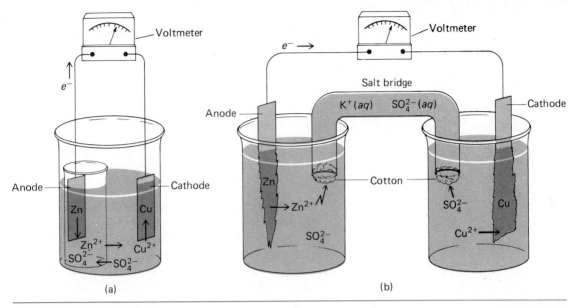

FIGURE 16.1

Two possible designs for a voltaic cell employing zinc and copper electrodes immersed in zinc sulfate and copper sulfate solutions, respectively. (a) A Zn electrode is immersed in a $ZnSO_4$ solution in a porous cup, which is immersed in a $CuSO_4$ solution. Zinc is oxidized to Zn^{2+} ions, which go into solution; the electrons that are lost flow through the wire to the Cu electrode. The electrons reduce Cu^{2+} ions to Cu metal. The circuit is completed by ion migration across the porous walls of the inner container. The voltmeter attests to the completed circuit. (b) The same oxidation of Zn, electron flow, and reduction of Cu^{2+} occurs in this setup. But the circuit is completed by the use of a so-called salt bridge through which ions migrate. Notice how the Zn electrode is eaten away as Zn is oxidized to Zn^{2+} and the Cu electrode built up as Cu^{2+} is reduced to Cu.

also shows how the zinc electrode is eaten away and the copper electrode grows as the reaction proceeds.

Before proceeding let us clarify the names of and reactions occurring at the two electrodes of a voltaic cell, and, in fact, in all electrochemical cells. By definition, *oxidation occurs at the anode, reduction at the cathode.* The names given to ions arise because of their migratory tendencies toward the two electrodes. Negative ions are called *anions* because they always move toward the *an*ode. Positive ions are called *cations* because they always move toward the *cat*hode. In summary:

Electrode	Reaction	Ions Which Move toward the Electrode
Anode	Oxidation	Anions
Cathode	Reduction	Cations

SAMPLE EXERCISE

16.12

Describe how you would construct a voltaic cell employing the spontaneous reaction discussed in Sample Exercise 16.11.

SOLUTION

Refer to Figure 16.1. The desired cell would have a magnesium electrode submerged in $Mg(NO_3)_2$ solution and a lead electrode submerged in $Pb(NO_3)_2$ solution. The anode is Mg and the cathode, Pb. The circuit is completed by a salt bridge or by the arrangement shown in Figure 16.1a.

Electrolytic cells *Non*spontaneous reactions can be forced to occur by the application of an electric current. For example, in Figure 16.2, by introducing a source of electric current, the electron flow in the Zn-Cu system has been reversed. By pumping in electrons the zinc electrode is made the cathode, and zinc ions are reduced there; copper metal is oxidized at the copper anode. This figure depicts an **electrolytic** (or **electrolysis**) **cell**, a cell in which a nonspontaneous reaction is made to occur by the application of electric energy.

The combination of Na metal and Cl_2 gas to form NaCl is a spontaneous reaction. Conversely, the decomposition of the compound NaCl is nonspontaneous. However, the nonspontaneous reaction can be made to occur by passing an electric current through molten NaCl (see Figure 16.3). The Na^+ cations are attracted to and reduced at the cathode. The Cl^- anions are attracted to and oxidized

FIGURE 16.2

An electrolytic cell. The electric current is applied in such a way that electrons are fed to the Zn electrode. These electrons reduce the Zn^{2+} ions to Zn metal. At the same time Cu is oxidized to Cu^{2+}. This process is the reverse of that shown in Figure 16.1b. Substitution of the power supply for the volmeter in Figure 16.1b reverses the reaction; the Zn electrode is built up and the Cu electrode is eaten away.

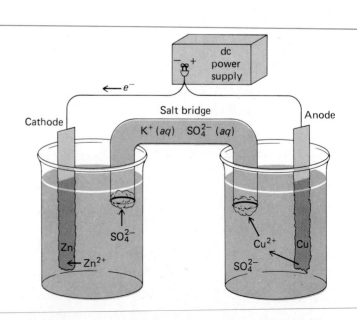

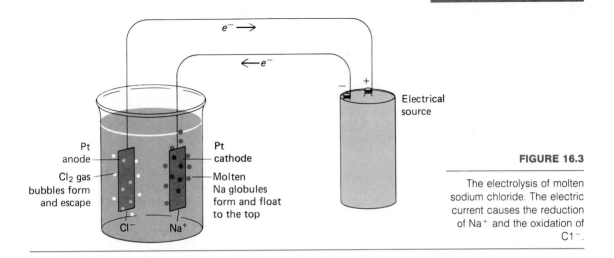

FIGURE 16.3

The electrolysis of molten sodium chloride. The electric current causes the reduction of Na^+ and the oxidation of Cl^-.

at the anode. The NaCl must be molten (liquid) so that the ions can flow and complete the circuit. Solid NaCl cannot be electrolyzed because the ions cannot move in the solid state.

A common practical application of electrolysis is **electroplating.** Figure 16.4 shows the silver plating of a fork. The fork cathode and silver-strip anode are immersed in a silver nitrate solution. Silver is plated onto the fork cathode as a result of the reduction of the Ag^+ ions in solution. Silver ions in solution are replenished through the oxidation of silver at the anode.

$$Ag^{+\cdot} + e^- \longrightarrow Ag \qquad \text{cathode: reduction}$$

$$Ag \longrightarrow Ag^+ + e^- \qquad \text{anode: oxidation}$$

FIGURE 16.4

Electroplating. Silver metal is plated onto the fork by reduction of Ag^+ ions in solution. The silver ions are replenished by the oxidation of the silver anode.

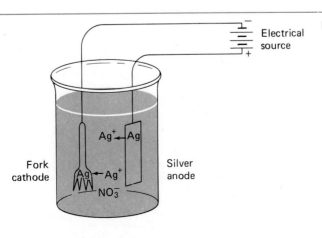

Chrome trim on your automobile is electroplated (chromplated) by a similar procedure. In the case of chromplating the cathode is the metallic piece to be plated and the anode is chromium metal.

SAMPLE EXERCISE

16.13

A cooking utensil is to be copper-plated. Describe the cathode and anode and write equations for the half-reactions occurring at the electrodes.

SOLUTION

In all electroplating procedures the object to be plated is the cathode and the metal to be deposited is the anode. In this case then, the cooking utensil is the cathode and copper metal is the anode. The *cathode reaction is always reduction* of the metal ion to the metal and the *anode reaction is always oxidation* of the metal to its cation. In this case

$$Cu^{2+} + 2e^- \longrightarrow Cu \qquad \text{cathode: reduction}$$

$$Cu \longrightarrow Cu^{2+} + 2e^- \qquad \text{anode: oxidation}$$

FIGURE 16.5

A flow of electrons (electrical current) is established when the electrons released in the oxidation half-reaction at the lead plates are transferred to the PbO_2 plates where reduction occurs.

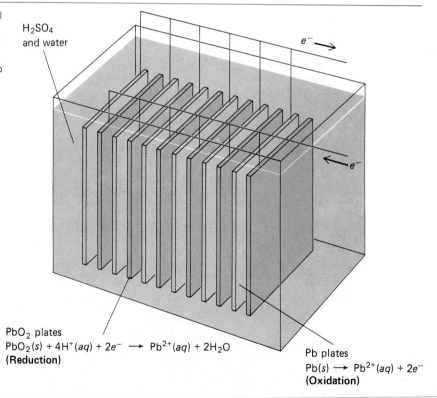

H_2SO_4 and water

$e^- \longrightarrow$

$\longleftarrow e^-$

PbO_2 plates
$PbO_2(s) + 4H^+(aq) + 2e^- \longrightarrow Pb^{2+}(aq) + 2H_2O$
(Reduction)

Pb plates
$Pb(s) \longrightarrow Pb^{2+}(aq) + 2e^-$
(Oxidation)

Lead storage batteries   The workings of your car battery are based on the principles of both voltaic and electrolytic cells. Figure 16.5 shows a typical automobile battery. The half-reactions cited are those that occur spontaneously and thereby produce the electric current to start the car, light the lights, or power the radio.

If the battery operated only as a voltaic cell, it would soon "run down" as the electrode plates were eaten away. A car's generator provides an electric current which reverses the battery reactions shown. The recharging of the battery by the generator is an electrolysis reaction.

REDOX IN THE BODY

16.11   The conversion of the food that we eat and the oxygen that we breathe into energy for body processes is mediated by redox reactions. Figure 16.6 shows a simplified skeletal diagram of the respiratory chain which is the central focus of energy transfer reactions in the body. The overall result of this very complicated stepwise process is the production of water from its elements, which is a highly exothermic reaction. Many of the individual steps are either oxidations or reductions. The oxidation of hydrogen and reduction of oxygen are labeled. In each step involving the cytochromes, iron-containing biological molecules, iron ions are either oxidized or reduced. For example, the electrons lost by hydrogen reduce two Fe^{3+} ions to Fe^{2+} ions in cytochrome b. Cytochromes a and a_3 lose electrons to oxygen and in the process Fe^{2+} ions are oxidized to Fe^{3+} ions.

Redox reactions, particularly biochemical examples, are often represented by diagrams with intersecting arrows. The point of intersection represents the electron transfer and consequent oxidation-reduction. For example, in Figure 16.6 one set of intersecting arrows depicts an electron being transferred from Fe^{2+} of cytochrome b, to Fe^{3+} of cytochrome c_1. The Fe^{2+} (cytochrome b) is oxidized to Fe^{3+} (cytochrome b), while the Fe^{3+} (cytochrome c_1) is reduced to Fe^{2+}

A skeletal view of the respiratory chain. Oxidation and reduction occur alternately along the chain as electrons are lost and gained. The overall process is the exothermic production of water. Through the mediation of the cytochromes, the electrons lost by hydrogen as it is oxidized reduce oxygen.

FIGURE 16.6

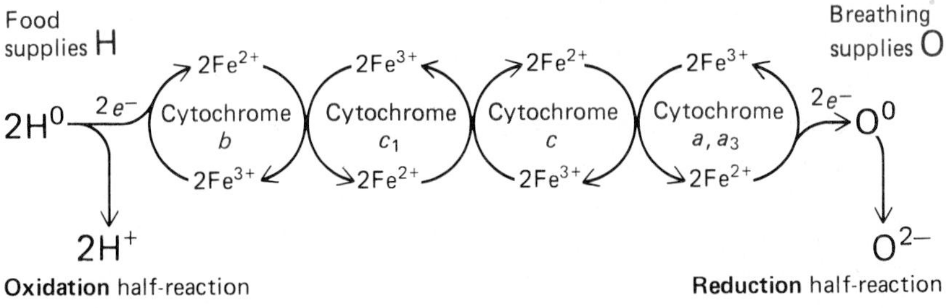

Food supplies H

Breathing supplies O

$2H^0 \xrightarrow{2e^-}$ Cytochrome b → $2Fe^{2+}$ ... $2Fe^{3+}$ → Cytochrome c_1 → $2Fe^{3+}$ ... $2Fe^{2+}$ → Cytochrome c → $2Fe^{2+}$ ... $2Fe^{3+}$ → Cytochrome a, a_3 → $2Fe^{3+}$ ... $2Fe^{2+}$ $\xrightarrow{2e^-}$ O^0

$2H^+$

O^{2-}

Oxidation half-reaction

Reduction half-reaction

Overall reaction: $2H^+ + O^{2-} \longrightarrow H_2O$ + Energy

(cytochrome c_1). The reaction between zinc and Cu^{2+}, which you have seen as

$$Zn + Cu^{2+} \longrightarrow Zn^{2+} + Cu$$

could be represented as

$$
\begin{array}{cc}
Zn & Zn^{2+} \\
& \\
Cu^{2+} & Cu
\end{array}
$$

We have presented just a few of the important redox processes. Indeed, all of chemistry involves electron transfer and therefore an understanding of these processes and manipulative skill with redox equations are important to acquire.

SUMMARY

Chemical reactions involve rearrangements of electron structure (Section 16.1). When electrons are transferred from one atom to another, the loss of electrons by one atom is called **oxidation** and the gain of electrons by the other is called **reduction.** Electron-transfer reactions are also called *redox* reactions (Section 16.2).

Every net ionic equation corresponding to a redox reaction can be separated into an oxidation half-reaction and a reduction half-reaction. The material that is oxidized is also referred to as the **reducing agent**; similarly, the material reduced is the **oxidizing agent** (Section 16.3).

By using a set of rules determined by convention, one can assign an oxidation number to any atom regardless of its state of combination (Section 16.4). Oxidation and reduction can also be defined in terms of oxidation number (Section 16.5).

Equations for redox reactions can be balanced by using to advantage the fact that as the reaction occurs, the total number of electrons gained must be equal to the total number of electrons lost. The Ion-Electron Method of balancing follows a set of illustrated guidelines (Section 16.6). For redox reactions done in basic solution, certain modifications of the guidelines for the Ion-Electron Method of balancing must be applied (Section 16.7). An alternate balancing procedure is the Oxidation-Number Method (Section 16.8).

The basis of the Activity Series is the relative ease of oxidation of the various metal atoms to cations. Spontaneous reactions occur when metals high in the series are oxidized at the same time that metallic cations lower in the series are reduced (Section 16.9). Spontaneous redox reactions can be used to produce electricity via a voltaic cell.

Conversely, electricity can be employed to drive a nonspontaneous reaction in an electrolytic cell (Section 16.10).

Electron-transfer reactions are also very important in body chemistry (Section 16.11).

CHAPTER ACCOMPLISHMENTS

After completing this chapter you should be able to

16.1 Introduction

16.2 Electron-transfer reactions

1. Define the terms "oxidation" and "reduction" in terms of electron transfer.
2. State the relationship between electrons gained and electrons lost during any chemical reaction.
3. Given an oxidation half-reaction and a reduction half-reaction, add them to obtain a balanced chemical equation.

16.3 Writing half-reactions

4. Recognize half-reactions as oxidation or reduction.
5. Define oxidizing and reducing agents in terms of materials undergoing oxidation and reduction.
6. Given an ionic equation, write the oxidation and reduction half-reactions, and identify the oxidizing and reducing agents.

16.4 Oxidation numbers

7. Given the formula of a chemical species, such as a compound or polyatomic ion, determine the oxidation number of each element in that species.

16.5 Definitions revisited

8. Distinguish between oxidation and reduction in terms of changes in oxidation number.
9. Given a chemical equation, indicate the substances oxidized and reduced, and identify the oxidizing and reducing agents.

16.6 Balancing redox reactions

10. Given an unbalanced equation for an oxidation-reduction reaction in acidic or neutral solution, write a balanced ionic equation using the Ion-Electron Method.

16.7 Basic redox reactions

11. Given an unbalanced equation for an oxidation-reduction reaction in basic solution, write a balanced ionic equation using the Ion-Electron Method.

16.8   An alternate method

12.   Given an unbalanced equation for an oxidation-reduction reaction, write a balanced equation using the Oxidation-Number Method.

16.9   Activity Series revisited

13.   State the relationships among the activity of a metal, the ease with which it is oxidized, and its strength as a reducing agent.

14.   Given an activity series, designate reactions as spontaneous or nonspontaneous based on the strengths of the oxidizing and reducing agents.

16.10   Uses of redox

15.   Distinguish between voltaic cells and electrolytic cells.

16.   Given the equation for a spontaneous single replacement reaction, describe the construction of a voltaic cell employing that reaction.

17.   State the name of the reaction that occurs at the anode and the one that occurs at the cathode in all electrochemical cells.

18.   State the names of the ions which move toward the anode and the cathode in electrochemical cells.

19.   Write equations for the half-reactions that occur during the electrolysis of a given molten ionic compound.

20.   Given the metal to be electroplated, write equations for the oxidation and reduction half-reactions that occur during the process.

16.11   Redox in the body

21.   Recognize oxidation-reduction reactions in the intersecting arrow representation.

PROBLEMS

16.2   Electron-transfer reactions

16.8   Explain why oxidation-reduction reactions can be thought of as electron-transfer reactions.

16.9   Oxidation and reduction occur simultaneously in a chemical reaction. Explain why this must be true.

16.10   **a.** $Mg \rightarrow Mg^{2+} + 2e^-$
b. $Br_2 + 2e^- \rightarrow 2Br^-$
Add the preceding two half-reactions together. How many electrons are lost in part (a)? How many electrons are gained in part (b)?

16.11 $Fe \rightarrow Fe^{3+} + 3e^-$
$Br_2 + 2e^- \rightarrow 2Br^-$
a.   Will the electrons lost equal the electrons gained if we simply add the preceding two half-reactions?
b.   How can we add the two half-reactions so that the electron change will cancel out in the sum?

16.3   Writing half-reactions

16.12   Identify the following half-reactions as oxidation or reduction:
a. $K^+ + e^- \rightarrow K$
b. $Zn \rightarrow Zn^{2+} + 2e^-$
c. $MnO_4^- + 8H^+ + 5e^- \rightarrow Mn^{2+} + 4H_2O$
d. $3SO_2 + 6H_2O \rightarrow 3HSO_4^- + 9H^+ + 6e^-$

16.13   **a.**   Give a definition of oxidizing and reducing agents.

b. For each of the half-reactions in Problem 16.12, indicate whether the reactant is an oxidizing or reducing agent.

16.14 Identify the following half-reactions as oxidation or reduction and identify the reactant as an oxidizing or reducing agent:

a. $Br_2 + 2e^- \rightarrow 2Br^-$
b. $Pb^{2+}(aq) + 2e^- \rightarrow Pb(s)$
c. $SO_3^{2-} + H_2O \rightarrow SO_4^{2-} + 2H^+ + 2e^-$
d. $Fe^{2+} \rightarrow Fe^{3+} + e^-$

16.15 For each of the following ionic equations, write the oxidation-reduction half-reactions, and identify the oxidizing and reducing agents:

a. $2Fe^{3+}(aq) + Sn^{2+}(aq) \rightarrow$
$$2Fe^{2+}(aq) + Sn^{4+}(aq)$$
b. $Zn(s) + 2H^+(aq) \rightarrow Zn^{2+}(aq) + H_2(g)$
c. $2Fe^{3+}(aq) + 3Zn(s) \rightarrow 2Fe(s) + 3Zn^{2+}(aq)$

16.16 For each of the following unbalanced ionic equations, write the oxidation-reduction half-reactions and identify the oxidizing and reducing agents.

a. $Al(s) + Pb^{2+}(aq) \rightarrow Al^{3+}(aq) + Pb(s)$
b. $SO_4^{2-}(aq) + Zn(s) \rightarrow Zn^{2+}(aq) + SO_2(g)$
c. $I^-(aq) + NO_3^-(aq) \rightarrow I_2(s) + NO(g)$

16.4 Oxidation numbers

16.17 Determine the oxidation numbers of
a. Each element in K_2SO_4
e. N in NH_3
b. Cl in ClO_3^-
f. P in H_3PO_4
c. Cl in Cl_2
g. C in CH_4
d. N in NH_4^+

16.18 Determine the oxidation number of N in N_2, NO, NO_2, N_2O, N_2O_4, NO_2^-, NO_3^-, and NH_3

16.19 Determine the oxidation number of Cl in each of the following:
a. Cl_2
d. Cl_2O_7
b. Cl_2O
e. Cl_2O_5
c. Cl_2O_3

16.20 State the oxidation number of each element in
a. Na_2SO_4
c. H_2SO_4
b. $KMnO_4$
d. C_2H_6

16.21 Give the formula of a species in which the oxidation number of hydrogen is
a. +1

b. −1
c. 0

16.5 Definitions revisited

16.22 a. Give a definition of oxidation and reduction in terms of oxidation-number changes.
b. In the following unbalanced equations, determine the substances oxidized and reduced, and identify the oxidizing and reducing agents:
1. $HNO_3 + HI \rightarrow NO + I_2 + H_2O$
2. $Bi(OH)_3 + K_2SnO_2 \rightarrow Bi + K_2SnO_3 + H_2O$
3. $I_2O_5 + CO \rightarrow I_2 + CO_2$
4. $HCl + Mg \rightarrow MgCl_2 + H_2$
5. $H_2 + O_2 \rightarrow H_2O$

16.23 For each of the following equations, determine the substances oxidized and reduced and identify the oxidizing and reducing agents:
a. $NaNO_3 + Pb \rightarrow NaNO_2 + PbO$
b. $Na_2SO_4 + 4C \rightarrow Na_2S + 4CO$
c. $NH_4NO_2 \rightarrow N_2 + 2H_2O$
d. $As_2O_3 + Cl_2 + H_2O \rightarrow H_3AsO_4 + HCl$

16.6 Balancing redox reactions

16.24 Balance the following half-reactions (solutions are acidic or neutral):
a. $NO(g) \rightarrow NO_3^-(aq)$
b. $MnO_4^-(aq) \rightarrow Mn^{2+}(aq)$
c. $Cr_2O_7^{2-} \rightarrow Cr^{3+}$
d. $C_2O_4^{2-} \rightarrow CO_2$

16.25 Write a balanced net ionic equation for the following oxidation-reduction reactions (solutions are either acidic or neutral):
a. $Zn(s) + HCl(aq) \rightarrow H_2(g) + ZnCl_2(aq)$
b. $MnO_2(s) + HBr(aq) \rightarrow$
$$MnBr_2(aq) + Br_2(aq) + H_2O(l)$$
c. $HNO_3(aq) + I_2(s) \rightarrow$
$$NO_2(g) + H_2O(l) + HIO_3(aq)$$
d. $K_3AsO_4(aq) + HI(aq) \rightarrow$
$$K_3AsO_3(aq) + I_2(aq)$$
e. $H_2S(aq) + HNO_3(aq) \rightarrow$
$$S(s) + NO(g) + H_2O(l)$$
f. $KClO_3(aq) + HI(aq) \rightarrow KCl(aq) + I_2(aq)$
g. $Cu(s) + H_2SO_4(aq) \rightarrow$
$$CuSO_4(aq) + SO_2(g) + H_2O(l)$$

h. $FeSO_4(aq) + KMnO_4(aq) + H_2SO_4(aq) \rightarrow$
$Fe_2(SO_4)_3(aq) + MnSO_4(aq) + H_2O(l) + K_2SO_4$

i. $H_2(g) + FeCl_3(aq) \rightarrow HCl(aq) + FeCl_2(aq)$

j. $C_3H_8(g) + O_2(g) \rightarrow CO_2(g) + H_2O(l)$

k. $C_6H_{12}O_6(aq) + O_2(g) \rightarrow CO_2(g) + H_2O(l)$

16.26 Write a balanced net ionic equation for the following oxidation-reduction reactions (solutions are either acidic or neutral):

a. $Ag + HNO_3 \rightarrow AgNO_3 + NO + H_2O$

b. $HNO_2 + HI \rightarrow NO + I_2 + H_2O$

c. $Pb(NO_3)_2 + S + H_2O \rightarrow$
$Pb + H_2SO_3 + HNO_3$

d. $H_2O_2 + HI \rightarrow I_2 + H_2O$

16.27 Balance the following half-reactions (solutions are basic):

a. $Zn \rightarrow Zn(OH)_4{}^{2-}$

b. $Sn \rightarrow HSnO_2{}^-$

c. $BrO_4{}^- \rightarrow Br^-$

d. $CN^- \rightarrow CNO^-$

16.28 Write a balanced net ionic equation for the following oxidation-reduction reactions (solutions are basic):

a. $Zn(s) + KMnO_4(aq) \rightarrow$
$Zn(OH)_2(aq) + MnO_2(s) + KOH(aq)$

b. $Cd(s) + NiO_2(aq) \rightarrow$
$Cd(OH)_2(aq) + Ni(OH)_2(aq)$

c. $K_2S(aq) + I_2(aq) + KOH(aq) \rightarrow$
$K_2SO_4(aq) + KI(aq)$

d. $NaOH(aq) + Br_2(aq) \rightarrow$
$NaBrO_3(aq) + NaBr(aq)$

e. $Bi(OH)_3(aq) + K_2SnO_2(aq) \rightarrow$
$K_2SnO_3(aq) + Bi(s)$

f. $Al(s) + KNO_3(aq) \rightarrow NH_3(aq) + KAlO_2(aq)$

16.29 Write a balanced net ionic equation for the following oxidation-reduction reactions (solutions are basic):

a. $Cr(OH)_3 + NaClO + NaOH \rightarrow$
$Na_2CrO_4 + NaCl + H_2O$

b. $KMnO_4 + NaClO_2 \rightarrow MnO_2 + NaClO_4$

c. $Cr^{3+} + ClO^- \rightarrow CrO_4{}^{2-} + Cl^-$

d. $CoCl_2 + Na_2O_2 \rightarrow Co(OH)_3 + NaCl$

16.8 An alternate method

16.30 Balance the following reactions by the oxidation number method:

a. $Cu(s) + HNO_3(aq) \rightarrow$
$Cu(NO_3)_2(aq) + NO(g) + H_2O(l)$

b. $K_2Cr_2O_7(aq) + H_2O(l) + S(s) \rightarrow$
$SO_2(g) + KOH(aq) + Cr_2O_3(aq)$

c. $H_2O(l) + P_4(s) + HOCl(aq) \rightarrow$
$H_3PO_4(aq) + HCl(aq)$

d. $PbO_2(aq) + HI(aq) \rightarrow$
$PbI_2(aq) + I_2(aq) + H_2O(l)$

e. $HNO_3(aq) + H_2S(aq) \rightarrow$
$NO(g) + S(s) + H_2O(l)$

16.31 Balance the following reactions by the oxidation number method:

a. $Br_2 + H_2O + SO_2 \rightarrow HBr + H_2SO_4$

b. $KClO + H_2 \rightarrow KCl + H_2O$

c. $SnSO_4 + FeSO_4 \rightarrow Sn + Fe_2(SO_4)_3$

d. $HNO_2 + HI \rightarrow NO + I_2 + H_2O$

16.9 Activity series revisited

16.32 Use Table 16.1 to predict which of the following reactions will occur spontaneously:

a. $2Al(s) + 3CuSO_4(aq) \rightarrow$
$3Cu(s) + Al_2(SO_4)_3(aq)$

b. $3Ag(s) + Al(NO_3)_3(aq) \rightarrow$
$Al(s) + 3AgNO_3(aq)$

c. $Fe(s) + ZnSO_4(aq) \rightarrow Zn(s) + FeSO_4(aq)$

d. $Fe(s) + 2AgNO_3(aq) \rightarrow$
$2Ag(s) + Fe(NO_3)_2(aq)$

16.33 Write the half-reactions for the spontaneous reactions you identified in Problem 16.32.

16.34 Use Table 16.1 to

a. Give the name of a metal that is a better reducing agent than Mg

b. Give the names of two metals that will not spontaneously react with HCl

16.35 A gold ring can be immersed in a sodium chloride solution or even a concentrated hydrochloric acid solution without damage to the ring. Explain how Table 16.1 enables you to predict these facts.

16.10 Uses of redox

16.36 Describe the construction of voltaic cells that might be made employing appropriate reactions in Problem 16.32.

16.37 What is the function of a salt bridge in voltaic and electrolytic cells?

16.38 A student constructs a battery by immersing an aluminum strip in $Al(NO_3)_3$ solution and a tin strip in $Sn(NO_3)_2$ solution. A salt bridge is used and the set-up is wired in a manner similar to the apparatus in Figure 16.1*b*.

a. Which metal is the anode and which the cathode?

b. Write the half-reactions and the total cell reaction.

c. Which electrode becomes heavier as the battery runs?

16.39 A common flashlight battery works through the coupling of the two half-reactions:

$Zn(s) \rightarrow Zn^{2+}(aq) + 2e^-$
$MnO_2(s) + NH_4^+(aq) + e^- \rightarrow$
$$MnO(OH)(s) + NH_3(aq)$$

a. Which reaction occurs at the anode and which at the cathode?

b. What material is the anode made of?

16.40 Which reactions in Problem 16.32 could be induced to occur in an electrolytic cell?

16.41 The method by which aluminum metal is claimed from its ore involves the electrolysis of molten Al_2O_3.

a. Write the two half-reactions that occur during electrolysis.

b. Does aluminum metal form at the anode or cathode?

16.42 Given an ordinary metal bracelet, a strip of gold, and a solution of $Au(NO_3)_3$, describe how you would plate the bracelet with gold.

16.43 Write the anode and cathode reactions that would occur during the gold plating described in Problem 16.42.

16.44 Consult Figure 16.5.

a. The half-reactions shown in Figure 16.5 are those which occur spontaneously and allow the battery to supply current. Which plates are the anode and which the cathode?

b. Write equations for the half-reactions that occur when a car's generator recharges a battery.

c. During recharging which plates are the anode and which the cathode?

NUCLEAR CHEMISTRY CHAPTER

17

Cl

Chlorine

35.45

INTRODUCTION

17.1 The "ordinary" chemistry that we have discussed so far in this book involves changes only in the electron structure of atoms, ions, and molecules. The numbers and/or arrangements of electrons in atoms change as chemical reactions occur, but the nucleus containing protons and neutrons remains unaltered. However, the nuclei of some isotopes of many elements do undergo changes in their numbers of protons and/or neutrons during a process called radioactive decay or **radioactivity.** Such isotopes are called **radioisotopes.**

For example, carbon-14 is radioactive and its nucleus changes in such a way that it acquires an extra proton. This nuclear change converts the carbon atom (with six protons) to a nitrogen atom (with seven protons). The most abundant isotope of carbon, ^{12}C, has no tendency to undergo such a nuclear change. Carbon is typical of the common elements. The most abundant isotopes of the common elements are not radioactive, but usually a small amount of some radioisotope (generally of higher mass) does exist.

On the other hand, all isotopes of the element uranium are radioactive. This behavior is typical of the heavier elements (those of larger atomic numbers) such as uranium.

Nuclei, such as those of uranium, which undergo changes (nuclear reactions) do so in the same exact manner regardless of the nature of their surrounding electron structure. That is, the nuclei of free uranium metal atoms, which have the electron structure characteristic of that element, undergo nuclear reactions in the same way that chemically combined, and thus electronically altered, uranium atoms do. In considering nuclear chemistry, therefore, we will completely ignore the electrons of the elements we discuss and look only at the content of the nucleus—the protons and neutrons.

THE NUCLEUS REVISITED

17.2   Let us briefly review Sections 4.5 through 4.7. The atomic number of an element tells us the number of protons in the nuclei of atoms of that element. The sum of the number of protons and the number of neutrons in an atom is given by the mass number. Consequently,

Number of neutrons = mass number − atomic number

Information about the content of the nucleus is conveniently given by symbols of the form

$$\text{atomic number}^{\text{mass number}} Sy$$

for example, $^{12}_{6}C$, $^{90}_{38}Sr$, and $^{235}_{92}U$. Because the atomic number is identical for all atoms of a particular element, the atomic number is frequently omitted from these symbols, for example, ^{12}C, ^{90}Sr, and ^{235}U. Should the atomic number be needed, the symbol (C, Sr, or U) enables us to find the atomic number in the periodic table. Another common symbolism is carbon-12, strontium-90, and uranium-235, in which the mass number follows the name of the element.

Along with the symbol, the mass number of an atom must always be specified to identify that atom uniquely because whereas the number of protons in atoms of a particular element is invariable (and hence given by the symbol alone), the number of neutrons may vary. Atoms with the same atomic number (same number of protons) but different mass numbers (different numbers of neutrons) are called **isotopes.** For example, the most abundant isotope of uranium is ^{238}U (92 p^{+}, 146 n^{0}). The isotope of uranium that is most useful as a nuclear fuel is ^{235}U (92 p^{+}, 143 n^{0}).

Problem 17.1

Consider the isotopes $^{24}_{11}Na$, $^{27}_{13}Al$, ^{23}Na, ^{79}Br, and ^{206}Pb, and indicate the numbers of protons and neutrons in the nuclei of each.

RADIOACTIVITY

17.3   As you know, "ordinary" chemical reactions are accompanied by energy changes. Nuclear reactions are accompanied by much greater energy changes. The spontaneous emission of energy which accompanies nuclear change is called **radioactivity.**

It was the accidental detection of the emission of energy from a uranium salt that led ultimately to knowledge of nuclear reactions. In 1896 in France, Henri Becquerel happened to place a sample of a uranium salt on top of a photographic plate wrapped in opaque black paper in a dark drawer. Several days later he developed the photographic plate and found the image of the outlines of the uranium crystal on the plate. He concluded that it was the uranium sample that gave off radiant energy since the photographic plate had been subjected to no other source of radiation. Becquerel soon found that specifically uranium nuclei produced the phenomenon because the radiation occurred for free uranium and for various different uranium salts. Moreover, the amount of radiation was directly proportional to the amount of uranium only.

Marie (Sklodowska) and Pierre Curie took up the study of the mysterious emissions from uranium, and it was they who coined the term "radioactive" to describe elements which spontaneously give off radiation. In the course of their studies of radioactivity, they discovered two previously unknown elements, polonium (Po) (named for Marie's homeland, Poland) and radium (Ra). In 1903 the Curies shared the Nobel Prize in Physics with Henri Becquerel for their discoveries.

Meanwhile in England, Ernest Rutherford was also studying radiation, and he was able to establish that there were two types of radiation, which he called alpha (α) and beta (β). (Alpha and beta are the first two letters of the Greek alphabet.) His experiments also showed that these two types of nuclear radiation involve the emission of particles from the nucleus.

Another scientist demonstrated the existence of a third type of radiation which does not involve particles. This is γ-radiation, named for the third letter of the Greek alphabet, gamma. An experimental setup that identifies the three types of radiation is shown in Figure 17.1. When a beam of nuclear radiation is passed between electric or magnetic poles, the beam splits into three separate beams. One beam, α, bends toward the negative plate; another, β, bends toward the positive plate; and the third, γ, is unaffected by the electric poles and passes along the same line as the original emission.

The existence of three types of radiation can be demonstrated by passing a beam from a radioactive source between electrical plates. The radiation that is unaffected is called **gamma** (γ). That bent toward the positive plate is called **beta** (β). That bent toward the negative plate is called **alpha** (α). The fact that β particles are deflected more than α particles indicates that β particles must be lighter than α particles.

FIGURE 17.1

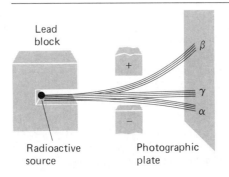

Lead
block

β

$+$

γ

$-$

α

Radioactive
source

Photographic
plate

PROPERTIES OF α-, β-, and γ-RADIATION

17.4 The experiment shown in Figure 17.1 clearly shows that α-radiation is positive because it is attracted toward the negative plate, β-radiation is negative because it is attracted to the positive plate, and γ-radiation is electrically neutral since it is attracted to neither plate. The magnitudes of these charges were also determined by experiment and found to be α, $+2$; β, -1; and γ, 0, using the same relative scale as for the subatomic particles. These and other properties are summarized in Table 17.1.

Alpha (α) radiation

Alpha radiation involves the emission of particles from a radioactive nucleus. From a determination of the relative charge ($+2$) and relative mass (4 amu) of α-particles, scientists were able to conclude that α-particles are identical to helium nuclei. Identity is symbolized $\equiv$.

$$\alpha\text{-Particle} \qquad {}^{4}_{2}\text{He nucleus}$$

$$\left(\begin{matrix} 2\,p^{+} \\ 2\,n^{0} \end{matrix}\right) \equiv \left(\begin{matrix} 2\,p^{+} \\ 2\,n^{0} \end{matrix}\right)$$

Because α-particles are the heaviest of the three radiations, they are the slowest-moving and least-penetrating of the three types. They are readily stopped by a few pieces of paper or a layer of human skin. Therefore, α-radiation is usually not hazardous to living organisms unless swallowed or inhaled. Figure 17.2 shows the relative penetrating power of the three radiation forms.

Beta (β) radiation

Beta radiation is also particulate. The mass and charge of the β-particle

TABLE 17.1   CHARACTERISTICS OF THE THREE TYPES OF NUCLEAR RADIATION

Type of Radiation	Symbol	Composition	Charge	Relative Mass, amu	Velocity*	Penetrating Power
Alpha	α, ${}^{4}_{2}\text{He}$	Identical to the He nucleus	$+2$	4	90% c	Low
Beta	β, $-{}^{0}_{1}e$	Identical to the electron	-1	1/1840	Variable, as fast as 99% c	Moderate
Gamma	γ, ${}^{0}_{0}\gamma$	High-energy radiation	0	0	c	High

* c is the standard symbol for the speed of light; $c = 3 \times 10^{10}$ cm/s.

FIGURE 17.2

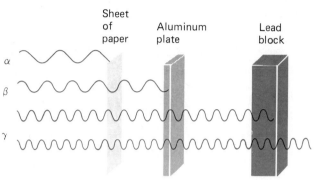

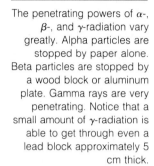

Sheet
of
paper

Aluminum
plate

Lead
block

α

β

γ

The penetrating powers of α-, β-, and γ-radiation vary greatly. Alpha particles are stopped by paper alone. Beta particles are stopped by a wood block or aluminum plate. Gamma rays are very penetrating. Notice that a small amount of γ-radiation is able to get through even a lead block approximately 5 cm thick.

indicate that it is identical to an electron:

$$\beta\text{-Particle} \equiv e^-$$

Because β-particles are smaller and faster-moving, they have about 100 times the penetrating power of α-particles. An aluminum plate, a block of wood, or heavy protective clothing is necessary to stop β-radiation. Although most β-radiation is not sufficiently energetic to reach the internal organs of the body, it goes deep enough within the outer layers of skin to cause damage (similar to severe sunburn) and represents a special hazard to eyes.

Gamma (γ) radiation

Gamma radiation is pure energy; γ-rays are not particles. This high-energy radiation, which is similar to x-rays, travels at the speed of light and can be stopped only by a block of several layers of lead. Gamma radiation easily penetrates the skin and can cause severe internal damage. We assume you have an idea of the properties of x-rays and know about their ability to travel through space (or the body) from your contact with them in medical and dental diagnosis. Your notions are sufficient for an understanding of the concepts being presented.

IONIZING RADIATION

17.5 Nuclear radiation, like x-radiation, is also referred to as **ionizing radiation.** This is so because the interaction of α-, β-, and γ-radiation and x-rays often causes ionization of an atom or molecule. When α-, β-, γ-, or x-radiation collides with an atom or molecule, some of the energy of the radiation is transferred to the particle that has been hit. At the very least, electrons in the atom or molecule are promoted to higher-energy, less-stable states. Often electrons are kicked out of atoms or molecules, producing ions and breaking bonds (see Figure 17.3).

Within the cells of the body, these ionization effects can be devastating. Smooth-working body chemistry depends on the presence of particular substances participating in certain reactions. Alteration of these substances through ionization interferes with necessary cell reactions and leads to other undesirable reactions. The result is a total disruption of cell activity.

The ionizing power of nuclear radiation is inversely related to its penetrating power; that is, γ-radiation is most penetrating but produces the fewest ions per unit volume. On the other hand, the least penetrating radiation, α produces the largest number of ions, and β-radiation is moderate in penetrating power and intermediate in ionizing ability. Many of the units which measure the power or biological effectiveness of a radioactive source relate to ionizing ability. See Table 17.2 for the definition of units which are used to measure radiation.

Radiation intensity decreases rapidly with increasing distance from the source of radiation. Doubling your distance from a radiation source ensures that the radiation felt will be one-fourth of the original value; at triple the distance, the radiation is one-ninth the original value. Quantitatively, this is known as the inverse square law: intensity I is

FIGURE 17.3

Radiant energy knocks electrons out of atoms or molecules. This produces ions. Bond breakage often occurs in molecular ions.

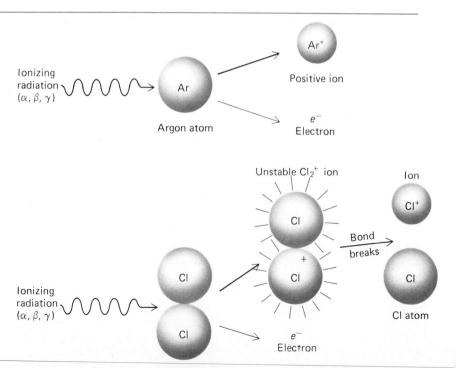

UNITS OF RADIATION MEASUREMENT **TABLE 17.2**

Unit	What It Measures	Description	Example
Curie (Ci)	Activity	An amount of radioactive material that undergoes 3.7×10^{10} disintegrations per second	^{60}Co γ-ray sources delivered to hospitals for radiation therapy are typically in the millicurie range
Roentgen (R)	Intensity of exposure	An amount of x- or γ-radiation that produces 1 electrostatic unit of charge ($\sim$ 20 billion ion pairs) in 1 mL of dry air at 0°C and 1 atm of pressure	The output from an x-ray machine is described in terms of roentgens
Radiation absorbed dose (rad)	Absorbed dose	An amount of radiation which when absorbed by tissue delivers 100 ergs per gram of tissue (1 erg $= 2.4 \times 10^{-8}$ cal); for x- and γ-rays, 1 R delivers 1 rad	Radiation therapists quote dosages in rads
Relative biological effectiveness (rbe)	Biological effectiveness	A quality factor which accounts for the fact that the same amount of exposure to different forms of radiation produces different amounts of biological damage	X-, γ-, and β-rays all have approximately the same effectiveness and are given an rbe = 1; α-rays are much more damaging and are rated at 10 rbe
Roentgen equivalent for human (rem)	Dose equivalent weighted for different kinds of radiation	An amount of radiation which produces the same damage to tissue as 1 R of x-radiation (dose in rems = dose in rads × rbe); dosages in rems are additive; thus cumulative effects are expressed by this unit	Preferred dosage statement because it compensates for the differences in effectiveness among the forms of radiation

inversely proportional to the square of the distance from the source.

$$I \propto \frac{1}{d^2}$$

This diminishment of radiation with distance offers the simplest protection against radiation. Distance and shielding are the principal safety measures practiced around the radiation laboratory and clinic.

DETECTION DEVICES

17.6 Many of the devices for detecting and measuring radioactive emissions depend on the fact that nuclear radiation produces ions.

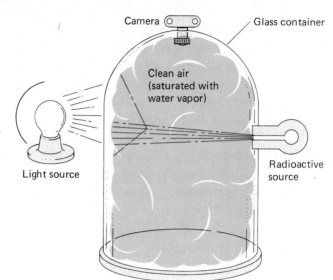

Camera   Glass container

Clean air
(saturated with
water vapor)

Light source   Radioactive
source

Cloud chamber. The chamber is supersaturated with water vapor. Radioactive emissions ionize the air through which they pass and cause the condensation of visible water droplets that leave a trail behind the traveling emitted ray. The trail can be photographed—notice the camera.

For example, consider the cloud chamber shown in Figure 17.4. The cloud chamber contains air supersaturated with water vapor. As radiation passes through this air and ionizes molecules in the air, water vapor condenses on the ions. This condensation is visible. Therefore, the radiation leaves a trail as it travels, in the same way that a jet leaves a vapor trail.

Probably the best known device for detecting and measuring radiation is the Geiger-Müller tube or the Geiger counter. This instrument is a tube filled with a gas such as argon (see Figure 17.5). Entering radiation ionizes the gas, and an electric current is produced as electrons go to the anode (+ electrode) and argon (+) ions go to the cathode (− electrode). The amount of current is proportional to the amount of radiation. Frequently, the electric current is used to produce audible clicks, the speed and intensity of which increase with increasing radiation.

Of course, photographic film detects radiation, as we saw in the case of the discovery of radioactivity by Becquerel. People who work with radioactive materials wear "film badges" which are developed periodically to monitor the extent of their exposure.

PHYSIOLOGICAL EFFECTS OF RADIATION

17.7 Ionizing radiation is most damaging to the *nuclei* of living cells. The cell nucleus contains the "blueprints" for producing more identical cells. Because the cell nucleus directs division and replication, cells that are dividing most rapidly are the first to show the effects of

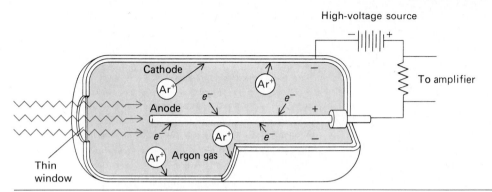

High-voltage source

To amplifier

Cathode
Anode
Argon gas
Thin window

FIGURE 17.5

Geiger-Müller counter. Ionizing radiation enters the Geiger tube through the thin window at the left of the apparatus. As the radiation passes through the gas inside the tube, it ionizes argon atoms along its path by kicking out electrons (see Figure 17.3). The Ar^+ cations are attracted to the negatively charged walls. The electrons are attracted to the central rod, which is positively charged. This produces an electrical current in the circuit which is amplified and recorded or heard as a click.

radiation. Genetic damage frequently occurs because the principal genetic material resides in the cell nucleus. The cells most susceptible to radiation damage are those in the lymphatic system, bone marrow, intestinal tract, reproductive organs, and lens of the eye.

The body can tolerate exposure to small amounts of ionizing radiation without apparent symptoms. Thus, background radiation (from the soil and outer space) and medical x-rays produce no noticeable harm. However, small doses may be cumulative, and it is important for persons who work near sources of radiation to monitor their exposure. A common device employed for this purpose is the film badge which makes use of the fact that photographic film detects radiation. Film badges are developed periodically to determine the extent of exposure.

One source of background radiation in homes is the gas radon. Specialists fear it may be appearing in many homes in amounts greater than previously believed. The source of the radon is soil or rocks on which a house is built or the actual building materials.

Rocks, especially granite, contain minute amounts of radium-226, a solid which, as such, stays in the rock. However, radium-226 emits α-particles and is thereby converted to radon. (You will see this conversion in Sample Exercise 17.1.) The radioactive gas radon goes into the air. Outdoors the radon atoms are quickly dispersed and are harmless. Indoors, on the other hand, they may accumulate.

In some homes, background radiation from radon may provide more unwanted exposure to radiation in the course of a year than the medical x-rays a person may have. Increased incidence of lung cancer from this exposure is especially likely, because radon is a gas, and a source of very damaging alpha radiation can therefore be inhaled into the lungs.

Even in the absence of noticeable symptoms of illness, excessive low-level exposure could lead to sterility or birth defects because reproductive cells and fetal tissue are especially sensitive to radiation.

TABLE 17.3   EFFECTS ON HUMAN BEINGS OF SHORT-TERM WHOLE-BODY
RADIATION EXPOSURE

Dose, rems	Effects
0–25	No detectable clinical effects
25–100	Slight short-term reduction in number of some blood cells; disabling sickness not common
100–200	Nausea and fatigue, vomiting if dose is greater than 125 rems, longer-term reduction in number of some blood cells
200–300	Nausea and vomiting first day of exposure, up to a 2-week latent period followed by appetite loss, general malaise, sore throat, pallor, diarrhea, and moderate emaciation; recovery in about 3 months unless complicated by infection or injury
300–600	Nausea, vomiting, and diarrhea in first few hours; up to a 1-week latent period followed by loss of appetite, fever, and general malaise in the second week, followed by hemorrhage, inflammation of mouth and throat, diarrhea, and emaciation; some deaths in 2–6 weeks; eventual death for 50% if exposure is above 450 rems; others recover in about 6 months
≥ 600	Nausea, vomiting, and diarrhea in first few hours; rapid emaciation and death as early as second week; eventual death of nearly 100%

From *Medical Aspects of Radiation Accidents*, E. L. Saenger, ed. (Washington, D.C.: United States Atomic Energy Commission, 1963), page 9.

The first clinical symptom for higher levels of exposure is a drop in white blood cell count. White blood cells have short life spans; therefore, damage to tissue producing these cells shows up rapidly. A reduction in white blood cell count increases susceptibility to infection because a person's natural resistance is lowered. Red blood cells are also affected, and anemia may result.

Higher doses of radiation cause symptoms such as nausea, vomiting, and diarrhea because of damage to cells in the intestinal tract.

The highest doses may produce burns to the skin, clouding of the eye lens (cataracts), and frequently death because of damage to so many essential bodily functions. If a person does survive massive exposure to radiation, the likelihood of developing cancer, particularly leukemia or blood cancer, is greatly increased. Table 17.3 lists some of the effects on human beings of short-term radiation exposure.

NUCLEAR REACTIONS

17.8 Let's get back to the fact that the emission of radiation accompanies a change in the nucleus of the emitting radioactive atom. The change in the nucleus produces an atom of a new element. For example, in Section 17.1, the conversion of $^{14}_{6}C$ to $^{14}_{7}N$ was described. The change from one element to another is called **transmutation.** Because radioactive elements disappear as they emit radiation and are transmuted into other elements, they are said to disintegrate or *decay.* Because we know the composition of α- and β-particles, we can predict the identity of the new element formed from the emitting element as it decays. Using symbols, we can record this conversion, or nuclear reaction, by a nuclear equation of the form:

$$\text{Emitting element} \longrightarrow \text{emitted particle} + \text{new element}$$

α-Emission

Uranium-238 is an α-emitter. This means $^{238}_{92}U$ nuclei lose α-particles, which we know to be composed of two protons and two neutrons, i.e., the helium nucleus, $^{4}_{2}He$. The nuclear equation for this event begins with the uranium-238 emitter and shows the emitted α-particle represented by its nuclear symbol, $^{4}_{2}He$.

$$^{238}_{92}U \longrightarrow \, ^{4}_{2}He + \, ?$$

Because of the loss of two protons in the α-particle, after emission the original nucleus contains only 90 protons (92 − 2). Because the number of protons characterizes an atom, this nucleus must be the element thorium, which we determine by looking up atomic number 90 in the periodic table. The mass number of the Th atom is 4 less than the original U atom because a total of four particles, each of mass 1, have been lost. The nuclear equation is completed by showing the symbol for the thorium isotope.

$$^{238}_{92}U \longrightarrow \, ^{4}_{2}He + \, ^{234}_{90}Th$$

In this, as in all nuclear equations, there is a balance (that is, their sums are equal) of both atomic numbers (92 = 2 + 90) and mass numbers (238 = 4 + 234) on the two sides of the arrow, which is like an equals sign. Nothing is lost; all protons and neutrons can be accounted for. Figure 17.6 illustrates the reaction and equation.

Another α-emitter is $^{218}_{84}Po$. The new element formed upon α-emission is an isotope of lead.

$$^{218}_{84}Po \longrightarrow \overbrace{\underbrace{^{4}_{2}He + \, ^{214}_{82}Pb}_{84}}^{218}$$

SAMPLE EXERCISE

17.1

Write the nuclear equation for the change that occurs in radium-226 when it emits an α-particle. Radium-226 was the first radioisotope used to treat cancer.

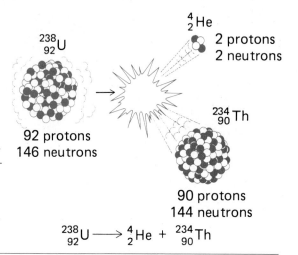

FIGURE 17.6

Alpha emission from uranium-238. Two protons and two neutrons leave the uranium-238 nucleus, leaving behind 90 protons and 144 neutrons. A nucleus with 90 protons is the element thorium.

$$^{238}_{92}\text{U} \longrightarrow ^{4}_{2}\text{He} + ^{234}_{90}\text{Th}$$

SOLUTION

Step 1   Write the symbol of the emitter, including atomic number and mass number, on the left-hand side of the equation. In this case, the mass number is given, and you must find radium's atomic number in the periodic table.

$$^{226}_{88}\text{Ra} \longrightarrow$$

Step 2   Write the symbol for the α-particle which shows that it is a helium nucleus on the right side of the equation:

$$^{226}_{88}\text{Ra} \longrightarrow ^{4}_{2}\text{He} + ?$$

Step 3   Complete the equation by writing a symbol for an isotope that has an atomic number 2 less than the original isotope ($88 - 2 = 86$), and a mass number 4 less than the original ($226 - 4 = 222$). Use the periodic table to ascertain that atomic number 86 identifies radon, Rn.

$$^{226}_{88}\text{Ra} \longrightarrow ^{4}_{2}\text{He} + ^{222}_{86}\text{Rn}$$

Step 4   Check the equation to see that the mass numbers and atomic numbers are balanced, that is, that the totals on each side of the arrow are equal.

$$226 = 4 + 222$$
$$88 = 2 + 86$$

Problem 17.2

Write a nuclear equation for α-emission from $^{239}_{94}\text{Pu}$.

β-Emission

Because you know that β-particles are electrons, you may be wondering how an electron emerges from the nucleus which contains only protons and neutrons. The answer is that, in effect, a neutron is transformed into a proton which stays in the nucleus and an electron which leaves the nucleus. The overall result is that the number of protons in the nucleus increases by 1 (atomic number increases by 1), the number of neutrons decreases by 1, and the mass number remains the same, since the sum of the number of protons and neutrons is not altered.

For example, $^{234}_{90}$Th is a β-emitter. One of its 144 neutrons is transformed into a proton and an electron, which is emitted. Thus the original nucleus contained 144 neutrons and 90 protons, and the new nucleus contains 143 neutrons and 91 protons. The element with atomic number 91 is protactinium. The mass number of the new element is 234 because 143 neutrons and 91 protons has that mass. This reaction is illustrated in Figure 17.7 and can be summarized by the nuclear equation

$$^{234}_{90}\text{Th} \longrightarrow {}^{0}_{-1}e + {}^{234}_{91}\text{Pa}$$

The use of the symbol $_{-1}^{0}e$ for the β-particle enables us to check that the nuclear equation is balanced.

$$234 = 0 + 234$$

$$90 = -1 + 91$$

Another β-emitter is $^{210}_{83}$Bi. The new nucleus formed has 84 protons and the same mass number:

$$^{210}_{83}\text{Bi} \longrightarrow {}^{0}_{-1}e + {}^{210}_{84}\text{Po}$$

FIGURE 17.7

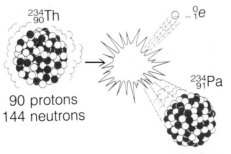

234
90Th

$_{-1}^{0}e$

234
91Pa

90 protons
144 neutrons

91 protons
143 neutrons

$$^{234}_{90}\text{Th} \longrightarrow {}^{0}_{-1}e + {}^{234}_{91}\text{Pa}$$

Beta emission from thorium-234. One of the thorium neutrons decomposes to an electron and a proton. The electron leaves the nucleus as a β particle. The nucleus now contains one more proton (91) and one less neutron (143). A nucleus with 91 protons is the element Pa.

SAMPLE EXERCISE

17.2

Write the nuclear equation for the change that occurs in cobalt-60 when it emits a β-particle.

SOLUTION

Step 1 Write the symbol of the emitter, including atomic number and mass number, on the left-hand side of the equation. From the periodic table, the atomic number of cobalt is 27. The mass number 60 is given.

$$_{27}^{60}\text{Co} \longrightarrow$$

Step 2 Write the symbol for the β-particle which shows that it is an electron on the right side of the equation.

$$_{27}^{60}\text{Co} \longrightarrow \ _{-1}^{0}e + ?$$

Step 3 Complete the equation by writing a symbol for an isotope that has an atomic number 1 greater than the original isotope ($27 + 1 = 28$; atomic number 28 indicates Ni) and the same mass number:

$$_{27}^{60}\text{Co} \longrightarrow \ _{-1}^{0}e + \ _{28}^{60}\text{Ni}$$

Step 4 Check the equation by seeing if the mass numbers and atomic numbers are balanced.

$$60 = 0 + 60$$
$$27 = -1 + 28$$

Problem 17.3

Write a nuclear equation for β-emission from $_{89}^{227}\text{Ac}$.

γ-Emission

Virtually all α- and β-emissions are accompanied by γ-emission. Because γ-radiation produces no change in nuclear contents, it is usually ignored in writing nuclear equations. For example, γ-emission from cobalt-60 is used in the treatment of cancerous tumors. However, in Sample Exercise 17.2, only β-emission was mentioned. We can include the fact that γ-emission accompanies β-emission in the nuclear equation. Notice that the symbol $_{0}^{0}\gamma$ does not alter the balanced equation.

$$_{27}^{60}\text{Co} \longrightarrow \ _{-1}^{0}e + \ _{28}^{60}\text{Ni} + \ _{0}^{0}\gamma$$

RADIOACTIVE DECAY SERIES

17.9 Because radioactive elements disappear as they emit radiation and are changed into other elements, they are said to **disintegrate** or

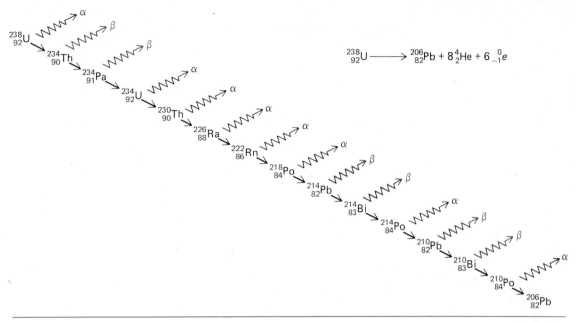

$$^{238}_{92}U \longrightarrow ^{206}_{82}Pb + 8\,^4_2He + 6\,^0_{-1}e$$

FIGURE 17.8

The main pathway of the uranium-238 radioactive decay series. This series could also proceed through the following series of emissions beginning with polonium-218: β, α, α, β, β, α, β. The sequence shown is the one followed by most atoms.

decay. Very often the product of radioactive decay is also radioactive, and so it will decay also. For example, we have seen $^{238}_{92}U$ decay to $^{234}_{90}Th$ by α-emission. We have also seen that $^{234}_{90}Th$ is radioactive and decays to $^{234}_{91}Pa$ by β-emission. It turns out that $^{234}_{91}Pa$ is also radioactive and decays by β-emission:

$$^{234}_{91}Pa \longrightarrow ^{\ 0}_{-1}e + ^{234}_{92}U$$

$^{234}_{92}U$ is radioactive, as are all isotopes of uranium.

What we see developing here is a radioactive decay series, which is summarized in Figure 17.8. When decay of a "parent" nucleus produces a radioactive "daughter" nucleus, the daughter will also decay, and so the series continues. A radioactive decay series stops when a stable *non*radioactive isotope is produced. In the example given (Figure 17.8), the series stops when nonradioactive $^{206}_{82}Pb$ is formed.

HALF-LIFE $t_{\frac{1}{2}}$

17.10 The rate at which radioactive elements decay varies greatly. Rate of radioactive decay is usually measured in terms of an element's **half-life,** symbolized $t_{\frac{1}{2}}$, the time required for the decay of one-half of a radioactive sample. For example, the half-life of ^{51}Cr, a radioisotope used to study blood volumes, is 28 days. This means that if you have 100 mg of ^{51}Cr today, then 28 days from today you would have only

50 mg because one-half the sample (50 mg) would have decayed. During the following 28 days, half the 50 mg would decay, so that only 25 mg would remain, and the process would continue.

$$100 \text{ mg } {}^{51}\text{Cr} \xrightarrow{\text{28 days}} 50 \text{ mg } {}^{51}\text{Cr} \xrightarrow{\text{28 days}} 25 \text{ mg } {}^{51}\text{Cr} \xrightarrow{\text{28 days}}$$

$$12.5 \text{ mg } {}^{51}\text{Cr} \xrightarrow{\text{28 days}} \cdots$$

The listing of half-lives of other commonly used radioisotopes in Table 17.4 points out the wide variation in half-life, from a few hours to millions of years.

Figure 17.9 shows the meaning of half-life graphically. Each time a half-life period elapses, half the amount of material present at the beginning of that period disintegrates. Figure 17.9a shows the principle of half-life in general. In Figures 17.9b and 17.9c, examples of the decay of specific amounts of specific radioisotopes are represented.

TABLE 17.4   HALF-LIVES AND USES OF SELECTED RADIOISOTOPES

Name	Symbol	Half-life	Radiation Emitted	Usefulness
Carbon-14	${}^{14}_{6}\text{C}$	5720 years	β	Radioactive dating and labeling
Sodium-24	${}^{24}_{11}\text{Na}$	15 h	β, γ	Blood-circulation studies
Phosphorus-32	${}^{32}_{15}\text{P}$	14 days	β	Fertilizer uptake monitor
Sulfur-35	${}^{35}_{16}\text{S}$	88 days	β	Protein label for metabolic studies
Potassium-42	${}^{42}_{19}\text{K}$	12 h	β	Plant and animal nutrition studies
Calcium-45	${}^{45}_{20}\text{Ca}$	165 days	β	Animal nutrition studies
Iron-59	${}^{59}_{26}\text{Fe}$	45 days	β	Red blood cell studies
Cobalt-60	${}^{60}_{27}\text{Co}$	5.3 years	β, γ	Radiation therapy
Yttrium-90	${}^{90}_{39}\text{Y}$	64 h	β	Pituitary implant radiation therapy
Technetium-99	${}^{99}_{43}\text{Tc}$	6 h	γ	Brain scans
Iodine-123	${}^{123}_{53}\text{I}$	13 h	γ	Thyroid radiation therapy
Iodine-131	${}^{131}_{53}\text{I}$	8 days	β, γ	Thyroid activity studies
Uranium-235	${}^{235}_{92}\text{U}$	710 million years	α, γ	Nuclear reactors

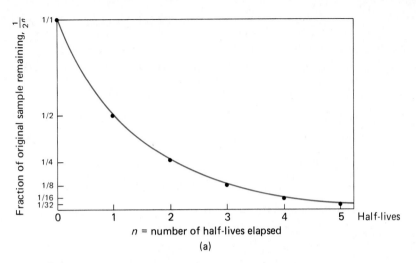

(a)

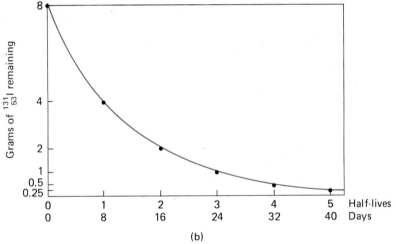

(b)

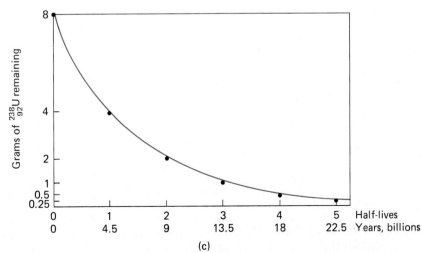

(c)

FIGURE 17.9

(a) Half-life decay curve for any radioactive isotope. During each half-life interval, half of the material present at the beginning of the interval decays. (b) Half-life decay curve for an 8-g sample of $_{53}^{131}I$, which has a half-life equal to 8 days. (c) Half-life decay curve from an 8-g sample of $_{92}^{238}U$, which has a half-life equal to 4.5 billion years.

Comparison of Figure 17.9*b* and *c* points out the idea that long-lived radioisotopes such as ^{238}U present a radioactive-waste storage problem because they "hang around" almost forever, whereas short-lived radioisotopes "dispose" of themselves. In these examples, a sample of iodine-131 would diminish from 8 to 0.25 g in only 40 days, whereas 22.5 billion years would be required for the decay of uranium-238 to this extent. Radioactive waste from nuclear power plants is mostly long-lived uranium isotopes which must be shielded from the environment for thousands of centuries before the uranium decays sufficiently to be harmless.

Calculations relating amounts of radioactive material and $t_{\frac{1}{2}}$ can be done by using the format shown earlier for the decay of ^{51}Cr, or more compactly, we can recognize that the graphic displays in Figure 17.9 give us an equation for this relationship:

$$\frac{\text{Quantity remaining}}{\text{after } n \text{ half-lives}} = \frac{\text{original}}{\text{quantity}} \times (\tfrac{1}{2})^n \qquad (17.1)$$

For ^{131}I, if we begin with 8.0 g, after 5 half-lives ($n = 5$), 0.25 g remain, as the graph told us and the equation tells us:

$$\begin{aligned}\text{Quantity remaining} &= 8.0 \text{ g originally} \times (\tfrac{1}{2})^5 \\ \text{after 5 half-lives} & \qquad\qquad\qquad\qquad (17.1)\end{aligned}$$

$$= 8.0 \text{ g} \times \frac{1}{2 \times 2 \times 2 \times 2 \times 2}$$

$$= 0.25 \text{ g}$$

SAMPLE EXERCISE

17.3

The half-life of ^{24}Na is 15 h. If you have a 240-mg sample of this radioisotope at noon on a Monday, how many milligrams will remain at 3 P.M. on Thursday?

SOLUTION

In order to use Equation (17.1) it is necessary to determine the total time elapsed and from that the number (n) of half-lives that have elapsed.

Monday $\xrightarrow{24\text{ h}}$ Tuesday $\xrightarrow{24\text{ h}}$ Wednesday $\xrightarrow{24\text{ h}}$ Thursday $\xrightarrow{3\text{ h}}$ Thursday
noon   noon   noon   noon   3 P.M.

Over the times specified, 75 h elapse:

$$75 \text{ h} \times \frac{1 \text{ half-life}}{15 \text{ h}} = 5.0 \text{ half-lives}$$

Now use Equation (17.1):

$$\text{Quantity remaining} = 240 \text{ mg} \times \frac{1}{2 \times 2 \times 2 \times 2 \times 2}$$

$$= 7.5 \text{ mg}$$

Problem 17.4

The half-life of ^{214}Bi is 20 min. How much of an original 16-g sample remains after 2 h 20 min?

The half-life concept can be used to determine the age of objects which contain radioisotopes. The age of museum relics can often be determined by carbon-14 dating. This procedure is based on the fact that all living things—plants and animals alike—are composed principally of the element carbon. Furthermore, whereas most carbon atoms are the nonradioactive $^{12}_{6}$C isotopes, a small percentage of carbon atoms are radioactive $^{14}_{6}$C isotopes.

While a plant is alive, the ratio of ^{14}C to ^{12}C isotopes is a constant because although ^{14}C is continually decaying in the plant, it is being replaced by ^{14}C from $^{14}CO_2$ in the atmosphere. Similarly, the ^{14}C to ^{12}C ratio is constant in animals because they continually eat fresh plants and inhale some $^{14}CO_2$. When a plant or animal dies, there is no longer a continual replacement of decaying ^{14}C, and the ratio diminishes.

The Dead Sea Scrolls were dated by examining the ^{14}C to ^{12}C ratio in their paper (paper is made from trees). It was found that the ratio in these scrolls was only 79.5 percent of the ratio in a living plant. From this and the known $t_{\frac{1}{2}}$ of carbon-14 (5720 years), archaeologists calculated that the scrolls are approximately 1900 years old.

Similar dating procedures are applied with other radioisotopes. For example, rocks can be dated from their uranium-238 or rubidium-87 content.

SAMPLE EXERCISE

17.4

A fragment of animal-skin clothing from an archaeological dig was found to have a carbon-14 to carbon-12 ratio only 0.25 times that of the ratio in a living animal's fur. The half-life of carbon-14 is 5720 years. How old is the clothing?

SOLUTION

Let the original ratio of carbon-14 to carbon-12 be represented by r. Then every time a half-life elapses, the ratio will be cut in half, because the amount of carbon-14 will be halved.

$$r \xrightarrow{\text{5720 y}} \tfrac{1}{2}r \xrightarrow{\text{5720 y}} \tfrac{1}{4}r$$

The given ratio is $0.25r$ or $\tfrac{1}{4}r$. Therefore, two half-lives, or 11,440 years (2 × 5720), must have elapsed since the animal skin was part of a living animal.

Problem 17.5

In the same archaeological dig as described in Sample Exercise 17.4, a wooden cart displayed a ^{14}C to ^{12}C ratio 0.5 times that in living trees. How old is the cart?

USES OF RADIOISOTOPES

17.11   Radioisotopes are very useful in chemistry, biology, industry, and especially medicine because of their following properties:

1.   They are easily detected in even minute amounts.

2.   The chemical reactivity of radioisotopes (that is, electron activity as described in Section 17.1) is identical to that of nonradioactive isotopes.

3.   Radiation damages cells, particularly those which divide rapidly.

The first two properties are used in medical diagnosis, chemical and industrial applications. Some medicinal treatments employ the latter two properties. Other properties desirable in radioisotopes used medicinally are a short half-life, so that exposure is not long-term, and high energy, so that they are intense and easily detected.

Radioactive labeling

Radioisotopes react and combine with other elements just like nonradioactive elements. For example, the compound sodium iodide (NaI), which can be used to study thyroid activity, is ordinarily composed of nonradioactive ^{23}Na and ^{127}I. However, ^{23}Na can equally readily combine with radioactive ^{131}I. This sodium iodide ($^{23}Na^{131}I$) is radioactively **labeled.** If it is introduced into the body, such a labeled compound can be readily detected and followed, that is, **traced** throughout the body. Tracing is the basis of medical diagnostic techniques.

Medical diagnosis

The rate at which a thyroid gland absorbs ^{131}I from a $Na^{131}I$ solution a patient has drunk can be readily monitored by a Geiger counter. This rate of absorption indicates whether the thyroid is working properly or is underactive or overactive.

A particularly useful radioisotope with several applications is $^{99}_{43}Tc$. For example, labeled sodium pertechnetate ($NaTcO_4$), a combination of Na^+ ions and TcO_4^- ions, is often used for brain studies. Ordinarily, an ion such as TcO_4^- cannot pass the blood-brain barrier, the body's mechanism for protecting brain cells. However, certain tumors or other abnormalities seem to ignore this barrier and hence radioactive TcO_4^- can enter brain tissue. The resulting accumulation of technetium-99 in brain tissue indicates to doctors that brain abnormalities are present.

Technetium-99 combined in other forms has an affinity for tissue types other than brain tissue. Technetium pyrophosphate, $Tc_2P_2O_7$, selectively collects in bone tissue; a ^{99}Tc-sulfur combination is taken up preferentially by cells of the liver, spleen, and bone marrow. Studies of the lung and kidneys are also possible using technetium radiopharmaceuticals.

Other radioisotopes are also used for medical diagnosis. Iron-59 is used to study the formation of red blood cells because the compound hemoglobin in red blood cells contains iron. Sodium-24 is commonly used to study blood circulation. A ^{24}Na sample is injected into the bloodstream, and its course throughout the body is followed by a Geiger tube.

Medical treatment

Radioisotopes have been used for medical treatment almost from the very first moment of the discovery of radium. Malignant tissues are irradiated with γ-radiation, usually from a cobalt-60 source. Ionizing radiation is more damaging to fast-growing cancer cells than it is to normal tissues, so it is possible to kill the cancer cells while leaving normal cells relatively unharmed. The rays are directed as much as possible toward the tumor, while normal tissue is shielded.

Another treatment technique is to use radioisotopes that show selectivity for certain kinds of tissues. Thus ^{131}I can be used to treat cancerous thyroid glands because iodine is absorbed only by thyroid tissue and accumulates there. ^{123}I may also be used. It has two advantages over ^{131}I: its half-life is shorter (13 h) and it emits only γ radiation.

Radioactive iodine is also sometimes used to destroy healthy (nonmalignant) thyroid tissue. In Graves' disease, the thyroid is dangerously overactive. One treatment involves ingestion of a pill containing ^{131}I. The radioactive iodine accumulates in the thyroid gland, emits radiation, and destroys some thyroid tissue, thus diminishing the activity of the gland.

Leukemia can be treated by phosphorus-32, which becomes part of bone structure. Bones normally incorporate phosphorus from dietary sources into their makeup. Phosphorus-32 in bone tissue emits beta radiation; the excess white blood cells in the leukemia patient's bone marrow are thus exposed to radiation and many are killed.

Chemical uses

Many of the important concepts and theories of chemistry can be demonstrated or proved through the use of radioactive tagging or radioactive labeling. For example, it can readily be demonstrated that there is indeed a dynamic equilibrium present in a saturated solution of $PbCl_2$ by the use of radioactive **^{212}Pb.** The experiment would proceed as follows:

1. Begin with a saturated solution of unlabeled $PbCl_2$ in which we assume the following equilibrium exists:

$$PbCl_2(s) \rightleftharpoons \underset{\text{Nonradioactive}}{^{206}Pb^{2+}(aq)} + Cl^-(aq)$$

2. Add a few milliliters of tagged $^{212}Pb(NO_3)_2$ solution, which supplies radioactive lead ions. The solution equilibrium can now be represented by:

Solid lead(ll)chloride $\rightleftharpoons$
$$^{206}Pb^{2+}(aq) + {}^{212}Pb^{2+}(aq) + Cl^-(aq) + NO_3^-(aq)$$

3. If there is truly a dynamic equilibrium, ^{212}Pb will appear in the solid after a period of time. In forming $PbCl_2$, Cl^- does not care whether it teams up with ^{206}Pb or ^{212}Pb. If there is no equilibrium reaction, the solid will not acquire radioactivity. The solid *does* become radioactive as solid $^{212}PbCl_2$ forms; thus the exchange between solid and ions in solution is demonstrated and can be represented:

$$^{206}PbCl_2(s) + {}^{212}PbCl_2(s) \rightleftharpoons$$
$$^{206}Pb^{2+}(aq) + {}^{212}Pb^{2+}(aq) + Cl^-(aq) + NO_3^-(aq)$$

Another example of the use of radioactive labels can be seen in a study of photosynthesis:

$$6CO_2 + 6H_2{*}O \xrightarrow{\text{Photosynthesis}} \underset{\text{Glucose}}{C_6H_{12}O_6} + 6{*}O_2$$

The * indicates that the oxygen of water is radioactively tagged, whereas the oxygen of CO_2 is not. The appearance of the tag only in the product O_2 demonstrates that all the oxygen in glucose comes from CO_2 and all the oxygen in water becomes free oxygen gas.

Industrial uses

The harnessing of nuclear energy (Section 17.13) to generate electric power is probably the most significant industrial use of radioisotopes. Tagging or labeling techniques, i.e., the replacement of minute numbers of inactive atoms by radioisotopes, are used for wear and corrosion tests. Smoke detection devices contain small amounts of radioactive materials; the detection is based on the interaction of smoke particles and radiation which produces an electric current and sounds the alarm (Figure 17.10).

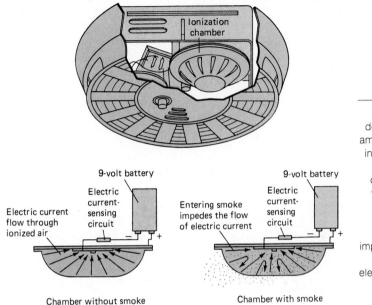

FIGURE 17.10

Diagram of a home smoke detector. A small quantity of americium-241 ionizes the air in the ionization chamber so that the ions in the air conduct an electric current when an electric voltage is applied. When smoke particles pass through the ionization chamber, they impede the flow of electricity, which is detected by electronic circuitry monitoring the electric current and signals an alarm.

NEW ELEMENTS THROUGH BOMBARDMENT

17.12   Radioactive decay is a natural process that occurs spontaneously. However, it is also possible to induce nuclear reactions by bombarding stable nuclei with nuclear-sized "bullets." For example, if a high-speed neutron is shot at the stable isotope of aluminum $^{27}_{13}Al$, a radioactive aluminum isotope is produced:

$$^{27}_{13}Al + {}^{1}_{0}n \longrightarrow {}^{28}_{13}Al \qquad \textbf{neutron bombardment}$$

In this case the force of the collision has simply caused neutron capture: that is, the neutron is taken into the aluminum nucleus. The $^{28}_{13}Al$ is a β-emitter and decays:

$$^{28}_{13}Al \longrightarrow {}^{0}_{-1}e + {}^{28}_{14}Si \qquad \textbf{decay}$$

Neutron bombardment also occurs naturally and produces other results besides simple capture. Carbon-14 exists in the atmosphere because of high-speed collisions between neutrons from high-energy cosmic rays in the upper atmosphere and the common isotope of nitrogen:

$$^{14}_{7}N + {}^{1}_{0}n \longrightarrow {}^{14}_{6}C + {}^{1}_{1}H$$

In this case the bombardment knocks a proton ($^{1}_{1}H$) out of the nucleus as the neutron enters.

Subatomic particles other than neutrons can be involved in bombardment. For example, bombardment of $^{14}_{7}N$ by an α-particle produces a nonradioactive isotope of oxygen:

$$^{14}_{7}N + {}^{4}_{2}He \longrightarrow {}^{17}_{8}O + {}^{1}_{1}H$$

Notice that these nuclear equations are balanced; that is, the sum of the mass numbers and the sum of the atomic numbers are identical on both sides of the equation.

SAMPLE EXERCISE

17.5

Complete the following nuclear equations, which represent bombardment reactions:

a. $^{35}_{17}Cl + {}^{1}_{0}n \longrightarrow {}^{34}_{16}S + ?$
b. $^{23}_{11}Na + ? \longrightarrow {}^{23}_{12}Mg + {}^{1}_{0}n$
c. $^{238}_{92}U + {}^{4}_{2}He \longrightarrow ? + {}^{1}_{0}n$

SOLUTION

The sum of the mass numbers and the sum of atomic numbers must be identical on each side of a nuclear equation. By establishing this equality, the identity of the unknown species can be determined.

a. Because the sum of the mass numbers on the left side is 36, the mass number of the unknown must be 2 (36 = 34 + 2). Similarly, the atomic number of the unknown must be 1 (17 = 16 + 1). Atomic number 1 characterizes H.

$$^{35}_{17}Cl + {}^{1}_{0}n \longrightarrow {}^{34}_{16}S + {}^{2}_{1}H$$

b.
$$^{23}_{11}Na + {}^{1}_{1}H \longrightarrow {}^{23}_{12}Mg + {}^{1}_{0}n$$

c. Because the sum of the mass numbers on the left is 242, the mass number of the unknown must be 241. Similarly, the atomic number must be 94. Plutonium is the element with atomic number 94.

$$^{238}_{92}U + {}^{4}_{2}He \longrightarrow {}^{241}_{94}Pu + {}^{1}_{0}n$$

Perhaps the most exciting application of bombardment reactions is in the preparation of non-naturally-occurring elements. Indeed, most of the radioisotopes routinely used in medicine do not occur naturally and are made artificially. The name *technetium*, given to the first element made artificially, is derived from a Greek word meaning "artificial."

Elements with atomic numbers greater than 92 are called **transuranium elements;** these elements do not occur naturally, but rather have been synthesized by bombardment reactions. For example, in Sample Exercise 17.5 you saw the synthesis of $^{241}_{94}Pu$ by α-bombardment of $^{238}_{92}U$. Other bombardment reactions lead to other trans-uranium elements. Also, the decay of these elements, all of which are radioactive, often leads to new elements. For example, the β-decay of $^{241}_{94}Pu$ produces americium-241:

$$^{241}_{94}Pu \xrightarrow{\text{decay}} {}^{241}_{95}Am + {}^{0}_{-1}e$$

Americium is used commercially in smoke detectors (Figure 17.10).

Problem 17.6

What radioisotope is produced when $^{242}_{96}Cm$ is bombarded by an α-particle and a neutron is ejected, that is,

$$^{242}_{96}Cm + {}^{4}_{2}He \longrightarrow {}^{1}_{0}n + ?$$

NUCLEAR ENERGY

17.13 Radioisotopes and nuclear reactions have many other practical applications besides medical uses. For example, they provide an important energy source. The nuclear power plants in operation today all employ **nuclear fission** reactions. Nuclear fission produces a great deal of heat energy, which is used to generate steam to drive turbines and produce electricity in much the same way that heat energy from fossil fuels is used (Figure 17.11).

FIGURE 17.11

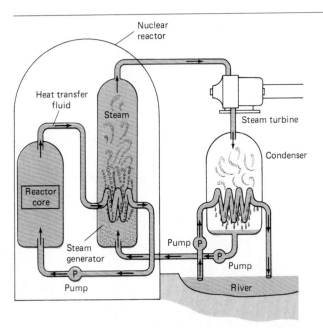

Diagram of a nuclear reactor. The heat source is the core of the reactor. The heat produced by the core is transferred by a closed loop of liquid sodium or liquid water under pressure to a steam generator. The steam runs a steam turbine, which produces electricity. Steam from the turbine is cooled by water from a nearby source, such as a river, and is pumped back into the steam generator.

Nuclear fission is the splitting of an atom upon bombardment. For example,

$$^{235}_{92}U + ^1_0n \longrightarrow ^{141}_{56}Ba + ^{92}_{36}Kr + 3\,^1_0n$$

Because more neutrons (three) are produced than are required for the fission to occur, a self-sustaining **chain reaction** will occur as long as there is a sufficient mass of uranium-235 present for the new neutrons to interact with (see Figure 17.12). This sufficient mass is called the **critical mass.** The chain reaction is self-sustaining because the product neutrons can react with the uranium-235. After the first fission, no additional energy need be added. Indeed, a great deal of energy is released.

FIGURE 17.12

The fission of one uranium-235 nucleus by one neutron can set off a chain reaction if there are sufficient numbers of the uranium-235 nuclei present (critical mass) to be split by the neutrons generated by the original fission.

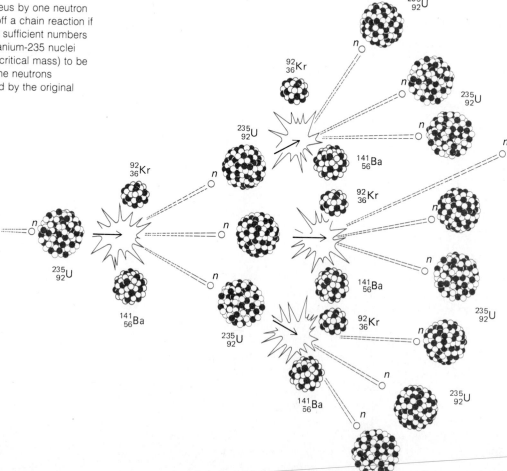

A major disadvantage of the use of nuclear fission in power plants is the fact that many of the products of the fission are themselves radioactive. For example, the barium-141 and krypton-92 shown in the fission equation are both radioactive and decay to other radioactive species. This presents a problem for radioactive-waste disposal; the products as well as the reactants are radioactive. The leakage of radiation following the accident at the Chernobyl nuclear power plant in the USSR dramatizes the need for safety precautions in harnessing nuclear energy.

Much more energy can be generated without the production of radioactive waste through nuclear **fusion,** which is the combination of two small nuclei into one larger nucleus. For example, the energy of the sun comes about through a series of nuclear fusion reactions, the overall result of which is

$$4\,{}^{1}_{1}\text{H} \longrightarrow {}^{4}_{2}\text{He} + 2\,{}^{0}_{1}\beta + \text{energy}$$

where ${}^{0}_{1}\beta$, a positively charged particle with the mass of an electron, is called a **positron.**

Fusion is preferable to fission as an energy source because (1) fusion reactions produce more energy per amount of starting material than do fission reactions, (2) the starting materials for fusion are more abundant than fissionable isotopes, and (3) the products of fusion are safe, i.e., not radioactive. However, there are extreme technological problems to be overcome before nuclear fusion can be a practical means of energy production. Most notably, fusion reactions require extremely high temperatures (1 to 5 million °C) for initiation, and no known substance can withstand these temperatures and act as a container for the reaction. However, research continues, and it is hoped that fusion will meet the energy needs of the twenty-first century.

SUMMARY

Whereas ordinary chemical reactions involve changes in electron structure with nuclei remaining unaltered, nuclear reactions involve changes in the nucleus, i.e., changes in the numbers of protons and neutrons (Section 17.1). To understand nuclear reactions it is necessary to review the meaning of the symbols used to indicate nuclear content (Section 17.2).

Radioactivity is the spontaneous emission of energy which accompanies nuclear reactions. There are three types of radiation: alpha (α), beta (β), and gamma (γ) (Section 17.3). Alpha-particles are helium nuclei, ${}^{4}_{2}\text{He}^{2+}$; β-particles are electrons, ${}^{0}_{-1}e$; and γ-radiation is pure energy, ${}^{0}_{0}\gamma$ (Section 17.4).

When nuclear radiation interacts with matter, ions are produced and bonds are broken. The more penetrating γ-radiation produces fewer ions than the less penetrating α- and β-radiations (Section 17.5).

Detection devices for radiation are frequently designed around the fact that radioactivity is ionizing radiation (Section 17.6). Ionization within body cells is the mechanism by which radiation damage occurs. There are a range of symptoms of exposure to excess radiation (Section 17.7).

During α or β emission, transmutation occurs; i.e., one element is changed into another. Alpha emission produces daughter nuclei with atomic numbers 2 less and mass numbers 4 less than the parent nuclei. Beta emission produces daughter nuclei with the same mass, but with atomic numbers 1 greater than the parent nuclei. Gamma-radiation usually accompanies α- and β-radiation (Section 17.8). Natural radioisotopes emit α- and β-particles and thereby decay in a series of reactions until a stable, nonradioactive isotope is formed (Section 17.9).

The rate of radioactive decay is usually measured in terms of the half-life of an element, the time required for the decay of one half of a radioactive sample. The half-life concept has a practical application in the dating of ancient objects (Section 17.10). There are a large number of other applications of radioisotopes in medicine, chemistry, and industry (Section 17.11).

Nuclear reactions can be induced by the bombardment of a nucleus with nuclear-sized "bullets" such as fast-moving neutrons or α-particles. The non-naturally-occurring transuranium elements have been made in this way (Section 17.12). Nuclear bombardment can also lead to nuclear fission, the splitting of large nuclei into smaller nuclei. A great deal of energy is released upon fission, and this energy can be harnessed to produce electricity. Nuclear fusion, the combination of smaller nuclei into larger ones, produces even more energy per quantity of reactant than fission. Hopefully this energy will be harnessed by the beginning of the twenty-first century (Section 17.13).

CHAPTER ACCOMPLISHMENTS

After completing this chapter you should be able to

17.1 Introduction
1. Distinguish between "ordinary" chemical reactions and nuclear reactions.

17.2 The nucleus revisited
2. Given a periodic table and the symbol for a nucleus which includes the mass number, indicate the number of protons and neutrons in the nucleus.

17.3 Radioactivity
3. Define radioactivity.

4. Name the three types of nuclear radiation.
5. Define radioisotope.

17.4 Properties of α-, β-, and γ-radiation

6. Write symbols which indicate the composition of α-, β-, and γ-radiation.
7. State the charge and relative mass of each of the three types of radiation.
8. State the relative penetrating powers of α-, β-, and γ-radiation.

17.5 Ionizing radiation

9. State the effect on matter of interaction with α-, β-, and γ-, or x-radiation.
10. State the relative ionizing abilities of α-, β-, and γ-radiation.

17.6 Detection devices

11. Name three ways to detect radiation.
12. State the property of radiation on which the design of the cloud chamber and Geiger counter are based.

17.7 Physiological effects of radiation

13. Describe some of the physiological effects of exposure to nuclear radiation.

17.8 Nuclear reactions

14. Define "transmutation."
15. Write a nuclear equation representing the emission of an α-particle from some specified radioisotope.
16. Write a nuclear equation representing the emission of a β-particle from some specified radioisotope.

17.9 Radioactive decay series

17. Explain why radioisotopes decay in a series of steps.
18. State the event that terminates a radioactive decay series.

17.10 Half-life $t_{\frac{1}{2}}$

19. Define "half-life."
20. Given the amount of a radioactive sample and the half-life of the radioisotope, calculate the amount of sample remaining after some specified time interval.
21. Explain how carbon-14 dating works.

17.11 Uses of radioisotopes

22. State at least one each of the medical, chemical, and industrial uses of radioisotopes.

17.12 New elements through bombardment

23. Complete nuclear equations representing bombardment reactions.
24. Define "transuranium" elements.

17.13 Nuclear energy

25. Define "fission."
26. Define "critical mass."
27. State one disadvantage of the use of nuclear fission in electric power plants.
28. Define "fusion."
29. State three advantages of nuclear fusion over fission as an energy source.
30. State the technological problem that prohibits the practical use of fusion today.

PROBLEMS

17.1 Introduction

17.7 What are the differences between "ordinary" chemical reactions and nuclear reactions in terms of:

- **a.** Subatomic particles participating
- **b.** Amount of energy change
- **c.** Effect of the state of chemical combination of an element

17.2 The nucleus revisited

17.8 Tell the number of protons and the number of neutrons in each of the following:

- **a.** ^{241}Am
- **b.** ^{63}Cu
- **c.** ^{3}H
- **d.** ^{40}Ar

17.9 Write nuclear symbols corresponding to

- **a.** Magnesium-25
- **b.** Gold-197
- **c.** Iodine-131
- **d.** Lead-210

17.10 Write nuclear symbols for two isotopes of carbon-13.

17.3 Radioactivity

17.11 What is radioactivity?

17.12 How many types of nuclear radiation are there? Name them.

17.13 Describe the difference between the terms "radioisotope" and "isotope."

17.4 Properties of α-, β-, and γ-radiation

17.14 Describe the mass and charge characteristics of α-, β-, and γ-radiation.

17.15 Write the proper nuclear symbols for α- and β-particles.

17.16 If only a heavy-cloth curtain stood between you and a radioactive source, would you be endangered more by α- or γ-radiation? Explain.

17.17 Describe the type of shielding necessary for protection against:

- **a.** α-particles
- **b.** β-particles
- **c.** γ-rays
- **d.** x-rays

17.5 Ionizing radiation

17.18 What is the effect of nuclear radiation on matter?

17.19 **a.** Which type of radiation is most penetrating?
b. Which type of radiation produces the greatest amount of ionization?

c. Which radiation measurement unit gives an amount of radiation weighted for different types of radiation?

17.6 Detection devices

17.20 Explain how radiation "tracks" form in a cloud chamber.

17.21 Explain how radiation causes a Geiger counter to click.

17.22 What actually exposes the film badges worn by workers who work with radioactive materials?

17.7 Physiological effects of radiation

17.23 What part of a living cell is most susceptible to radiation damage?

17.24 What is the first clinical symptom of exposure to higher levels of radiation?

17.25 Besides immediate clinical symptoms, what are the possible long-term effects of radiation exposure?

17.8 Nuclear reactions

17.26 Give an example of transmutation.

17.27 Why does α-emission lead to transmutation?

17.28 Why does β-emission lead to transmutation?

17.29 Write nuclear equations for α-emission from
 a. $^{222}_{86}Rn$
 b. $^{227}_{89}Ac$
 c. $^{235}_{92}U$

17.30 Write nuclear equations for β-emission from
 a. $^{40}_{19}K$
 b. $^{131}_{53}I$
 c. $^{14}_{6}C$

17.31 Based on the following partial equations, decide whether the emitting isotopes are α- or β-emitters:
 a. $^{3}_{1}H \rightarrow {}^{3}_{2}He + ?$ **c.** $^{211}_{83}Bi \rightarrow {}^{207}_{81}Tl + ?$
 b. $^{218}_{84}Po \rightarrow {}^{214}_{82}Pb + ?$ **d.** $^{35}_{16}S \rightarrow {}^{35}_{17}Cl + ?$

17.32 Identify X in each of the following nuclear reactions:
 a. $^{15}_{8}O \rightarrow {}^{15}_{7}N + X$
 b. $^{219}_{86}Rn \rightarrow {}^{4}_{2}He + X$
 c. $X \rightarrow {}^{41}_{20}Ca + {}_{-1}^{0}\beta$

17.33 Why is the emission of γ-radiation often ignored in writing nuclear equations?

17.9 Radioactive decay series

17.34 Fill in the blanks in the following partial radioactive decay series:

$$^{224}_{88}Ra \rightarrow {}^{4}_{2}He + \underline{\quad} \rightarrow {}^{4}_{2}He + \underline{\quad} \rightarrow$$

$$^{4}_{2}He + {}^{212}_{82}Pb \rightarrow {}_{-1}^{0}e + \underline{\quad} \rightarrow$$

$$\underline{\quad} + {}_{-1}^{0}e \rightarrow {}^{208}_{82}Pb + {}^{4}_{2}He$$

17.35 The series shown in Problem 17.34 terminates with the formation of $^{208}_{82}Pb$. Would you conclude that this isotope is radioactive? Explain.

17.10 Half-life, $t_{\frac{1}{2}}$

17.36 Compare sodium-24 and carbon-14 in Table 17.4. Why is sodium-24 a better choice for use in blood-circulation studies than carbon-14?

17.37 In an experiment, 1.6 g of ^{90}Sr ($t_{\frac{1}{2}}$ = 28 years) was buried in 1952. How much of this sample will be left in 2036?

17.38 A nuclear chemist needs 12 mg of $^{68}_{29}Cu$ ($t_{\frac{1}{2}}$ = 32 s) to do a particular experiment. At 10 A.M. a colleague brings him 750 mg of ^{68}Cu which he has just synthesized. The chemist is distracted and does not begin to work until 10:10 A.M. Is there still sufficient ^{68}Cu for the experiment?

17.39 A hospital has a 24-g supply of ^{131}I($t_{\frac{1}{2}}$ = 8 days) on January 1. Even if none is used by the staff, by what date will the sample have diminished to only 3 g?

17.40 Two ancient scraps of paper are found. Scrap A has a carbon-14–carbon-12 ratio twice as large as that for scrap B. Which piece of paper is older?

17.41 The carbon-14–carbon-12 ratio in a deeply buried fossil is only 0.125 times that in a living plant. How old is the fossil?

17.42 Phosphorus-32 ($t_{\frac{1}{2}}$ = 14.3 days) can be used to locate brain tumors because when injected into the body it is preferentially absorbed by diseased brain tissue. Technetium-99 ($t_{\frac{1}{2}}$ = 6 h) works just as well for this purpose. Why is the newer isotope technetium-99 used more prevalently?

17.43 The half-life of a particular radioisotope is 50 years.

    a. Will all the radioisotope be gone in 100 years?

    b. If your answer to part (a) is "no," what fraction of the original sample will be left after 100 years?

17.44 The element selenium occurs in crops in certain regions and is toxic to animals. The radioisotope selenium-75 can be used to label plants, and the uptake of selenium by animals can then be measured readily. The half-life of selenium-75 is 120 days. If an animal ingested 0.1 mg on January 1, 1985, in what year will there be less than 1 μg (1×10^{-6} g) in the animal (assuming that none is excreted)?

17.11 Uses of radioisotopes

17.45 Describe one way in which radioisotopes are used in a medical diagnostic procedure.

17.46 Describe one way in which radioisotopes are used in a medical treatment.

17.47 What is the function of radiation in the treatment of cancer?

17.48 How might ^3HCl (i.e., HCl in which the hydrogen is the radioactive isotope tritium ^3_1H) be used to demonstrate the equilibria in water:

$$HCl + H_2O \rightleftharpoons H_3O^+ + Cl^-$$

$$H_3O^+ \rightleftharpoons H^+ + H_2O$$

17.12 New elements through bombardment

17.49 What is the nuclear "bullet" in each of the following bombardment reactions?

    a. $^{242}_{96}Cm + ? \rightarrow ^{245}_{98}Cf + ^1_0n$

    b. $^6_3Li + ? \rightarrow ^4_2He + ^3_2He$

    c. $^{27}_{13}Al + ? \rightarrow ^{28}_{14}Si$

    d. $^{113}_{48}Cd + ? \rightarrow ^0_0\gamma + ^{114}_{48}Cd$

17.50 Complete the following bombardment reactions:

    a. $^{197}_{99}Au + ^1_1H \rightarrow ? + ^1_0n$

    b. $^{253}_{99}Es + ^4_2He \rightarrow ^1_0n + ?$

    c. $^{27}_{13}Al + ^1_0n \rightarrow ^4_2He + ?$

    d. $^{12}_6C + ^2_1H \rightarrow ^{13}_7N + ?$

17.51 Complete the following nuclear equations:

    a. $^{90}_{37}Rb \rightarrow ^{90}_{38}Sr + ?$

    b. $^{98}_{42}Mo + ^2_1H \rightarrow ^{99}_{43}Tc + ?$

    c. $^{216}_{85}At \rightarrow ^{212}_{83}Bi + ?$

    d. $^{39}_{19}K + ? \rightarrow ^{36}_{18}Ar + ^4_2He$

    e. $^{23}_{11}Na + ^2_1H \rightarrow ? + ^4_2He$

    f. $^3_1H \rightarrow _{-1}^0e + ?$

    g. $^{228}_{90}Th \rightarrow ^4_2He + ?$

17.52 Which of the equations in Problem 17.51 represent the following?

    a. α-emissions

    b. β-emissions

    c. Bombardment reactions

17.53 Give the symbols of all the transuranium elements shown in your periodic table.

17.13 Nuclear energy

17.54 Identify the following as fission or fusion reactions:

    a. $^{235}_{92}U + ^1_0n \rightarrow ^{94}_{38}Sr + ^{139}_{54}Xe + 3^1_0n$

    b. $^3_1H + ^1_1H \rightarrow ^4_2He$

    c. $^{13}_6C + ^1_0n \rightarrow ^{10}_4Be + ^4_2He$

17.55 Give at least two advantages for the fusion process compared to fission.

17.56 What is one practical problem preventing the commercial use of fusion?

ORGANIC AND BIOLOGICAL MOLECULES
CHAPTER

18

Ar

Argon

39.95

INTRODUCTION

18.1 As you know, your body is made up of chemical compounds. As you sit here reading, a multitude of chemical reactions are taking place in your body. The branches of chemistry that deal with body chemistry are *organic chemistry* and *biochemistry*.

Prior to about 1830, scientists believed that living matter was distinctly different from nonliving matter. They classified compounds originating from living organisms as **organic compounds** and compounds from nonliving sources as **inorganic compounds.** It was also observed that all organic compounds contained the element carbon.

Chemists in the early part of the nineteenth century believed that organic compounds could be formed only from living organisms because such organisms possessed a "vital force" that was necessary for the synthesis of organic chemicals. However, in 1828 Friedrich Wöhler was able to obtain the organic compound urea by simply heating the inorganic compound ammonium cyanate:

$$NH_4^+ \ {}^-N{=}C{=}O \xrightarrow{\Delta} \underset{\substack{H_2N \qquad NH_2}}{\overset{\overset{\textstyle O}{\|}}{C}}$$

Ammonium cyanate   Urea
Inorganic   **Organic**

Despite reluctance scientists were forced to give up the Vital Force Theory, because it had been proven wrong by Wöhler's experiment in which an organic compound was made from a nonliving source.

The classification of compounds as organic or inorganic was maintained and is based on the observation that most compounds that contain the element carbon are very different in properties from those which do not. **Organic chemistry** is the study of carbon compounds.

Today the field of organic chemistry encompasses all the compounds of carbon, except carbonates, metallic carbides, cyanides, and cyanates.

In contrast to only a few hundred thousand known inorganic compounds, there are well over 3 million known organic compounds. Many occur naturally, but a large number have been prepared synthetically in the laboratory. Students often ask why there are so many organic compounds. The answer can be found in the nature of the chemical bonds that carbon forms.

BONDING IN CARBON COMPOUNDS

18.2 The predominant type of bonding between carbon atoms and between carbon and other elements in organic compounds is covalent or polar covalent (Section 6.16). Carbon is unique among the elements in its ability to form strong bonds *to itself* as well as to other elements such as hydrogen and oxygen.

The fact that carbon-carbon bonds are strong means that chains of carbon atoms can form and will be stable. The chains can be of varying length, leading to literally an infinite number of compounds differing in chain length.

Figure 18.1 shows carbon compounds with three carbons and with six carbons and a segment of polyethylene. The common plastic polyethylene is constructed of huge molecules in which thousands of carbon atoms are joined in a chain. In addition, carbon atoms can also be joined together to form rings. Furthermore, both chains and rings can have "branches" of one or more carbons.

Thus, because of the chain- and ring-forming abilities of carbon, we see that a great variety of organic compounds arise just through different arrangements of different numbers of carbon and hydrogen atoms. However, carbon can also bond to other elements. Most notably, in organic compounds we find the elements oxygen, nitrogen, phosphorus, and sulfur in addition to carbon and hydrogen. Each different structural arrangement of the atoms, which can be represented by a structural formula (Lewis structure), corresponds to a different compound with a unique set of physical and chemical properties. All the foregoing leads to a very large number of possible compounds with a wide spectrum of properties and reactivities.

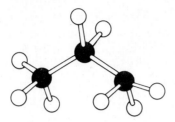

Three-carbon chain (propane)

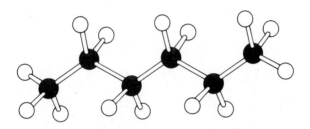

Six-carbon chain (hexane)

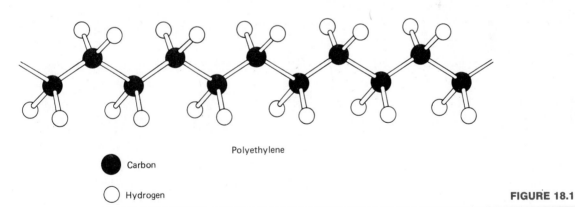

Polyethylene

● Carbon

○ Hydrogen

FIGURE 18.1

Carbon chains can be of any length, from very short to literally infinite. We can show only a short segment of polyethylene because it is a chain of thousands of carbon atoms. The zigzag relationship of adjacent carbon-carbon bonds arises from the tetrahedral geometry around each carbon.

STRUCTURAL FORMULAS FOR ORGANIC MOLECULES

18.3 Organic molecules are almost always represented by structural formulas (Lewis structures) rather than molecular formulas. A structural formula shows the kinds and numbers of atoms in a molecule and the way in which they are bonded together. In Section 6.13 you learned how to write structural formulas for *any* molecular compound. For organic molecules there are some general bonding patterns, the recognition of which offers a simpler approach to writing organic structural formulas than the more general guidelines of Section 6.13. In organic compounds carbon forms four bonds:

| Four single bonds to C | One double bond and two single bonds to each C | One triple bond and one single bond to each C |

Hydrogen forms only one bond, as shown in all examples. Oxygen forms two bonds:

| Two single bonds to O | One double bond to O |

Nitrogen forms three bonds:

| Three single bonds to N | One double bond and one single bond to N | One triple bond to N |

Table 18.1 summarizes these bonding patterns. In writing structural formulas, satisfy the bonding requirement of each atom as shown in Table 18.1 and then fill in nonbonding electrons so that each atom has an octet (H requires only a duet). The total number of electrons shown must equal the number of valence electrons. Sample Exercises 18.1 to 18.3 demonstrate this method of writing structural formulas.

SAMPLE EXERCISE

18.1

Write a structural formula for the molecular formula CH_4O.

SOLUTION

Begin with the atom that forms the largest number of bonds, namely, carbon:

TYPICAL COVALENT BONDING PATTERNS FOR THE COMMON ELEMENTS **TABLE 18.1**

Lewis Dot Structure	Periodic A Group	Number of Bonds Usually Formed*	Bonding Patterns
$\cdot \overset{\cdot}{\underset{\cdot}{C}} \cdot$	IV	4	$-\overset{\vert}{\underset{\vert}{C}}-$ four single bonds
			$-\underset{\vert}{C}=$ two single, one double bond
			$-C\equiv$ one single, one triple bond
H·	I	1	—H one single bond
$:\overset{\cdot\cdot}{\underset{\cdot\cdot}{X}}\cdot^{\dagger}$	VII	1	$:\overset{\cdot\cdot}{\underset{\cdot\cdot}{X}}-$ one single bond
$:\overset{\cdot\cdot}{O}:^{\ddagger}$	VI	2	$-\overset{\cdot\cdot}{\underset{\cdot\cdot}{O}}-$ two single bonds
			$:\overset{\cdot\cdot}{O}=$ one double bond
$\cdot\overset{\cdot\cdot}{N}\cdot^{\S}$	V	3	$-\overset{\vert}{\underset{\cdot\cdot}{N}}-$ three single bonds
			$-\underset{\cdot\cdot}{N}=$ one single, one double bond
			$:N\equiv$ one triple bond

* These patterns do not take into account the possibility of coordinate covalent bonding. Hence, for example, N usually forms three bonds, but occasionally we see it forming four bonds, as in the case of ammonium, NH_4^+. This is discussed in Section 6.14.
† X represents the halogens chlorine, fluorine, bromine, and iodine.
‡ Sulfur is in the same periodic group as oxygen; therefore it follows the same bonding patterns.
§ Phosphorus theoretically can use the same patterns as nitrogen. In fact, P most often forms three single bonds or it is pentavalent (i.e., forms five bonds), as in the phosphates.

$$-\overset{\vert}{\underset{\vert}{C}}-$$

Next, consider oxygen which forms two bonds. "Attach" it (i.e., bond it) to carbon and see how many more bonds must be formed.

$$-\overset{\vert}{\underset{\vert}{C}}-\overset{\cdot\cdot}{\underset{\cdot\cdot}{O}}-$$

Carbon requires three more bonds and oxygen one more. There are four hydrogens, each of which forms one bond to complete the structure:

Check that each atom (except H) has an octet. If we had neglected to include the lone pairs on oxygen, this oversight would be revealed in the check.

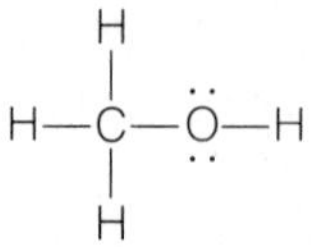

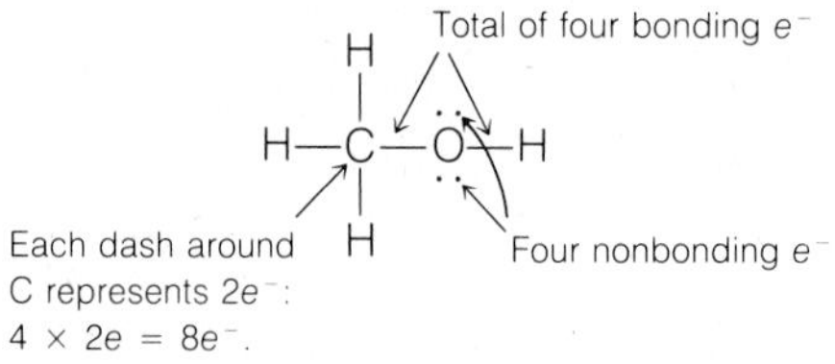

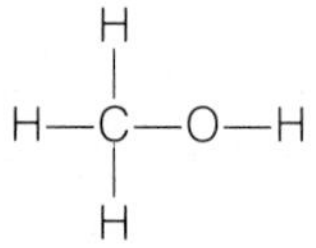

Incorrect; oxygen lacks an octet.

SAMPLE EXERCISE 18.2

Construct a structural formula for formaldehyde (CH_2O), a preservative for biological specimens.

SOLUTION

Begin as in Sample Exercise 18.1 and arrive at

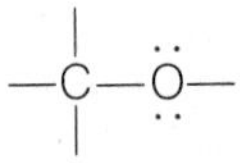

In this case we do not have four atoms to complete the structure as before. We have only two H atoms. This tells us to consider a bonding pattern with multiple bonds; i.e., carbon and oxygen can be double-bonded:

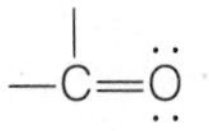

Now we see that only two hydrogens are required to complete the structure:

Check that each atom (except H) has an octet.

SAMPLE EXERCISE 18.3

What is the structural formula for propane, C_3H_8?

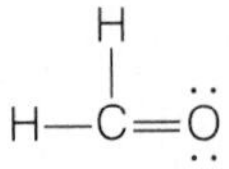

SOLUTION

For compounds with many carbon atoms, always begin by linking the given numbers of carbons together:

$$-\overset{\displaystyle |}{\underset{\displaystyle |}{C}}-\overset{\displaystyle |}{\underset{\displaystyle |}{C}}-\overset{\displaystyle |}{\underset{\displaystyle |}{C}}-$$

Then count the number of bonding slots to be filled. In this case, there are eight slots and eight hydrogens to fill them:

$$\begin{array}{ccc} H & H & H \\ | & | & | \\ H-C-&C-&C-H \\ | & | & | \\ H & H & H \end{array}$$

An insufficient number of atoms to fill the slots, would indicate that a multiple-bonding pattern was necessary.

Problem 18.1

Use Table 18.1 to write a structural formula for CH_5N.

CONDENSED STRUCTURAL FORMULAS

18.4 So far we have used only *full* structural formulas which show individual symbols for each atom and all connecting bonds. For example, in the case of propane, C_3H_8, in Sample Exercise 18.3, we came to the structure

$$\begin{array}{ccc} H & H & H \\ | & | & | \\ H-C-&C-&C-H \\ | & | & | \\ H & H & H \end{array}$$

The three carbons, eight hydrogens, and all bonds are shown.

To save space and time of writing, *condensed* structural formulas are more often employed. In these formulas most carbons are individually represented, but hydrogens on the same carbon are added up, and the bonding dashes are not shown. For example, the condensed structural formula for propane is $CH_3CH_2CH_3$. Note carefully in the full structural formula that the three carbons in propane are bonded in a chain with no intervening hydrogens. It is merely the convention employed in writing condensed structures to list atoms bonded to each carbon after each carbon.

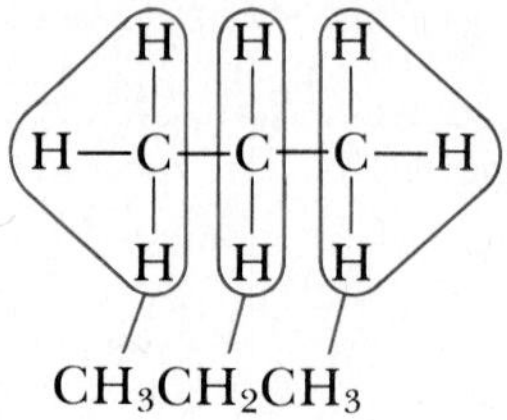

$$CH_3CH_2CH_3$$

Branches off of a main carbon chain are handled by parentheses in condensed structural formulas:

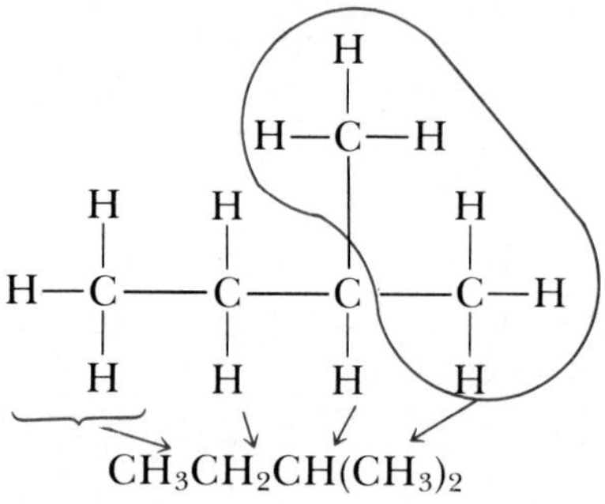

$$CH_3CH_2CH(CH_3)_2$$

The condensed structure $CH_3CH_2C(CH_3)_3$ has the full structural formula

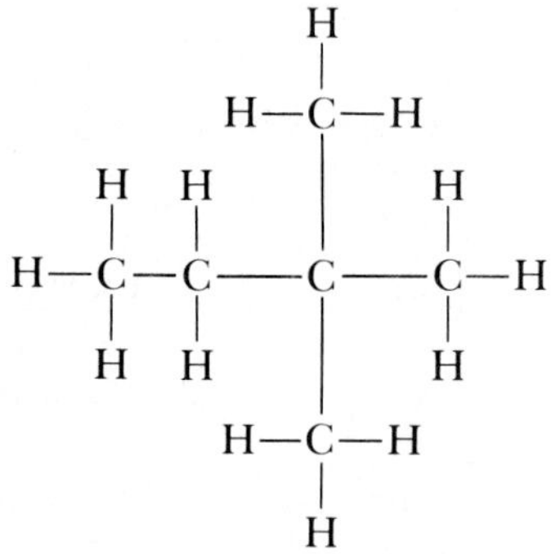

Problem 18.2

Write full structural formulas from the following condensed structures:

a. CH_2ClCH_3

b. $CH_3CH_2CH(CH_3)CH_2CH_3$

c. CH_3CH_2OH

Compounds with rings, i.e., **cyclic** compounds, are represented by regular polygons. Each apex of the polygon represents a carbon atom. Although H's are not written, it is understood that a number of H's are attached to each carbon atom such that each carbon forms four bonds.

For example, compare the full and condensed formulas for rings:

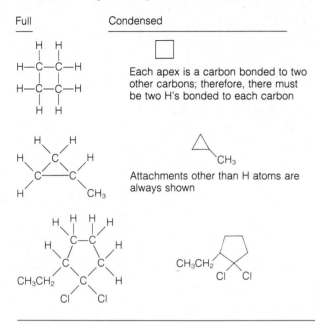

Full   Condensed

Each apex is a carbon bonded to two
other carbons; therefore, there must
be two H's bonded to each carbon

Attachments other than H atoms are
always shown

FUNCTIONAL-GROUP CONCEPT

18.5 Because of the large number of organic compounds, a classification system for organizing them is needed. Organic compounds are classified according to the presence of so-called functional groups in the compound. A **functional group** is a specific group of atoms, such as the —OH group or the —NH$_2$ group, that gives a compound a particular set of physical and chemical properties. Table 18.2 lists the common functional groups.

For example, a compound with an —OH group is an alcohol. An alcohol with one carbon (methanol) and an alcohol with two carbons (ethanol) have very similar chemistry. The functional group, and not the number of carbons and hydrogens, is usually the key factor in determining the chemical properties of an organic compound.

$$
\begin{array}{cc}
\quad\ \text{H} & \quad\ \text{H}\quad\text{H} \\
\quad\ | & \quad\ |\quad\ | \\
\text{H}\!-\!\text{C}\!-\!\text{OH} & \text{H}\!-\!\text{C}\!-\!\text{C}\!-\!\text{OH} \\
\quad\ | & \quad\ |\quad\ | \\
\quad\ \text{H} & \quad\ \text{H}\quad\text{H}
\end{array}
$$

One-carbon alcohol   Two-carbon alcohol

TABLE 18.2 COMMON FUNCTIONAL GROUPS

Structural Feature*	Name of Class
	Alkane
	Alkene
—C≡C—	Alkyne
or †	Aromatic
—C—OH	Alcohol
—C—NH₂	Amine
—C—O—C—	Ether
—C—X	Alkyl halide
C=O (—CHO)	Aldehyde
—C—C—C—	Ketone
—C—OH (—COOH)	Carboxylic acid
—C=O (—CONH₂) NH₂	Amide
—C—O—C— (—COOC—)	Ester
—C=O X	Acyl halide
—C—O—C—	Acid anhydride

* Where other atoms are not shown on the bonds to carbon, hydrogens or carbons are normally found bonded at these positions. Forms in parentheses are the common abbreviations of these structural features. Halogens (F, Cl, Br, I) are represented by X.
† The meaning of this representation will be clarified in Section 18.8.

The carbons and hydrogens act as a backbone or skeleton on which the functional groups are placed. In order to make predictions concerning the properties and reactions of specified organic compounds, it is necessary to be able to identify the functional group(s) present in the compound.

SAMPLE EXERCISE

18.4

Indicate the functional groups in the following structures:

a. 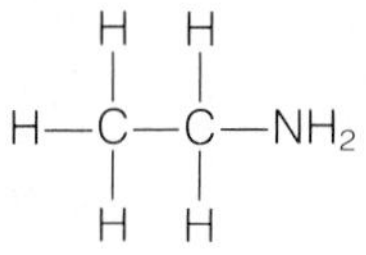
$$
\begin{array}{ccc}
& \overset{\displaystyle H}{\underset{\displaystyle H}{|}} & \overset{\displaystyle H}{\underset{\displaystyle H}{|}} \\
H{-}C{-}C{-}NH_2 \\
\end{array}
$$

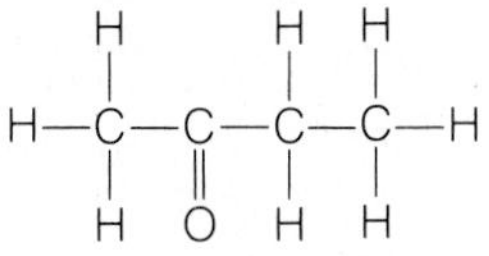

b. 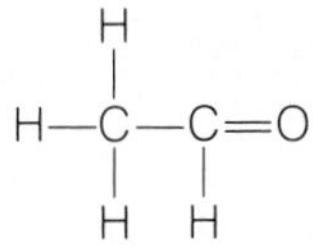
$$
\begin{array}{cc}
\overset{\displaystyle H}{|} & \\
H{-}C{-}C{=}O \\
\underset{\displaystyle H}{|}\ \underset{\displaystyle H}{|}
\end{array}
$$

c.
$$
H{-}\overset{H}{\underset{H}{C}}{-}\overset{}{\underset{O}{C}}{-}\overset{H}{\underset{H}{C}}{-}\overset{H}{\underset{H}{C}}{-}H
$$

d. 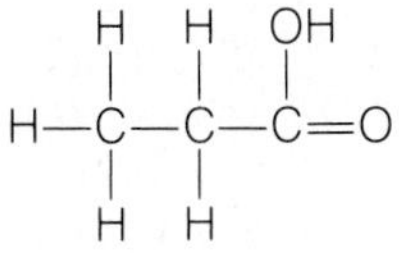
$$
H{-}\overset{H}{\underset{H}{C}}{-}\overset{H}{\underset{H}{C}}{-}\overset{OH}{C}{=}O
$$

SOLUTION

Refer to Table 18.2 to identify the functional groups.

a. The presence of the —NH_2 group indicates an **amine.**
b. The presence of the —C=O group indicates an **aldehyde.**
$$\underset{\displaystyle H}{|}$$

c. The C—C—C indicates a **ketone.** Notice the difference between this group
$$\underset{\displaystyle O}{\|}$$
and the aldehyde group, which has an H attached directly to the C that is double-bonded to O.

d. —C=O indicates the **carboxylic acid** group. Do not confuse this group
$$\underset{\displaystyle OH}{|}$$
with the alcohol functional group. Alcohols have —O—H and no C=O; acids have —OH attached to C=O. Also compare amines with amides, ethers with esters, and alkyl with acyl halides.

ISOMERISM

18.6 Throughout the foregoing we have been using **structural** formulas to represent organic molecules and will continue to do so. It is essential that we use full or condensed structural formulas in organic chemistry rather than **molecular** formulas, which simply give the numbers of each type of atom in the compound. This is so because often more than one structure can be written for a particular molecular formula. For example, consider the molecular formula C_2H_6O. Two correct Lewis structures can be written for C_2H_6O:

$$
\begin{array}{cc}
\text{H} \quad\;\; \text{H} & \text{H} \;\; \text{H} \\
\;| \qquad | & | \;\; | \\
\text{H—C—}\overset{..}{\underset{..}{\text{O}}}\text{—C—H} & \text{H—C—C—}\overset{..}{\underset{..}{\text{O}}}\text{H} \\
\;| \qquad | & | \;\; | \\
\text{H} \quad\;\; \text{H} & \text{H} \;\; \text{H}
\end{array}
$$

Structure 1   Structure 2

The fact that two structural formulas can be written for C_2H_6O tells us that there are two distinctly different compounds with the same molecular formula. Compounds which have the same molecular formula but different structural formulas are called **isomers.**

Isomers are completely different compounds, each with its own unique set of properties. In the example just given, compound 1 is an ether and has the characteristic properties of an ether, whereas compound 2 is an alcohol with the properties characteristic of that functional group. Compound 1 is a gas at room temperature (bp = $-23°C$), whereas compound 2 is a liquid which boils at 78°C.

Given a molecular formula, one can write corresponding structural formulas by using the skills developed in Section 6.13 or 18.3.

Problem 18.3

Write structural formulas for isomers with molecular formula $C_2H_4Br_2$.

Often it is necessary to be able to recognize whether two given structures are related as isomers. This can be done according to the following guidelines.

Guidelines for recognizing structures as identical, isomeric, or unrelated

1. Determine the molecular formula of the given structures. Isomers (and identical compounds) must have the same molecular formula. If the molecular formulas of the given structures are *not* the same, the two structures are neither identical nor isomers. If the molecular formulas are the same, proceed to 2.

2. Examine the bonding arrangement of the atoms in the structure. Isomers will have their atoms arranged differently, whereas identical compounds will exhibit the same bonding arrangement.

SAMPLE EXERCISE

18.5

For each of the following pairs of structures, decide whether the two structures are identical, isomeric, or unrelated:

a.
$$
\begin{array}{cc}
\text{H H} & \text{H H} \\
\text{HO—C—C—H} & \text{and} \quad \text{H—C—C—OH} \\
\text{H H} & \text{H H}
\end{array}
$$

b.
$$
\begin{array}{cc}
\text{H} \quad\quad \text{H} & \text{H H} \\
\text{H—C—C—C—H} & \text{and} \quad \text{H—C—C—C}{=}\text{O} \\
\text{H O H} & \text{H H H}
\end{array}
$$

c.
$$
\begin{array}{cc}
\text{H} & \text{H H} \\
\text{H—C—H} & \text{and} \quad \text{H—C—C—H} \\
\text{H} & \text{H H}
\end{array}
$$

SOLUTION

a. *Identical.* The two structures have the same molecular formula, C_2H_6O. Although one is written as the reverse of the other, a careful examination will show that the atoms are bonded together in exactly the same arrangement in both structures.

b. *Isomers.* The two structures have the same molecular formula, C_3H_6O. However, an examination of the bonding arrangement will show that the structural formulas are different. In this case, recognition of one compound as a ketone and the other as an aldehyde emphasizes the difference in structure.

c. *Unrelated.* The molecular formula for the compound on the left is CH_4. The one on the right has the formula C_2H_6. Therefore, these are neither identical nor isomeric.

GEOMETRY AROUND CARBON ATOMS

18.7 The spatial arrangement around a particular carbon atom follows directly from its Lewis structure and VSEPR theory (see Section 12.2). The geometry, or arrangement, of bonded atoms around a particular carbon atom can readily be determined by the carbon's bonding pattern. Four single bonds are arranged tetrahedrally; when a double bond is present, the array is planar triangular; the presence of a triple bond produces a linear array. Figure 18.2 shows these arrangements, and they can be summarized as follows:

Bonding Pattern		Geometry
Four singles	$-\overset{\displaystyle\vert}{\underset{\displaystyle\diagup}{C}}\diagdown$	Tetrahedral
One double and two singles	$\overset{\diagdown}{\underset{\diagup}{C}}{=}$	Planar triangular
One triple and one single	$-C\equiv$	Linear

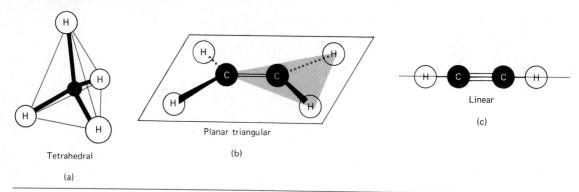

Tetrahedral

(a)

Planar triangular

(b)

Linear

(c)

FIGURE 18.2

Geometry around carbon atoms. (a) Four single bonds to carbon are arranged in three dimensions so that they point to the corners of a tetrahedron; CH_4 offers a simple example. (b) When a carbon is involved in a double bond, the carbon and all of the atoms bonded to it lie in the same plane. The full description of the geometry is planar triangular because the atoms bonded to the doubly bonded carbon can be conceived to form the apexes of a triangle. (c) When a carbon is involved in a triple bond, the carbon and all the atoms bonded to it lie along a line, a linear array.

The tetrahedral arrangement around carbon is the most commonly encountered geometry because carbon atoms that form four single bonds are most prevalent. In a connecting chain of all single-bonded carbons, the tetrahedral shape about each carbon leads to a zigzag array for the chain, as you have already seen in Figure 18.1. A chain of four carbons (butane) looks like

As we have said, organic chemistry is set off as a separate branch of chemistry. But it is useful to remember that the principles of general chemistry presented in earlier chapters apply to carbon compounds as well. Structural formulas, geometry, functional groups, and isomerism are all consequences of the basic Lewis ideas of proper formulas for molecular compounds.

Physical and chemical property trends among organic compounds follow the same rules concerning polarity and intermolecular forces described previously. Organic compounds are traditionally studied by functional group, and such a presentation begins here.

HYDROCARBONS

18.8 Compounds containing only carbon and hydrogen are called **hydrocarbons.** The main sources of hydrocarbons are natural gas, which is principally methane, and petroleum. Petroleum is refined (distilled into fractions with different boiling points) to give hydrocarbons containing from 5 to 40 carbons. Some of these hydrocarbons are used in commercial products, such as gasoline, jet fuels, and lubricating oils; others are consumed in the chemical industry in the production of more complicated chemicals. The hydrocarbons are subdivided into classes depending on the type of carbon-carbon bonding and whether the particular compound is an open chain or contains a ring of carbons. Figure 18.3 summarizes this classification scheme.

Physical properties of hydrocarbons

Because their elemental composition is the same (C and H), the physical properties of all classes of hydrocarbons are very similar. Consequently, we can discuss trends in physical properties of hydrocarbons in general. The trends apply equally well to alkanes, alkenes, alkynes, and aromatics.

Since all C—C and C—H bonds are nonpolar, the only attractive force in hydrocarbons is the dispersion force (Section 12.4); this attractive force generally increases with increasing molecular weight, leading to an increase in boiling point and melting point in the higher-molecular-weight hydrocarbons.

Hydrocarbons of similar molecular weight exhibit a decrease in boiling point with increased branching on the main carbon chain. This is a particular example of the general concept that for any given molecular weight, a molecule that has the shape of a spaghetti strand will have greater dispersion forces than a molecule which is ball-shaped (see Section 12.4 and particularly Figure 12.5).

Hydrocarbons are not soluble in water. This can be deduced from

FIGURE 18.3

Hydrocarbons are compounds containing only the elements carbon and hydrogen. They are classified according to bond type and the presence or absence of rings. A representative example is shown for each class.

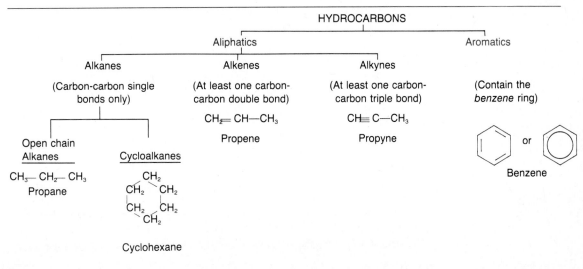

the observation that they float on water and do not diffuse through water. Hydrocarbons dissolve only in nonpolar solvents. This is another example of the general principle "like dissolves like."

Alkanes

Hydrocarbons with only single bonds between the carbons are known as **alkanes.** Alkanes all have a molecular formula C_nH_{2n+2}, where n is the number of carbon atoms in the molecule. Table 18.3 lists the structures and names of the first 10 alkanes.

The alkanes form a **homologous series;** that is, starting with any alkane formula, the next higher member in the series may be formed by adding a CH_2 group. Notice that the fragment $—CH_2—$ is the only difference between any two adjacent members in the series. The chemical properties of any homologous series remain fairly constant throughout the series. This means that if we know the reactions of one alkane, we can project the properties and reactions of all alkanes. The low-molecular-weight alkanes (methane through butane) are gases and are used as fuels for home heating and cooking. Their use as fuels in the combustion reaction with oxygen represents the most important chemical reaction of alkanes.

$$CH_4 + 2O_2 \longrightarrow CO_2 + 2H_2O + heat$$

Alkanes can be used to prepare other organic compounds by substitution reactions. Substitution in the alkanes occurs with replacement of a hydrogen atom by another group such as a halogen:

TABLE 18.3   ROOT NAMES FOR CONTINUOUS CARBON CHAINS AND NAMES OF THE FIRST 10 STRAIGHT-CHAIN ALKANES

Number of Carbons in a Continuous Chain	Root Name	Alkane Structure	Alkane Name
1	Meth-	CH_4	Methane
2	Eth-	CH_3CH_3	Ethane
3	Prop-	$CH_3CH_2CH_3$	Propane
4	But-	$CH_3CH_2CH_2CH_3$	Butane
5	Pent-	$CH_3CH_2CH_2CH_2CH_3$	Pentane
6	Hex-	$CH_3CH_2CH_2CH_2CH_2CH_3$	Hexane
7	Hept-	$CH_3(CH_2)_5CH_3$	Heptane
8	Oct-	$CH_3(CH_2)_6CH_3$	Octane
9	Non-	$CH_3(CH_2)_7CH_3$	Nonane
10	Dec-	$CH_3(CH_2)_8CH_3$	Decane

$$
\underset{\text{Ethane}}{\text{H}-\overset{\overset{\displaystyle H}{|}}{\underset{\underset{\displaystyle H}{|}}{C}}-\overset{\overset{\displaystyle H}{|}}{\underset{\underset{\displaystyle H}{|}}{C}}-\text{H}} + \underset{\text{Chlorine}}{\text{Cl}_2} \xrightarrow[\text{light}]{\text{UV}} \underset{\text{Ethyl chloride}}{\text{H}-\overset{\overset{\displaystyle H}{|}}{\underset{\underset{\displaystyle H}{|}}{C}}-\overset{\overset{\displaystyle H}{|}}{\underset{\underset{\displaystyle H}{|}}{C}}-\text{Cl}} + \underset{\text{Hydrogen chloride}}{\text{HCl}}
$$

The alkanes play no important role in biological systems.

Nomenclature of alkanes   The name of an organic compound must specify a unique structure; that is, from the written name for a compound it should be possible to construct a structural formula. A systematic nomenclature provides a means of constructing names so that there is a unique correspondence between name and structure.

The International Union of Pure and Applied Chemistry (IUPAC) has developed a systematic approach for organic nomenclature known as the **IUPAC system of rules.** In the IUPAC system the root of the name for a given compound is based on the name of the longest continuous chain of carbons in the compound. A suffix is added to the root to indicate the functional-group family. Other connected substituent groups are named as branches on the longest chain. The basic concepts developed for the naming of alkanes are carried through in the nomenclature of all other functional groups.

Before you can proceed further, you will need to learn the root names of the first ten straight carbon chains in Table 18.3. The suffix which denotes the alkane family is *-ane*. Notice that the names in the table are constructed by combining the root name (indicating the number of carbons in the chain) with *-ane*.

In order to name alkanes with branched chains, you must learn the names of some alkyl groups. An **alkyl group** (R group) is the fragment of a molecule that remains if one atom is removed from a carbon. For example, when one of the hydrogens of ethane is removed, the fragment that remains is called the ethyl group.

$$
\underset{\text{Ethane}}{\text{H}-\overset{\overset{\displaystyle H}{|}}{\underset{\underset{\displaystyle H}{|}}{C}}-\overset{\overset{\displaystyle H}{|}}{\underset{\underset{\displaystyle H}{|}}{C}}-\text{H}} \xrightarrow{-\,\text{H}\cdot} \underset{\text{Ethyl}}{\text{H}-\overset{\overset{\displaystyle H}{|}}{\underset{\underset{\displaystyle H}{|}}{C}}-\overset{\overset{\displaystyle H}{|}}{\underset{\underset{\displaystyle H}{|}}{C}}-}
$$

The grouping CH_3CH_2- is not a compound (one carbon has only three bonds) and must be bonded to another grouping or atom. You will find such alkyl groups as substituents on the longest (root) continuous chain of carbons. For example,

$$CH_3—CH_2—CH_2—CH_2—CH_2—CH_3$$

Hexane

$$\begin{array}{c} CH_3 \\ | \\ CH_2 \\ | \\ CH_3—CH_2—CH—CH_2—CH_2—CH_3 \end{array}$$

Hexane with an ethyl substituent or branch

Table 18.4 lists the structure and names of some alkyl groups. Note that the name of an alkyl group is constructed:

$$\text{Root name} + \text{suffix } \textit{-yl}$$

The two-carbon fragment is ethyl; the one-carbon fragment CH_3— is methyl.

The alkane propane yields two different alkyl groups depending on which hydrogen is removed. Removal of a hydrogen from either of the end carbons gives a straight-chain alkyl group, called normal propyl, n-propyl.

$$\begin{array}{ccc} & H\;\;H\;\;H \\ & |\;\;\;\;|\;\;\;\;| \\ H—C—C—C—H & \xrightarrow{-H\cdot} & H—C—C—C— \\ & |\;\;\;\;|\;\;\;\;| \\ & H\;\;H\;\;H \end{array}$$

Removal of any one of the six colored hydrogens gives the same group. n-Propyl

Removal of a hydrogen from the middle carbon gives the isopropyl group.

TABLE 18.4   NAMES OF COMMON ALKYL GROUPS

CH_3—	Methyl
CH_3CH_2—	Ethyl
$CH_3CH_2CH_2$—	n-Propyl
CH_3CHCH_3 |	Isopropyl
$CH_3CH_2CH_2CH_2$—	n-Butyl
$CH_3CH_2CHCH_3$ |	sec-Butyl
$\begin{array}{c} CH_3 \\ \| \\ CH_3CHCH_2— \end{array}$	Isobutyl
$\begin{array}{c} CH_3 \\ \| \\ CH_3CCH_3 \\ \| \end{array}$	t-Butyl

$$\underset{\substack{|\\H}}{\overset{\substack{H\ \ H\ \ H\\|\ \ \ |\ \ \ |}}{H-C-C-C-H}} \xrightarrow{-H\cdot} \underset{\substack{H\ \ H\ \ H}}{\overset{\substack{H\ \ \ \ \ \ H\\|\ \ \ |\ \ \ |}}{H-C-C-C-H}}$$

Removal of one of the two colored   Isopropyl
hydrogens gives the same group.

The basic idea of how root names and alkyl group names are used can be demonstrated by a simple example:

$$\underset{\substack{|\\CH_3-CH-CH_2CH_3}}{CH_3}$$

is named 2-methylbutane. The longest continuous chain of four carbons provides the root name; hence the parent chain is a butane. There is a methyl substituent or branch on the chain. Methyl is connected to the *second* carbon along the butane chain, hence the number 2 to designate its position. As structures become more complex, so must the name, but the basic idea always involves finding a parent chain and recognizing the attached substituents.

Let us lay out a stepwise procedure for naming alkanes and demonstrate the procedure by naming

$$\underset{CH_3-CH-CH-CH-CH_2-CH_2-CH_3}{\overset{CH_3\ \ CH_3\ \ CH_2CH_3}{|\ \ \ \ \ |\ \ \ \ \ |}}$$

Step 1   *Find the longest continuous chain of carbons to establish the parent name.*

Seeing how many carbons you can pencil through without lifting the pencil off the paper is one way of accomplishing this. In this case the longest continuous chain happens to occur straight across the horizontal, but you should be aware that this is not always the case. The longest continuous chain may go "around corners." See, for example, Sample Exercise 18.6.

$$\underset{\substack{1\ \ \ \ \ 2\ \ \ \ \ 3\ \ \ \ \ 4\ \ \ \ \ 5\ \ \ \ \ 6\ \ \ \ \ 7}}{\overset{\substack{CH_3\ \ CH_3\ \ CH_2CH_3\\|\ \ \ \ \ |\ \ \ \ \ |}}{CH_3-CH-CH-CH-CH_2-CH_2-CH_3}}$$

Because the longest chain is seven carbons, this is a heptane.

Step 2   *Identify the substituents on the longest chain.*

In our example the branches are two methyl groups and one ethyl group:

$$\overset{\displaystyle \overset{\textstyle (CH_3)}{|} \; \overset{\textstyle (CH_3)}{|} \; \overset{\textstyle (CH_2CH_3)}{|}}{CH_3\text{—}CH\text{—}CH\text{—}CH\text{—}CH_2\text{—}CH_2\text{—}CH_3}$$

Step 3 Use appropriate prefixes to indicate more than one of a particular substituent.

In the example there are two methyl groups. This can be indicated by saying *dimethyl*. Recall that the prefix *di-* means "two." Table 7.3 lists other numerical prefixes. Only *di-* and *tri-* are used commonly. There is only one ethyl group, so no prefix is needed.

Step 4 Number the chain so that the lowest possible numbers are assigned to the substituent positions.

In the example, number the chain from left to right because that gives the substituent positions the numbers 2, 3, and 4, rather than 4, 5, and 6, which would be assigned if numbering began at the right end of the chain. Thus we have 2,3-dimethyl and 4-ethyl substituents.

$$
\begin{array}{c}
\overbrace{}^{\text{2,3-Dimethyl}} \quad \overbrace{\phantom{CH_2\text{—}CH_3}}^{\text{4-Ethyl}} \\
\overset{CH_3}{|} \quad \overset{CH_3}{|} \quad \overset{CH_2\text{—}CH_3}{|} \\
\underset{1}{CH_3}\text{—}\underset{2}{CH}\text{—}\underset{3}{CH}\text{—}\underset{4}{CH}\text{—}\underset{5}{CH_2}\text{—}\underset{6}{CH_2}\text{—}\underset{7}{CH_3}
\end{array}
$$

Step 5 Use commas between substituent numbers and dashes between numbers and prefixes.

Unlike substituent groups or different branching positions are separated by a dash. The final substituent group and the base name are written together as one word. It is preferred that the substituent names be written in alphabetical order without consideration of numerical prefixes. The assembled name for the sample compound is

$$\underbrace{\text{4-ethyl-}}\underbrace{\text{2,3-dimethyl}}\underbrace{\text{heptane}}$$

| An ethyl on carbon 4 | Two methyl groups on carbons 2 and 3 | A 7-carbon chain |

SAMPLE EXERCISE

18.6

Give the IUPAC name for

$$
\begin{array}{c}
\overset{CH_3}{|} \\
\overset{CH_3}{|} \quad \overset{CH_2}{|} \\
CH_3\text{—}CH\text{—}CH\text{—}CH_3
\end{array}
$$

SOLUTION

The longest chain, as shown, is five carbons, giving the parent name "pentane."

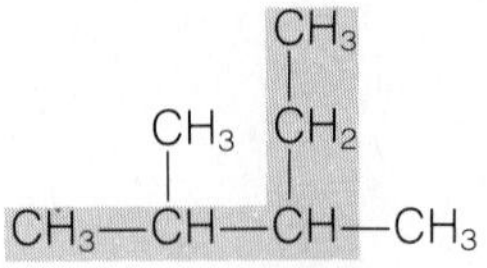

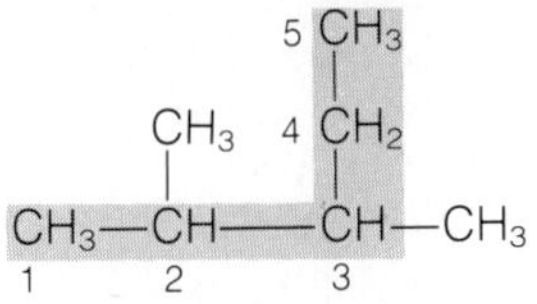

$$
\begin{array}{c}
\qquad\qquad\ \text{CH}_3 \\
\qquad \text{CH}_3\ \ \text{CH}_2 \\
\text{CH}_3\!-\!\text{CH}\!-\!\text{CH}\!-\!\text{CH}_3
\end{array}
$$

$$
\begin{array}{c}
\qquad\qquad 5\ \text{CH}_3 \\
\qquad \text{CH}_3\ \ 4\ \text{CH}_2 \\
\text{CH}_3\!-\!\text{CH}\!-\!\!-\!\text{CH}\!-\!\text{CH}_3 \\
\ \ 1 \qquad 2 \qquad\ \ 3
\end{array}
$$

There are methyl groups branched on the chain in the 2 and 3 positions. The full name is 2,3-dimethylpentane.

Problem 18.4

Give IUPAC names for

a. 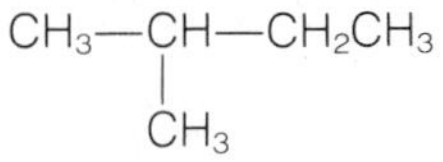 CH$_3$—CH—CH$_2$CH$_3$
         |
        CH$_3$

b. 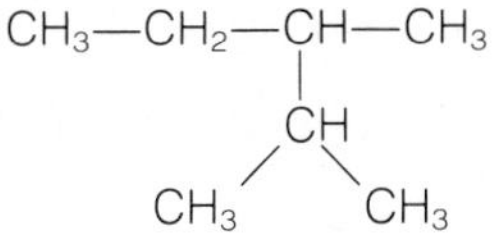 CH$_3$—CH$_2$—CH—CH$_3$
              |
              CH
           /   \
       CH$_3$ CH$_3$

Alkenes

Alkenes are hydrocarbons with at least one carbon-carbon double bond in the structure. The general formula for alkenes with one double bond is C_nH_{2n}. An alkene of a given number of carbons has two less hydrogens than the corresponding alkane with the same number of carbons. The alkene is said to be **unsaturated** because it does not contain as many hydrogens as the number of carbons can accommodate by all single bonds (the **saturated** state). Table 18.5 gives the structures and names of a few straight-chain alkenes.

The systematic naming of alkenes follows closely that of the alkanes, except that the suffix is *-ene* instead of *-ane,* and:

1. The longest chain must contain the double bond.

2. The position of the double bond is indicated by a number before the base name.

Review the names in Table 18.5 and the following example:

TABLE 18.5   STRUCTURE AND NAMES OF THE SMALLEST ALKENES

Molecular Formula	Structure	Name
C_2H_4	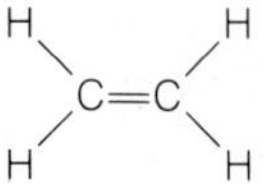	Ethene
C_3H_6	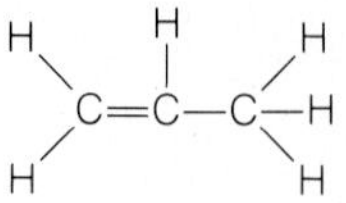	Propene
C_4H_8	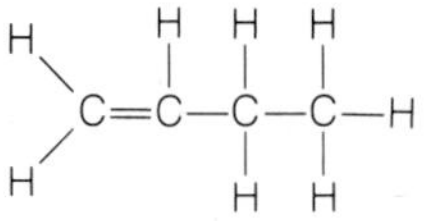	1-Butene
C_4H_{10}	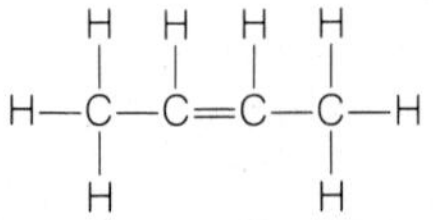	2-Butene

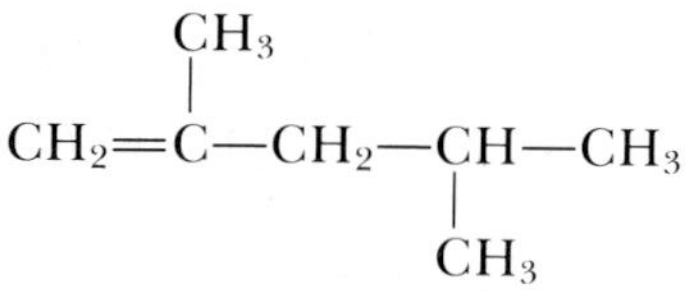

2,4-dimethyl-1-pentene

Only one structural formula each can be written for the molecular formulas C_2H_4 and C_3H_6. However, for C_4H_8 we can write two different four-carbon chains corresponding to the compounds 1-butene and 2-butene.[1] In this case, we have an example of isomerism resulting from the position of the double bond in a chain.

The unsaturated alkenes can be converted into the saturated alkanes by the addition of hydrogen to the carbon-carbon double bond:

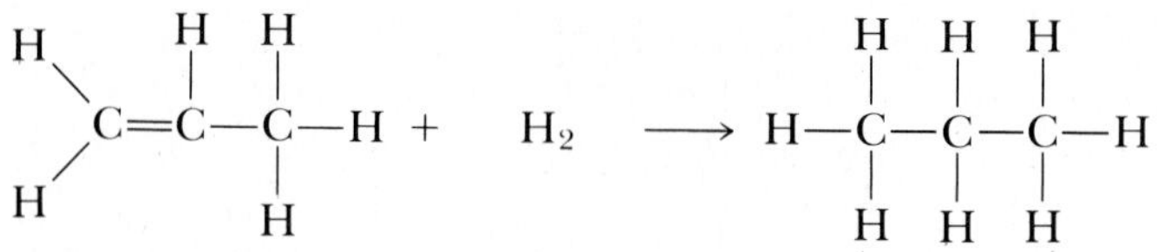

Propene   Hydrogen   Propane

---

[1] Later in your career you will learn that 2-butene can be written as a pair of geometric isomers cis-2-butene and trans-2-butene.

This reaction is an example of a type known as **addition reactions**—the attacking reagent adds to both carbons of the double bond, converting the double bond into a single bond.

Although there are many bonds in the propene molecule, the only bond affected in the reaction is the double bond. This behavior is typical of organic reactions in general. Often within a very large molecule a reaction will take place at only a limited small site, namely, the functional group; the rest of the molecule remains unreacted.

Under the appropriate conditions, alkenes can also undergo a type of addition reaction known as **polymerization.** Small molecules such as ethylene combine with each other to form large chains called **polymers.** The repeating unit in a polymer is known as a **monomer.**

$$\underset{/}{\overset{\backslash}{}}C=C\underset{\backslash}{\overset{/}{}} \;+\; \underset{/}{\overset{\backslash}{}}C=C\underset{\backslash}{\overset{/}{}} \;+\; \underset{/}{\overset{\backslash}{}}C=C\underset{\backslash}{\overset{/}{}} \;\longrightarrow\; \cdots-\overset{|}{\underset{|}{C}}-\overset{|}{\underset{|}{C}}-\overset{|}{\underset{|}{C}}-\overset{|}{\underset{|}{C}}-\overset{|}{\underset{|}{C}}-\overset{|}{\underset{|}{C}}-\cdots$$

Polymer

$$\text{or} \quad \left(-\overset{|}{\underset{|}{C}}-\overset{|}{\underset{|}{C}}-\right)_n$$

Monomer

Polymer

A common polymer such as polyethylene may be made up of thousands of monomer units ($n > 1000$). Many interesting biological molecules, such as proteins, DNA, RNA, and carbohydrates (e.g., starch) are also polymers. In many of these latter cases, however, the monomer units involved are not all identical.

Alkynes

Hydrocarbons with at least one triple bond are classified as alkynes. Alkynes have the general formula C_nH_{2n-2}. Naming of alkynes follows closely the naming of alkenes, discussed above, with the exception that the suffix for alkynes is *-yne.* For example,

$$CH_3-\overset{\overset{\textstyle CH_3}{|}}{\underset{\underset{\textstyle H}{|}}{C}}\rule{2em}{0pt}\overset{\overset{\textstyle H}{|}}{\underset{\underset{\textstyle H}{|}}{C}}-C\equiv C-H$$

is called 4-methyl-1-pentyne.

Alkynes, like alkenes, undergo the addition reaction, but in the case of the alkynes, 1 mole of alkyne can react with 2 moles of the adding reagent:

$$H-C\equiv C-\underset{\underset{H}{|}}{\overset{\overset{H}{|}}{C}}-H \ + \ 2H_2 \ \longrightarrow \ H-\underset{\underset{H}{|}}{\overset{\overset{H}{|}}{C}}-\underset{\underset{H}{|}}{\overset{\overset{H}{|}}{C}}-\underset{\underset{H}{|}}{\overset{\overset{H}{|}}{C}}-H$$

Propyne   Hydrogen   Propane

The most important alkyne is acetylene, C_2H_2 (ethyne), which is used commercially as a fuel in welding torches.

The triple bond is rarely found in biological molecules. However, it is interesting to note that it is this functional group that is used to modify the structure of natural female hormones so that they may be ingested orally as contraceptives.

Aromatic hydrocarbons

Chemists originally used the term **aromatic** to designate compounds that had a low hydrogen to carbon ratio and were very fragrant. Many such compounds were found in natural products, such as bitter almonds, clove oil, and vanilla beans.

August Kekulé (1829–1896) was the first to propose that all compounds previously called aromatic had a benzene ring common to their structures. Benzene itself is a compound with molecular formula C_6H_6. Kekulé pictured benzene as having the structure:

Kekulé benzene

where each apex of the ring represents a carbon atom. It would appear from the preceding structure that the carbon-carbon bonding in benzene involves alternating single and double bonds. However, experimentally it is found that all the carbon-carbon bonds in benzene are identical. Furthermore, benzene does *not* undergo any of the addition reactions of other compounds containing carbon-carbon double bonds.

To account for these properties, chemists now picture the benzene molecule as a six-carbon ring with single bonds between the carbons and, in addition, each carbon participates in a bond which extends over all six carbons. This latter bond is known as a **cyclic-π bond** because it extends in a cycle (ring) over the six carbons. In this cyclic-π bond, instead of electrons being localized between two atoms, as in all previously discussed covalent bonds, they are delocalized over the six carbon atoms.

A modern pictorial representation of benzene is given by

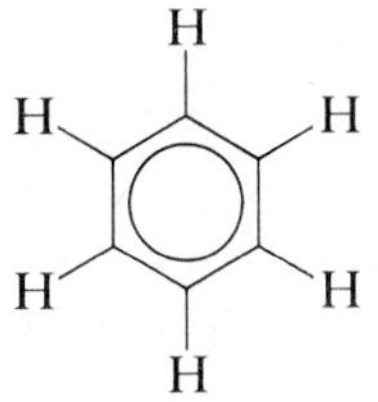

Very often the H's are not written in, but they are understood to be present at every apex at which no other bond is specifically shown.

The circle in the middle of the hexagon represents the cyclic-π bond. The fact that the six π electrons are attracted to all six carbon nuclei makes the π bond a low-energy, highly stable bond. Any addition reactions would break up the cyclic-π bond. Therefore, these reactions do not normally occur on the benzene ring.

Substitution for a hydrogen on a benzene ring is the most common reaction of aromatic compounds. In the following reaction, a methyl group replaces a hydrogen:

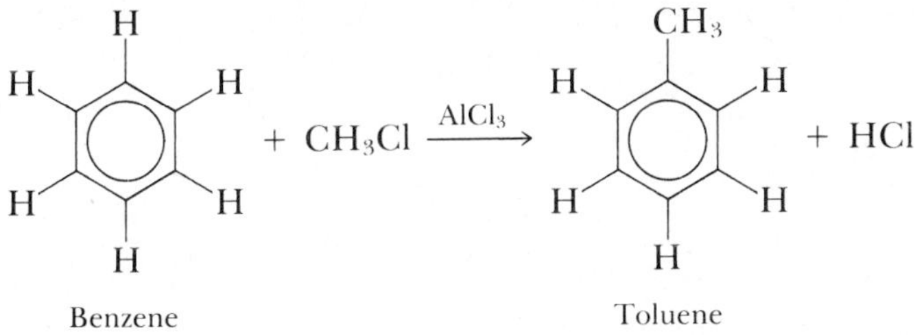

Benzene   Toluene

A variety of other types of groups may be substituted as well.

More than one substituent can be bonded to a benzene ring. This leads to positional isomerism. For example, there are three dimethylbenzenes:

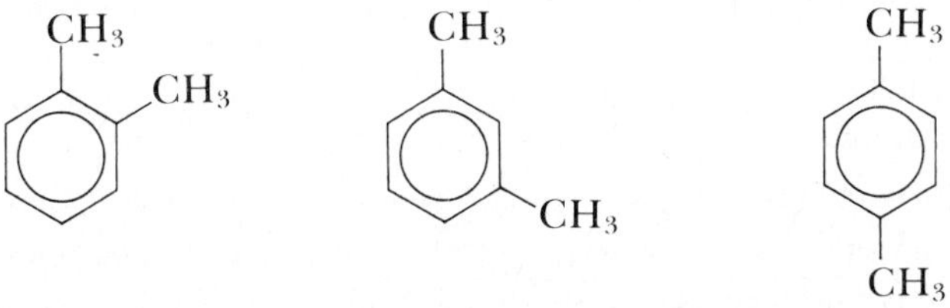

1,2-Dimethylbenzene   1,3-Dimethylbenzene   1,4-Dimethylbenzene

Many naturally occurring aromatic compounds have benzene rings fused together:

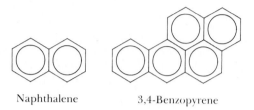

Naphthalene   3,4-Benzopyrene

Naphthalene is commonly used as a moth repellent. 3,4-benzopyrene is a known carcinogen, i.e., a cancer-producing chemical. It is formed during the burning of a cigarette and apparently also during the charcoal broiling of meat.

ORGANIC COMPOUNDS CONTAINING OXYGEN

18.9 In order to achieve a stable octet, oxygen commonly forms two covalent bonds, i.e., participates in sharing two pairs of electrons. Two arrangements are possible: either oxygen forms two single bonds, as found in alcohols, ethers, and one part of the acid and ester functional groups, or oxygen may be involved in a carbon-oxygen double bond, as in aldehydes, ketones, and another part of the acid and ester functional groups (see Table 18.2).

Alcohols

Compounds containing the —OH (hydroxyl) group bonded to an alkyl-type carbon are alcohols. In the systematic naming of alcohols the suffix employed is *-ol*, and the position of the alcohol group on the longest chain is shown by a number before the base name. For example,

$$
\begin{array}{c}
\qquad\quad CH_3 \\
\qquad\quad | \\
CH_3-C-CH_2CH_2-CH-CH_3 \\
\qquad\quad | \qquad\qquad\quad | \\
\qquad\quad H \qquad\qquad\quad OH
\end{array}
$$

is called 5-methyl-2-hexanol.

Because of the polar nature of the —OH group, alcohols can hydrogen-bond to themselves and water molecules. This leads to high boiling points and a high degree of solubility for those compounds in water. A few common alcohols are listed in Table 18.5.

Ethanol or ethyl alcohol is the alcohol found in all alcoholic bev-

COMMONLY ENCOUNTERED ALCOHOLS **TABLE 18.5**

Formula	Structure	Name	Use
CH_3OH	H | H—C—OH | H	Methanol	Used as solvent; causes blindness and death if taken internally
CH_3CH_2OH	H H | | H—C—C—OH | | H H	Ethanol	The alcohol in alcoholic beverages
CH_2OHCH_2OH	H H | | H—C—C—H | | OH OH	Ethylene glycol	Antifreeze component
$CH_2OHCHOHCH_2OH$	H H H | | | H—C—C—C—H | | | OH OH OH	Glycerol	Lubricant; formed as a breakdown product of fat metabolism

erages. The ethanol in these beverages is produced by the action of yeast on sugars found in barley and hops (beer), grapes (wine), and grains such as rye and corn (whiskey). In the chemical process known as **fermentation,** the biological catalyst or the enzyme called zymase, found in yeast, catalyzes the conversion of the sugar glucose into ethanol and carbon dioxide.

$$\text{Complex sugars} \xrightarrow{\text{hydrolysis}} C_6H_{12}O_6 \xrightarrow{\text{zymase}} 2CH_3CH_2OH + 2CO_2$$

Ethers

The ether functional group is the arrangement $-\overset{\displaystyle|}{\underset{\displaystyle|}{C}}-\overset{\displaystyle..}{\underset{\displaystyle..}{O}}-\overset{\displaystyle|}{\underset{\displaystyle|}{C}}-$.

Unlike alcohols, ethers do not have a hydrogen connected to an electronegative oxygen and therefore do not engage in hydrogen bonding with themselves. Consequently, the boiling points of ethers are much lower than boiling points of alcohols with corresponding molecular weights.

The most useful feature of ethers is their inertness to most organic substances. Because of their low reactivity, ethers can be used as solvents in many organic reactions without fear of their interference in the desired reaction.

Ethers are named by naming the alkyl groups bonded to oxygen followed by the word *ether;* for example, the ether most commonly identified as an anesthetic is the compound diethyl ether which has the structure:

$$H-\underset{\underset{H}{|}}{\overset{\overset{H}{|}}{C}}-\underset{\underset{H}{|}}{\overset{\overset{H}{|}}{C}}-\overset{..}{\underset{..}{O}}-\underset{\underset{H}{|}}{\overset{\overset{H}{|}}{C}}-\underset{\underset{H}{|}}{\overset{\overset{H}{|}}{C}}-H$$

Aldehydes and ketones

Aldehydes and ketones are organic compounds containing the fragment $>C=O$, the carbonyl group, bound to hydrocarbon fragments. If a hydrogen is bonded to the carbonyl carbon, we have an **aldehyde.** If only alkyl groups are linked to the carbonyl carbon, the compound is called a **ketone.** Aldehydes and ketones cannot form hydrogen bonds with themselves. Therefore, they have lower boiling points than do alcohols of similar molecular weight. Because aldehydes and ketones are highly polar, they do show appreciable water solubility.

In the systematic nomenclature, aldehydes are identified by an -*al* suffix, whereas ketones display an -*one* suffix and a number before the base name indicating the position of the carbonyl group. The lower-molecular-weight aldehydes and ketones are more often called by their common names, and we will use those common names in Table 18.6.

TABLE 18.6   COMMON ALDEHYDES AND KETONES

Name	Structure	Use or Occurrence
Formaldehyde		Preservative for biological specimens
Acetaldehyde		Key component of smog
Acetone		Nail-polish remover
Benzaldehyde		Extracted from bitter almonds
Cinnamaldehyde		Flavor of cinnamon

Aldehydes can be formed by the oxidation of alcohols which have at least two H atoms on the carbon attached to the OH group:

$$H-\overset{\displaystyle H}{\underset{\displaystyle H}{C}}-\overset{\displaystyle H}{\underset{\displaystyle H}{C}}-\overset{\displaystyle H}{\underset{\displaystyle H}{C}}-OH \ + \ [O]^* \ \longrightarrow \ H-\overset{\displaystyle H}{\underset{\displaystyle H}{C}}-\overset{\displaystyle H}{\underset{\displaystyle H}{C}}-\overset{\displaystyle H}{C}{=}O$$

Carboxylic acids

Carboxylic acids are generally weak acids and contain the functional group, $-C\overset{\displaystyle O}{\underset{\displaystyle OH}{\diagup}}$. These acids ionize slightly in aqueous solution to yield H$^+$ ions:

$$H-\overset{\displaystyle H}{\underset{\displaystyle H}{C}}-C\overset{\displaystyle O}{\underset{\displaystyle OH}{\diagup}} \ \rightleftharpoons \ H-\overset{\displaystyle H}{\underset{\displaystyle H}{C}}-C\overset{\displaystyle O}{\underset{\displaystyle O^-}{\diagup}} \ + \ H^+$$

Acetic acid (HC$_2$H$_3$O$_2$)

This equation gives the structural formula for acetic acid, which we have previously written simply as HC$_2$H$_3$O$_2$.

The systematic naming of carboxylic acids is based on the longest continuous chain dictating the base name and a characteristic suffix for the functional group. The suffix for carboxylic acids is -*oic*, followed by the word "acid." Thus CH$_3$CH$_2$CH$_2$CH$_2$—C=O is pentanoic acid
$\qquad\qquad\qquad\qquad\qquad\qquad\qquad\qquad$ |
$\qquad\qquad\qquad\qquad\qquad\qquad\qquad\qquad$ OH
because there is a five-carbon chain in which is included the acid functional group. The acid group —C=O is always at the end of the
$\qquad\qquad\qquad\qquad\qquad\qquad\qquad\qquad\qquad$ |
$\qquad\qquad\qquad\qquad\qquad\qquad\qquad\qquad\qquad$ OH
chain and thus is position 1. The most familiar acids (acetic and formic) and many physiologically important acids are generally referred to by their common names.

Carboxylic acids tend to have exceptionally high boiling points and are thereby all liquids or solids at room temperature. Acids have even higher boiling points than do alcohols of similar molecular weight.

---

* [O] is a symbol used to represent an oxidizing agent, which in this reaction could be K$_2$Cr$_2$O$_7$ or KMnO$_4$.

This is a consequence of the excellent opportunity for hydrogen bonding between acid molecules. Because of hydrogen bonding, a greater amount of heat energy must be supplied in order for molecules to move independently of one another and enter the gas phase (review Section 12.7). Because of the "double" H-bond formation between the acidic hydrogens and carbonyl

$$
\begin{array}{c}
\overset{\delta-}{O}\cdots\overset{\delta+}{H}-O \\
R-C \qquad\qquad C-R \\
O-H\cdots O \\
\overset{\delta+}{}\quad\overset{\delta-}{}
\end{array}
$$

oxygens, many carboxylic acids essentially exist as dimers; that is, they move together as two associated molecules. Because of this, their boiling-point behavior is like that of molecules that are twice as large as their apparent molecular weight would indicate.

Carboxylic acids with chains of less than five carbons are water-soluble because of H bonding between the acid and water molecules.

$$
\begin{array}{c}
\overset{\delta-}{O}\cdots\overset{\delta+}{H} \\
R-C \qquad\qquad O-H \\
O-H \qquad \overset{\delta-}{} \\
\overset{\delta+}{}
\end{array}
$$

Carboxylic acids can be prepared in the laboratory by the oxidation of aldehydes. Acids are often formed in metabolic reactions. Lactic acid, for example, is formed in muscles during exercise. Unless it is broken down by reactions requiring oxygen, it builds up in the muscles and leads to pain, which causes the exercise to halt.

Acids in which the alkyl group consists of a long chain of an even number of carbons are called **fatty acids** because they are found in naturally occurring fats and oils. Two examples are stearic and oleic acids:

$$
\underset{\text{Stearic acid}}{CH_3(CH_2)_{16}-\overset{\overset{\textstyle O}{\|}}{C}-OH} \qquad\qquad \underset{\text{Oleic acid}}{CH_3(CH_2)_7-\overset{\textstyle H}{C}=\overset{\textstyle H}{C}-(CH_2)_7\overset{\overset{\textstyle O}{\|}}{C}-OH}
$$

Stearic acid is a saturated fatty acid because all the carbon-carbon bonds are single bonds, whereas oleic is an unsaturated acid because there is a carbon-carbon double bond in the chain. When the materials that we commonly call oils or fats are formed from unsaturated fatty acids, they have a lower melting point than do those made up of saturated fatty acids. This leads to the different terminology, oils versus fat. The unsaturated fats are more likely to be liquids or oils, such as vegetable oils. In contrast, the saturated fats are solids, such as lard. When dieticians talk of unsaturated fats, they mean fats composed predominantly of unsaturated fatty acids. Oils (unsaturated) can be converted to fats (saturated) by the addition of hydrogen to double bonds.

Carboxylic acids react with bases to form salts. When the acid involved is a long-chain fatty acid, the salt formed is called a **soap:**

$$CH_3(CH_2)_{16}-C\overset{O}{\underset{OH}{\diagup}} + NaOH \longrightarrow CH_3(CH_2)_{16}-C\overset{O}{\underset{O^-}{\diagup}} Na^+ + H_2O$$

Stearic acid   Sodium stearate, a soap

Grease, the substance that binds dirt to clothing, skin, etc., is nonpolar in nature. Because it is a polar substance, water cannot dissolve grease. Soap molecules are unique in their solubility properties. Soaps have a nonpolar end, consisting of a long chain of carbons, which can dissolve in the grease and a polar end capable of dissolving in water (Figure 18.4). The result of this is the cleansing action of soap in water.

Esters

Carboxylic acids react with alcohols to form esters. The ester function

is characterized by a $-\overset{O}{\overset{\|}{C}}-O-\overset{|}{C}-$ linkage.

FIGURE 18.4

The cleansing action of soap. (a) Grease and water are immiscible. (b) The hydrocarbon end of soap can dissolve in grease; the ionic end can dissolve in water. (c) The large grease globule is broken up into small droplets called **micelles.** (d) Micelles are dispersed in water.

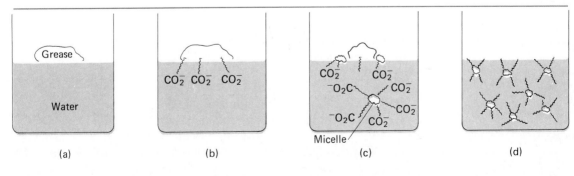

(a)   (b)   (c)   (d)

$$H-\underset{\underset{H}{|}}{\overset{\overset{H}{|}}{C}}-\underset{\underset{H}{|}}{\overset{\overset{H}{|}}{C}}-\overset{\overset{O}{\|}}{C}-\boxed{O-H} + \boxed{H}-O-\underset{\underset{H}{|}}{\overset{\overset{H}{|}}{C}}-\underset{\underset{H}{|}}{\overset{\overset{H}{|}}{C}}-H \longrightarrow H-\underset{\underset{H}{|}}{\overset{\overset{H}{|}}{C}}-\underset{\underset{H}{|}}{\overset{\overset{H}{|}}{C}}-\overset{\overset{O}{\|}}{C}-O-\underset{\underset{H}{|}}{\overset{\overset{H}{|}}{C}}-\underset{\underset{H}{|}}{\overset{\overset{H}{|}}{C}}-H + H_2O$$

Acid
fragment

Alcohol
fragment

Salicylic acid reacts with methanol to produce the ester methyl salicylate commonly known as "oil of wintergreen."

$$\text{(18.1)}$$

Salicylic acid   Methanol   Methyl salicylate

Salicylic acid is an example of a compound with two functional groups. In Equation (18.1) the acid group of salicylic acid was utilized to form an ester; in a reaction with acetic acid salicylic acid uses its alcoholic type —OH group to form the ester acetylsalicylic acid, the active component of aspirin.

$$+ H_2O \qquad \text{(18.2)}$$

Salicylic acid   Acetic acid   Acetylsalicylic acid
(aspirin)

Esters are named by identifying and naming the alkyl group from the alcohol and then naming the acid fragment. The -*ic* ending of the acid name is changed to -*ate*. Note the examples in Table 18.7.

Esters lack an OH group and thus cannot form hydrogen bonds among themselves, esters have lower boiling points than do alcohols or acids of comparable molecular weights.

Esters supply the pleasant odors of fruits and flowers. They are isolated from natural sources or synthesized for use in perfumes, cos-

TABLE 18.7

FLAVORS OR FRAGRANCES OF SOME ESTERS

Name	Structure	Flavor or Fragrance
Ethyl formate	$\overset{\displaystyle O}{\overset{\displaystyle \|}{H-C}}-O-CH_2CH_3$	Rum
Isobutyl formate	$\overset{\displaystyle O}{\overset{\displaystyle \|}{H-C}}-O-CH_2-\overset{\displaystyle CH_3}{\overset{\displaystyle \|}{CH}}-CH_3$	Raspberries
Methyl butyrate	$CH_3CH_2CH_2-\overset{\displaystyle O}{\overset{\displaystyle \|}{C}}-O-CH_3$	Apple
Ethyl butyrate	$CH_3CH_2CH_2-\overset{\displaystyle O}{\overset{\displaystyle \|}{C}}-O-CH_2CH_3$	Pineapple
Amyl* acetate	$CH_3-\overset{\displaystyle O}{\overset{\displaystyle \|}{C}}-O-CH_2CH_2CH_2CH_2CH_3$	Banana
Isoamyl acetate	$CH_3-\overset{\displaystyle O}{\overset{\displaystyle \|}{C}}-OCH_2CH_2-\overset{\displaystyle CH_3}{\overset{\displaystyle \|}{CH}}-CH_3$	Pear
Amyl* butyrate	$CH_3CH_2CH_2-\overset{\displaystyle O}{\overset{\displaystyle \|}{C}}-O-CH_2CH_2CH_2CH_2CH_3$	Apricot
Octyl acetate	$CH_3-\overset{\displaystyle O}{\overset{\displaystyle \|}{C}}-OCH_2CH_2CH_2CH_2CH_2CH_2CH_2CH_3$	Orange
Methyl salicylate	$\underset{\displaystyle OH}{\overset{\displaystyle \bigcirc}{}}\overset{\displaystyle O}{\overset{\displaystyle \|}{C}}-O-CH_3$	Wintergreen
Benzyl acetate	$\bigcirc-CH_2O-\overset{\displaystyle O}{\overset{\displaystyle \|}{C}}-CH_3$	Jasmine

* Amyl = pentyl.

metics, and flavorings. Table 18.7 shows the esters responsible for many fruit and spice flavorings.

ORGANIC COMPOUNDS CONTAINING NITROGEN

18.10

Amines

Amines can be thought of as organic derivatives of ammonia (NH_3) in which one or more hydrogens of NH_3 is replaced by an alkyl group or an aromatic ring:

$$\begin{array}{c} H \\ | \\ :N-H \\ | \\ H \end{array} \qquad \begin{array}{c} H \\ | \\ :N-CH_3 \\ | \\ H \end{array} \qquad \begin{array}{c} CH_3 \\ | \\ :N-CH_3 \\ | \\ H \end{array} \qquad \begin{array}{c} CH_3 \\ | \\ CH_2 \\ | \\ :N-CH_2CH_3 \\ | \\ CH_2 \\ | \\ CH_3 \end{array}$$

Ammonia   Methylamine   Dimethylamine   Triethylamine

Amines are most often named by simply listing the names of the organic groups attached to nitrogen and then adding the word *amine*. The prefixes *di-* and *tri-* are used to indicate the multiple presence of any organic group. Aromatic amines have an aromatic ring directly bonded to nitrogen. They are named as derivatives of aniline:

Aniline   4-Methylaniline

Like ammonia, amines act as bases:

$$\begin{array}{c} H\ \ H \\ |\ \ \ | \\ H-C-N: \\ |\ \ \ | \\ H\ \ H \end{array} + HCl \longrightarrow \begin{array}{c} H\ \ H \\ |\ \ \ | \\ H-C-N^{\pm}-H \\ |\ \ \ | \\ H\ \ H \end{array} + Cl^-$$

Methyl ammonium chloride

Because of the presence of the N—H bonds, amines can hydrogen-bond to each other and to water. Their boiling points are higher than the boiling points of alkanes with similar molecular weights, and the simpler amines are all water soluble.

When an atom other than carbon occupies a ring position, the ring is called a **heterocycle.** Many physiologically active compounds contain a nitrogen atom as an amine group in a heterocyclic ring.

Nicotine

Amides

Carboxylic acids react with amines to yield amides. The **amide** function

is characterized by a $-\overset{\overset{\displaystyle O}{\|}}{C}-N\overset{/}{\diagdown}$ linkage.

| Carboxylic acid | Amine | Amide |

Amides have very strong hydrogen bonding from the H of a N—H bond to the O of a C=O group.

Consequently, most amides are solids.

Nylon is a polyamide, a long-chain compound formed by the reaction of diacids and diamines.

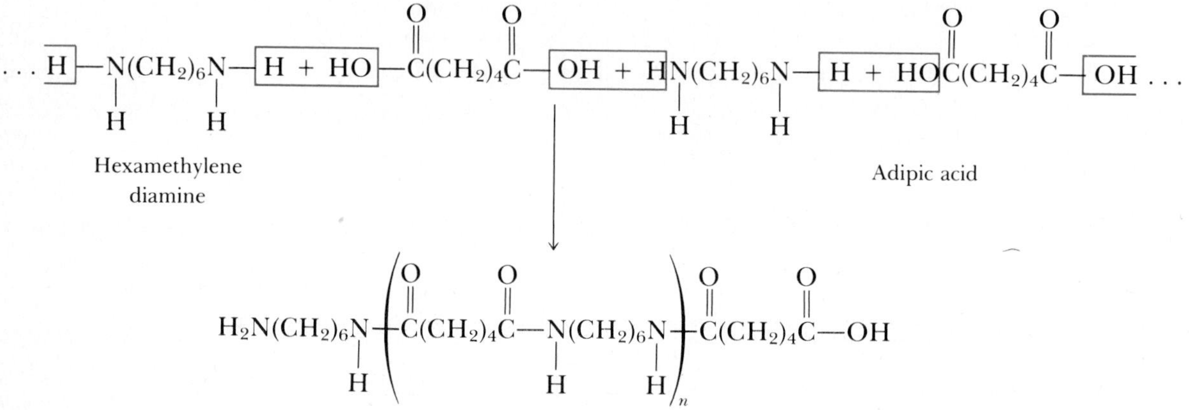

$$\ldots \boxed{H} - N(CH_2)_6N - \boxed{H + HO} - \overset{\overset{O}{\|}}{C}(CH_2)_4\overset{\overset{O}{\|}}{C} - \boxed{OH + H}N(CH_2)_6N - \boxed{H + HO}\overset{\overset{O}{\|}}{C}(CH_2)_4\overset{\overset{O}{\|}}{C} - \boxed{OH} \ldots$$

Hexamethylene diamine

Adipic acid

$$H_2N(CH_2)_6N\left(\!\!\begin{array}{c}\overset{O}{\|}\\C\end{array}\!\!(CH_2)_4\overset{\overset{O}{\|}}{C} - N(CH_2)_6N\!\!\right)\!\!\begin{array}{c}\overset{O}{\|}\\C\end{array}\!\!(CH_2)_4\overset{\overset{O}{\|}}{C} - OH$$

Nylon-6,6

Urea, one of the metabolic breakdown products in the human body, is a diamide with structure

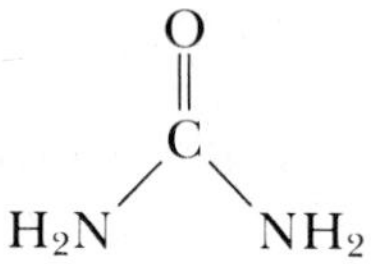

$$\underset{H_2N}{}\overset{\overset{\displaystyle O}{\|}}{C}\underset{NH_2}{}$$

Urea

The amide linkage is the key to an understanding of proteins, which form from a combination of amino acids.

ORGANIC MOLECULES OF BIOLOGICAL SIGNIFICANCE

18.11

Carbohydrates

Carbohydrates are naturally occurring compounds which structurally are polyhydroxyaldehydes or polyhydroxyketones. *Poly-* is a prefix meaning "many." Glucose (blood sugar) exemplifies a polyhydroxy-aldehyde and is a member of the carbohydrate subclass called **mono-saccharides.** Monosaccharides in general are the simplest carbohy-drates from which more complicated carbohydrates are constructed.

The monosaccharide glucose in solution is an equilibrium mixture of an open-chain structure and a cyclic structure:

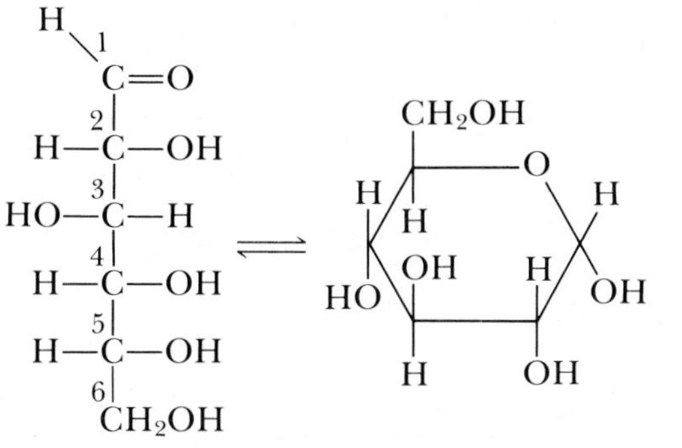

Glucose

Open form   Cyclic form

The open form clearly shows the presence of both the alcohol and aldehyde functional groups. The ring forms by the reaction of the aldehyde group and the alcohol group at carbon-5. The relative positional relationship of the hydroxyl groups in glucose is quite complicated and distinguishes glucose from other six-carbon monosaccharides containing five alcohol groups and an aldehyde function. For example, mannose, a compound distinct from glucose, has as its open-chain structure:

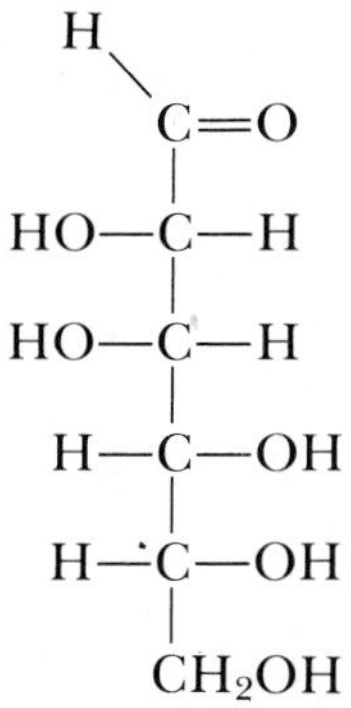

Mannose

Mannose is an isomer of glucose. Fructose is also an isomer of glucose and is a polyhydroxyketone.

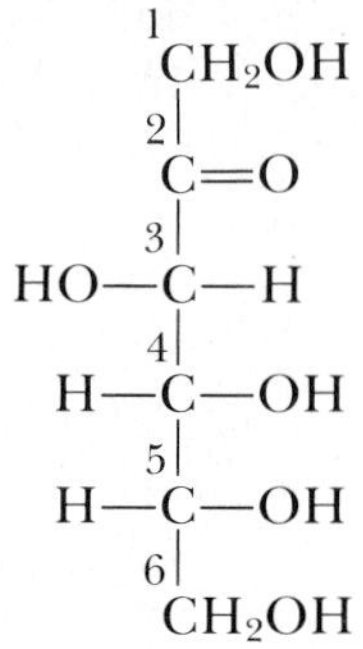

$$
\overset{1}{C}H_2OH
$$
$$
\overset{2}{C}=O
$$
$$
HO\overset{3}{-C-}H
$$
$$
H\overset{4}{-C-}OH
$$
$$
H\overset{5}{-C-}OH
$$
$$
\overset{6}{C}H_2OH
$$

Fructose

Another member of the carbohydrate family is sucrose, common table sugar, which is a disaccharide. **Disaccharides** are constructed of two monosaccharide units hooked together. In the case of sucrose, the two monosaccharides are glucose and fructose. **Polysaccharides,** such as starch, are made up of thousands of monosaccharide units. Starch is composed of glucose units.

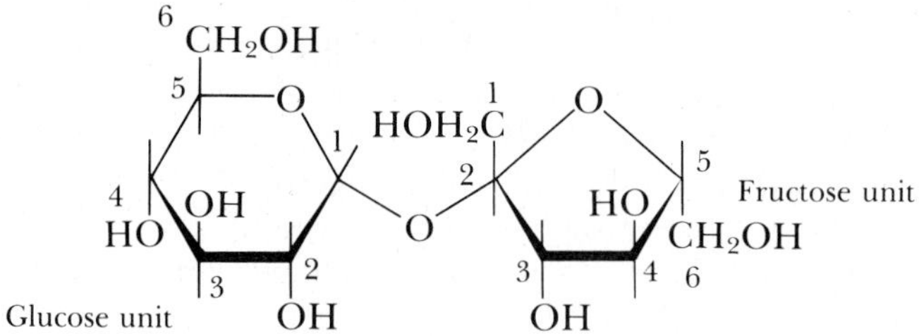

Sucrose

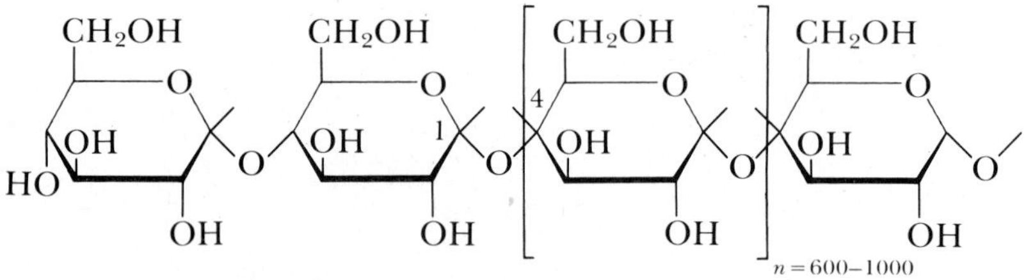

Starch

The first step in the digestion of carbohydrates is the breakdown of disaccharides and polysaccharides to monosaccharide units by *hydrolysis reactions,* i.e., by reactions with water:

$$\text{Sucrose}(aq) + H_2O \longrightarrow \text{glucose}(aq) + \text{fructose}(aq)$$
$$\text{Starch}(aq) + H_2O \longrightarrow n \text{ glucose units}(aq)$$

Commercially, carbohydrates are a key food source. Glucose is one of the body's main sources of energy. Cotton and linen are largely composed of cellulose, the most abundant polysaccharide. Wood and other plant materials have cellulose as their principal structural component.

Triglycerides

The fats and oils previously mentioned in the discussion of carboxylic acids are esters of long-chain fatty acids and the alcohol glycerol. The chemical name for compounds with the structure of fats and oils is **triglyceride**:

$$
\begin{array}{c}
\text{H}\qquad\quad\text{O}\\
|\qquad\quad\;\|\\
\text{H}-\text{C}-\text{O}-\text{C}-(\text{CH}_2)_{16}\text{CH}_3\\
|\\
\\
\text{H}-\text{C}-\text{O}-\text{C}-(\text{CH}_2)_{16}\text{CH}_3\\
|\qquad\quad\;\|\\
\qquad\quad\;\text{O}\\
\\
\text{H}-\text{C}-\text{O}-\text{C}-(\text{CH}_2)_{16}\text{CH}_3\\
|\qquad\quad\;\|\\
\text{H}\qquad\quad\text{O}
\end{array}
$$

Stearin, a triglyceride

Triglycerides are broken down during the digestion process to fatty acids and glycerol.

$$
\begin{array}{c}
\text{H}\qquad\text{O}\\
|\qquad\;\|\\
\text{H}-\text{C}-\text{O}-\text{C}-(\text{CH}_2)_{16}\text{CH}_3\\
|\\
\text{H}-\text{C}-\text{O}-\text{C}-(\text{CH}_2)_{16}\text{CH}_3 + 3\text{H}_2\text{O}\\
|\qquad\;\|\\
\qquad\;\text{O}\\
\text{H}-\text{C}-\text{O}-\text{C}-(\text{CH}_2)_{16}\text{CH}_3\\
|\qquad\;\|\\
\text{H}\qquad\text{O}
\end{array}
\xrightarrow{\text{digestion}}
\begin{array}{c}
\text{H}\\
|\\
\text{H}-\text{C}-\text{OH}\\
|\\
\text{H}-\text{C}-\text{OH} + 3\text{CH}_3(\text{CH}_2)_{16}\overset{\displaystyle O}{\overset{\|}{\text{C}}}-\text{OH}\\
|\\
\text{H}-\text{C}-\text{OH}\\
|\\
\text{H}
\end{array}
$$

Glycerol   Fatty acid

Triglycerides serve as the storage form of fat in animals. They are a subclass of a larger group of compounds known as **lipids,** the part of biological matter which dissolves in nonpolar solvents. Subclasses such as steroids are nonester containing members of the lipid group. Cholesterol is an example of a steroid.

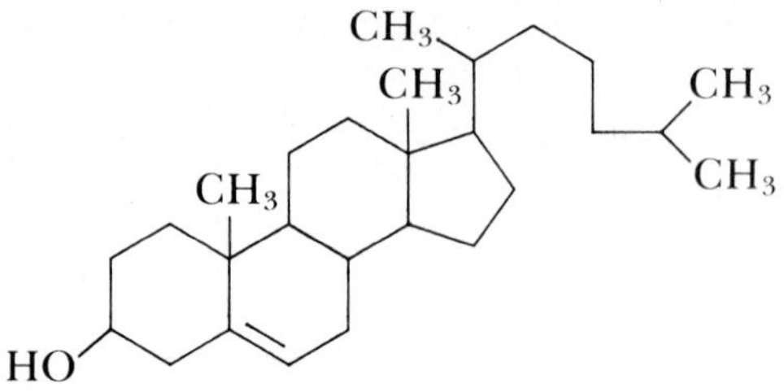

Cholesterol

Amino acids

Carboxylic acids containing an amine group on the carbon adjacent to the acid group are called *alpha* amino acids, or simply, amino acids. In general, an amino acid has the formula

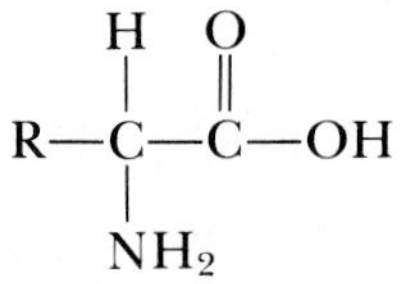

where R can be any alkyl group. There are 20 common naturally occurring amino acids.

Let us consider two amino acids, valine and alanine:

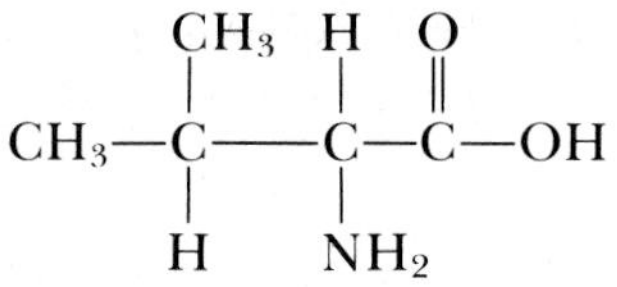

Valine 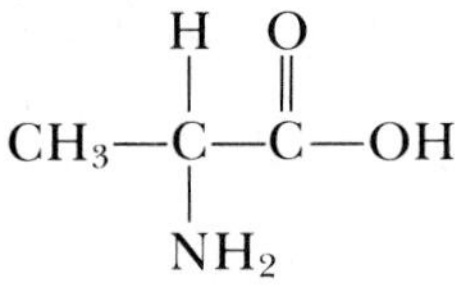   Alanine

Recall that acids can react with amine groups. What forms if we react the acid group of valine with the amino group of alanine?

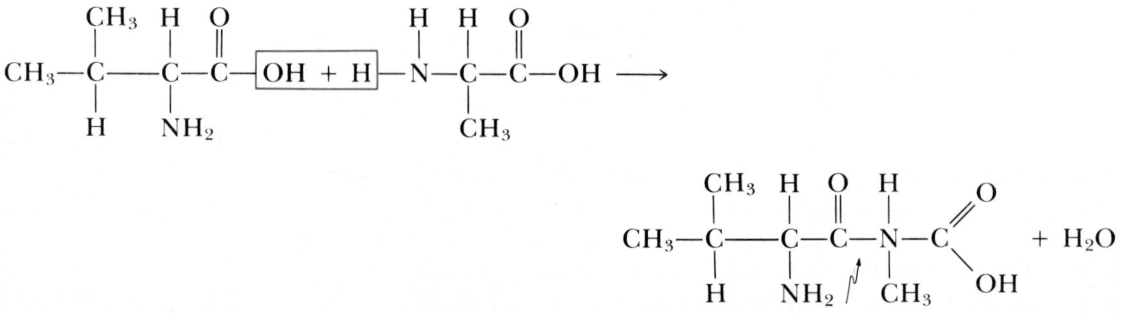

amide linkage

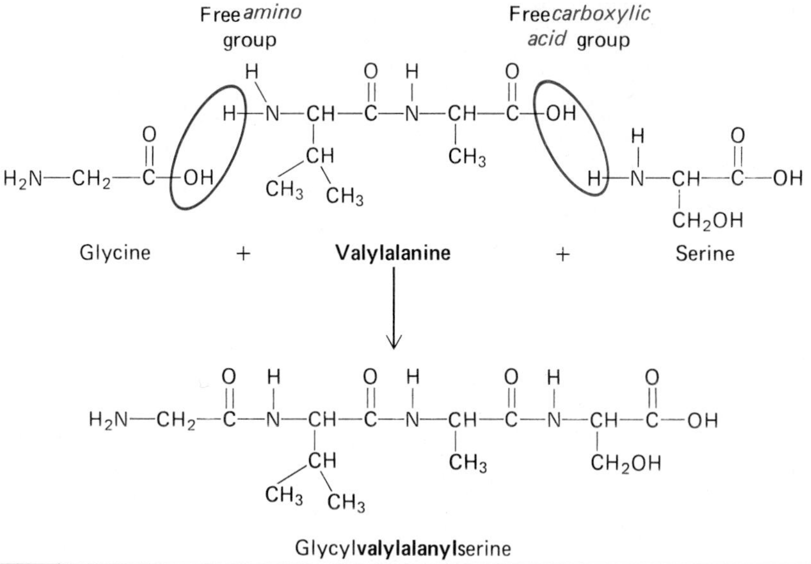

FIGURE 18.5

Peptide chains can be extended at either end because of the free amino and free carboxylic acid groups. In this case the dipeptide valylalanine is extended on the left by the reaction with glycine and on the right by reaction with serine. The result is a tetrapeptide.

An amide linkage forms, which in the case of combining amino acids is called a **peptide linkage.** Notice that the resulting product, valyl alanine, still has an acid group on one end and a free amino group on the other end. The acid end could react with the $-NH_2$ of another amino acid, and the free NH_2 group could react with the acid group of yet another amino acid. The final resulting compound would contain four amino acids bound together by peptide bonds (see Figure 18.5). Four amino acids bound together by peptide bonds is called a *tetrapeptide.*

Proteins

When the number of amino acids bound together is at least 50, we have a **protein.** All proteins are formed from the basic set of 20 amino acids; however, a common protein has 100 to 300 amino acids held together by peptide bonds. Clearly, amino acids are repeated in the chain. It should be noted that the order of putting the amino acids together is important. If in our original dipeptide example we had used the acid group of alanine and the amino group of valine, we would have obtained a different compound, namely, alanyl valine.

SAMPLE EXERCISE

18.7

Write the structure of alanyl valine.

SOLUTION

The name alanyl valine tells us that the first amino acid in the chain is alanine, and that alanine's acid group is part of the peptide linkage. Since alanine's acid group reacts, it must be valine's amino group that reacts.

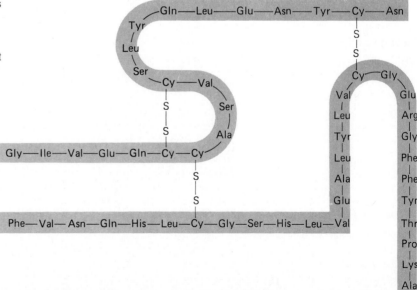

Alanine   Valine   Alanyl valine

Figure 18.6 shows the structure of insulin. If the order of attachment of amino acids in the chain is altered in any way, the compound would no longer be insulin, and it would not have the expected biological activity.

The word "protein" is derived from the Greek word *protos,* meaning "the first." In many ways proteins are first in importance among the three classes of biomolecules: carbohydrates, lipids, and proteins. In most organisms the mass of protein is greater than the combined mass of carbohydrates and lipids. Moreover, proteins serve in a wider variety of biological functions than do the other biomass components. These functions range from structural support and movement,

FIGURE 18.6

The sequence of amino acids in beef insulin. The amino acids are commonly represented by three-letter abbreviations, usually the first three letters of the name. For example, Ala = alanine, Gly = glycine, and Val = valine.

because proteins are the material of skin, ligaments, tendons, cartilage, and muscle, to biological catalysis, because enzymes are largely proteinaceous. The oxygen transporters, hemoglobin and myoglobin, are proteins, as are the defense and regulatory agents of the body, the antibodies and hormones.

SUMMARY

Organic compounds were so named because originally it was believed that they could be synthesized only by living organisms. Today the term "organic," is applied to carbon compounds (Section 18.1). Carbon is unique among the elements in its ability to form strong covalent bonds with itself as well as with other elements. This property leads to huge numbers of carbon compounds (Section 18.2).

Structural formulas for organic molecules can be written by recognizing the typical bonding patterns for the common elements found in these compounds (Section 18.3). Condensed structural formulas are used most often because it is more convenient to write them (Section 18.4). In order to organize the study of organic compounds, they are classified according to functional groups (Section 18.5).

Isomers are compounds which have the same molecular formulas but different structural formulas. The phenomenon of isomerism is another factor leading to the large numbers of organic molecules in the world (Section 18.6).

The relationship between bonding patterns and the geometry of molecular compounds was first introduced in Chapter 12. The ideas are reviewed briefly in this chapter as they apply to carbon compounds. Structural formulas do not show geometry, but you can always predict the arrangement around an atom by using VSEPR theory. The tetrahedral array around carbon is encountered most often (Section 18.7).

Compounds containing only carbon and hydrogen are called hydrocarbons. The physical properties of hydrocarbons are dictated by the fact that there is essentially no difference in electronegativity between carbon and hydrogen. Hence, dispersion forces are the only operative intermolecular forces in hydrocarbons. There are four classes of hydrocarbons based on the type of bonding between carbons: alkanes, alkenes, alkynes, and aromatics. For the first three classes, the concept of the homologous series, a set of compounds differing only in chain length, is useful. Hydrocarbons do not play an important role in biological systems, but they are useful as fuels and in the chemical industry (Section 18.8).

The IUPAC nomenclature system is based on identifying the longest continuous carbon chain in a molecule. Suffixes to the root chain name denote the family or functional group class of the compound. Alkanes are signified by the suffix -ane (Section 18.8).

There are a wide variety of organic compounds containing oxygen which are subdivided into classes based on the functional group concept: alcohols, ethers, aldehydes, ketones, carboxylic acids, and esters (Section 18.9).

The element nitrogen is encountered in the organic functional groups, amines and amides. Polyamides form by the reaction of diacids and diamines (Section 18.10).

The three classes of biologically significant organic molecules are carbohydrates, lipids, and proteins. Carbohydrates are important foodstuffs that are chemically distinguished by the fact that they contain many alcohol functional groups and either the aldehyde or ketone function. Examples of lipids are triglycerides, i.e., fats and oils, which are esters, and steroids such as cholesterol. Proteins are the foodstuffs which bring the element nitrogen into the body in usable form. Proteins are polymers of amino acids (Section 18.11).

CHAPTER ACCOMPLISHMENTS

After completing this chapter you should be able to

18.1 Introduction
1. Distinguish between organic and inorganic chemistry with respect to atomic composition and numbers of known compounds.

18.2 Bonding in carbon compounds
2. State the predominant type of bonding in organic compounds.
3. Give at least two reasons why there is a very large number of organic compounds.

18.3 Structural formulas for organic molecules
4. Write a structural formula for a given molecular formula.

18.4 Condensed structural formulas
5. Write a full structural formula from a given condensed formula.

18.5 Functional group concept
6. State the usefulness of the functional group concept.
7. Identify the functional groups present in an organic compound given the structural formula.

18.6 Isomerism
8. State the relationship between compounds with the same molecular formula but different structural formulas.

9. Given a set of structural formulas, distinguish among isomers, identical compounds, and unrelated compounds.
10. Write structural formulas for isomers of a given molecular formula.

18.7 Geometry around carbon atoms

11. State the relationship between the number of atoms around a carbon and the geometry around that carbon.

18.8 Hydrocarbons

12. Define hydrocarbon.
13. Describe the solubility of hydrocarbons.
14. Describe what is meant by the term homologous series.
15. Name and describe two chemical reactions of alkanes.
16. Give the IUPAC name for a given alkane structural formula.
17. Give the IUPAC name for a given alkene structural formula.
18. Describe the relationship between the position of a double bond and the concept of isomerism.
19. Name and describe the typical chemical reaction of alkenes and alkynes.
20. Give an IUPAC name for a given alkyne structural formula.
21. Given a structural formula, recognize an aromatic compound.

18.9 Organic compounds containing oxygen

22. Recognize an alcohol, ether, aldehyde, ketone, carboxylic acid, or ester from a given structural formula or name.
23. Draw structural formulas which show the hydrogen bonding interaction between organic molecules or between an organic molecule and water.
24. Describe a method of preparation for aldehydes and ketones.
25. Given the structural formula of a carboxylic acid, write an equation for the acid ionization equilibrium.
26. Describe the characteristic structural feature and natural origin of fatty acids.
27. Distinguish between saturated and unsaturated acids.
28. Given the structural formula of a carboxylic acid, write the equation for the reaction of the acid with an inorganic base.
29. Given the structural formula of a carboxylic acid and the structural formula of an alcohol, write the equation for the preparation of an ester.

18.10 Organic compounds containing nitrogen

30. Recognize an amide or amine from a given structural formula or name.
31. State the structural relationship between organic amines and ammonia.

32. Given the structural formula of an amine, write the equation for the reaction of the amine with a strong acid.
33. Compare the physical properties of amines to those of alkanes.
34. Define heterocycle.
35. Given the structural formula of a carboxylic acid and the structural formula of an amine, write the equation for the preparation of an amide.

Organic molecules of biological significance

36. State the type of functional groups commonly present in the open form of carbohydrates.
37. Distinguish among monosaccharides, disaccharides, and poly-saccharides.
38. Recognize a triglyceride.
39. Write structural formulas for the products of hydrolysis of a given triglyceride.
40. Recognize an amino acid.
41. Write the structural formula of the product formed by combining two or more given amino acids in a given sequence.

PROBLEMS

18.1 Introduction

18.5 a. Describe the major compositional difference between organic and inorganic compounds.
b. What element besides carbon is common to almost all organic compounds.

18.2 Bonding in carbon compounds

18.6 Give reasons for the large number of organic compounds in existence.

18.7 What other element in the periodic table would you expect to show bonding properties similar to those of carbon and hence possibly form a large number of compounds.

18.3 Structural formulas for organic molecules

18.8 Write a structural formula for a compound with molecular formula:
a. CH_4
b. CH_2Cl_2
c. C_2H_4O
d. C_2H_4
e. CH_2O_2
f. $C_2H_4O_2$
g. CH_4S

18.4 Condensed structural formulas

18.9 Write full structural formulas from the following condensed structures:
a. $CH_3CH_2CH_2CH_3$
b. $CH_3CH(CH_3)_2$
c. $CH(CH_3)_3$
d. ⬠—CH_3
e. $CH_3CHClCH_2CH_2Br$
f. $(CH_3)_2CH(CH_2)_6CH_3$

18.10 Convert the following structural formulas into condensed structures:

a.
$$H\!-\!\underset{\underset{H}{|}}{\overset{\overset{H}{|}}{C}}\!-\!\underset{\underset{OH}{|}}{\overset{\overset{H}{|}}{C}}\!-\!\underset{\underset{H}{|}}{\overset{\overset{H}{|}}{C}}\!-\!\underset{\underset{H}{|}}{\overset{\overset{H}{|}}{C}}\!-\!H$$

b.
$$\begin{array}{c} H \qquad\quad H \\ | \qquad\quad / \\ H\!-\!C\!-\!C\!-\!H \\ | \qquad | \\ H\!-\!C\!-\!C\!-\!Cl \\ | \qquad | \\ H \qquad H \end{array}$$

c. [chemical structure: H—C—C—C—C—OH with Br, H, H, H substituents and a branched CH₃ group]

b. [chemical structure: H—C—C—C—C—OH with NH₂ and O substituents]

c. [chemical structure: benzene ring with C(=O)—C—H group]

d. [chemical structure: H—C—C—N with O]

18.5 Functional-group concept

18.11 Classify the following compounds by functional group:

a. [chemical structure: H—C—H]

g. [chemical structure: H—C—O—C—H]

b. [chemical structure: H—C—C—C—H with O]

h. [chemical structure: H—C—C—O—C—H with O]

e. [chemical structure: H—C—C—C with OH, OH, O]

f. [chemical structure: benzene ring with C—OH (O) and O—C—CH₃ (O)]

18.6 Isomerism

18.13 In your own words, carefully distinguish among isomers, identical compounds, and unrelated compounds.

18.14 For each of the following pairs of structures, indicate whether the two are identical, isomeric, or unrelated.

c. [chemical structure: H—C—C—C—H with Cl]

i. [chemical structure: H—C—N with H, H]

d. [chemical structure: H—C≡C—C—H]

j. [chemical structure: H—C—C=C with H atoms]

a. [chemical structure: H—C—C—C—OH] and [chemical structure: H—C—C—C—H with OH]

e. [chemical structure: benzene ring with C—H]

k. [chemical structure: H—C—N with O, H, H]

b. [chemical structure: N—C—C—H] and [chemical structure: H—C—C—N with H, H]

l. [chemical structure: H—C—C—C with O, H]

c. [chemical structure: C=C—C—C—H] and [chemical structure: H—C—C=C—C—H]

f. [chemical structure: H—C—C=O with OH]

d. [chemical structure: H—C—C—O—C—C—H] and [chemical structure: H—C—C—C—C—OH]

18.12 Circle and label the functional groups present in each compound:

a. [chemical structure: H—C—C—C—OH with OH, O]

e. [chemical structure: H—C—N with H, H] and [chemical structure: H—C—N with O, H, H]

f. 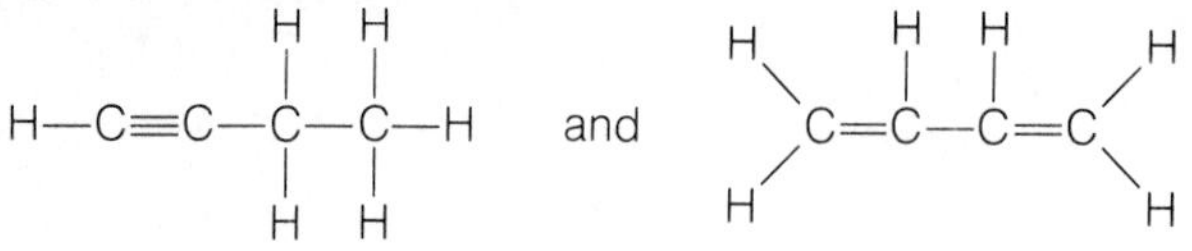 and

c.

g. 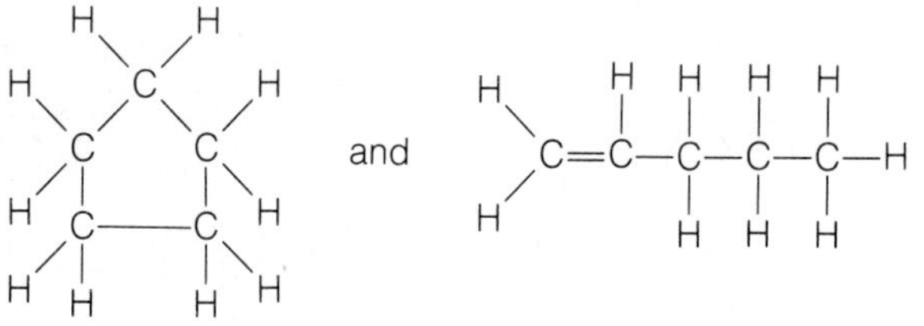 and

d.

18.15 Write structural formulas for isomers of molecular formula C_5H_{12}. (You should find three.)

18.16 Write structural formulas for isomers with molecular formula C_3H_7Cl. (There are two.)

Write structural formulas for isomers with molecular formula C_4H_9Br. (You should find four.)

e.

f.

18.7 Geometry around carbon atoms

18.18 a. List the relationships between the number of atoms around a given carbon and the geometric arrangement around the carbon atom.
b. What is the most common geometric arrangement of covalent bonds around a carbon?

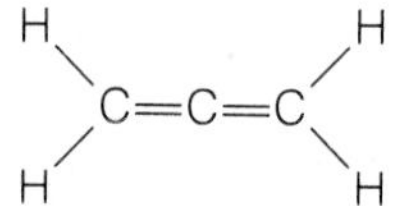

18.8 Hydrocarbons

18.21 Explain why hydrocarbons have low boiling points and are insoluble in water.

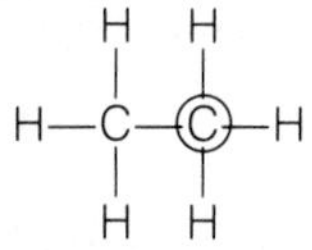

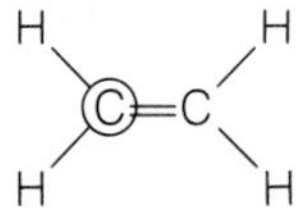

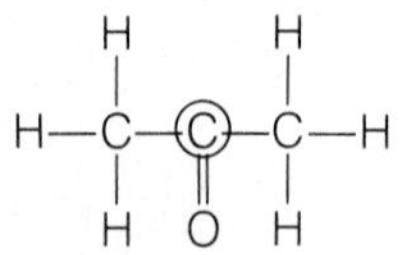

18.22 Arrange the following set of compounds in order of increasing boiling point:
a. CH_3CH_3 **c.** $CH_3CH_2CH_2CH_3$
b. $CH_3(CH_2)_5CH_3$

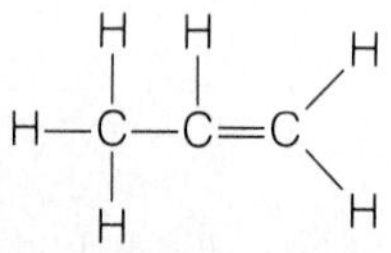

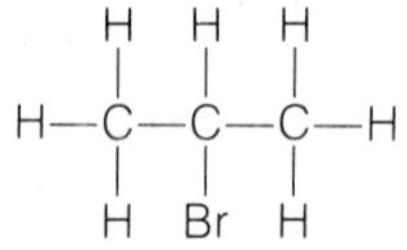

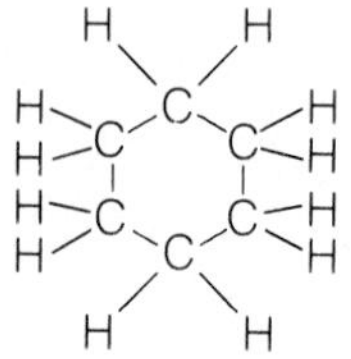

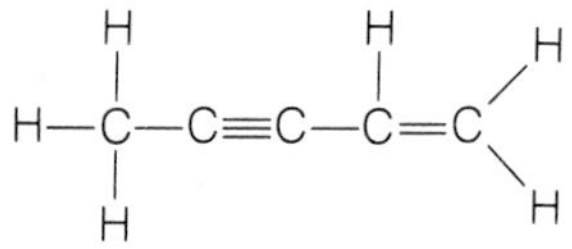

18.19 Give the geometry around each of the circled carbons:

18.23 Predict the compound with the highest boiling point in the following set:

a. 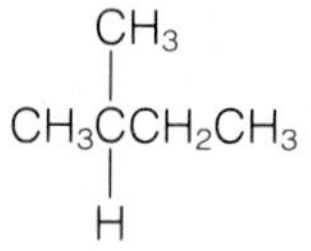   **c.**

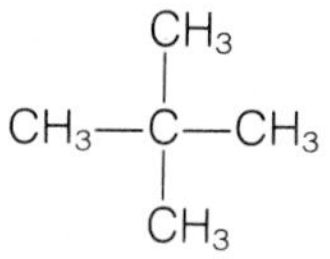

b.   **d.**

a.   **c.**

b. $CH_3CH_2CH_2CH_2CH_3$

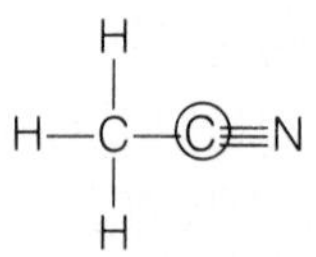

18.20 Give the shape about each carbon atom in each of the following structures:
a. $CH_3—CH_3$

18.24 How could you tell whether a colorless, odorless liquid is a water solution or a hydrocarbon.

b.

18.25 Classify the following substances according to the scheme of Figure 18.3:
a. $CH_3(CH_2)_5CH_3$
b. $CH_3—C{\equiv}C—H$

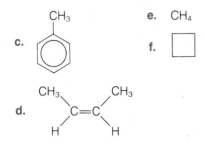

e. CH_4

f. □

c. $CH_3-\overset{\overset{\displaystyle CH_3}{|}}{\underset{\underset{\displaystyle CH_3}{|}}{C}}-CH=CHCH_3$

d. $CH_3CH_2\overset{\overset{\displaystyle}{}}{\underset{\underset{\displaystyle CH_2}{\|}}{C}}CH_2CH_3$

18.26 A hydrocarbon has a molecular formula C_3H_8. What would be the molecular formula of another hydrocarbon in the same homologous series?

18.27 **a.** What is the most important commercial chemical reaction of alkanes?
b. Give two specific uses of alkanes that involve the reaction in part (*a*).

18.28 What is the difference between an alkane molecule and an alkyl group? Give a specific example of each.

18.29 Write the structures and names of the two alkyl groups that can be derived from propane, $CH_3CH_2CH_3$.

18.30 Give an unambiguous name for each of the following:

a. $CH_3-\overset{\overset{\displaystyle CH_3}{|}}{\underset{\underset{\displaystyle CH_3}{|}}{C}}-CH_3$

b. $CH_3CH_2CH_2\overset{\overset{\displaystyle CH_3}{|}}{C}HCH_2CH_3$

c. $CH_3\overset{\overset{\displaystyle CH_2-CHCH_2CH_3}{|}}{C}H\overset{\overset{\displaystyle CH_3}{}}{\underset{\underset{\displaystyle CH_3}{|}}{\underset{\underset{\displaystyle CH_3}{|}}{CH_2}}}$

d. $CH_3CH_2\overset{\overset{\displaystyle}{}}{\underset{\underset{\displaystyle CH_3CHCH_3}{|}}{C}}HCH_3$

18.31 Write a structure corresponding to each of the following names:
a. 2-Methylpentane
b. 2,2-Dimethyl-5-ethyloctane
c. 4-Isopropylheptane

18.32 Give an unambiguous name for the following alkenes.
a. $H_2C=CH_2$
b. $H_2C=CHCH_2CH_3$

18.33 What is the monomer unit in the following partial structure of a polymer chain?

$-CH_2-CH-CH_2-CH-CH_2-CH-CH_2-CH-$

18.34 Give an unambiguous name to each of the following alkynes:
a. $HC\equiv CCH_2CH_3$
b. $CH_3C\equiv CCH_3$

c. $CH_3CH_2\overset{\overset{\displaystyle CH_3}{|}}{C}HC\equiv CH$

18.35 Explain why benzene rings tend to undergo substitution reactions rather than addition reactions.

18.36 Write the structural formulas of the three diethylbenzene isomers.

18.9 Organic compounds containing oxygen

18.37 Draw a figure showing hydrogen bonding between two methanol (CH_3OH) molecules.

18.38 What chemical property allows ethers to be used as common solvents?

18.39 Write the structural formula of an ether that would be an isomer of

$$H-\overset{\overset{\displaystyle H}{|}}{\underset{\underset{\displaystyle H}{|}}{C}}-\overset{\overset{\displaystyle H}{|}}{\underset{\underset{\displaystyle H}{|}}{C}}-\overset{\overset{\displaystyle H}{|}}{\underset{\underset{\displaystyle H}{|}}{C}}-OH$$

18.40 Although aldehyde molecules cannot

hydrogen-bond to themselves, they can hydrogen-bond to water molecules. Show how this is possible.

18.41　**a.**　Write the structural formula for the organic product of the following reactions:

$$\underset{\substack{| \quad | \quad | \quad |\\ H \;\; OH \; H \;\; H}}{H-\overset{\displaystyle H}{\underset{\displaystyle H}{C}}-\overset{\displaystyle H}{C}-\overset{\displaystyle H}{C}-\overset{\displaystyle H}{C}-H} + [O] \longrightarrow$$

$$\underset{\substack{| \quad | \quad |\\ H \;\; H \;\; H}}{HO-\overset{\displaystyle H}{C}-\overset{\displaystyle H}{C}-\overset{\displaystyle H}{C}-H} + [O] \longrightarrow$$

b.　Identify the products of the reactions as aldehydes or ketones.

18.42　**a.**　Write an equilibrium equation for the ionization of propionic acid:

$$CH_3CH_2-C\overset{\displaystyle O}{\underset{\displaystyle OH}{\Big\langle}}$$

b.　Which side is favored in the equilibrium? Explain.

18.43　**a.**　What are the relative magnitudes of the boiling points of carboxylic acids and aldehydes of approximately equal molecular weight?
b.　What structural feature is present in carboxylic acids that will account for your answer in part (a)?

18.44　Distinguish between a fat and an oil in terms of chemical structure and physical properties.

18.45　Vegetable oils are commercially converted into solid fats such as margarine. Describe the chemical reaction for this process.

18.46　**a.**　Write the structural formula for the organic product in the reaction between oleic acid and sodium hydroxide.
b.　What kind of substance is the organic product of this reaction?

18.47　**a.**　Explain why water cannot dissolve grease.

b.　Describe how a soap can accomplish the task of dissolving grease in water.

18.48　Write the structural formula of the ester that could form from the reaction of propionic acid with methanol.

$$CH_3CH_2-C\overset{\displaystyle O}{\underset{\displaystyle OH}{\Big\langle}} \qquad\qquad CH_3OH$$

Propionic acid　　　　　　　Methanol

18.49　From the given name identify each of the following compounds as an alcohol, ether, aldehyde, ketone, carboxylic acid, or ester:
　a.　Ethanoic acid　**f.**　Ethyl formate
　b.　2-Pentanone　**g.**　Dimethyl ether
　c.　2-Pentanol　**h.**　3-Bromobenz-
　d.　Pentanal　　　　　aldehyde
　e.　n-Propyl　　　**i.**　Isopropyl alcohol
　　　acetate　　　**j.**　2-Methylbutanoic
　　　　　　　　　　　acid

18.10　Organic compounds containing nitrogen

18.50　In what way does the structure of methyl ethyl amine differ from that of ammonia? In what way is it similar?

$$\underset{\substack{| \quad\quad\;\; | \quad |\\ H \quad\quad\; H \;\; H}}{H-\overset{\displaystyle H}{C}-\overset{\displaystyle H}{\underset{\displaystyle ..}{N}}-\overset{\displaystyle H}{C}-\overset{\displaystyle H}{C}-H}$$

Methyl ethyl amine

18.51
　a.　Write the equation for the acid-base transfer reaction between HCl and methyl ethyl amine.
　b.　Will the product be more or less soluble in water than the starting amine? Explain.

18.52　**a.**　Compare the relative boiling points and water solubility of methyl ethyl amine and an alkane of similar molecular weight.
　b.　What structural feature is present in methyl ethyl amine that could account for your answer in part (a)?

18.53 Write the structural formula for the amide that would form from methanoic acid and ethyl amine.

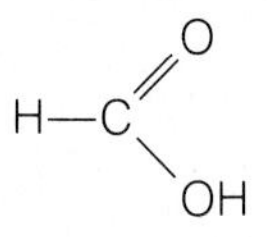

Methanoic acid 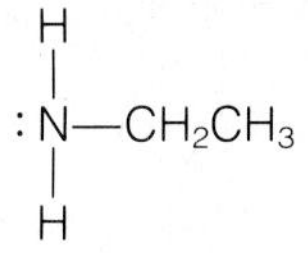   Ethyl amine

18.54 Circle and label the amide linkage in the following structure:

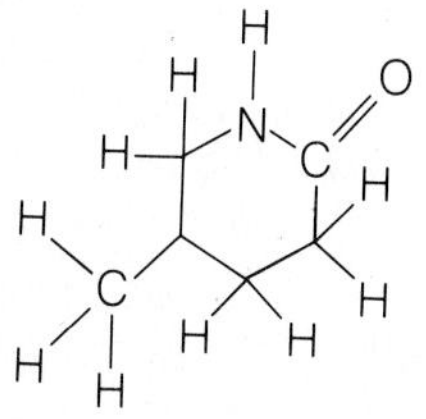

18.11 Organic molecules of biological significance

18.55 Circle and label the functional groups present in fructose:

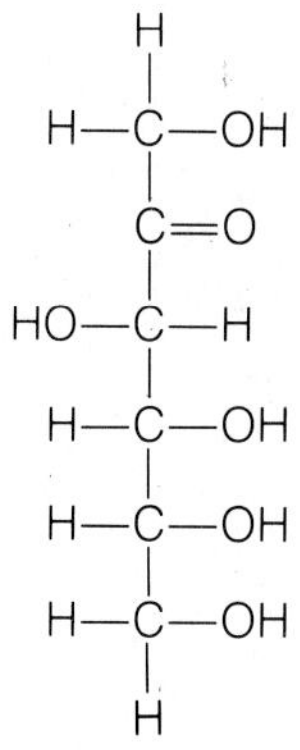

Fructose

18.56 **a.** What two monosaccharides make up sucrose?

b. What monosaccharide is starch composed of?

18.57 Do you predict carbohydrates to have stronger or weaker intermolecular attractive forces than simple aldehydes or alcohols?

18.58 Write the structure of the triglyceride that would form from glycerol and oleic acid.

18.59 What functional groups are present in an amino acid?

18.60 Write the structure of an amino acid made from a four-carbon carboxylic acid.

18.61 Write the structural formula of the dipeptide that could form from the amino acids alanine and glycine using the acid group of alanine and the amino group of glycine.

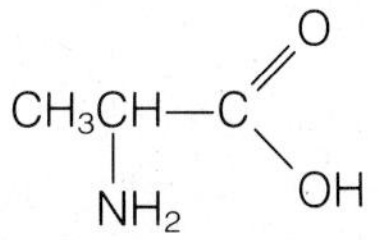

Alanine 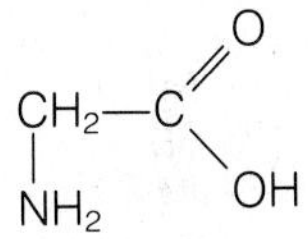   Glycine

18.62 **a.** Carry out Problem 18.61 using the acid group of glycine and the amino group of alanine.

b. Are the dipeptides formed in Problems 18.61 and 18.62, part (a) the same?

18.63 Form a tripeptide by using the acid group of the dipeptide you formed in Problem 18.61 and the amino group of valine.

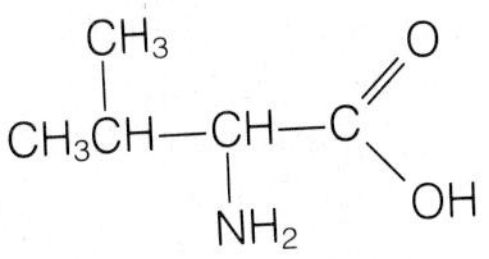

Valine

18.64 Form a tripeptide by using the acid group of the dipeptide you formed in Problem 18.62 and the amino group of valine.

BASIC ARITHMETIC REVIEW

APPENDIX 1

A PLACE VALUES

The decimal system uses the ten digits, 0, 1, 2, 3, 4, 5, 6, 7, 8, 9, to indicate the value of a particular place and thereby give the magnitude of a number. For example, a number indicating several common places is shown below:

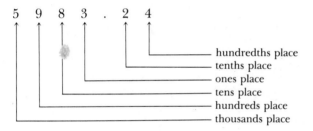

The digits zero through nine are used to indicate the value of each place. Thus the number above is made up of 5 thousands, 9 hundreds, 8 tens, 3 ones, 2 tenths, and 4 hundredths.

Notice that the place values are always one-digit numbers. Two digits in a place indicate that one digit should be carried over to the next higher place. For example, 17 in the ones column should be written as 1 in the tens column and 7 in the ones column.

B ADDITION OF POSITIVE DECIMAL NUMBERS

The key to the successful addition of decimal numbers is to line up the numbers so that digits of the same place value are found in the same vertical column.

SAMPLE EXERCISE

1A.1

Add 12.25 + 5.14 + 6.3.

SOLUTION

Step 1 Write the numbers one under the other so that the digits of the same place value fall into the same vertical column. Note that 6.3 has no digit in the hundredths column.

$$
\begin{array}{r}
12.25 \\
5.14 \\
6.3 \\
\end{array}
$$

Step 2   Add the numbers in each column, "carrying over" any value of ten or greater in any column as one or more in the next column to the left.

$$
\begin{array}{r}
12.25 \\
5.14 \\
\underline{6.3} \\
23.69 \\
\end{array}
$$

    Note that the decimal point is brought straight down

The value in the ones column was 13; 10 ones were carried to the tens column as 1 ten.

SAMPLE EXERCISE

1A.2

Add 8.93 + 0.0014 + 117.

SOLUTION

Step 1   Line up the numbers.

$$
\begin{array}{r}
8.93 \\
0.0014 \\
\underline{117} \\
\end{array}
$$

Step 2   Add.

$$
\begin{array}{r}
8.93 \\
0.0014 \\
\underline{117} \\
125.9314 \\
\end{array}
$$

Problem A.1

Add:

a.   1.79 + 5.32

b.   7.301 + 11.3 + 2.0903

C   SUBTRACTION OF POSITIVE DECIMAL NUMBERS

The key again in subtraction is to line the place values up properly. Decimal point should lie underneath decimal point. In any place value, if you are trying to subtract a larger digit from a smaller digit, you must borrow ten units from the next place value to the left. This is illustrated in the exercise below.

Do the subtraction:

$$635.49 - 173.52$$

SOLUTION

Step 1 Line up numbers by place value, putting decimal point underneath decimal point.

$$
\begin{array}{r}
635.49 \\
-173.52
\end{array}
$$

Step 2 Subtract, borrowing where necessary.

Here we borrowed 10 tens (= 1 hundred) from the hundreds column, making this 13 tens.

$$
\begin{array}{r}
5\ 4 \\
6\cancel{3}5.49 \\
-173.52 \\
\hline
461.97
\end{array}
$$

Note: We borrowed 10 tenths (= one) from the ones column, making this 14 tenths.

Problem A.2

Subtract:

a. $18.93 - 11.76$

b. $109.40 - 52.87$

D MULTIPLICATION OF POSITIVE DECIMAL NUMBERS

The decimal numbers are first multiplied using the same procedure as is used with whole numbers. The significant difference is that you must now locate the decimal point in the product. Add up the number of decimal places (digits to the right of the decimal point) in the numbers being multiplied, and place the decimal point in the product so that the number of decimal places in the product will be equal to this sum. Zeros are added to the left of the product answer when there are not enough digits for the required number of decimal places.

Multiply 18.4×0.9.

SOLUTION

Step 1 Multiply the numbers as if they were ordinary whole numbers.

$$
\begin{array}{r}
18.4 \\
\times 0.9 \\
\hline
16\,56
\end{array}
$$

Step 2 Add up decimal places in multiplied numbers and place decimal point in product so that the number of decimal places in product equals this sum.

$$18.4 \longleftarrow \text{One decimal place}$$
$$\underline{\times 0.9} \longleftarrow \text{One decimal place}$$
$$16.56 \qquad \text{Requires two decimal places}$$
$$\text{in the product}$$

SAMPLE EXERCISE

1A.5

Multiply 1.7×0.04.

SOLUTION

$$1.7 \longleftarrow \text{One decimal place}$$
$$\underline{\times 0.04} \longleftarrow \text{Two decimal places}$$

Simple multiplication gives $\longrightarrow$ 68 $\longleftarrow$ Product requires three decimal places

A zero must be added to the left of 68 to hold the tenths' place.

$$1.7$$
$$\underline{\times 0.04}$$
$$.068 \longleftarrow \text{Final answer}$$

Problem A.3

Multiply:

a. 78.45×3.2

b. 19.01×0.05

c. 0.14×0.25

E DIVISION OF POSITIVE DECIMAL NUMBERS

The key step in the division of decimal numbers is to move the decimal point in the *divisor* (denominator of a fraction) as many places to the right as necessary to make the divisor a whole number. The decimal point in the *dividend* (numerator of a fraction) is then moved to the right an equal number of places with zeros being added if necessary to hold a decimal place.

$$\frac{148.2}{11.1} = 148.2 \div 11.1$$

$$11.\underset{\smile}{1} \overline{\smash{)}148.\underset{\smile}{2}}$$

Move one place to right. Move decimal in dividend one place to right.

SAMPLE EXERCISE

1A.6

Divide:

$$\frac{17.5}{.35}$$

$$\frac{17.5}{.35} = 17.5 \div .35$$

Step 1   Move decimal point in divisor to obtain a whole number.

$$.35\overline{)17.5}$$

Step 2   Move decimal point in dividend an equal number of places to the right, filling in zeros to hold places.

$$35.\overline{)17.50}$$ Decimal point was moved two places to right with a zero filled in.

Step 3   Locate decimal point in *quotient* (answer) immediately above its current position in dividend.

$$35.\overline{)1750.}$$

Step 4   Zeros may be placed to the right of the decimal in the dividend and the division carried out to as many decimal places as desired.

$$
\begin{array}{r}
50.0 \\
35.\overline{)1750.0} \\
\underline{175} \\
000
\end{array}
$$

SAMPLE EXERCISE

1A.7

Divide:

1.482 by .111

SOLUTION

Step 1 $.111\overline{)1.482}$

Step 2 $111.\overline{)1.482}$

Step 3 $111.\overline{)1482.}$

Step 4
$$
\begin{array}{r}
13.35 \\
111.\overline{)1482.00} \\
\underline{111} \\
372 \\
\underline{333} \\
390 \\
\underline{333} \\
570 \\
\underline{555} \\
15
\end{array}
$$

Problem A.4

Divide:

a. $78.3 \div 1.1$

b. $12.90 \div .638$

c. $0.384 \div .15$

F   DIRECT AND INVERSE PROPORTIONALITY

Direct proportionality   There are many experiments in chemistry in which one measures two variable properties of a sample and finds that an increase or decrease in one variable leads to an increase or decrease in the other variable. For example, an increase in the volume of liquid leads to an increase in the mass of the liquid. (Volume and mass are properties of matter that are defined and discussed in Section 3.2.)

If an increase (or decrease) by a given multiplying factor in one variable produces an increase (or decrease) by the same factor in the other variable, then the two variables are said to be directly proportional. Experiments show that if we double the volume of a liquid, we find that the mass doubles. If we triple the volume, we triple the mass. If we cut the volume in half, we find that the mass is also cut in half. Because this is true, volume and mass are said to be directly proportional. The symbol $\propto$ is shorthand for the words *is proportional to*. We can write:

$$M \propto V$$

The direct proportionality can be converted into an equation by inserting a proportionality constant k. Thus

$$M \propto V \quad \text{can also be expressed} \quad M = kV$$

SAMPLE EXERCISE

1A.8

Solve for k in $M = kV$.

SOLUTION

Divide both sides by V.

$$\frac{M}{V} = \frac{k\cancel{V}}{\cancel{V}} = k$$

$$k = \frac{M}{V}$$

In Section 3.4 the proportionality constant in this problem is given the name density.

If any two variables x and y are directly proportional, then the equation relating them will have the form $x = ky$, where k is constant. As we did in Sample Exercise 1A.8, we can rearrange this equation to read $k = x/y$. This tells us that if two variables are directly proportional, their ratio is a constant.

Given the following data, are x and y directly proportional?

x	y
3	6
6	9
9	12

SOLUTION

If two variables are directly proportional, then their ratio is a constant. When we examine the ratio y/x, we find $\frac{6}{3} = \frac{2}{1}$, $\frac{9}{6} = \frac{3}{2}$, $\frac{12}{9} = \frac{4}{3}$. The ratio y/x is not constant, and therefore x and y are *not* directly proportional. Another way of identifying directly proportional variables is to graph them (Section 2.7). Such a graph is always a straight line through the origin.

The directly proportional concept is particularly important in understanding Boyle's law problems (Section 11.5).

Inverse proportionality   If two variables are related such that an increase (decrease) in one by a given factor produces a decrease (increase) in the other by the same factor, then the two variables are said to be inversely proportional. For example, average exam score and difficulty of a test may be inversely proportional. If we make the test twice as difficult (increase by a factor of 2), the average exam score will be cut in half (decrease by a factor of 2).

The inverse proportionality between two variables x and y can be represented as follows:

$$x \propto \frac{1}{y}$$

We can convert the inverse proportionality into an equation by inserting a proportionality constant. Thus

$$x \propto \frac{1}{y} \quad \text{can also be expressed} \quad x = \frac{k}{y}$$

Solve for k in $x = \dfrac{k}{y}$.

SOLUTION

Multiply each side of the equation by y.

$$xy = k$$

Sample exercise 1A.10 shows that if two variables x and y are inversely proportional, then their product must be equal to a constant; that is, $xy = k$. This means that if x is doubled, y must be cut in half in order for the product xy to still be a constant. The equation of two variables x and y that are inversely proportional is always of the form $xy = k$.

SAMPLE EXERCISE

1A.11

Given the following data, are x and y inversely proportional?

x	y
16	12
8	24
4	48

SOLUTION

If two variables are inversely proportional, then their product is a constant. When we examine the product xy, we find $(16)(12) = 192$, $(8)(24) = 192$, $(4)(48) = 192$. The product xy is a constant, and therefore x and y are inversely proportional.

The inversely proportional concept is particularly important in understanding Charles' law problems (Section 11.6).

CHEMICAL ARITHMETIC USING A HAND CALCULATOR

APPENDIX 2

INTRODUCTION

Using an electronic hand calculator can speed up solving numerical problems. In this review we summarize those operations which apply to problems in this text. Calculators use an operating system based on either algebraic notation or reverse polish notation. Most beginning students have the first type, and we limit our discussion to that type.

ADDITION/SUBTRACTION

Before doing a new calculation, press the clear key to delete any previous data or operation. The $(+)$ key is used for addition and $(-)$ for subtraction. The $(+/-)$ key changes the sign of a number in the display. Pressing the $(=)$ button provides the result of the operations in the order in which they were entered. Operations are entered by reading calculations from left to right. Let's look at a few examples.

SAMPLE EXERCISE

2A.1

$$8.94 + 0.39 - 3 =$$

1. Enter the number 8.94.

2. Press $(+)$ for addition.

3. Enter the number 0.39. [At this point the calculator will have added 0.39 to 8.94. You could check this by pressing $(=)$.]

4. Press $(-)$ for subtraction.

5. Enter 3.

6. Press $(=)$. The calculator will display the result 6.33.

SAMPLE EXERCISE

2A.2

$$-4.95 - (-2.1) + 0.93 =$$

1. Enter the number 4.95.

2. Press $(+/-)$ to change the sign of 4.95 to -4.95.

3. Press $(-)$ for subtraction.

4. Enter 2.1.

5. Press $(+/-)$ to change the sign of 2.1 to -2.1.

6. Press $(+)$ for addition.

7. Enter 0.93.

8. Press $(+)$ to display the result -1.92.

MULTIPLICATION/DIVISION

The $(\times)$ sign is pressed for multiplication, and $(\div)$ is pressed for division.

SAMPLE EXERCISE

2A.3

$$7.21 \times 4.01 \div 6.35 =$$

1. Enter the number 7.21.

2. Press $(\times)$.

3. Enter 4.01.

4. Press $(\div)$.

5. Enter 6.35.

6. Press $(=)$ to display the result 4.55.

SAMPLE EXERCISE

2A.4

$$-8.50 \div 2.30 \times -1.75 =$$

1. Enter the number 8.50.

2. Press $(+/-)$ to change the sign of 8.50 to -8.50.

3. Press $(\div)$.

4. Enter 2.30.

5. Press $(\times)$.

6. Enter 1.75.

7. Press $(+/-)$ to change the sign of 1.75 to -1.75.

8. Press $(=)$ to display the result 6.47.

Note that in multiplication and division, it is not necessary to change signs when numbers are entered. The appropriate sign in the answer can be determined as described in Section 2.2.

COMBINED ADDITION/SUBTRACTION AND MULTIPLICATION/DIVISION

In combining addition/subtraction with multiplication/division operations you need to be careful about the order in which the operations are carried out. For example, a chain of operations such as $3 \times 5 + 2 \times 4$, shown without parentheses, could be carried out as

$$(3 \times 5) + (2 \times 4) = 23 \quad \text{or}$$
$$((3 \times 5) + 2) \times 4 = 68 \quad \text{or}$$
$$3 \times (5 + 2) \times 4 = 84 \quad \text{or}$$
$$3 \times (5 + (2 \times 4)) = 39$$

Some calculators using the algebraic operating system use a priority system of operations. Multiplications and divisions are carried out before addition and subtractions. With such a calculator the result 23 is obtained.

Other calculators simply apply the operations in order from left to right. Such a calculator would yield 68 by first multiplying $3 \times 5 = 15$, then adding 2 to obtain 17, and then multiplying 17×4 to obtain 68.

If your calculator has parenthesis keys, you can have it carry out a chain operation in any order you wish. For example, using parentheses before 5 and after 2, the example reads $3 \times (5 + 2) \times 4 =$. The calculator will first carry out the operations within parentheses, in this case $5 + 2$ yielding 7, then multiply 3×7 to obtain 21, and then multiply by 4 to display 84.

SCIENTIFIC NOTATION

Many calculators have an EE or Exp (exponential) key which allows decimal numbers to be entered in scientific notation. Numbers in scientific notation can be used in arithmetic operations with other numbers in scientific notation or ordinary decimal numbers.

SAMPLE EXERCISE

2A.5

Enter 6.02×10^{23}.

1. Enter the number 6.02.

2. Press EE or Exp.

3. Enter 23.

SAMPLE EXERCISE

2A.6

Enter 1.67×10^{-24}.

1. Enter the number 1.67.
2. Press EE or Exp.
3. Enter 24.
4. Press $(+/-)$ to change the exponent 24 to -24.

SAMPLE EXERCISE

2A.7

$$(6.02 \times 10^{23}) \times (1.67 \times 10^{-24})$$

1. Enter 6.02×10^{23} as in Sample Exercise 2A.5.
2. Press $(\times)$.
3. Enter 1.67×10^{-24} as in Sample Exercise 2A.6.
4. Press $(=)$ to display the result.

In using results from your calculator, remember that the calculator does not necessarily yield an answer to the correct number of significant figures. In fact, it rarely does. You must consider the measured quantities and the arithmetic operations and round to the appropriate number of sig figs as described in Sections 3.8 and 3.9.

VAPOR PRESSURE OF WATER AT VARIOUS TEMPERATURES

APPENDIX 3

Temperature (°C)	Vapor pressure (torr)	Temperature (°C)	Vapor pressure (torr)
5	6.5	34	40.0
10	9.2	35	42.2
15	12.8	36	44.6
16	13.6	37	47.1
17	14.5	38	49.7
18	15.5	39	52.4
19	16.5	40	55.3
20	17.5	45	71.9
21	18.6	50	92.5
22	19.8	55	118.0
23	21.1	60	149.4
24	22.4	65	187.5
25	23.8	70	233.7
26	25.2	75	289.1
27	26.7	80	355.1
28	28.3	85	433.6
29	30.0	90	525.8
30	31.8	95	633.9
31	33.7	100	760.0
32	35.7		

ANSWERS TO SELECTED PROBLEMS

CHAPTER 1

1.1 a. (1) Solid   (2) Liquid   (3) Liquid   (4) Gas
  b. (1) Solid   (2) Solid   (3) Semisolid   (4) Gas

1.2 a. Mixture   **b.** Mixture   **c.** Pure substance   **d.** Mixture

1.3 a. Chemical   **b.** Physical   **c.** Physical   **d.** Chemical
  e. Chemical

1.4 Laboratory glassware is odorless, colorless, transparent, brittle, solid at room temperature, and a poor conductor of electricity.

1.5 Table 1.3 lists the boiling point of carbon tetrachloride as 76.5°C. Since the boiling point of chloroform is given as 61°C, an experimental measurement of the boiling points of the two unlabeled liquids would distinguish between them.

1.6 Try to ignite the samples. Methane burns. Oxygen does not.

1.7 a. Mixture   **b.** Compound   **c.** Element   **d.** Compound

1.8 The elements and compounds are pure substances.

1.9 a. Homogeneous   **b.** Homogeneous   **c.** Homogeneous
  d. Homogeneous

1.11 Matter is anything that occupies space and has mass.

1.13 A gas will expand to fill any size container.

1.15 Gases and liquids. (Solids can also be poured in the powdered form.)

1.17 a. Lower its temperature to its freezing point.
  b. Raise its temperature to its boiling point.

1.19 a. Melting   **b.** Condensation   **c.** Freezing   **d.** Sublimation
  e. Evaporation

1.21 Three physical states: gas—air; liquid—water; solid—ice, wood

1.23 a. Mixture   **b.** Mixture   **c.** Mixture   **d.** Mixture (an alloy).
  e. Pure substance   **f.** Pure substance (water); dirty snow is a mixture.
  g. Pure substance.

1.25 a. Observe escaping gas and consequent "flatness" acquired. Heat the cola to evaporate the solvent. The syrup remaining is thicker than the original cola at room temperature.
  b. Let the dry ice sublime in a closed evacuated container. Decrease the temperature to the freezing point and the solid will be the same as the original solid.

1.27 a. A lump of sugar is ground to a powder (physical change) and then heated in air. It melts (physical change), then darkens (chemical change), and finally bursts into flames and burns (chemical change).
  b. Gasoline is sprayed into the carburetor, mixed with air (physical change), converted to vapor (physical change), and burned (chemical change), and the combustion products expand (physical change) in the cylinder.

1.29 a. Physical   **b.** Chemical   **c.** Chemical   **d.** Physical

1.31 An element

1.33 The pure substances in Problem 1.18 are all compounds.

1.35 a. Solid   **b.** Gas   **c.** Liquid   **d.** Solid   **e.** Liquid

1.37 Odor, miscibility, melting point, boiling point, and flammability

1.39 a. Physical   **b.** Physical   **c.** Chemical   **d.** Physical

1.41 They are similar in that the final condition is the liquid state in both cases. They are different in that melting converts pure solid sugar into pure liquid sugar whereas dissolving converts pure solid sugar into a homogeneous mixture of sugar and water.

1.43 a. Club soda = CO_2 gas (compound) in water (compound) with a pinch of salt (compound).
  b. Many of the components such as oxygen, nitrogen, and argon are elements. Some such as CO_2 and water vapor are compounds.
  c. Aerated water = water (compound) with air—nitrogen (element) and oxygen (element)—dissolved in it.

1.45 3 g of A will combine with 1 g of B to form 4 g of AB. 1 g of A will remain.

1.47 Homogeneous refers to uniform matter. Heterogeneous refers to nonuniform matter.

1.49 Water is a compound. Pepsi Cola is a mixture containing water, sugar, CO_2, and other substances.

1.51 Br, N, Hg, Ag, and Au.

1.53 a. The "L" should be lowercase: Cl.
  b. The "3" and "6" should be subscripts: C_3H_6.
  c. Correct.
  d. The "f" should be a capital: F.
  e. Correct.

1.55 a. 4, 6   **b.** 1   **c.** 2,5   **d.** 2, 6   **e.** 5

1.57 Potential and kinetic

1.59 a. The energy stored in a compressed spring, or the energy stored in a book held above the floor.
  b. The chemical energy stored in a compound; e.g., gasoline, TNT, or nitroglycerine.
  c. Yes. The composition of a compound does not change with its position.

1.61 Although energy can change from one form to another, the total amount of energy in the universe is constant.

1.63 Excess chemical energy is stored as fat.

CHAPTER 2

2.1 a. –   **b.** –

2.2 a. –2.7   **b.** 7.4   **c.** 36   **d.** –36

2.3 a. -3.0 **b.** 3.0   **c.** -3.0

2.4 a. 11.2   **b.** 3.6   **c.** -11.2 **d.** 3.6   **e.** -11.2 **f.** -3.6

2.5 a. 1.73×10^2 **b.** 2.9×10^{-3} **c.** 1.31982×10^5 **d.** 4.01×10^{-6}

   e. 1.64×10^1 **f.** 8.1×10^0 or 8.1

2.6 a. 7310   **b.** 0.000192   **c.** 638,000   **d.** 0.0000836

2.7 a. 2.30×10^7 **b.** 1.65×10^2 **c.** 3.84×10^{-14} **d.** 3.16×10^2

2.8 a. 2.0×10^4 **b.** 2.73×10^{-10} **c.** 2.01×10^8 **d.** 9.47×10^{-2}

2.9 a. 4.07×10^4 **b.** 2.84×10^2 **c.** 8.25×10^{-3} **d.** 5.09×10^{-3}

2.10 a. -1 **b.** 3   **c.** $\dfrac{13}{A}$ **d.** -22

2.11 a. 0.045   **b.** 0.33   **c.** 0.00092

2.12 6.23 lb

2.13 56%

2.14 a.

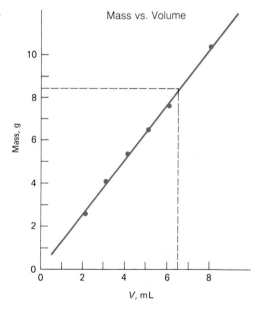

b. Mass = 8.4 g

2.15

	Magnitude	Sign
a.	13.3	$+$
b.	3.13	$-$
c.	0.0184	$-$
d.	90.5	$+$
e.	42.3	$+$
f.	4	$-$

2.17 a. -4 **b.** 0.25   **c.** -4.3 **d.** -3.9 **e.** 3.9

2.19 a. -22 **b.** 61   **c.** -36.6 **d.** -5.3

2.21 If $\dfrac{2}{17} = \dfrac{6}{51}$, then (6)(17) must = (2)(51); 102 = 102.

2.23 3/8 = 0.375

2.25 a. 0.222   **b.** 2.0   **c.** -0.889 **d.** 27   **e.** 0.667
    f. -3.67

2.27 a. 6.0×10^{11} **b.** 2.69×10^{0} or 2.69   **c.** 2.97×10^{-15}
    d. 2.72×10^{-6} **e.** 4.32×10^{16}

2.29 a. 5.7×10^{4} **b.** 8.06×10^{4} **c.** 8.29×10^{4} **d.** 4.18×10^{7}

2.31 $x = 4$

2.33 $F = (9/5)C + 32$

2.35 $n = \dfrac{PV}{RT}$

2.37 $x = 3$

2.39 $\Delta t = 2.5°C$

2.41 10.3 lb

2.43 3.25%

2.45 0.349

2.47 a.

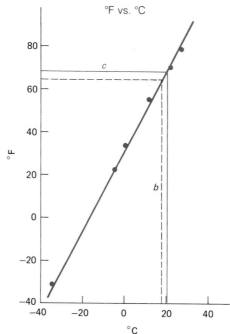

b. 18°C = 64°F
c. 68°F = 20°C

2.49 a.

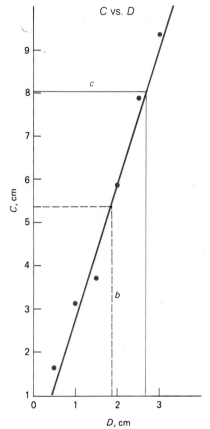

C vs. *D*

b. D = 1.8; C = 5.4 cm
c. C = 8.0; D = 2.7 cm

CHAPTER 3

3.1 a. $\dfrac{2.2 \text{ lb}}{1 \text{ kg}}$ and $\dfrac{1 \text{ kg}}{2.2 \text{ lb}}$

b. $\dfrac{1 \text{ liter}}{1000 \text{ mL}}$ and $\dfrac{1000 \text{ mL}}{1 \text{ liter}}$

3.2 439 mL

3.3 549 g

3.4 0.90 g/mL

3.5 53 mL

3.6 22°C

3.7 60. cal

3.8 a. 4   **b.** 3   **c.** 4   **d.** 4

3.9 23

3.10 1032.91 g

3.11 Measurement is the comparison of some quantity to a reference standard.

3.13 a. 1 g = 0.001 kg or 1 kg = 1000 g
b. 1 liter = 1000 mL or 1 mL = 0.001 liter
c. 1 m = 100 cm or 1 cm = 0.01 m
d. 1 μm = 0.000001 m or 1 m = 1,000,000 μm
e. 1 mL = 1 cm^3

3.15 a. Centimeter < meter < kilometer
b. Microgram < milligram < gram
c. Milliliter = cubic centimeter < liter

3.17 a. 0.0189 km **b.** 0.375 m **c.** 145 mL **d.** 0.452 liter
e. 645 mL **f.** 89 g **g.** 1900 mL **h.** 38,400 mg

3.19 a. 5300 cm **b.** 0.000385 km **c.** 0.00319 kg **d.** 8.2 cm

3.21 a. 84,000 cm **b.** 0.106 gal **c.** 43 g

3.23 109 yd

3.25 212 mi

3.27 41 cm

3.29 18.4 m

3.31 a. 1500 m (1.5 × 10^3 m) **b.** 0.93 mi

3.33 30.2 cents/liter

3.35 11 km/liter

3.37 a. 88 km/h **b.** 9.6 km/liter

3.39 a. 55 mi/h **b.** 23 mi/gal

3.41 a. 3.85 × 10^8 m **b.** 1.28 s

3.43 0.7016 g/mL

3.45 a. 23 g **b.** 999 g **c.** 17 g

3.47 486 g

3.49 a. 2.2 lb **b.** 1.4 lb

3.51 11.3 g/cm^3

3.53 1.9 × 10^5 g

3.55 a. 24°C **b.** −21°C **c.** 64°F **d.** −110°F (−1.1 × 10^2) **e.** −9.194°C

3.57 116°C

3.59 265°C

3.61 43,500 cal (4.35 × 10^4 cal)

3.63 3.2 kcal

3.65 227 g

3.67 0.106 $\dfrac{\text{cal}}{\text{g} \times \text{°C}}$

3.69 b. The aluminum sample is coldest.

3.71 a. 3 **b.** 1 **c.** 4 **d.** 2 **e.** 4 **f.** 1 **g.** 2 **h.** 3
i. 4 **j.** 3

3.73 a. 9.4   **b.** 9   **c.** 0.002   **d.** 5.1×10^{-3} **e.** 0.191

3.75 a. 7.015 lb   **b.** 7.01 lb

3.77 184 mL

CHAPTER 4

4.1 Compounds are pure substances with a constant composition because atoms are combined in simple whole-number ratios. These ratios lead to a fixed proportion by weight. Chemical changes can disrupt the combination of atoms.

4.2 The diameter of a Cs atom is 5.24×10^{-8} cm. The number of Cs atoms along a 1-mm line is 1.91×10^6 atoms. The number of Cs atoms in a 1-cm cube is 6.95×10^{21} atoms.

4.3 a. Mg, 12; S, 16; Ag, 47
　　b. Mg, 12; S, 16; Ag, 47

4.4 a. Chlorine   **b.** Calcium

4.5 Al, 13; K, 19; Cu, 29

4.6 a. 3, 18, 53
　　b. Lithium, argon, iodine

4.7 12 protons

4.8 a. 8 protons, 8 electrons, and 9 neutrons
　　b. 15 protons, 15 electrons, and 16 neutrons
　　c. 27 protons, 27 electrons, and 33 neutrons

4.9 a. $^{32}_{16}Y$ contains 16 p^+, 16 e^-, 16 n^0.
　　$^{32}_{15}Y$ contains 15 p^+, 15 e^-, 17 n^0.
　　$^{127}_{53}Y$ contains 53 p^+, 53 e^-, 74 n^0.
　　$^{31}_{15}Y$ contains 15 p^+, 15 e^-, 16 n^0.
　　$^{130}_{53}Y$ contains 53 p^+, 53 e^-, 77 n^0.
　　b. $^{32}_{16}Y$ is sulfur. $^{32}_{15}Y$ is phosphorus. $^{127}_{53}Y$ is iodine.
　　$^{31}_{15}Y$ is phosphorus. $^{130}_{53}Y$ is iodine.
　　c.

　　d. $^{32}_{15}Y$ and $^{31}_{15}Y$ are isotopes of phosphorus.
　　$^{127}_{53}Y$ and $^{130}_{53}Y$ are isotopes of iodine.

4.10 Q + 3 (cation)
　　R + 2 (cation)
　　T − 1 (anion)

4.11 16 amu

4.12 6.93 amu

4.13 Fluorine and oxygen exist as diatomic molecules.

4.15 All atoms have protons and neutrons in the nucleus and electrons outside the nucleus.

4.17

Dense nucleus
Fluffy electron cloud

4.19 Electron

4.21 Proton $\approx$ neutron $\approx$ 1 amu
Electron $\approx$ 0

4.23 Protons

4.25 Protons and neutrons are in the nucleus of the atom. Electrons move in a region of space around the nucleus.

4.27 The atomic number equals the number of protons. The number of electrons may or may not be equal to this number.

4.29 a. Nitrogen   **b.** Sodium   **c.** Argon   **d.** Helium

4.31 Isotopes of the same element contain equal numbers of protons. They differ in the number of neutrons they contain.

4.33 The number of protons must be equal.

4.35 a. 23 p^+, 28 n^0, 23 e^-
b. 14 p^+, 14 n^0, 14 e^-
c. 9 p^+, 10 n^0, 9 e^-

4.37

Symbol	Protons	Neutrons	Electrons
^9_4Be	4	5	4
$^{127}_{53}\text{I}$	53	74	53
$^{31}_{15}\text{P}$	15	16	15
$^{40}_{18}\text{Ar}$	18	22	18

4.39 a and f; b and d.

4.41 a. $^{39}_{19}\text{K}$ **b.** $^{40}_{19}\text{K}$

4.43 $^{15}_7\text{N}$

4.45 Yes. No. Carbon-12 is $^{12}_6\text{C}$; i.e., it has 6 n^0, whereas $^{13}_6\text{C}$ and carbon-13 both represent atoms with 7 n^0.

4.47 a. 38 p^+:50 n^0:36 e^-
b. 11 p^+:12 n^0:10 e^-
c. 7 p^+: 7 n^0:10 e^-
d. 53 p^+:74 n^0:54 e^-

4.49 a. $^{31}_{15}\text{P}^{3-}$ **b.** $^{39}_{19}\text{K}^+$ **c.** $^1_1\text{H}^+$ **d.** $^{19}_9\text{F}^-$ **e.** $^{85}_{37}\text{Rb}^+$

4.51 Anions are formed when a neutral atom gains electrons.

4.53 a. F^- **b.** Mg^{2+} **c.** Na^+ **d.** O^{2-} **e.** N^{3-}

4.55 60 amu

4.57 2:1

4.59 0.950 sulfur-32:0.0076 sulfur-33:0.0422 sulfur-34

4.61 20.2 amu

4.63 Group IA: sodium, potassium, lithium
Group VIA: oxygen, sulfur
Group VA: phosphorus, arsenic

4.65 a. Be   **b.** Te   **c.** F   **d.** Kr

4.67 b, c, and d

4.69 Boron is the most nonmetallic. Thallium is the most metallic.

CHAPTER 5

5.1 Less stable

5.2 Case A

5.3 a. Two electrons in the 1s, two in the 2s, and six in the 2p; two electrons in the 3s and four in the 3p
   b. Two electrons in the 1s, two in the 2s, and six in the 2p; two electrons in the 3s and three in the 3p
   c. Two electrons in the 1s and two in the 2s; six electrons in the 2p; and two electrons in the 3s

5.4 $1s^2 2s^2 2p^6 3s^2 3p^2$

5.5 Three

5.6 4p sublevel

5.7

	n	ℓ	m	S
e_1	1	0	0	$-\frac{1}{2}$
e_2	1	0	0	$+\frac{1}{2}$
e_3	2	0	0	$-\frac{1}{2}$
e_4	2	0	0	$+\frac{1}{2}$
e_5	2	1	-1	$-\frac{1}{2}$
e_6	2	1	0	$-\frac{1}{2}$
e_7	2	1	1	$-\frac{1}{2}$
e_8	2	1	-1	$+\frac{1}{2}$
e_9	2	1	0	$+\frac{1}{2}$
e_{10}	2	1	1	$+\frac{1}{2}$
e_{11}	3	0	0	$-\frac{1}{2}$
e_{12}	3	0	0	$+\frac{1}{2}$

5.8 a. 6s ($n + \ell = 6$)
     4f ($n + \ell = 7$)
     $n + \ell$ is lower for 6s.
   b. 4f ($n + \ell = 7$)
     5d ($n + \ell = 7$)
     Same value of $n + \ell$ but lower value for n for 4f sublevel

5.9 $1s^2 2s^2 2p^6 3s^2 3p^6 4s^2 3d^{10} 4p^6 5s^2 4d^{10} 5p^6 6s^1$

5.10 Two for helium; eight for all the other elements in Group VIIIA

5.11 $2s^2 2p^4$

5.12 Na· ·N̈· :B̈r· ·C̈· :N̈e: Mg·

5.13 a. Group IA   **b.** Alkali metals

5.14 Metals. The atomic size decreases from left to right across a period, and the metals are on the left of each period.

5.15 a. Helium   **b.** Francium

5.17 a. The apple is in a lower potential energy state on the ground.
   b. Water seeks a lower energy state.
   c. Close approach of negatively charged e^- to the positively charged nucleus is a low-energy condition.

5.19 a. Electron-proton attractions.
  b. Electron-electron repulsions.
  c. Energy is required.
  d. Greater. According to the experimental result in part c, the force in part a must be greater than that in part b.

5.21 a. Force between the electron and the proton.
  b. Force between the electron of one atom and the proton of that same atom, and between the electron of the first atom and the proton of the second atom.
  c. Electron$_1$: electron$_2$
    Proton$_1$: proton$_2$
  d. As the distance between the atoms is made smaller, the attractive and repulsive forces will become larger. A minimum energy could and does exist. A hydrogen molecule.

5.23 s is spherical; p is shaped like a long balloon tied in the middle.

5.25 Electrons in an orbital must be of opposite spin to be in the lowest-energy state.

5.27 s sublevel

5.29 a. $1s^2\ 2s^22p^6\ 3s^23p^6\ 4s^2$
  b. $1s^2\ 2s^22p^6\ 3s^23p^6\ 4s^23d^{10}4p^3$
  c. $1s^2\ 2s^22p^6\ 3s^23p^6\ 4s^1$
  d. $1s^2\ 2s^22p^6\ 3s^23p^6\ 4s^23d^{10}4p^5$
  e. $1s^2\ 2s^22p^6\ 3s^23p^6\ 4s^23d^{10}4p^6\ 5s^24d^{10}5p^1$

5.31 a. Na

$1s^2$	$2s^2$	$2p^6$			$3s^1$
↑↓	↑↓	↑↓	↑↓	↑↓	↑

  b. Si

$1s^2$	$2s^2$	$2p^6$			$3s^2$	$3p^2$	
↑↓	↑↓	↑↓	↑↓	↑↓	↑↓	↑	↑

  c. Be

$1s^2$	$2s^2$
↑↓	↑↓

  d. C

$1s^2$	$2s^2$	$2p^2$	
↑↓	↑↓	↑	↑

  e. F

$1s^2$	$2s^2$	$2p^5$	
↑↓	↑↓	↑↓	↑↓	↑

  f. Co

$1s^2$	$2s^2$	$2p^6$			$3s^2$	$3p^6$			$4s^2$
↑↓	↑↓	↑↓	↑↓	↑↓	↑↓	↑↓	↑↓	↑↓	↑↓

$3d^7$

↑↓	↑↓	↑	↑	↑

5.33 The electrons fill the $2p$ orbitals singly, with the same spin, before any pairing takes place. This is the lowest-energy arrangement for three electrons in the p orbitals.

5.35 a. $n = 4$; $n^2 = 16$ orbitals
  b. $4s$, $4p$, $4d$, $4f$
  c. $2\,n^2 = 32\ e^-$

5.37 a. $2s$ **b.** $4d$ **c.** $3p$ **d.** $2p$ **e.** $2p$

5.39 $6f\ (n + \ell = 9)$
    $7d\ (n + \ell = 9)$
  $6f$ is lower in energy.

5.41 No

5.43 Absorption spectra are obtained by measuring, quantitatively, the amount of *energy absorbed* by a sample in promoting its electrons to higher levels, while emission spectra are obtained by measuring the intensity of *light emitted* when the excited electrons of a sample fall to the ground state.

5.45 Light energy is emitted as excited electrons fall back to the ground state. The wavelength of light energy corresponds to red light.

5.47 They are all in Group VA; valence configuration is ns^2np^3.

5.49 a. B **b.** K **c.** Co **d.** Ga **e.** Kr

5.51 a. $3s^23p^4$ **b.** $3s^23p^3$ **c.** $5s^25p^5$ **d.** $4s^24p^6$

5.53 a. O, Se, Te, or Po **b.** Be, Mg, Ca, Ba or Ra **c.** Cl, Br, I, or At
d. Li, K, Rb, Cs, or Fr **e.** Ne, Ar, Kr, Xe, or Rn

5.55 Li· ·Si· :Ï· :Är: :P· ·Ba·

5.57 a. To the left
b. To the right, nonmetals
c. Inert gases; highly stable

5.59 a. C **b.** F **c.** Cs **d.** Cs

5.61 Transition metals are those elements found in the B groups of the periodic table.

5.63 Atomic number, atomic weight, relative size, relative ionization energy, electron configuration, nature of element (metal or nonmetal), physical and chemical properties

CHAPTER 6

6.1 a. 3 e^-; Ne; **b.** 2 e^-; Ar; **c.** Al^{3+} cation, S^{2-} anion

6.2 a.

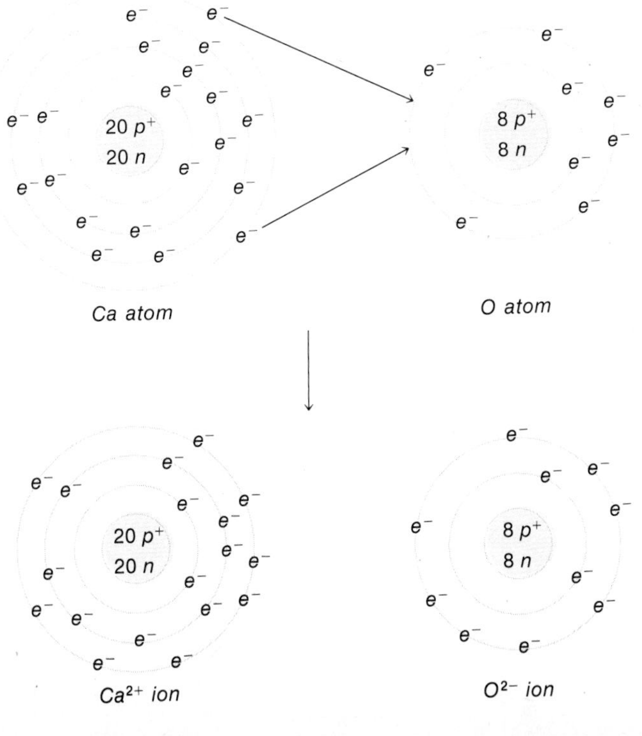

Ca atom   O atom

Ca^{2+} ion O^{2-} ion

b.

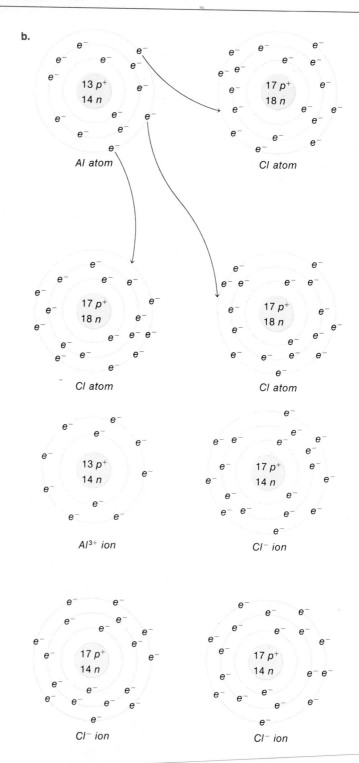

6.3 3^-

6.4 a. NaI **b.** $(NH_4)_2S$ **c.** $Al(NO_3)_3$ **d.** $Al_2(CO_3)_3$
e. $AlPO_4$ **f.** $(NH_4)_3PO_4$

6.5 a. 1:1 **b.** 1:2 **c.** 3:1 **d.** 1:1 **e.** 3:2 **f.** 1:1

6.6 a. MgS **b.** $CrCl_3$ **c.** Ca_3N_2 **d.** PbO_2

6.7 a. NaI **b.** $(NH_4)_2S$ **c.** $Al(NO_3)_3$ **d.** $Al_2(CO_3)_3$
e. $AlPO_4$ **f.** $(NH_4)_3PO_4$

6.8 a. Cl_2 **b.** I_2 **c.** Atoms

6.9 H:B̈r: H:F̈: H:Ï:

6.10 a.

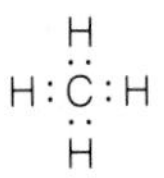

H
H:C̈:H
H

b.

:Ö:
|
H:Ö—P—Ö:H
|
:O:
H

c.

$$\left[\begin{array}{c} :Ö: \\ | \\ :Ö—S—Ö: \\ | \\ :O: \end{array} \right]^{2-}$$

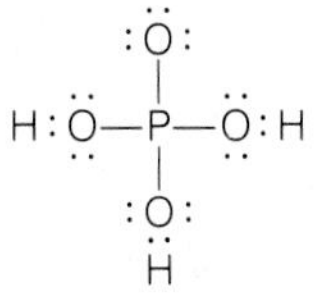

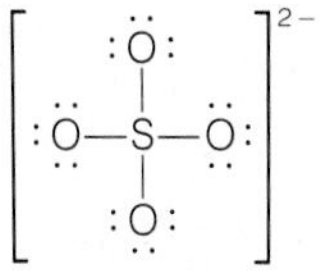

d.

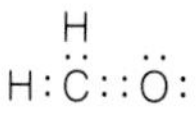

H
H:C̈::Ö:

6.11 H:Ö: + H^+ ⟶ $\left[\begin{array}{c} H:Ö:H \\ H \end{array} \right]^+$
H

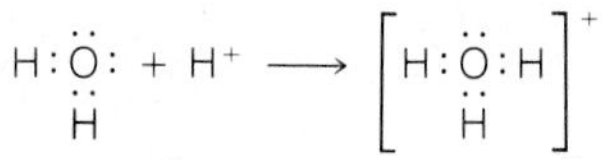

6.12 C_2H_2 24:2 = 12:1

6.13 H_2S 2:32 = 1:16

6.14 MgS 24.3:32.1

6.15 The force that holds elements together in compounds.

6.17 Filled valence electron levels, that is, $1s^2$ duet for He and octet (ns^2np^6) for other noble gases.

6.19 Transfer or sharing of electrons.

6.21 Loss of two electrons from Ca produces the Ar configuration. Loss of a third electron would destroy that.

6.23 a. Two **b.** Xe

6.25 a. Krypton **b.** Helium **c.** Krypton

6.27 Gain of one electron makes them isoelectronic with a noble gas.

6.29 a. $1s^2\ 2s^22p^6\ 3s^23p^6$
b. Argon

6.31 a. Krypton **b.** Argon **c.** Xenon

6.33 a.

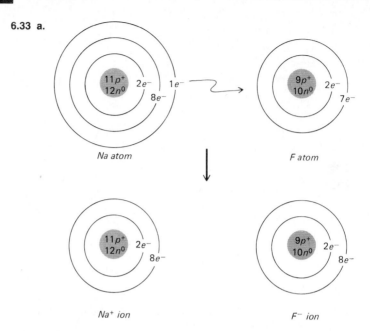

b. Na needs to lose only 1 e^- and F needs to gain only 1 e^- in order for them both to attain the Ne noble-gas configuration. Then the $+1$ charge on Na^+ is exactly neutralized by F^-.

6.35 a.

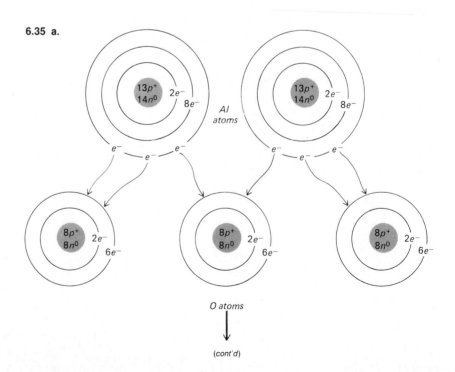

(cont'd)

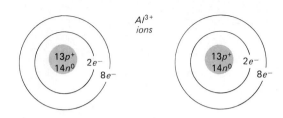

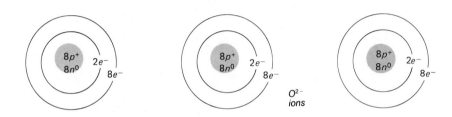

b. Al atoms needs to lose 3 e^- to attain a Ne noble-gas configuration. Each O atom needs only 2 e^- to achieve the Ne configuration. For equal numbers of e^- to be lost and gained, 2 Al atoms lose 3 e^- each (6 e^-), and 3 O atoms gain 2 e^- each (6 e^-). The total positive charge of $+6$ (2 Al^{3+}) is exactly neutralized by the total negative charge of -6 (3 O^{2-}).

6.37 a. CO_3^{2-} **b.** HCO_3^- **c.** PO_4^{3-} **d.** NH_4^+ **e.** SO_3^{2-}

6.39 a. Li_2O **b.** $BaCl_2$ **c.** Na_2S **d.** Ni_3N_2 **e.** AgBr   **f.** FeO
   g. Fe_2O_3 **h.** $SnCl_2$ **i.** $SnCl_4$ **j.** Ga_2S_3 **k.** CaI_2

6.41 X_2Y

6.43 The attraction between cations and anions

6.45 a. 3   **b.** 5   **c.** 4   **d.** 3

6.47 Ionic bonds are formed through the transfer of electrons from metal to nonmetal. Covalent bonds are formed through the sharing of electrons between nonmetals.

6.49 Helium has a filled outer level; therefore, it has no need to share electrons. Each hydrogen atom has one valence electron and requires one more electron to obtain a noble-gas configuration. If two hydrogen atoms share their electrons, they both obtain noble-gas configurations.

6.51 a. Six   **b.** Two

6.53 a. High-energy content
   b. High-energy content
   c. Minimum-energy content

6.55 The chlorine molecule is more stable because energy must be added to the molecule in order to separate the atoms.

6.57 a. S   **b.** C   **c.** S   **d.** N

6.59 a. $\ddot{O}=C=\ddot{O}$ **b.** $H-\ddot{O}-\overset{\displaystyle \cdot}{\underset{\displaystyle :\ddot{O}:}{Cl}}-\ddot{O}:$ **c.** $:C:::O:$ **d.** $H:\overset{H}{\underset{H}{C}}:\overset{H}{\underset{H}{C}}:H$

e. $:\ddot{F}:\ddot{O}:\ddot{F}:$ **f.** $H:\ddot{O}:\ddot{O}:H$ **g.** $H:\overset{\displaystyle H}{\underset{\displaystyle H}{\ddot{C}}}:\ddot{S}:H$ **h.** $H:\ddot{O}:\underset{\displaystyle :\ddot{O}:}{C}:\ddot{O}:H$

i. $:\ddot{O}-\overset{\displaystyle :O:}{S}-\ddot{O}:$

6.61 a. $:\ddot{O}-\ddot{S}=\ddot{O}$ $:\ddot{O}-\ddot{O}=\ddot{O}$

b. Each has the same numbers of valence electrons. Each central atom forms a single bond and a double bond.

c. Yes, because the alternate Lewis structure can be written:

$\ddot{O}=\ddot{S}-\ddot{O}:$ $\ddot{O}=\ddot{O}-\ddot{O}:$

which means that each bond can be regarded as "$1\frac{1}{2}$," i.e., intermediate between a single and double bond.

6.63 a. $:\ddot{F}:\overset{\displaystyle }{\underset{\displaystyle :\ddot{F}:}{B}}:\ddot{F}:$ **b.** $\ddot{N}H_3$ or $H_2\ddot{O}:$ **c.** $:\ddot{F}:\overset{\displaystyle H:\overset{\displaystyle H}{N}:H}{\underset{\displaystyle :\ddot{F}:}{B}}:\ddot{F}:$

6.65 Electron-deficient materials

6.67 Electronegativity is a measure of the attractive force that an atom exerts on a shared pair of electrons in a chemical bond.

6.69 a. To the right and up   **b.** To the left and down

6.71 IBr; P—Br; N—H; H—O

6.73 a. $F-\overset{\displaystyle F}{\underset{\displaystyle F}{C}}-F$ **b.** $F-\overset{\displaystyle F}{\underset{\displaystyle F}{C}}-F$ **c.** On the F   **d.** On the C

6.75 a. Polar   **b.** Ionic   **c.** Ionic   **d.** Ionic   **e.** Polar
f. Nonpolar

6.77 Compound A is likely to be ionic.
Compound B is likely to be covalent.

6.79 $H_2Te < H_2Se < H_2S < H_2O$

6.81 Atoms combine in a fixed ratio to form compounds. The fixed ratio leads to a definite composition by weight which reflects the relative weight of combining atoms.

6.83 No, atoms only combine in whole-number ratios.

CHAPTER 7

7.1 Sodium chloride, calcium bromide, barium oxide

7.2 Iron(II) oxide, iron(III) oxide, copper(II) oxide

7.3 Potassium carbonate, ammonium nitrate, lead(II) phosphate

7.4 a, b, c, g

7.5 Carbon dioxide, carbon tetrachloride, phosphorus trichloride, dinitrogen tetroxide, oxygen difluoride, hydrogen fluoride

7.6 SO_3, HBr, N_2O_4, SF_4

7.7 Carbonic acid, acetic acid, sulfurous acid, chloric acid, perchloric acid

7.8 a. $HClO_2$ **b.** HF   **c.** HNO_3

7.9 Magnesium chloride, hydrobromic acid, sulfuric acid, sulfurous acid, calcium phosphate, oxygen difluoride

7.10 $SO_2 = 64.06$ amu
$N_2O_4 = 92.02$ amu
$H_2SO_4 = 98.08$ amu

7.11 NaOH $= 40.00$ amu
$CaCl_2 = 110.98$ amu
$Al_2(CO_3)_3 = 233.99$ amu

7.12 11.2% hydrogen; 88.8% oxygen

7.13 a. Sodium chloride   **b.** Sodium hydroxide
c. Sodium hydrogen carbonate (or sodium bicarbonate)

7.15 a, d, f

7.17 a. Potassium oxide   **b.** Aluminum chloride   **c.** Calcium bromide
d. Silver oxide   **e.** Copper(II) iodide   **f.** Copper(I) chloride
g. Tin(II) bromide   **h.** Tin(IV) bromide   **i.** Gallium sulfide
j. Barium iodide   **k.** Nickel nitride

7.19 a. Calcium bicarbonate   **b.** Lithium phosphate   **c.** Tin(II) chloride
d. Iron(III) hydroxide   **e.** Magnesium chlorate

7.21 Periodate, iodate, iodite, hypoiodite

7.23

	OH^-	SO_4^{2-}	PO_4^{3-}
NH_4^+	Ammonium hydroxide	Ammonium sulfate	Ammonium phosphate
Fe^{2+}	Iron(II) hydroxide	Iron(II) sulfate	Iron(II) phosphate
Al^{3+}	Aluminum hydroxide	Aluminum sulfate	Aluminum phosphate

7.25 a. Phosphorus pentachloride   **b.** Sulfur dioxide
c. Diphosphorus pentoxide   **d.** Hydrogen fluoride
e. Carbon tetrabromide   **f.** Dichlorine heptoxide

7.27 Dinitrogen monoxide, nitrogen monoxide, dinitrogen trioxide, nitrogen dioxide, dinitrogen tetroxide, dinitrogen pentoxide

7.29 a. SF_4 **b.** Nitrogen trichloride   **c.** BCl_3 **d.** N_2O_3
e. Carbon tetrabromide

7.31 a. Oxyacid   **b.** Nonoxyacid   **c.** Oxyacid   **d.** Nonoxyacid
e. Nonoxyacid   **f.** Oxyacid

7.33 a. Sulfuric acid   **b.** Sulfurous acid   **c.** Nitric acid
d. Phosphoric acid   **e.** Chloric acid   **f.** Perchloric acid
g. Carbonic acid   **h.** Acetic acid

7.35 a. Potassium chromate   **b.** Sodium fluoride   **c.** Iron(II) oxide
d. Copper(I) carbonate   **e.** Lithium hydroxide   **f.** Carbon monoxide
g. Hydrogen chloride   **h.** Mercury(II) chloride   **i.** Chlorous acid
j. Dichlorine oxide

7.37 a. 253.80 amu **b.** 17.03 amu **c.** 96.09 amu **d.** 174.26 amu
e. 182.90 amu **f.** 149.10 amu **g.** 82.08 amu **h.** 100.46 amu
i. 158.04 amu **j.** 162.12 amu

7.39 63.50%

7.41 88.1% Sn; 11.9% O

7.43 79.88% Cu; 20.12% O

CHAPTER 8

8.1 Al = 26.98 g; F = 19.00 g; Br = 79.90 g

8.2 80.94 g

8.3 8

8.4 110.98 g

8.5 1.65 moles

8.6 12.75 moles K^+; 4.25 moles PO_4^{3-}

8.7 a. KCl **b.** C_2H_5Cl

8.8 HO

8.9 Pair, gross, ream

8.11 $\dfrac{Ne}{H} = 20.02$

8.13 a. No **b.** Sodium

8.15 a. 4.003 g **b.** 78.96 g **c.** 19.00 g **d.** 197.0 g **e.** 6.941 g
f. 72.59 g

8.17 24.31 g

8.19 a. 263.0 g **b.** 263.0 g

8.21 4.2×10^{25}. This is seven times greater than the mass of the earth.

8.23 127 g

8.25 0.213 mole

8.27 0.0392 mole

8.29 1.75 moles

8.31 9.99 g

8.33

Element	Number of moles	Number of atoms	Mass, g
Lithium	1.50	9.03×10^{23}	10.4
Calcium	1.000	6.022×10^{23}	40.08
Phosphorus	0.415	2.50×10^{24}	12.9
Silicon	3.16	1.90×10^{24}	88.8
Zinc	3.90	2.35×10^{24}	255

8.35 $\dfrac{1 \text{ H atom}}{1 \text{ HNO}_3 \text{ molecule}} ; \dfrac{1 \text{ N atom}}{1 \text{ HNO}_3 \text{ molecule}} ; \dfrac{3 \text{ O atoms}}{1 \text{ HNO}_3 \text{ molecule}}$

8.37 $\dfrac{\text{3 C atoms}}{\text{1 C}_3\text{H}_7\text{NO molecule}} : \dfrac{\text{7 H atoms}}{\text{1 C}_3\text{H}_7\text{NO molecule}} : \dfrac{\text{1 N atom}}{\text{1 C}_3\text{H}_7\text{NO molecule}} :$
$\dfrac{\text{1 O atom}}{\text{1 C}_3\text{H}_7\text{NO molecule}}$

8.39 6 Al^{3+} ions; 18 Cl^- ions

8.41 a. 74.55 g   **b.** 80.06 g   **c.** 98.08 g   **d.** 68.14 g   **e.** 84.01 g
f. 262.87 g

8.43 a. 1.50 moles   **b.** 2.00 moles; 508 g

8.45 141.94 g

8.47 a. 19.0 g   **b.** 19.7 g   **c.** 0.426 g   **d.** 2.64×10^3 g
e. 12.97 g   **f.** 125 g

8.49 a. 7.48 g   **b.** 6.231 g   **c.** 851 g   **d.** 241 g

8.51 96.4 mL

8.53 a. 5.25 moles K^+; 5.25 moles I^-
b. 4.00 moles Al^{3+}; 12.0 moles Cl^-
c. 1.10×10^{-3} moles Ca^{2+}; 2.20×10^{-3} moles NO_3^-
d. 0.750 mole Ba^{2+}; 0.500 mole PO_4^{3-}

8.55 a. 64.06 g
b. 6.022×10^{23} molecules
c. 1.064×10^{-22} g

8.57 The molecular formula gives the actual number of atoms in one molecule. The empirical formula gives the smallest whole-number ratio of the atoms in one molecule.

8.59 a. NO   **b.** CH   **c.** CH_4 **d.** NCl_3 **e.** C_3H_8O **f.** $PbSO_4$
g. KNO_2 **h.** Na_3PO_4

8.61 FeS

8.63 C_2H_6

8.65 $C_2H_4O_2$

8.67 $C_6H_9N_3O_2$

8.69 $C_6H_6N_2$

CHAPTER 9

9.1 $Pb + O_2 \longrightarrow PbO_2$

9.2 a. Not balanced   **b.** Not balanced   **c.** Balanced

9.3 $H_2 + Br_2 \longrightarrow 2 HBr$

9.4 a. Balanced as written
b. $2 Al + 3 Cl_2 \longrightarrow 2 AlCl_3$
c. $2 Ag_2O \longrightarrow O_2 + 4 Ag$

9.5 $CaBr_2 + 2 AgNO_3 \longrightarrow Ca(NO_3)_2 + 2 AgBr$

9.6 a. $2 K + 2 H_2O \longrightarrow 2 KOH + H_2$
b. $2 NO + O_2 \longrightarrow 2 NO_2$
c. $Na_2CO_3 + Fe(NO_3)_2 \longrightarrow 2 NaNO_3 + FeCO_3$

9.7 a. No reaction.   **b.** Reaction will occur.

9.9 $\longrightarrow$

9.11 a. Zinc and hydrogen chloride (hydrochloric acid)
    b. Zinc chloride and hydrogen

9.13 a. Barium chloride reacts with sodium sulfate to yield sodium chloride and barium sulfate.
    b. Silicon reacts with oxygen to yield silicon dioxide.
    c. Potassium reacts with iodine to yield potassium iodide.
    d. Sulfuric acid reacts with calcium hydroxide to yield calcium sulfate and water.

9.15 a. Balanced   **b.** Not balanced   **c.** Not balanced   **d.** Balanced
    e. Not balanced   **f.** Not balanced

9.17 Coefficients; subscripts

9.19 a. $2 KI + Cl_2 \longrightarrow 2 KCl + I_2$
    b. $2 Cu + O_2 \longrightarrow 2 CuO$
    c. $Li_2O + H_2O \longrightarrow 2 LiOH$
    d. $2 SO_2 + O_2 \longrightarrow 2 SO_3$
    e. $3 H_2 + N_2 \longrightarrow 2 NH_3$
    f. $2 Na + ZnSO_4 \longrightarrow Na_2SO_4 + Zn$

9.21 a. $CH_4 + 2 O_2 \longrightarrow CO_2 + 2 H_2O$
    b. $2 FeS_2 + 5 O_2 \longrightarrow 2 FeO + 4 SO_2$
    c. $P_4O_{10} + 6 H_2O \longrightarrow 4 H_3PO_4$
    d. $Cl_2 + H_2O \longrightarrow HCl + HClO$
    e. $(NH_4)_2Cr_2O_7 \longrightarrow Cr_2O_3 + N_2 + 4 H_2O$
    f. $2 Al + 3 CuSO_4 \longrightarrow Al_2(SO_4)_3 + 3 Cu$

9.23 a. $C + 2 Cl_2 \longrightarrow CCl_4$
    b. $6 K + N_2 \longrightarrow 2 K_3N$
    c. $Ba(NO_3)_2 + H_2SO_4 \longrightarrow BaSO_4 + 2 HNO_3$
    d. $Ca(OH)_2 \longrightarrow CaO + H_2O$
    e. $4 P + 5 O_2 \longrightarrow 2 P_2O_5$

9.25 a. $2 Na + I_2 \longrightarrow \underline{2 NaI}$
    b. $2 Ag_2O \longrightarrow 4 Ag + \underline{O_2}$
    c. $Pb(NO_3)_2 + 2 KCl \longrightarrow \underline{PbCl_2} + \underline{2 KNO_3}$
    d. $\underline{Si} + O_2 \longrightarrow \underline{SiO_2}$

9.27 a. $2 Al + 3 H_2SO_4 \longrightarrow \underline{Al_2(SO_4)_3} + 3 H_2$
    b. $N_2 + 3 H_2 \longrightarrow 2 \underline{NH_3}$
    c. $BaCl_2 + Na_2SO_4 \longrightarrow 2 \underline{NaCl} + BaSO_4$
    d. $2 Mg + O_2 \longrightarrow 2 \underline{MgO}$

9.29 (s) = solid; (ℓ) = liquid; (g) = gas; (aq) = aqueous; Δ = heat; $\downarrow$ = precipitate formed; $\uparrow$ = gas formed.

9.31 a. $2 Li (s) + 2 H_2O (\ell) \longrightarrow 2 LiOH (aq) + H_2 \uparrow$
    b. $SnCO_3 (s) \xrightarrow{\Delta} SnO (s) + CO_2 \uparrow$
    c. $KCl (aq) + AgNO_3 (aq) \longrightarrow AgCl \downarrow + KNO_3 (aq)$

9.33 Combination, decomposition, single replacement, double replacement

9.35 Combination: 9.12a, c; 9.13b, c; 9.15b, d, f; 9.19b, c, d, e; 9.20a, c, e, f, j; 9.21c; 9.23a, b, e; 9.24d; 9.25a, d; 9.26b, c, d; 9.27b, d; 9.28d, e; 9.30a; 9.32a

Decomposition: 9.15e; 9.20i; 9.21e; 9.23d; 9.25b; 9.30b; 9.31b; 9.32c

Single replacement: 9.11; 9.15a, c; 9.19a, f; 9.20b, h; 9.21f; 9.22e; 9.24b; 9.27a; 9.28h; 9.30d; 9.31a; 9.32b, d

Double replacement: 9.12b, d; 9.13a, d; 9.20d; 9.22d, f; 9.23c; 9.24a, c; 9.25c; 9.26a; 9.27c; 9.28a, f; 9.30c; 9.31c

9.37 . . . a single product (a compound)

9.39 a. Calcium bromide; $Ca + Br_2 \longrightarrow CaBr_2$
b. Sulfurous acid; $SO_2 + H_2O \longrightarrow H_2SO_3$
c. Magnesium nitride; $3\ Mg + N_2 \longrightarrow Mg_3N_2$
d. Calcium hydroxide; $CaO + H_2O \longrightarrow Ca(OH)_2$

9.41 a. Yes   **b.** No   **c.** Yes   **d.** No   **e.** Yes   **f.** Yes
g. Yes   **h.** No

9.43 a. $2\ Li + CuO \longrightarrow Li_2O + Cu$
b. $2\ Li + 2\ HNO_3 \longrightarrow 2\ LiNO_3 + H_2$

9.45 a. $K_2SO_4 + Ba(NO_3)_2 \longrightarrow 2\ KNO_3 + BaSO_4$
b. $(NH_4)_2CO_3 + MgCl_2 \longrightarrow 2\ NH_4Cl + MgCO_3$
c. $2\ (NH_4)_3PO_4 + 3\ Ca(NO_3)_2 \longrightarrow 6\ NH_4NO_3 + Ca_3(PO_4)_2$
d. $3\ FeCl_2 + 2\ K_3PO_3 \longrightarrow Fe_3(PO_4)_2 + 6\ KCl$
e. $Na_2S + Ni(NO_3)_2 \longrightarrow 2\ NaNO_3 + NiS$

9.47 a. $3\ Li + AuCl_3 \longrightarrow 3\ LiCl + Au$; single replacement
b. $2\ Al(NO_3)_3 + 3\ K_2CO_3 \longrightarrow Al_2(CO_3)_3 + 6\ KNO_3$; double replacement
c. $Ba + F_2 \longrightarrow BaF_2$; combination
d. $Ba + SnF_2 \longrightarrow BaF_2 + Sn$; single replacement
e. $3\ Mg + 2\ P \longrightarrow Mg_3P_2$; combination
f. $3\ SnCl_2 + 2\ Na_3PO_4 \longrightarrow Sn_3(PO_4)_2 + 6\ NaCl$; double replacement

9.49 $2\ NaHCO_3\ (s) \overset{\Delta}{\longrightarrow} CO_2\ (g) + H_2O\ (g) + Na_2CO_3\ (s)$

9.51 $ZnO + H_2 \longrightarrow Zn + H_2O$

9.53 $2\ C_8H_{18} + 25\ O_2 \longrightarrow 16\ CO_2 + 18\ H_2O$

9.55 $2\ SO_2 + O_2 \longrightarrow 2\ SO_3$
$SO_3 + H_2O \longrightarrow H_2SO_4$
$H_2SO_4 + CaCO_3 \longrightarrow CaSO_4 + CO_2 + H_2O$

CHAPTER 10

10.1 a. Two moles of $Al(OH)_3$ plus three moles of H_2SO_4 yields one mole of $Al_2(SO_4)_3$ plus six moles of H_2O.
b Four moles of Li plus one mole of O_2 yields two moles of Li_2O.

10.2 $\dfrac{2\ \text{moles Na}}{1\ \text{mole } H_2}$

10.3 $\dfrac{1\ \text{mole Sn}}{1\ \text{mole } H_2}$

10.4 $2\ K + Br_2 \longrightarrow 2\ KBr$; 15.0 moles KBr

10.5 224 g KCl

10.6 231 g Li_2S

10.7 4.50 g SO_3

10.8 a. Endothermic   **b.** Exothermic   **c.** Exothermic

10.9 58.5 kcal

10.11 The chemical equation must be balanced.

10.13 a. Two moles of P plus three moles of H_2 yields two moles of PH_3.
b One mole of HBr plus one mole of KOH yields one mole of KBr plus one mole of H_2O.

c. Two moles of CO plus one mole of O_2 yields two moles of CO_2.

d. One mole of C plus two moles of Cl_2 yields one mole of CCl_4.

e. One mole of Mg plus two moles of HCl yields one mole of $MgCl_2$ plus one mole of H_2.

10.15 No. The amount of matter must be conserved in a chemical reaction. In the example cited, 1 g of matter would have been lost. Also, equal weights of substances do not ensure equal numbers of particles.

10.17 a. $\dfrac{2 \text{ moles P}}{3 \text{ moles } H_2}$; $\dfrac{2 \text{ moles P}}{2 \text{ moles } PH_3}$; $\dfrac{3 \text{ moles } H_2}{2 \text{ moles } PH_3}$

b. $\dfrac{1 \text{ mole Mg}}{2 \text{ moles HCl}}$; $\dfrac{1 \text{ mole Mg}}{1 \text{ mole } MgCl_2}$; $\dfrac{1 \text{ mole Mg}}{1 \text{ mole } H_2}$; $\dfrac{2 \text{ moles HCl}}{1 \text{ mole } MgCl_2}$; $\dfrac{2 \text{ moles HCl}}{1 \text{ mole } H_2}$; $\dfrac{1 \text{ mole } MgCl_2}{1 \text{ mole } H_2}$

10.19 The molar ratios are equal to the particle ratios. The total number of atoms of each element is equal regardless of how the balanced equation is interpreted.

10.21 a. 0.600 mole CS_2 **b.** 1.20 moles SO_2
c. 28.0 moles CO   **d.** 7.50 moles C

10.23 a. $4 \text{ Al} + 3 \text{ O}_2 \longrightarrow 2 \text{ Al}_2O_3$; 3.0 moles Al_2O_3
b. 4.5 moles O_2

10.25 a. 0.45 mole H_2SO_4 **b.** 0.345 mole H_2O **c.** 6.94 moles KOH

10.27 a. 8.0 g N_2H_4 **b.** 7.0 g N_2

10.29 a. 10.2 g Cl_2 **b.** 3.48 g HCl   **c.** 4.44 g H_2O

10.31 a. 466 g of sand   **b.** 2.58 moles CO_2 **c.** 0.720 mole $CaCO_3$

10.33 a. 12.3 g Fe   **b.** 0.330 mole Br_2

10.35 a. 299 g O_2 **b.** 168 g H_2O

10.37 a. 2.43 kg NH_3 **b.** 1.51 kg H_2

10.39 a. 3.00 moles Cl_2 **b.** 1.50 moles CH_4 **c.** 0.50 mole Cl_2

10.41 1.00 mole $CHCl_3$

10.43 20. g HF

10.45 17 g N_2

10.47 101 g CO_2

10.49 2.3 g PbS

10.51 a. The actual yield is the amount of product which is actually collected in the lab.
b. No.

10.53 64%

10.55 a. 101 g Fe   **b.** 92%

10.57 a. Wood burning   **b.** Melting ice

10.59 a. 383 kcal   **b.** 57.4 kcal

10.61 a. 7.0 kcal   **b.** 27 g $CaCO_3$

10.63 a. 9.5 kcal   **b.** 88°C

CHAPTER 11

11.1 Original density = 1.9 g/liter
Final density = 0.080 g/liter

11.2 760 torr

11.3 3.0 atm

11.4 750. mmHg

11.5 a. 273 K   **b.** 298 K   **c.** 373 K

11.6 a. $\dfrac{9.6 \times 10^2 \text{ cm}^3}{263 \text{ K}} = 3.7 \text{ cm}^3/\text{K}$

$\left.\begin{array}{l}\dfrac{11.5 \times 10^2 \text{ cm}^3}{313 \text{ K}} = 3.67 \text{ cm}^3/\text{K} \\[2mm] \dfrac{13.4 \times 10^2 \text{ cm}^3}{363 \text{ K}} = 3.69 \text{ cm}^3/\text{K}\end{array}\right\}$ Constant 3.7 to two sig figs.

b. $\dfrac{9.6 \times 10^2 \text{ cm}^3}{-10.0°\text{C}} = -96 \text{ cm}^3/°\text{C}$

$\dfrac{11.5 \times 10^2 \text{ cm}^3}{40.0°\text{C}} = 28.7 \text{ cm}^3/°\text{C}$

11.7 2.6 mL

11.8 0.76 mole hydrogen

11.9 0.179 g/liter

11.10 83.2 liter

11.11 0.0499 mole

11.12 46.5 g/mole

11.13 15.7 liter Cl_2

11.14 10.7 g SF_4

11.15 (1) Gases can be easily compressed.
(2) Gases expand to fill the entire volume of their container.
(3) Gases have indefinite densities.
(4) Gases have low densities.
(5) Gases can diffuse rapidly through each other.
(6) Gases exert a pressure on the walls of any container or surface that they touch.
(7) Gases expand when heated and contract when cooled.

11.17 Any gas not removed from the container would affect the measurement.

11.19 Since the atmosphere is composed of gases, it exerts pressure on any surface it touches.

11.21 The atmospheric pressure acting on the surface of the reservoir of mercury is sufficient to support the column of mercury in the barometer.

11.23 a. 580. mmHg   **b.** 0.763 atm

11.25 887 mmHg or 1.167 atm

11.27 Gases flow spontaneously from a region of higher pressure to a region of lower pressure. Gas 1 will flow into mixture A, and gas 2 will flow into mixture B. Eventually the partial pressures of each gas will be equal on both sides of the barrier.

11.29 a. $V \propto 1/P$
  b. $PV = k$
  c. Breathing; increasing P on a piston decreases V in a cylinder.

11.31 a. 2 atm   **b.** 6 liters·atm   **c.** 6 liters·atm   **d.** 6 liters·atm

11.33 Volume increase is proportional to the height or length increase of trapped air. Temperature markings in °C are equally spaced and can be used as a measure of length (height).

11.35 Warm air molecules are moving faster, and therefore are better able to overcome gravitational attraction and thus rise up over colder air.

11.37 a. No.
  b. The volume of a gas is directly related to the Celsius temperature, but not directly proportional. An increase in Celsius temperature leads to an increase in volume but not by the same factor.

11.39 1.13 liters

11.41 No, volume of a gas varies with pressure.

11.43 754 K or 481°C

11.45 2.77 atm

11.47 2.43 liters

11.49 68.1 mL

11.51 Yes

11.53 67 liters

11.55 1.25 g/liter

11.57 26.0 g/mole

11.59 a. No   **b.** It must be at STP (0°C and 1 atm).

11.61 1.52 moles

11.63 375 g

11.65 144 g

11.67 0.609 g/liter

11.69 (1) Gases are made up of small particles which are constantly moving in random, straight-line motion.
  (2) The distance between particles is very large compared with the size of the particles. A gas is mostly empty space.
  (3) There are no attractive forces between particles. The particles move independently of each other.
  (4) The particles collide with each other and with the walls of the container without incurring a loss of energy.
  (5) The average kinetic energy of the particles is directly proportional to the kelvin temperature of the gas.

11.71 The increased motion of the gas inside the tire increases the rate and force at which the gas particles collide with the inside wall of the tire, thus increasing the pressure.

11.73 At constant pressure, a given mass of a gas increases in volume as its temperature rises. Since density is inversely proportional to volume, the density decreases.

11.75 273°C

11.77 1.5 liters O_2

11.79 23.7 liters

CHAPTER 12

12.1 a. Linear, linear   **b.** Planar, bent   **c.** Tetrahedral, tetrahedral

12.2 b and c

12.3 b, c

12.4 a. NH_3 **b.** H_2O **c.** Br_2

12.5 12.2 kcal

12.7 a. Tetrahedral   **b.** Planar triangular   **c.** Tetrahedral
d. Tetrahedral

12.9 a. This configuration minimizes the repulsive forces between pairs.
b. Since an angle of 200° would bring the pairs closer than an angle of 180° (i.e., 160°), the electron repulsion would prevent their moving to 200°.

12.11 a.

$$:\overset{\overset{\displaystyle ..}{:}\overset{\displaystyle Cl:}{|}}{\underset{\underset{\displaystyle :Cl:}{|}}{\overset{..}{Cl}}}-\overset{}{C}-\overset{..}{Cl}:\qquad H-\overset{\overset{\displaystyle :\overset{..}{Cl}:}{|}}{\underset{\underset{\displaystyle :\overset{..}{Cl}:}{|}}{C}}-H$$

b. CCl_4 is nonpolar because the bond dipoles cancel. CH_2Cl_2 is polar because the bond dipoles and geometry lead to − and + ends of the molecule.

12.13 a. $:\overset{..}{Cl}-Be-\overset{..}{Cl}:$

$$H\overset{\overset{\textstyle \overset{..}{S}}{\diagup \diagdown}}{}H$$

$$H\overset{\overset{\textstyle \overset{..}{O}}{\diagup \diagdown}}{}H$$

b. $BeCl_2$ is linear and symmetrical; therefore the bond dipoles cancel.
c. H_2O.

12.15 Place the compound between two charged plates. An observable change in the voltage between the plates indicates that polar molecules have lined themselves up with the electric field.

12.17 b

12.19 NH_3. N is more electronegative than P.

12.21 a. Dispersion forces, dipole-dipole interactions, and hydrogen bonds.
b. All molecules display dispersion forces. Dipole-dipole interactions occur between molecules with permanent dipole moments, which require polar bonds and noncanceling geometry. Hydrogen bonds exist when a hydrogen atom bonded to a N, O, or F in a molecule is attracted to a nonbonding electron pair on N, O, or F in another molecule.

12.23 a. Dispersion forces   **b.** Hydrogen bonding
c. Dipole-dipole interactions   **d.** Hydrogen bonding

12.25 a. Hydrogen bonding   **b.** Dipole-dipole interactions
c. Dispersion forces

12.27 The number of electrons decreases, resulting in weaker dispersion forces.

12.29 b, d, f, h

12.31 $C_{10}H_{22}$. The greater number of atoms means more electrons.

12.33 More heat is required to overcome stronger intermolecular attractions.

12.35 $He < CH_4 < CH_3F < CH_2F_2 < CH_3CH_2CH_2OH$

12.37 In a solid, the molecules are very close, and the intermolecular forces are very strong. The molecules in a liquid are not as close together, and the forces between them are weak enough to allow them to move past one another.

12.39 The molecules of gases are much farther apart than those of a liquid or solid. This permits gases to be compressed more easily.

12.41 a. The molecules with higher kinetic energy escape into the gas phase, leaving behind the molecules with lesser kinetic energy on the average in the liquid phase.
  b. No

12.43 a. CH_4 **b.** (H₂C=O structure)   **c.** (propane structure H—C—C—C—H with H's)

12.45 Increasing T increases the average kinetic energy of molecules so that more can escape into the gas phase.

12.47

Evaporation	Boiling
Occurs only at the liquid's surface.	Bubbles of gas form throughout the liquid.
Occurs independently of the external pressure.	Occurs only when vapor $P =$ external P.

12.49 a. The vapor pressure of water equals 1 atm at 100°C.
  b. Yes. It can boil at lower temperatures by reducing the external pressure, or it can boil at higher temperatures by increasing the external pressure.

12.51 a. Molar heat of vaporization.
  b. Gas. The energy comes from the heat of vaporization.
  c. Liquid.
  d. Energy required to overcome the stronger intermolecular attractive forces in the liquid.

12.53 16.8 kcal

12.55 Decreased surface tension

12.57 To prevent the escape of volatile components of the cheese which give it its aroma and flavor. Also, retards loss of water vapor, which would dry out the cheese.

12.59 a. 93°C   **b.** Longer

12.61 Heat it to determine if it had a definite melting point.

12.63 a. Oppositely charged ions   **b.** Molecules   **c.** Atoms
  d. Metallic cations

12.65 a. Ionic—electrostatic forces; molecular—dipole-dipole interactions, dispersion forces, and hydrogen bonds; covalent—covalent bonds; metallic—electrostatic attraction between metal cations and mobile electrons
  b. Molecular

12.67 Hydrogen iodide is a molecular compound in which the strongest intermolecular force is dipole-dipole interaction. This is much weaker than the electrostatic forces in the ionic crystal, NaCl.

12.69 a. Molar heat of fusion.
  b. Liquid. The energy comes from the heat of fusion.

   c. Solid.

   d. Energy required to overcome the strongest attractive forces in the solid.

12.71 The attractive forces between molecules differs much more between the gas and liquid states than they do between the liquid and solid states.

12.73 Molecules of gas which sublimed from the solid

12.75 19.7 kcal

CHAPTER 13

13.1 a. Unsaturated   **b.** Unsaturated   **c.** Unsaturated

13.2 Dilute

13.3 31.6% w/w

13.4 Dissolve 160. g KBr in 340. g H_2O.

13.5 700 mL (7.0×10^2)

13.6 12.2 g

13.7 4.22 liters

13.8 0.25 M

13.9 Ba^{2+} $(aq) + SO_4{}^{2-}$ $(aq) \longrightarrow BaSO_4 \downarrow$

13.10 No. No precipitate forms since both KCl and $Cu(NO_3)_2$ are soluble.

13.11 A solution is a homogeneous mixture.

13.13 (1) Oxygen gas in water, (2) oil in gasoline, (3) sugar in water

13.15 a, b, c, e, f

13.17 Dusty air is not homogeneous; dust particles settle out.

13.19 a. No, if solubility is low, solution can be saturated but dilute.

   b. Silver acetate has a solubility limit of 1.04 g/100 g H_2O; this solution would be saturated but dilute.

13.21 a. Unsaturated   **b.** Unsaturated   **c.** Unsaturated   **d.** Saturated

13.23 Dissolve 50. g of KNO_3 in 100 g of H_2O at 60°C. Carefully cool the solution to 20°C; the solution will now be supersaturated.

13.25 a. Mixtures of liquids with liquids

   b. Ethylene glycol and water—antifreeze mixture

   c. Turpentine and water

13.27 a. Solutes tend to dissolve in solvents which have similar intermolecular forces. Polar solutes dissolve in polar solvents; nonpolar solutes in nonpolar solvents.

   b. Polar solutes.

   c. Nonpolar solutes.

13.29 a. Opening the bottle relieves the pressure on the solution. The solubility of CO_2 in water decreases as pressure decreases.

   b. No, solubility of a gas in water decreases with increasing temperatures.

13.31 a. Stirring establishes better contact between the solute and solvent particles.

   b. No, the limit of solubility is unaltered.

13.33 Solubility refers to the maximum amount of solute that can dissolve in a given amount of solvent at a given temperature. For example, a maximum of 36 g NaCl can dissolve in 100 g H_2O at 20°C. Concentration is a statement of the amount of solute dissolved in a given amount of solvent or solution. 10 g NaCl in 90 g H_2O produces a 10% solution. Concentration is variable.

13.35 a. 22.1% w/w   **b.** 139 g H_2O

13.37 Dissolve 12.5 g $BaCl_2$ in 237.5 g water.

13.39 24.0% w/w

13.41 13.5 g sugar and 76.5 g water

13.43 38.7% w/v

13.45 31 mL

13.47 0.01% w/v

13.49 Percent weight per volume relates grams of solute to milliliters of solution. Molarity relates moles of solute to liters of solution.

13.51 0.0125 mole

13.53 Place 26.3 g NaCl in a 500-mL volumetric flask. Add some water, and swirl the flask to dissolve the NaCl. Add water to the etched mark which gives 0.500 liter of solution.

13.55 4.25 g

13.57 0.568 M

13.59 1810 mL (1.81×10^3)

13.61 0.200 M

13.63 480. mL

13.65 A solution of NaCl conducts electricity.

13.67 Some covalent materials react with H_2O to form many ions in solution (e.g., HBr). Other covalent materials form only a few ions (e.g., $HC_2H_3O_2$).

13.69 HCl reacts with H_2O to form ions. In benzene no ions are produced.

13.71 Strong acids: hydrochloric acid, HCl; hydrobromic acid, HBr; hydroiodic acid, HI; nitric acid, HNO_3; sulfuric acid, H_2SO_4; perchloric acid, $HClO_4$.
Weak acids: acetic acid, $HC_2H_3O_2$; boric acid, H_3BO_3; carbonic acid, H_2CO_3; hydrofluoric acid, HF; hydrocyanic acid, HCN. See Tables 15.1 and 15.3.

13.73 a. $Ba^{2+}(aq) + SO_4^{2-}(aq) \longrightarrow BaSO_4 \downarrow$
      b. $Zn(s) + 2 H^+(aq) \longrightarrow Zn^{2+}(aq) + H_2 \uparrow$
      c. $Ag^+(aq) + Cl^-(aq) \longrightarrow AgCl \downarrow$
      d. $Pb^{2+}(aq) + S^{2-}(aq) \longrightarrow PbS \downarrow$
      e. $Fe^{3+}(aq) + 3 OH^-(aq) \longrightarrow Fe(OH)_3 \downarrow$
      f. $2 H^+(aq) + CO_3^{2-}(aq) \longrightarrow H_2O + CO_2 \uparrow$

13.75 a. Na^+ and SO_4^{2-} ions   **b.** FeS $\downarrow$ **c.** K^+ and OH^- ions
      d. $Ca_3(PO_4)_2 \downarrow$ **e.** Mg^{2+} and NO_3^- ions   **f.** NH_4^+ and CO_3^{2-} ions

13.77 a. Water   **b.** Glucose solution   **c.** Water   **d.** Glucose solution

13.79 a. Net water flow from A to B
      b. Net water flow from A to B
      c. Net water flow from B to A
      d. No net flow

13.81 a. Hypertonic   **b.** Hypotonic   **c.** Hypertonic   **d.** Hypertonic
      e. Isotonic

13.83 0.600 M

13.85 7.60 g

13.87 1.78 g

CHAPTER 14

14.1 a. $K_{eq} = \dfrac{[HI]^2}{[H_2][I_2]}$ **b.** $K_{eq} = \dfrac{[NH_3]^2}{[N_2][H_2]^3}$

14.2 a. $K_{eq} = \dfrac{1}{[CO][H_2]^2}$ **b.** $K_{eq} = [NH_3][HCl]$ **c.** $K_{eq} = \dfrac{[H_2]^4}{[H_2O]^4}$

14.3 a. $K_{sp} = [Ca^{2+}][CO_3^{2-}]$ **b.** $K_{sp} = [Pb^{2+}][Cl^-]^2$ **c.** $K_{sp} = [Ba^{2+}]^3[PO_4^{3-}]^2$

14.4 a. More products than reactants at equilibrium
  b. Significant amounts of products and reactants at equilibrium
  c. Essentially all reactants at equilibrium

14.5 a. Shift right ($\longrightarrow$)   **b.** Shift left ($\longleftarrow$)   **c.** Shift right ($\longrightarrow$)
  d. Shift right ($\longrightarrow$)

14.6 a. Shift left ($\longleftarrow$)   **b.** Shift right ($\longrightarrow$)   **c.** No effect

14.7 At the melting point, melting and freezing occur at the same rate. Solid and liquid exist in dynamic equilibrium. In a closed soda bottle, CO_2 gas enters and leaves the liquid at the same rate. See also Section 14.1.

14.9 a. Water molecules leave the liquid state and enter the gaseous state; after a while some gaseous water molecules are recaptured into the liquid state. Eventually the rate gas $\longrightarrow$ liquid is equal to liquid $\longrightarrow$ gas, and an equilibrium condition is present.
  b. No, because water molecules can leave the surroundings of the liquid and not be recaptured into the liquid.

14.11 Equal concentrations requires $[R] = [P]$. Constant concentration requires $[R]$ remains the same, e.g., 0.5, and $[P]$ remains the same, e.g., 0.36, but $[R]$ and $[P]$ do not have to be equal.

14.13 Rate (forward) equals rate (reverse).

14.15 $K_{eq} = \dfrac{k}{k'} = \dfrac{[B]}{[A]^2}$

14.17 a. $\dfrac{[H_2][I_2]}{[HI]^2}$
  b. When the color of the system remains constant.
  c. 10.

14.19 a. $K_{eq} = \dfrac{[CO][H_2]}{[H_2O]}$ **b.** $K_{eq} = \dfrac{[H_2]^4}{[H_2O]^4}$ **c.** $K_{eq} = [H_2]^2[O_2]$
  d. $K_{eq} = \dfrac{[MnCl_2][Cl_2]}{[HCl]^4}$ **e.** $K_{eq} = \dfrac{[CO_2]}{[H_2CO_3]}$

14.21 a. $K_{eq} = \dfrac{[\beta\text{-D-glucose}]}{[\alpha\text{-D-glucose}]}$ **b.** $K_{eq} = \dfrac{[CH_3COOCH_3]}{[CH_3OH][CH_3COOH]}$
  c. $K_{eq} = \dfrac{[oxyhemoglobin]}{[hemoglobin][O_2]}$ **d.** $K_{eq} = \dfrac{[\text{single-stranded DNA}]^2}{[\text{double-stranded DNA}]}$

14.23 a. $K_{sp} = [Ag^+][Br^-]$ **b.** $K_{sp} = [Cu^{2+}][OH^-]^2$ **c.** $K_{sp} = [Al^{3+}][OH^-]^3$
  d. $K_{sp} = [Mg^{2+}]^3[PO_4^{3-}]^2$ **e.** $K_{sp} = [Bi^{3+}][S^{2-}]^3$

14.25 a. Essentially all products at equilibrium
  b. More reactions than product at equilibrium

c. Significant amounts of reactants and products at equilibrium

d. Essentially all reactants at equilibrium

14.27 a. 2   **b.** 3

14.29 a. Shift right ($\longrightarrow$)   **b.** Shift left ($\longleftarrow$)   **c.** Shift right ($\longrightarrow$)
         d. Shift left ($\longleftarrow$)   **e.** Shift right ($\longrightarrow$)

14.31 a. Decrease pressure.   **b.** Increase pressure.

14.33 a. Shift to $Ca_3(PO_4)_2(s)$ **b.** Shift to $Ca_3(PO_4)_2(s)$
         c. Shift to product ions.   **d.** No change in equilibrium

14.35 In pure water. In sulfuric acid, the sulfate ion shifts the equilibrium toward the solid (common ion effect).

14.37 $[HCO_3^-]$ is lowered.

14.39 a. $Ag^+(aq) + Br^-(aq) \longrightarrow AgBr\downarrow$
         b. $H^+(aq) + OH^-(aq) \longrightarrow H_2O$
         c. NR
         d. $Ba^{2+}(aq) + 2\,OH^-(aq) + 2\,H^+(aq) + SO_4^{2-}(aq) \longrightarrow BaSO_4\downarrow + 2\,H_2O$
         e. $3\,Ca^{2+}(aq) + 2\,PO_4^{3-}(aq) \longrightarrow Ca_3(PO_4)_2\downarrow$

14.41 The product precipitate is being removed from the system. This causes the shift in equilibrium toward the products.

CHAPTER 15

15.1 $KOH(s) \rightleftharpoons K^+(aq) + OH^-(aq)$
       $Ca(OH)_2(s) \rightleftharpoons Ca^{2+}(aq) + 2\,OH^-(aq)$

15.2 $\underset{\text{Base}}{OH^-} + \underset{\text{Acid}}{HNO_3} \rightleftharpoons \underset{\substack{\text{Conjugate} \\ \text{base}}}{NO_3^-} + \underset{\substack{\text{Conjugate} \\ \text{acid}}}{H_2O}$

15.3 $HNO_3 > HF > H_2CO_3 > H_3BO_3 > H_2O$

15.4 $NO_3^- < F^- < HCO_3^- < H_2BO_3^- < OH^-$

15.5 $3.1 \times 10^{-6}\ M$; basic

15.6 a. 5.00   **b.** 5.00

15.7 a. 5.17   **b.** 13.85

15.8 a. 1.5 M **b.** 0.4 M **c.** 0.45 M

15.9 $pH = pK_a + \log\dfrac{[HPO_4^{2-}]}{[H_2PO_4^-]}$

$pH = 7.21 + \log\dfrac{2.4 \times 10^{-3}}{1.5 \times 10^{-3}}$

$pH = 7.21 + \log(1.6)$

$pH = 7.21 + 0.20$

$pH = 7.41$

15.11 Refer to Table 15.1, e.g., acetic acid (vinegar), acetylsalicylic acid(aspirin), sodium hydroxide (lye), ammonia.

15.13 $HNO_3 + H_2O \rightleftharpoons H_3O^+ + NO_3^-$

15.15 a. Brønsted-Lowry acid = H^+ donor.
            Brønsted-Lowry base = H^+ acceptor.
         b. The definitions are essentially identical.
         c. Brønsted-Lowry base—H^+ acceptor (a broad, general definition).
            Arrhenius base—produces ^-OH in solution (a more limited definition).

15.17 $NH_4^+ + H_2O \rightleftharpoons H_3O^+ + NH_3$

15.19 a. OH^- **b.** F^- **c.** HPO_4^{2-} **d.** NH_3 **e.** HSO_4^- **f.** NH_2^-

15.21 $HCO_3^- \longrightarrow H^+ + CO_3^{2-}$
$HCO_3^- + H^+ \longrightarrow H_2CO_3$
HCO_3^- is an amphoteric substance since it is capable of both losing and accepting protons.

15.23 a. $NH_4^+ + NH_3 \rightleftharpoons NH_3 + NH_4^+$
      Acid   *Base*
b. $H_2SO_4 + H_2O \rightleftharpoons HSO_4^- + H_3O^+$
      Acid   *Base*
c. $HCl + CN^- \rightleftharpoons Cl^- + HCN$
      Acid   *Base*
d. $H_2CO_3 + NO_3^- \rightleftharpoons HCO_3^- + HNO_3$
      Acid   *Base*
e. $HSO_4^- + OH^- \rightleftharpoons SO_4^{2-} + H_2O$
      Acid   *Base*
f. $HSO_4^- + NO_3^- \rightleftharpoons SO_4^{2-} + HNO_3$
      Acid   *Base*

15.25 $ClO_4^- < SO_4^{2-} < H_2BO_3^- < CN^- < OH^-$

15.27 a. $\dfrac{[H_2SO_4][OH^-]}{[HSO_4^-]}$
b. The concentrations at equilibrium

15.29 a. $H_2CO_3 \rightleftharpoons H^+ + HCO_3^-$
$HCO_3^- \rightleftharpoons H^+ + CO_3^{2-}$
b. $H_3PO_4 \rightleftharpoons H^+ + H_2PO_4^-$
$H_2PO_4^- \rightleftharpoons H^+ + HPO_4^{2-}$
$HPO_4^{2-} \rightleftharpoons H^+ + PO_4^{3-}$

15.31 a. Acidic   **b.** Basic   **c.** Acidic   **d.** Basic
e. Acidic   **f.** Neutral

15.33 a. Acidic   **b.** Basic   **c.** Basic

15.35 a. Acidic   **b.** Acidic   **c.** Acidic

15.37 a. 2.30, acidic   **b.** 7.01, basic-neutral   **c.** 1.33, acidic
d. 0.041, acidic   **e.** 8.54, basic   **f.** 5.08, acidic
g. 12.49, basic   **h.** 9.30, basic

15.39 $[OH^-] = 0.625\ M$
$[H^+] = 1.6 \times 10^{-14}\ M$
$pH = 13.80$

15.41 The pH meter will read 7 when the titration has reached the equivalence point and $[OH^-] = [H^+]$.

15.43 a. $KOH + HCl \longrightarrow KCl + H_2O$
b. $Ba(OH)_2 + 2\ HNO_3 \longrightarrow Ba(NO_3)_2 + 2\ H_2O$
c. $3\ NaOH + H_3PO_4 \longrightarrow Na_3PO_4 + 3\ H_2O$
d. $Ca(OH)_2 + H_2SO_4 \longrightarrow CaSO_4 + 2\ H_2O$
e. $3\ Ca(OH)_2 + 2\ H_3PO_4 \longrightarrow Ca_3(PO_4)_2 + 6\ H_2O$

15.45 a. $HNO_3 + KHCO_3 \longrightarrow KNO_3 + H_2O + CO_2$
b. $2\ HCl + Li_2CO_3 \longrightarrow 2\ LiCl + H_2O + CO_2$
c. $Zn + 2\ HCl \longrightarrow ZnCl_2 + H_2$
d. $Na_2CO_3 + 2\ HC_2H_3O_3 \longrightarrow 2\ NaC_2H_3O_2 + H_2O + CO_2$
e. $Mg + H_2SO_4 \longrightarrow MgSO_4 + H_2$

15.47 0.0697 M

15.49 5.02 mL

15.51 Each mole of H_2SO_4 is capable of releasing two moles of H^+.

15.53 0.0350 mole

15.55 0.9009 M

15.57 A buffer is a weak acid–weak base pair that, by reacting with added amounts of a base or acid, can resist large changes in the solution's pH.

15.59 $H_2S + OH^- \rightleftharpoons HS^- + H_2O$
$HS^- + H^+ \rightleftharpoons H_2S$

15.61 $[H^+]$ increases; $[H_2CO_3]$ increases; $[HCO_3^-]$ decreases.

15.63 pH = 6.4

15.65 pH = 6.1

15.67 The coupling of the equilibria between H_2CO_3 and the unlimited supply of CO_2 in the lungs provides for the high buffering capacity toward base through appropriate shifts in equilibrium.

CHAPTER 16

16.1 a. Oxidation   **b.** Reduction   **c.** Oxidation   **d.** Reduction

16.2 $Ni(s) \longrightarrow Ni^{2+}(aq) + 2e^-$ Oxidation
$2H^+(aq) + 2e^- \longrightarrow H_2$ reduction
Oxidizing agent = H^+
Reducing agent = Ni

16.3 a. +5   **b.** +5   **c.** +4

16.4 a. Fe^{3+} is reduced to Fe^{2+}; S^{2-} is oxidized to S^0.
  b. Oxidizing agent = Fe^{3+}
   Reducing agent = S^{2-}

16.5 $3 SnBr_2 + KIO_3 + 6 HBr \longrightarrow 3 SnBr_4 + KI + 3 H_2O$

16.6 $3 Cl_2 + 6 KOH \longrightarrow KClO_3 + 5 KCl + 3 H_2O$

16.7 $10 HNO_3 + I_2 \longrightarrow 10 NO_2 + 2 HIO_3 + 4 H_2O$

16.9 Because when electrons are lost by one species (oxidation), they must be gained by another (reduction).

16.11 a. No.
  b. Multiply the first half-reaction by 2 and the second by 3 so that the sum is
   $2 Fe + 3 Br_2 + 6e^- \longrightarrow 2 Fe^{3+} + 6 Br^- + 6e^-$.

16.13 a. Oxidizing agent equals the material reduced. Reducing agent equals the material oxidized.
  b. K^+ is oxidizing agent; Zn is reducing agent; MnO_4^- is oxidizing agent; SO_2 is reducing agent.

16.15 a. Oxidation: $Sn^{2+}(aq) \longrightarrow Sn^{4+}(aq)$
         Reducing agent
      Reduction: $2 Fe^{3+}(aq) \longrightarrow 2 Fe^{2+}(aq)$
            Oxidizing agent
  b. Oxidation: $Zn(s) \longrightarrow Zn^{2+}(aq)$
        Reducing agent

Reduction: $2 H^+(aq) \longrightarrow H_2(g)$
  Oxidizing agent
  c. Oxidation: $3 Zn(s) \longrightarrow 3 Zn^{2+}(aq)$
    Reducing agent
  Reduction: $2 Fe^{3+}(aq) \longrightarrow 2 Fe(s)$
    Oxidizing agent

16.17 a. K: $+1$; S: $+6$; O: -2 **b.** $+5$ **c.** 0 **d.** -3
  e. -3 **f.** $+5$ **g.** -4

16.19 a. 0 **b.** $+1$ **c.** $+3$ **d.** $+7$ **e.** $+5$

16.21 a. H_2O **b.** NaH   **c.** H_2

16.23 a. NO_3^- is reduced and is the oxidizing agent. Pb is oxidized and is the reducing agent.
  b. SO_4^{2-} is reduced and is the oxidizing agent. C is oxidized and is the reducing agent.
  c. NH_4^+ is oxidized and is the reducing agent. NO_2^- is reduced and is the oxidizing agent.
  d. Cl_2 is reduced and is the oxidizing agent. As_2O_3 is oxidized and is the reducing agent.

16.25 a. $Zn(s) + 2 H^+(aq) \longrightarrow H_2(g) + Zn^{2+}(aq)$
  b. $MnO_2(s) + 4 H^+(aq) + 2 Br^-(aq) \longrightarrow Mn^{2+}(aq) + Br_2(aq) + 2 H_2O(\ell)$
  c. $8 H^+(aq) + 10 NO_3^-(aq) + I_2(s) \longrightarrow 10 NO_2(g) + 4 H_2O(\ell) + 2 IO_3^-(aq)$
  d. $AsO_4^{3-}(aq) + 2 H^+(aq) + 2 I^-(aq) \longrightarrow AsO_3^{3-}(aq) + I_2(aq) + H_2O(\ell)$
  e. $3 S^{2-}(aq) + 2 NO_3^-(aq) + 8 H^+(aq) \longrightarrow 3 S(s) + 2 NO(g) + 4 H_2O(\ell)$
  f. $ClO_3^-(aq) + 6 H^+(aq) + 6 I^-(aq) \longrightarrow Cl^-(aq) + 3 I_2(g) + 3 H_2O(\ell)$
  g. $Cu(s) + 4 H^+(aq) + SO_4^{2-}(aq) \longrightarrow Cu^{2+}(aq) + SO_2(g) + 2 H_2O(\ell)$
  h. $5 Fe^{2+}(aq) + MnO_4^-(aq) + 8 H^+(aq) \longrightarrow 5 Fe^{3+}(aq) + Mn^{2+}(aq) + 4 H_2O(\ell)$
  i. $H_2(g) + 2 Fe^{3+}(aq) \longrightarrow 2 H^+(aq) + 2 Fe^{2+}(g)$
  j. $C_3H_8(g) + 5 O_2(g) \longrightarrow 3 CO_2(g) + 4 H_2O(\ell)$
  k. $C_6H_{12}O_6(aq) + 6 O_2(g) \longrightarrow 6 CO_2(g) + 6 H_2O(\ell)$

16.27 a. $4 OH^- + Zn \longrightarrow Zn(OH)_4^{2-} + 2 e^-$
  b. $3 OH^- + Sn \longrightarrow HSnO_2^- + H_2O + 2 e^-$
  c. $8 e^- + 4 H_2O + BrO_4^- \longrightarrow Br^- + 8 OH^-$
  d. $2 OH^- + CN^- \longrightarrow CNO^- + H_2O + 2 e^-$

16.29 a. $10 OH^- + 2 Cr^{3+} + 3 ClO^- \longrightarrow 2 CrO_4^{2-} + 5 H_2O + 3 Cl^-$
  b. $2 H_2O + 4 MnO_4^- + 3 ClO_2^- \longrightarrow 4 MnO_2 + 3 ClO_4^- + 4 OH^-$
  c. $10 OH^- + 2 Cr^{3+} + 3 ClO^- \longrightarrow 2 CrO_4^{2-} + 5 H_2O + 3 Cl^-$
  d. $2 Co^{2+} + 2 H_2O + O_2^{2-} \longrightarrow 4 OH^- + 2 Co^{3+}$

16.31 a. $Br_2 + 2 H_2O + SO_2 \longrightarrow 2 HBr + H_2SO_4$
  b. $KClO + H_2 \longrightarrow KCl + H_2O$
  c. $SnSO_4 + 2 FeSO_4 \longrightarrow Sn + Fe_2(SO_4)_3$
  d. $2 HNO_2 + 2 HI \longrightarrow 2 NO + I_2 + 2 H_2O$

16.33

	Oxidation	Reduction
a.	$Al \longrightarrow Al^{3+} + 3 e^-$	$Cu^{2+} + 2 e^- \longrightarrow Cu$
b.	$Fe \longrightarrow Fe^{2+} + 2 e^-$	$Ag^+ + e^- \longrightarrow Ag$

16.35 Table 16.1 shows Au as the least reactive of the metals in the table. Gold cannot displace Na from NaCl or H_2 from HCl. Therefore, the gold does not react and is unharmed by the solutions.

16.37 The salt bridge allows anions to flow toward the anode and cations to flow toward the cathode.

16.39 a. Zn half-reaction at the anode; other reaction at the cathode.
     b. Zn.

16.41 a. $Al^{3+} + 3\,e^- \longrightarrow Al$
         $2\,O^{2-} \longrightarrow O_2 + 4\,e^-$
     b. Cathode

16.43 Anode $Au \longrightarrow Au^{3+} + 3\,e^-$
       Cathode $Au^{3+} + 3\,e^- \longrightarrow Au$

CHAPTER 17

17.1 $^{24}_{11}Na$ has 11 protons and 13 neutrons. $^{27}_{13}Al$ has 13 protons and 14 neutrons. ^{23}Na has 11 protons and 12 neutrons. ^{79}Br has 35 protons and 44 neutrons. ^{206}Pb has 82 protons and 124 neutrons.

17.2 $^{239}_{94}Pu \longrightarrow {}^4_2He + {}^{235}_{92}U$

17.3 $^{227}_{89}Ac \longrightarrow {}^{\,\,0}_{-1}e + {}^{227}_{90}Th$

17.4 0.13 g

17.5 5720 years

17.6 $^{245}_{98}Cf$

17.7

"Ordinary"	Nuclear
a. Only electrons.	Protons and neutrons.
b. Relatively small.	Very large.
c. Different state of combination leads to different reactions.	No effect.

17.9 a. $^{25}_{12}Mg$ **b.** $^{197}_{79}Au$ **c.** $^{131}_{53}I$ **d.** $^{210}_{82}Pb$

17.11 Spontaneous emission of energy

17.13 Isotopes are atoms with the same atomic number but different mass numbers. Radioisotopes are radioactive isotopes.

17.15 ^4_2He, $^{\,\,0}_{-1}e$

17.17 a. A few sheets of paper
     b. A sheet of aluminum, a block of wood, or heavy protective clothing
     c. Several layers of lead
     d. Several layers of lead

17.19 a. Gamma   **b.** Alpha   **c.** rem

17.21 Radiation ionizes the argon gas, allowing Ar^+ ions to move to the cathode and e^- to the anode. This movement of charged particles causes an electric current to be produced in the connected wires, and the current can be used to yield an audible click.

17.23 Nucleus of the cell

17.25 Sterility or birth defects in offspring

17.27 The number of protons in the emitting nucleus is changed, yielding a new element.

17.29 a. $^{222}_{86}Rn \longrightarrow {}^4_2He + {}^{218}_{84}Po$
     b. $^{227}_{89}Ac \longrightarrow {}^4_2He + {}^{223}_{87}Fr$
     c. $^{235}_{92}U \longrightarrow {}^4_2He + {}^{231}_{90}Th$

17.31 a. Beta   **b.** Alpha   **c.** Alpha   **d.** Beta

17.33 Gamma emission affects neither nuclear charge nor mass.

17.35 No, since the decay terminated, $^{208}_{82}\text{Pb}$ must be nonradioactive.

17.37 0.20 g

17.39 January 25

17.41 17,160 years

17.43 a. No   **b.** One-quarter

17.45 Use of ^{131}I to monitor activity of the thyroid gland is one example.

17.47 Radiation is more damaging to fast-growing cancer cells than to normal cells. Thus it is possible to kill the cancer cells, leaving the normal cells relatively unharmed.

17.49 a. $^{4}_{2}\text{He}$ **b.** $^{1}_{1}\text{H}$ **c.** $^{1}_{1}\text{H}$ **d.** $^{1}_{0}n$

17.51 a. $^{0}_{-1}e$ **b.** $^{1}_{0}n$ **c.** $^{4}_{2}\text{He}$ **d.** $^{1}_{1}\text{H}$ **e.** $^{21}_{10}\text{Ne}$ **f.** $^{3}_{2}\text{He}$ **g.** $^{224}_{88}\text{Ra}$

17.53 Np, Pu, Am, Cm, Bk, Cf, Es, Fm, Md, No, Lr, Rf, Ha

17.55 (1) Fusion reactions produce more energy than fission reactions.
      (2) The starting materials for fusion are more abundant than fissionable isotopes.
      (3) The products of fusion reactions are not radioactive.

CHAPTER 18

18.1

18.2

18.3

18.4 a. 2-Methylbutane   **b.** 2,3-Dimethylpentane

18.5 a. All organic compounds contain carbon.
    b. Hydrogen.

18.7 Silicon

18.9 a.

```
   H  H  H  H
   |  |  |  |
H—C—C—C—C—H
   |  |  |  |
   H  H  H  H
```

b.

```
   H  H  H
   |  |  |
H—C—C—C—H
   |  |  |
   H  C  H
      / \
     F  H  H
```

c.

```
   H  H  H
   |  |  |
H—C—C—C—H
   |  |  |
   H  C  H
     / \
    H  H  H
```

d.

```
      H      H   H
       \    /   /
        C        C
       / \      / \
  H   C       C    H
       \     /
    H   C—C   H
        |  |
        H  H
```

e.

```
   H  Cl H  H
   |  |  |  |
H—C—C—C—C—Br
   |  |  |  |
   H  H  H  H
```

f.

```
   H  H  H  H  H  H  H  H  H
   |  |  |  |  |  |  |  |  |
H—C—C—C—C—C—C—C—C—C—H
   |  |  |  |  |  |  |  |  |
   H  C  H  H  H  H  H  H  H
     / \
    H  H  H
```

18.11 a. Alkane   **b.** Ketone   **c.** Alkyl halide   **d.** Alkyne
       e. Aromatic   **f.** Carboxylic acid   **g.** Ether   **h.** Ester
       i. Amine   **j.** Alkene   **k.** Amide   **l.** Aldehyde

18.13 Isomers have the same molecular formulas but have different structures. Identical compounds have the same molecular formulas and structures. Unrelated compounds have different molecular formulas.

18.15

```
   H  H  H  H  H
   |  |  |  |  |
H—C—C—C—C—C—H
   |  |  |  |  |
   H  H  H  H  H
```

```
        H  H  H
        |  |  |
     H  C  H
        |
H—C—C—C—H
     |  |
     H  C  H
       / \
      H  H
```

```
   H  H  H  H
   |  |  |  |
H—C—C—C—C—H
   |  |  |  |
   H  C  H  H
     / \
    H  H  H
```

18.17

```
   H  H  H  H
   |  |  |  |
H—C—C—C—C—Br
   |  |  |  |
   H  H  H  H
```

```
   H  Br H
   |  |  |
H—C—C—C—H
   |  |  |
   H  C  H
     / \
    H  H  H
```

```
   H  Br H  H
   |  |  |  |
H—C—C—C—C—H
   |  |  |  |
   H  H  H  H
```

```
   H  H  H
   |  |  |
Br—C—C—C—H
   |  |  |
   H  C  H
     / \
    H  H  H
```

18.19 a. Tetrahedral   **b.** Planar triangular
    c. Planar triangular   **d.** Linear

18.21 They have low boiling points due to the weak (dispersion) intermolecular forces.
They are insoluble in water because they are nonpolar.

18.23 b

18.25 a. alkane (open chain)   **b.** alkyne   **c.** aromatic
    d. alkene   **e.** alkane (open chain)   **f.** cycloalkane

18.27 a. Combustion   **b.** Home heating fuel, gasoline

18.29

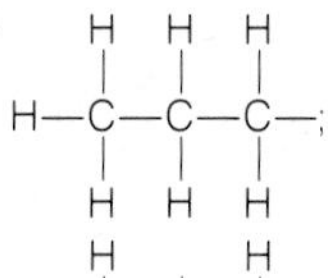

; *n*-propyl

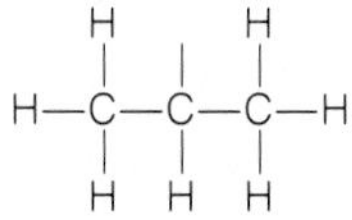

; isopropyl

18.31 a.

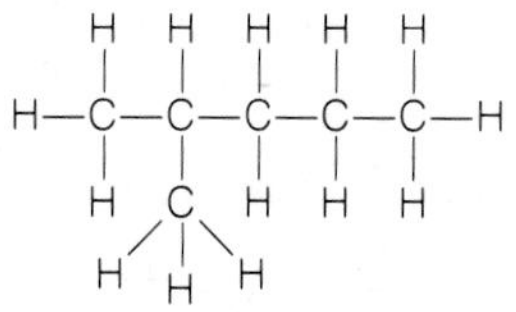

b.

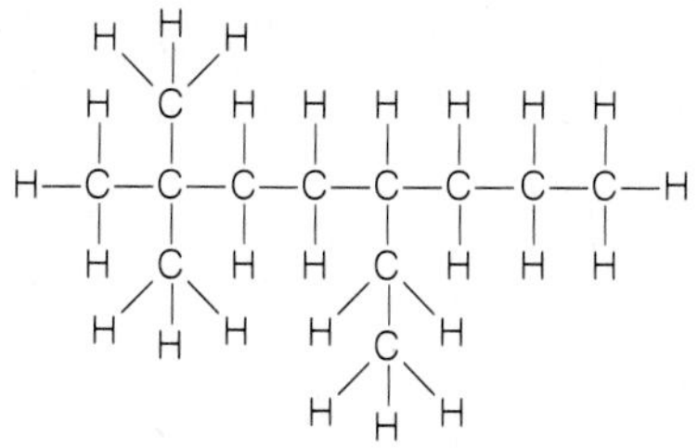

c.

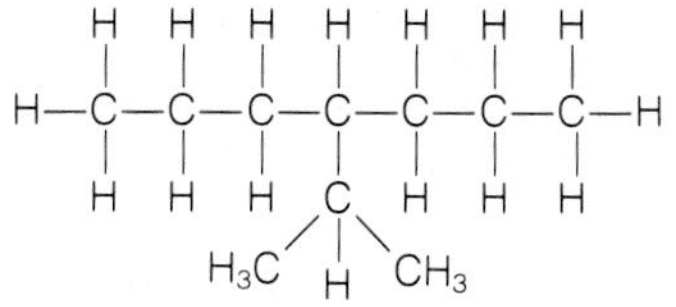

18.33   —CH—CH₂—

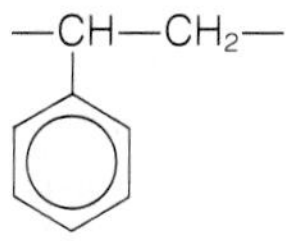

18.35 To preserve the cyclic π system

18.37

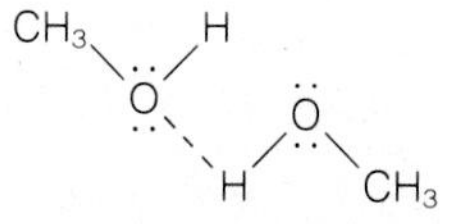

18.39

$$
\begin{array}{c}
\quad\;\; H \qquad\quad H \;\; H \\
\quad\;\; | \qquad\qquad | \;\;\; | \\
H-C-O-C-C-H \\
\quad\;\; | \qquad\qquad | \;\;\; | \\
\quad\;\; H \qquad\quad H \;\; H
\end{array}
$$

18.41 a. (1)

$$
\begin{array}{c}
\;\;\; H \qquad\quad H\;\; H \\
\;\;\; | \qquad\qquad |\;\;\; | \\
H-C-C-C-C-H \\
\;\;\; | \;\;\; \| \;\;\; |\;\;\; | \\
\;\;\; H \;\; O \; H\;\; H
\end{array}
$$

(2)

$$
\begin{array}{c}
\quad\;\; H\;\; H\;\; H \\
\quad\;\; |\;\;\; |\;\;\; | \\
O{=}C-C-C-H \\
\qquad\;\; |\;\;\; | \\
\qquad\;\; H\;\; H
\end{array}
$$

b. (1) Ketone   (2) Aldehyde

18.43 a. Carboxylic acids boil at higher temperatures.
  b. Hydrogen bonding of the type

18.45 Hydrogenation (alkene + $H_2 \longrightarrow$ alkane)

18.47 a. Grease is nonpolar; water is polar.
  b. Soap molecules have a nonpolar end which can dissolve in grease and an ionic end which can dissolve in water. The result is a breakdown of grease into small particles which can be dispersed in water. See Figure 18.4.

18.49 a. Carboxylic acid   **b.** Ketone   **c.** Alcohol   **d.** Aldehyde
  e. Ester   **f.** Ester   **g.** Ether   **h.** Aldehyde
  i. Alcohol   **j.** Carboxylic acid

18.51 a.

  b. More soluble because it is an ionic compound

18.53

$$
\begin{array}{c}
O\;\; H\;\; H\;\; H \\
\| \;\;\; |\;\;\; |\;\;\; | \\
H-C-N-C-C-H \\
\qquad\qquad |\;\;\; | \\
\qquad\qquad H\;\; H
\end{array}
$$

18.55

Fructose

18.57 Stronger

18.59 Carboxylic acid; amine

18.61

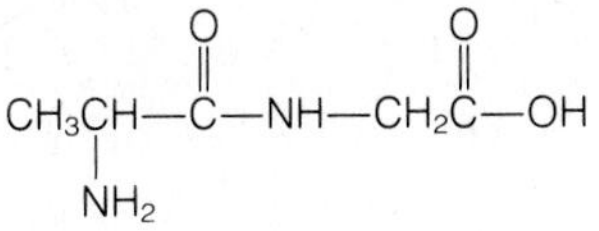

$$CH_3CH-\overset{O}{\underset{\underset{NH_2}{|}}{C}}-NH-CH_2\overset{O}{\overset{||}{C}}-OH$$

18.63

$$CH_3CHC-NH-CH_2-\overset{O}{\overset{||}{C}}-NH-CH-\overset{O}{\overset{||}{C}}-OH$$

with NH₂ below the first carbon and CH₃CHCH₃ below the CH of the third residue.

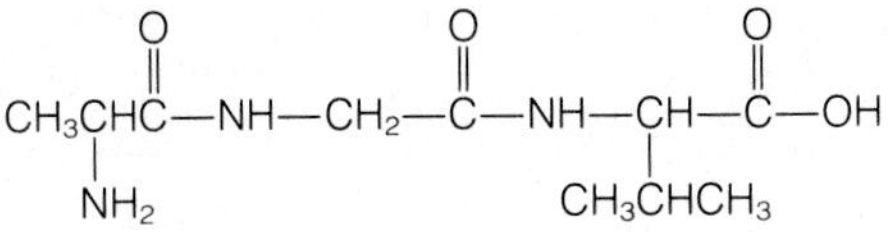

INDEX

Boldface entries and page numbers indicate defined words and their location in the text; page numbers in *italic* indicate illustrations or tables.

Table of atomic weights
(based on carbon-12)

	Symbol	Atomic number	Atomic weight
Actinium	Ac	89	[227]*
Aluminum	Al	13	26.9815
Americium	Am	95	[243]
Antimony	Sb	51	121.75
Argon	Ar	18	39.948
Arsenic	As	33	74.9216
Astatine	At	85	[210]
Barium	Ba	56	137.34
Berkelium	Bk	97	[247]
Beryllium	Be	4	9.01218
Bismuth	Bi	83	208.980
Boron	B	5	10.81
Bromine	Br	35	79.904
Cadmium	Cd	48	112.40
Calcium	Ca	20	40.08
Californium	Cf	98	[251]
Carbon	C	6	12.011
Cerium	Ce	58	140.12
Cesium	Cs	55	132.905
Chlorine	Cl	17	35.453
Chromium	Cr	24	51.996
Cobalt	Co	27	58.9332
Copper	Cu	29	63.546
Curium	Cm	96	[247]
Dysprosium	Dy	66	162.50
Einsteinium	Es	99	[254]
Erbium	Er	68	167.26
Europium	Eu	63	151.96
Fermium	Fm	100	[257]
Fluorine	F	9	18.9984
Francium	Fr	87	[223]
Gadolinium	Gd	64	157.25
Gallium	Ga	31	69.72
Germanium	Ge	32	72.59
Gold	Au	79	196.967
Hafnium	Hf	72	178.49
Hahnium	Ha	105	[262]
Helium	He	2	4.00260
Holmium	Ho	67	164.9304
Hydrogen	H	1	1.0079
Indium	In	49	114.82
Iodine	I	53	126.9045
Iridium	Ir	77	192.22
Iron	Fe	26	55.847
Krypton	Kr	36	83.80
Lanthanum	La	57	138.9055
Lawrencium	Lr	103	[260]
Lead	Pb	82	207.2
Lithium	Li	3	6.941
Lutetium	Lu	71	174.97
Magnesium	Mg	12	24.305
Manganese	Mn	25	54.9380
Mendelevium	Md	101	[258]

*A value given in brackets denotes the mass number of the longest-lived or best-known isotope.